Study Guide
with Selected Solutions
for

Linear Algebra with Applications
by Jeffrey Holt

Timothy J. Flaherty
Carnegie Mellon University

Jeffrey Holt
University of Virginia

Amy Yielding
Eastern Oregon University

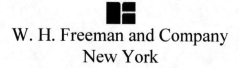

W. H. Freeman and Company
New York

© 2013 by W. H. Freeman and Company

ISBN-13: 978-1-4292-4088-8
ISBN-10: 1-4292-4088-1

Printed in the United States of America.

Third printing

W. H. Freeman and Company
41 Madison Avenue
New York, NY 10010
Houndmills, Basingstoke RG21 6XS, England
www.whfreeman.com

Contents

Part I: Study Guide

Chapter 1 1

Chapter 2 5

Chapter 3 8

Chapter 4 13

Chapter 5 17

Chapter 6 21

Chapter 7 25

Chapter 8 27

Chapter 9 30

Chapter 10 32

Chapter 11 34

Part II: Solutions to Odd-Numbered Exercises

1 Systems of Linear Equations.. 37

    1.1 Lines and Linear Equations.. 37

    1.2 Linear Systems and Matrices... 40

    1.3 Numerical Solutions.. 43

    1.4 Applications of Linear Systems.. 46

2 Euclidean Space.. 51

    2.1 Vectors... 51

    2.2 Span 58

    2.3 Linear Independence... 61

3 Matrices... 67

    3.1 Linear Transformations 67

    3.2 Matrix Algebra 72

    3.3 Inverses 79

    3.4 LU Factorization 84

    3.5 Markov Chains 90

4 Subspaces 95

4.1 Introduction to Subspaces 95

4.2 Basis and Dimension 99

4.3 Row and Column Spaces 103

5 Determinants 107

5.1 The Determinant Function 107

5.2 Properties of the Determinant 111

5.3 Applications of the Determinant 115

6 Eigenvalues and Eigenvectors 119

6.1 Eigenvalues and Eigenvectors 119

6.2 Approximation Methods 125

6.3 Change of Basis 128

6.4 Diagonalization 131

6.5 Complex Eigenvalues 135

6.6 Systems of Differential Equations 138

7 Vector Spaces 145

7.1 Vector Spaces and Subspaces 145

7.2 Span and Linear Independence 149

7.3 Basis and Dimension 152

8 Orthogonality 157

8.1 Dot Products and Orthogonal Sets 157

8.2 Projection and the Gram-Schmidt Process 161

8.3 Diagonalizing Symmetric Matrices and QR Factorization 167

8.4 The Singular Vale Decomposition 170

8.5 Least Squares Regression 175

9 Linear Transformations 183

9.1 Definition and Properties 183

9.2 Isomorphisms 186

9.3 The Matrix of a Linear Transformation 189

9.4 Similarity 193

10 Inner Product Spaces 197

10.1 Inner Products 197

10.2 The Gram-Schmidt Process Revisited 200

10.3 Applications of Inner Products.. 203

11 Additional Topics and Applications 207

11.1 Quadratic Forms... 207

11.2 Positive Definite Matrices 208

11.3 Constrained Optimization 211

11.4 Complex Vector Spaces.. 213

11.5 Hermitian Matrices.. 217

Part 1: Study Guide

1 Systems of Linear Equations

1.1 Lines and Linear Equations

Section Goals

1. Introduce systems of linear equations.

2. Solve systems of linear equations using elimination and back substitution.

3. Introduce triangular and echelon forms.

4. Linear algebra vocabulary: system of linear equations, solution for linear systems, solution set, hyperplane, consistent, inconsistent, free parameter, general solution, leading variable, and free variable.

Study Hints

1. There are three types of solution sets: a unique solution, infinitely many solutions, and no solution. A system can only have one type of solution set. For example, a given system cannot both have a unique solution and infinitely many.

2. Careful! There is a distinction between the vocabulary used for a solution to the system and the system itself:

The Solution Set to the System of Linear Equations	The System of Linear Equations
unique solution	system is consistent
infinitely many solutions	system is consistent
no solutions	system is inconsistent

3. A solution to a system of linear equations and a solution set can be viewed in many ways, such as:

A Solution	The Solution Set
a solution	the general solution
one solution to the system	every solution to the sytem
a point in common	every point in common
one point of intersection of these lines	every point of intersection of these lines
one point that satisfies the system	every point that satisfies the system

4. Keep in mind that triangular form is just a type of echelon form. Properties that distinguish triangular form from the general echelon form:

Triangular Form	Echelon Form
equal number of equations as unknowns	any number of equations and any number of unknowns
every variable must be a leading variable	not every variable has to be a leading variable
exactly one or no solution	can have any of the three types of solution sets
even sized "stair step" pattern	"stair step" pattern

1.2 Linear Systems and Matrices

Section Goals

1. Develop a systematic procedure for transforming any linear system into an echelon form.

2. Define row operations on a system of linear equations.

3. Introduce augmented matrices and implement elementary row operations.

4. Introduce Gaussian elimination as a method to transform a matrix into echelon form.

5. Introduce Gauss-Jordan elimination as a method to transform a matrix into reduced echelon form.

6. Begin the study of homogeneous systems and their solutions.

7. Linear algebra vocabulary: elementary row operations, equivalent systems, matrix, augmented matrix, equivalent matrices, zero row, zero column, Gaussian elimination, leading term, echelon form of a matrix, pivot positions, pivot columns, pivots, Gauss-Jordan elimination, reduced echelon form, forward phase, backward phase, homogeneous equation, homogeneous system, trivial solution, and nontrivial solution.

Study Hints

1. You need to master the process of moving from a system of linear equations to an augmented matrix and vice versa. This skill must become second nature for a rigorous study of this topic to continue.

2. It is common to write the elementary row operations performed above the $\sim$ symbol. This can prove useful when you are a beginner or are prone to making arithmetic mistakes. For example:

$$\begin{bmatrix} 1 & 2 & -1 \\ 2 & 3 & 4 \end{bmatrix} \begin{array}{c} R_2 - 2R_1 \Rightarrow R_2 \\ \sim \end{array} \begin{bmatrix} 1 & 2 & -1 \\ 0 & -1 & 6 \end{bmatrix}$$

3. When first learning how to implement row operations, it is recommended that you perform no more than two row operations at a time to avoid errors.

4. Keep in mind that equivalent systems will have equivalent augmented matrices and vice versa, all of which correspond to the same solution set.

5. Echelon form of a system of linear equations corresponds to an augmented matrix in echelon form and vice versa.

6. A leading variable in a system of linear equations corresponds to a pivot position in the augmented matrix.

7. Careful! You know you are performing a circular sequence of operations if you undo any zeros created in a pivot column while transforming the augmented matrix into echelon form. This is usually a clue that you are not sticking to the Gaussian/Gauss-Jordan elimination rules.

8. Gaussian elimination produces an augmented matrix in echelon form, whereas Gauss-Jordan elimination produces an augmented matrix in reduced echelon form. In both forms, the pivot positions are the same, even though the leading entries may not be equal.

9. Caution! If you are checking your work with a friend, check your reduced echelon forms. It is possible to attain two different echelon forms of the same system, but the reduced echelon form will always match. For example:

$$\begin{bmatrix} 1 & 2 & -1 \\ 2 & 3 & 4 \end{bmatrix} \sim \begin{bmatrix} 1 & 2 & -1 \\ 0 & -1 & 6 \end{bmatrix} \text{ and } \begin{bmatrix} 1 & 2 & -1 \\ 2 & 3 & 4 \end{bmatrix} \sim \begin{bmatrix} 2 & 4 & -2 \\ 0 & -1 & 6 \end{bmatrix},$$

and both are equivalent to $\begin{bmatrix} 1 & 0 & 11 \\ 0 & 1 & -6 \end{bmatrix}$.

10. Homogeneous systems are great since they always have a solution, namely the trivial solution. Since the last column of the matrix is a zero column, it is unaffected by row operations. Thus, some find it space saving to leave out the last column when implementing Gauss-Jordan elimination on homogeneous systems.

1.3 Numerical Solutions

Section Goals

1. Discover the shortcomings of the elimination methods we have discussed thus far.

2. Develop and discuss effectiveness of alternative solution methods used in computer computations.

3. Discuss the following types of direct\elimination methods:

   - Partial pivoting
   - Full pivoting

4. Discuss the following types of iterative methods:

   - Jacobi iteration
   - Gauss-Seidel iteration

5. Linear algebra vocabulary: partial pivoting, full pivoting, diverge, converge, Jacobi iteration, Gauss-Seidel iteration, diagonally dominant, sparse systems, and sparse matrix.

Study Hints

1. Careful! If you use the = symbol, you are claiming that the value is an exact answer and is perfectly accurate. Use :≈ to indicate that rounding has occurred.

2. Disadvantages of the direct\elimination methods:

   - The more arithmetic needed, the more rounding error effects your solution.
   - The more the coefficients differ in absolute value, the more rounding error effects your solution.
   - Accumulation of rounding errors occurs.

3. Advantages of the direct\elimination methods:

   - Always results in a solution.
   - Can be used on any system of linear equations.

4. Disadvantages of the iterative methods:

   - Does not always result in a solution.
   - May only be used on system of linear equations with the same number of equations as unknowns.

5. Advantages of the iterative methods:

   - Rounding errors do not accumulate.
   - Answers can be made as accurate as the user defines.

1.4 Applications of Linear Systems

Section Goals

1. Discuss a variety of applications of linear systems including:

   - Traffic flow
   - Team rankings
   - Orbital periods
   - Balancing chemical equations

Study Hints

1. It is essential that by this point you are able to freely move between systems of linear equations and augmented matrices. Review sections 1.1-1.3 before attempting this section.

2. You may find it useful to review basic linear word problems discussed in a College Algebra course. Techniques learned in such as course will be applicable to the word problems given in this section.

3. The problems given at the end of the section are very similar to the example problems. Thorough analysis of the examples will prove helpful before attempting the homework sets.

2 Euclidean Space

2.1 Vectors

Section Goals

1. Begin the study of vectors.

2. Define basic vector arithmetic.

3. Discuss algebraic properties of vectors.

4. Define vector equations and their connection to linear combinations.

5. Investigate the geometry of vectors in $\mathbf{R}^2$ and $\mathbf{R}^3$.

6. Linear algebra vocabulary: vector, $\mathbf{R}$, component, column vector, row vector, vector arithmetic, scalar, Euclidean space, zero vector, $\mathbf{0}$, linear combination, vector equation, vector form, tip and tail of vector, tip-to-tail rule, parallelogram rule, and scalar multiplication.

Study Hints

1. Some readers may have seen the idea of vectors in a Calculus or Physics course. The notation in Linear algebra can be different from what you have seen before, but rest assured that the concepts are the same. The following variety of notations all represent the same vector:

$$(\mathbf{R}^2) < 1, -2 >, \begin{bmatrix} 1 \\ -2 \end{bmatrix}, \mathbf{i} - 2\mathbf{j}$$

$$(\mathbf{R}^3) < 3, -1, 1 >, \begin{bmatrix} 3 \\ -1 \\ 1 \end{bmatrix}, 3\mathbf{i} - \mathbf{j} + \mathbf{k}$$

Note: We will discuss the third form in a later chapter, as well as why the first form is not commonly used in Linear algebra.

2. When working with vectors by hand you may want to implement the "hat" notation, $\vec{v}$ instead of $\mathbf{v}$. It is easy to mix up vectors and scalars without notation distinguishing the two and the bolded typeface used in the text is difficult to do by hand. This is especially important when attempting to prove various theorems given throughout the text.

3. Linear combinations will be an important concept revisited throughout the text. You are encouraged to thoroughly study the definition given on page 55.

4. Caution! Although the geometry of vectors in $\mathbf{R}^2$ and $\mathbf{R}^3$ is a great visual aid in understanding the properties and behaviors of vectors, we will expand the discussion of vectors to $\mathbf{R}^4$, $\mathbf{R}^5$, $\mathbf{R}^6$, and beyond. So it is important that you are able to work with vectors without the use of this visual aid, both computationally and theoretically.

2.2 Span

Section Goals

1. Define the span of a set of vectors.

2. Reinforce the concept of linear combination.

3. Explore the relationship between span and linear combination with augmented matrices, linear systems, and vector equations.

4. Introduce the $A\mathbf{x} = \mathbf{b}$ notation and its connection to linear combinations and linear systems.

5. Linear algebra vocabulary: span.

Study Hints

1. Holt has some fun, visual, and insightful examples to describe the span of set of vectors. The reader is encourage to look over the VecMobile II, VecMobile III, and pizza box plane examples.

2. The following results from this section are all equivalent:

   - $\mathbf{v}$ is an element of span$\{\mathbf{u}_1, \mathbf{u}_1, \ldots, \mathbf{u}_m\}$.
   - $\mathbf{v}$ is in span$\{\mathbf{u}_1, \mathbf{u}_1, \ldots, \mathbf{u}_m\}$.
   - $\mathbf{v}$ can be reached by VecMobileM.
   - $\mathbf{v}$ lies on the "pizza box" with strings $\mathbf{u}_1, \mathbf{u}_2, \ldots, \mathbf{u}_m$.
   - $\mathbf{v}$ is a linear combination of $\mathbf{u}_1, \mathbf{u}_2, \ldots, \mathbf{u}_m$.
   - The linear system represented by the augmented matrix $[\mathbf{u}_1 \mathbf{u}_2 \cdots \mathbf{u}_m \mathbf{v}]$ has at least one solution.
   - The linear system represented by the augmented matrix $[\mathbf{u}_1 \mathbf{u}_2 \cdots \mathbf{u}_m \mathbf{v}]$ is consistent.
   - The vector equation $x_1 \mathbf{u}_1 + x_2 \mathbf{u}_2 \cdots x_m \mathbf{u}_m = \mathbf{v}$ has at least one solution.
   - span$\{\mathbf{v}, \mathbf{u}_1, \mathbf{u}_2, \ldots, \mathbf{u}_m\}$ = span$\{\mathbf{u}_1, \mathbf{u}_2, \ldots, \mathbf{u}_m\}$.
   - If $U = [\mathbf{u}_1 \mathbf{u}_2 \cdots \mathbf{u}_m]$, then the equation $U\mathbf{x} = \mathbf{v}$ has at least one solution.

2.3 Linear Independence

Section Goals

1. Introduce linear independence and its connection to the homogeneous equation.

2. Discover the relationship between the solution set to $A\mathbf{x} = \mathbf{b}$ and the solution set to the associated $A\mathbf{x} = \mathbf{0}$.

3. Begin the development of The Big Theorem.

4. Linear algebra vocabulary: linear independence, nonhomogeneous, associated homogeneous systems, particular solution, and the Big Theorem

Study Hints

1. Let $\mathcal{A} = \{\mathbf{a}_1, \mathbf{a}_2, \ldots, \mathbf{a}_m\}$, the following table provides comparison between consequences of $\mathcal{A}$ being linear independent or dependent:

$\mathcal{A}$ Is Linearly Independent	$\mathcal{A}$ Is Linearly Dependent
$x_1 \mathbf{a}_1 + x_2 \mathbf{a}_2 + \cdots + x_m \mathbf{a}_m = \mathbf{0}$ has only the trivial solution	$x_1 \mathbf{a}_1 + x_2 \mathbf{a}_2 + \cdots + x_m \mathbf{a}_m = \mathbf{0}$ has more than the trivial solution
$x_1 \mathbf{a}_1 + x_2 \mathbf{a}_2 + \cdots + x_m \mathbf{a}_m = \mathbf{b}$ has at most one solution	$x_1 \mathbf{a}_1 + x_2 \mathbf{a}_2 + \cdots + x_m \mathbf{a}_m = \mathbf{b}$ has infinitely many solutions
$U\mathbf{x} = \mathbf{0}$ has only the trivial solution	$U\mathbf{x} = \mathbf{0}$ has more than the trivial solution
$U\mathbf{x} = \mathbf{b}$ at most one solution	$U\mathbf{x} = \mathbf{b}$ infinitely many solutions

2. To impress on you the importance of the Big Theorem, we restate it here:
   Let $\mathcal{A} = \{\mathbf{a}_1, \ldots, \mathbf{a}_n\}$ be a set of n vectors in $\mathbf{R}^n$, and let $A = [\mathbf{a}_1 \cdots \mathbf{a}_n]$. Then the following are equivalent:

    (a) $\mathcal{A}$ *spans* $\mathbf{R}^n$.

    (b) $\mathcal{A}$ *is linearly independent.*

    (c) $A\mathbf{x} = \mathbf{b}$ *has an unique solution for all* $\mathbf{b}$ *in* $\mathbf{R}^n$.

3. Warning! Do not be tempted to use the Big Theorem when you have an unequal number of equations as unknowns. Most of the results do not always hold otherwise. Here, we provide two such examples:

$\mathcal{A} = \{ \begin{bmatrix} 1 \\ 0 \end{bmatrix}, \begin{bmatrix} 0 \\ 1 \end{bmatrix}, \begin{bmatrix} 1 \\ 1 \end{bmatrix} \}$ spans $\mathbf{R}^2$ but is a linearly dependent set, and $A\mathbf{x} = \mathbf{b}$ has infinitely many solutions for every $\mathbf{b}$ *in* $\mathbf{R}^n$.

$\mathcal{A} = \{ \begin{bmatrix} 1 \\ 0 \\ 0 \end{bmatrix}, \begin{bmatrix} 0 \\ 1 \\ 0 \end{bmatrix} \}$ is linearly independent but does not span $\mathbf{R}^3$, and $A\mathbf{x} = \begin{bmatrix} 0 \\ 0 \\ 1 \end{bmatrix}$ has no solution.

4. We can extend our list from section 2.2 study hints to the following:

- $\mathbf{v}$ is an element of span$\{\mathbf{u}_1, \mathbf{u}_1, \ldots, \mathbf{u}_m\}$.
- $\mathbf{v}$ is in span$\{\mathbf{u}_1, \mathbf{u}_1, \ldots, \mathbf{u}_m\}$.
- $\mathbf{v}$ can be reached by VecMobileM.
- $\mathbf{v}$ lies on the "pizza box" with strings $\mathbf{u}_1, \mathbf{u}_2, \ldots, \mathbf{u}_m$.
- $\mathbf{v}$ is a linear combination of $\mathbf{u}_1, \mathbf{u}_2, \ldots, \mathbf{u}_m$.
- The linear system represented by the augmented matrix $[\mathbf{u}_1 \mathbf{u}_2 \cdots \mathbf{u}_m \mathbf{v}]$ has at least one solution.
- The linear system represented by the augmented matrix $[\mathbf{u}_1 \mathbf{u}_2 \cdots \mathbf{u}_m \mathbf{v}]$ is consistent.
- The vector equation $x_1 \mathbf{u}_1 + x_2 \mathbf{u}_2 \cdots x_m \mathbf{u}_m = \mathbf{v}$ has at least one solution.
- span$\{\mathbf{v}, \mathbf{u}_1, \mathbf{u}_2, \ldots, \mathbf{u}_m\}$ = span$\{\mathbf{u}_1, \mathbf{u}_2, \ldots, \mathbf{u}_m\}$.
- If $U = [\mathbf{u}_1 \mathbf{u}_2 \cdots \mathbf{u}_m]$, then the equation $U\mathbf{x} = \mathbf{v}$ has at least one solution.
- The set $\{\mathbf{v}, \mathbf{u}_1, \mathbf{u}_2, \ldots, \mathbf{u}_m\}$ is linearly dependent.
- If $W = [\mathbf{v}, \mathbf{u}_1 \mathbf{u}_2 \cdots \mathbf{u}_m]$, then the equation $W\mathbf{x} = \mathbf{0}$ has more than the trivial solution.
- The vector equation $x_1 \mathbf{u}_1 + x_2 \mathbf{u}_2 \cdots x_m \mathbf{u}_m + x_{m+1} \mathbf{v} = \mathbf{0}$ has more than the trivial solution.
- The linear system represented by the augmented matrix $[\mathbf{u}_1 \mathbf{u}_2 \cdots \mathbf{u}_m \mathbf{v} \ \mathbf{0}]$ has more than the trivial solution.

3 Matrices

3.1 Linear Transformations

Section Goals

1. Define linear transformations.

2. Establish properties of linear transformations.

3. Discuss the connections between linear transformations, consistent systems, linear combinations, homogeneous systems, and linear dependence.

4. Define $\mathbf{e_1}, \mathbf{e_2}, \ldots, \mathbf{e_m}$.

5. Discuss how to create the matrix of linear transformation.

6. Expand the Big Theorem.

7. Explore the geometry of linear transformations including:

    (a) Reflection.

    (b) Rotation.

    (c) Shear.

8. Linear algebra vocabulary: linear transformations, domain, codomain, image of $\mathbf{u}$ under T, range, $n \times m$ matrix, dimension, square matrix, one-to-one, and onto.

Study Hints

1. Some readers may find the difference between codomain and range confusing. Here are some college algebra examples that may help clarify the difference:

    (a) Consider $f(x) = x^2$ with $f : \mathbf{R}^1 \to \mathbf{R}^1$. Here, the domain is $\mathbf{R}^1$, the codomain is $\mathbf{R}^1$, and the range is $[0, \infty)$.

    (b) Consider $f(x) = x^3$ with $f : \mathbf{R}^1 \to \mathbf{R}^1$. Here, the domain is $\mathbf{R}^1$, the codomain is $\mathbf{R}^1$, and the range is $\mathbf{R}^1$.

    (c) Consider $f(x) = |x| - 3$ with $f : \mathbf{R}^1 \to \mathbf{R}^1$. Here, the domain is $\mathbf{R}^1$, the codomain is $\mathbf{R}^1$, and the range is $[-3, \infty)$.

2. Caution! Not all functions are linear. Consider: $f(x) = x^2$. $f(u + v) = (u + v)^2 = u^2 + 2uv + v^2 \neq u^2 + v^2 = f(u) + f(v)$, so $f(x)$ is not linear.

3. Some readers may find the concept of one-to-one and onto confusing. Here are some college algebra examples that may help:

    (a) $f(x) = x^2$ is neither one-to-one ($f(-2) = f(2)$) nor onto (there is no $x \in \mathbf{R}^1$ such that $f(x) = -1$).

    (b) $f(x) = \sqrt{x}$ is one-to-one and not onto (there is no $x \in \mathbf{R}^1$ such that $f(x) = -4$).

    (c) $f(x) = x(x - 3)(x + 2)$ is onto and not one-to-one ($f(3) = f(-2)$).

    (d) $f(x) = 2x$ is both one-to-one and onto.

4. $\mathbf{e_1}, \mathbf{e_2}, \ldots, \mathbf{e}_m$ will continually arise throughout our study of linear algebra, you are encouraged to become familiar with them.

5. Holt establishes many useful diagnostic tools in theorems 3.3, 3.5, 3.6, 3.7, and 3.9; you are encouraged to become familiar with these.

6. To impress on you the importance of the Big Theorem, we restate version 2 here:
   Let $\mathcal{A} = \{\mathbf{a}_1, \ldots, \mathbf{a}_n\}$ be a set of n vectors in $\mathbf{R}^n$, let $A = [\mathbf{a}_1 \cdots \mathbf{a}_n]$, and let $T : \mathbf{R}^n \to \mathbf{R}^n$ be given by $T(\mathbf{x}) = A\mathbf{x}$. Then the following are equivalent:

   (a) $\mathcal{A}$ spans $\mathbf{R}^n$.

   (b) $\mathcal{A}$ is linearly independent.

   (c) $A\mathbf{x} = \mathbf{b}$ has an unique solution for all $\mathbf{b}$ in $\mathbf{R}^n$.

   (d) T is onto.

   (e) T is one-to-one.

7. We can extend our list from section 2.2 study hints to the following:

   - $\mathbf{v}$ is an element of $\text{span}\{\mathbf{u}_1, \mathbf{u}_1, \ldots, \mathbf{u}_m\}$.
   - $\mathbf{v}$ is in $\text{span}\{\mathbf{u}_1, \mathbf{u}_1, \ldots, \mathbf{u}_m\}$.
   - $\mathbf{v}$ can be reached by VecMobileM.
   - $\mathbf{v}$ lies on the "pizza box" with strings $\mathbf{u}_1, \mathbf{u}_2, \ldots, \mathbf{u}_m$.
   - $\mathbf{v}$ is a linear combination of $\mathbf{u}_1, \mathbf{u}_2, \ldots, \mathbf{u}_m$.
   - The linear system represented by the augmented matrix $[\mathbf{u}_1 \mathbf{u}_2 \cdots \mathbf{u}_m \mathbf{v}]$ has at least one solution.
   - The linear system represented by the augmented matrix $[\mathbf{u}_1 \mathbf{u}_2 \cdots \mathbf{u}_m \mathbf{v}]$ is consistent.
   - The vector equation $x_1 \mathbf{u}_1 + x_2 \mathbf{u}_2 \cdots x_m \mathbf{u}_m = \mathbf{v}$ has at least one solution.
   - $\text{span}\{\mathbf{v}, \mathbf{u}_1, \mathbf{u}_2, \ldots, \mathbf{u}_m\} = \text{span}\{\mathbf{u}_1, \mathbf{u}_2, \ldots, \mathbf{u}_m\}$.
   - If $U = [\mathbf{u}_1 \mathbf{u}_2 \cdots \mathbf{u}_m]$, then the equation $U\mathbf{x} = \mathbf{v}$ has at least one solution.
   - The set $\{\mathbf{v}, \mathbf{u}_1, \mathbf{u}_2, \ldots, \mathbf{u}_m\}$ is linearly dependent.
   - If $W = [\mathbf{v}, \mathbf{u}_1 \mathbf{u}_2 \cdots \mathbf{u}_m]$, then the equation $W\mathbf{x} = \mathbf{0}$ has more than the trivial solution.
   - The vector equation $x_1 \mathbf{u}_1 + x_2 \mathbf{u}_2 \cdots x_m \mathbf{u}_m + x_{m+1} \mathbf{v} = \mathbf{0}$ has more than the trivial solution.
   - The linear system represented by the augmented matrix $[\mathbf{u}_1 \mathbf{u}_2 \cdots \mathbf{u}_m \mathbf{v}\ \mathbf{0}]$ has more than the trivial solution.
   - If $T(\mathbf{x}) = U\mathbf{x}$, then $\mathbf{v}$ is in the range of $T(\mathbf{x})$.
   - If $S(\mathbf{x}) = W\mathbf{x}$, then S is not one-to-one.

3.2 Matrix Algebra

Section Goals

1. Discuss properties of matrix algebra including:

   (a) Scalar multiplication.

   (b) Matrix addition.

   (c) Matrix equality.

   (d) Matrix multiplication.

2. Introduce the zero matrix 0_{nm} and the identity matrix I_m.

3. Define the transpose of a matrix and discuss some of its basic properties.

4. Investigate the powers of a matrix.

5. Introduce diagonal matrices and their powers.

6. Discuss partitioned matrices and the techniques of addition and multiplication via blocks.

7. Linear algebra vocabulary: equal matrices, addition of matrices, scalar multiplication of matrices, matrix multiplication, additive identity, identity matrix, zero matrix, transpose, symmetric matrix, diagonal matrix, diagonal of a matrix, upper triangular matrix, lower triangular matrix, triangular matrix, partitioned matrix, and blocks.

Study Hints

1. The matrix multiplication methods stated in definition 3.12 and example 4 will produce the same result, so whichever method you find easiest is the one you should use. Neither method is "best."

2. Careful! Stick to the properties listed in theorem 3.13. In general, matrices do not commute. Consider the following:

$$\begin{bmatrix} 1 & 0 \\ 2 & 1 \end{bmatrix} \begin{bmatrix} -1 & 3 \\ 0 & 5 \end{bmatrix} = \begin{bmatrix} -1 & 3 \\ -2 & 11 \end{bmatrix}$$

$$\begin{bmatrix} -1 & 3 \\ 0 & 5 \end{bmatrix} \begin{bmatrix} 1 & 0 \\ 2 & 1 \end{bmatrix} = \begin{bmatrix} 5 & 3 \\ 10 & 5 \end{bmatrix}$$

Sometimes, the multiplication is not even defined.

$$\begin{bmatrix} 1 & 1 \\ 2 & 1 \end{bmatrix} \begin{bmatrix} -1 & 1 & 3 \\ 1 & 2 & -1 \end{bmatrix} = \begin{bmatrix} 0 & 3 & 2 \\ -1 & 4 & 5 \end{bmatrix}$$

$$\begin{bmatrix} -1 & 1 & 3 \\ 1 & 2 & -1 \end{bmatrix} \begin{bmatrix} 1 & 1 \\ 2 & 1 \end{bmatrix} \text{ is not defined (a } 2 \times 3 \text{ and a } 2 \times 2\text{).}$$

3. Study theorem 3.14; many readers find these results surprising and counterintuitive.

3.3 Inverses

Section Goals

1. Define the inverse matrix and inverse transformation.

2. Develop the method for finding an inverse.

3. Introduce the quick formula for the inverse of a 2×2 matrix.

4. Expand the Big Theorem.

5. Discuss the inverse of a partitioned matrix.

6. Linear algebra vocabulary: inverse, invertible matrix, nonsingular, singular, block diagonal, and block lower triangular.

Study Hints

1. Practice makes perfect! The process for finding an inverse may seem overwhelming, but with practice you will master it.

2. Not every matrix has an inverse and that is OK.

3. Inverses are quite useful. You always have a solution to $A\mathbf{x} = \mathbf{b}$, namely $\mathbf{x} = A^{-1}\mathbf{b}$.

4. To impress on you the importance of the Big Theorem, we restate version 3 here:
   Let $\mathcal{A} = \{\mathbf{a}_1, \ldots, \mathbf{a}_n\}$ be a set of n vectors in $\mathbf{R}^n$, let $A = [\mathbf{a}_1 \cdots \mathbf{a}_n]$, and let $T : \mathbf{R}^n \to \mathbf{R}^n$ be given by $T(\mathbf{x}) = A\mathbf{x}$. Then the following are equivalent:

    (a) $\mathcal{A}$ *spans* $\mathbf{R}^n$.

    (b) $\mathcal{A}$ *is linearly independent.*

    (c) $A\mathbf{x} = \mathbf{b}$ *has an unique solution for all* $\mathbf{b}$ *in* $\mathbf{R}^n$.

    (d) T *is onto.*

    (e) T *is one-to-one.*

    (f) $\mathcal{A}$ *is invertible.*

5. We can extend our list from section 3.1 study hints to the following:

   - $\mathbf{v}$ is an element of span$\{\mathbf{u}_1, \mathbf{u}_1, \ldots, \mathbf{u}_m\}$.
   - $\mathbf{v}$ is in span$\{\mathbf{u}_1, \mathbf{u}_1, \ldots, \mathbf{u}_m\}$.
   - $\mathbf{v}$ can be reached by VecMobileM.
   - $\mathbf{v}$ lies on the "pizza box" with strings $\mathbf{u}_1, \mathbf{u}_2, \ldots, \mathbf{u}_m$.
   - $\mathbf{v}$ is a linear combination of $\mathbf{u}_1, \mathbf{u}_2, \ldots, \mathbf{u}_m$.
   - The linear system represented by the augmented matrix $[\mathbf{u}_1 \mathbf{u}_2 \cdots \mathbf{u}_m \mathbf{v}]$ has at least one solution.
   - The linear system represented by the augmented matrix $[\mathbf{u}_1 \mathbf{u}_2 \cdots \mathbf{u}_m \mathbf{v}]$ is consistent.
   - The vector equation $x_1\mathbf{u}_1 + x_2\mathbf{u}_2 \cdots x_m\mathbf{u}_m = \mathbf{v}$ has at least one solution.
   - span$\{\mathbf{v}, \mathbf{u}_1, \mathbf{u}_2, \ldots, \mathbf{u}_m\}$ = span$\{\mathbf{u}_1, \mathbf{u}_2, \ldots, \mathbf{u}_m\}$.
   - If $U = [\mathbf{u}_1 \mathbf{u}_2 \cdots \mathbf{u}_m]$, then the equation $U\mathbf{x} = \mathbf{v}$ has at least one solution.
   - The set $\{\mathbf{v}, \mathbf{u}_1, \mathbf{u}_2, \ldots, \mathbf{u}_m\}$ is linearly dependent.
   - If $W = [\mathbf{v}, \mathbf{u}_1 \mathbf{u}_2 \cdots \mathbf{u}_m]$, then the equation $W\mathbf{x} = \mathbf{0}$ has more than the trivial solution.
   - The vector equation $x_1\mathbf{u}_1 + x_2\mathbf{u}_2 \cdots x_m\mathbf{u}_m + x_{m+1}\mathbf{v} = \mathbf{0}$ has more than the trivial solution.
   - The linear system represented by the augmented matrix $[\mathbf{u}_1 \mathbf{u}_2 \cdots \mathbf{u}_m \mathbf{v} \ \mathbf{0}]$ has more than the trivial solution.
   - If $T(\mathbf{x}) = U\mathbf{x}$, then $\mathbf{v}$ is in the range of $T(\mathbf{x})$.
   - If $S(\mathbf{x}) = W\mathbf{x}$, then S is not one-to-one.
   - If $S(\mathbf{x}) = W\mathbf{x}$, then S is not invertible.

3.4 LU Factorization

Section Goals

1. Introduce LU-factorization.

2. Illustrate implementation of the LU-factorization algorithm.

3. Discuss conditions for which no LU-factorization is possible.

4. Investigate a modification of the LU-factorization, the LDU-factorization.

5. Introduce elementary matrices.

6. Linear algebra vocabulary: LU-factorization, LDU-factorization, and elementary matrix.

Study Hints

1. Keep in mind that; you can solve $A\mathbf{x} = \mathbf{b}$ without using LU-factorization, the convenience is that LU-factorization allows for back substitution.

2. Having L and U makes it easier to solve the system.

3. The advantage of the LU-factorization appears when you are asked to solve many systems that all have the same coefficient matrix.

4. Holt provides step-by-step instructions for finding L and U in example 2.

5. As seen in example 4, if at any point in the algorithm a row interchange is required, stop. The matrix does not have an LU-factorization.

6. There can be more than one LU-factorization for a given matrix. So use caution when checking your answer with a fellow student.

7. Holt provides step-by-step instructions for finding L, D, and U in example 7.

3.5 Markov Chains

Section Goals

1. Introduce Markov chains.

2. Introduce stochastic matrices.

3. Discuss properties of stochastic matrices and regular stochastic matrices.

4. Linear algebra vocabulary: stochastic matrix, doubly stochastic matrix, initial state vector, probability vector, transition matrix, state vector, markov chain, steady-state vector, and regular matrix.

Study Hints

1. The reader is encouraged to study the step-by-step examples Holt provides throughout the section.

4 Subspaces

4.1 Introduction to Subspaces

Section Goals

1. Introduce subspaces and discuss their connection to spanning sets and linear combinations.

2. Describe a general procedure for determining if a subset is a subspace.

3. Provide examples of common subspaces including:

   (a) Null spaces.
   (b) Kernel.
   (c) Range.

4. Expand the Big Theorem.

5. Linear algebra vocabulary: subspace, closed under addition, scalar multiplication, subspace spanned, subspace generated, null space, kernel, and range.

Study Hints

1. When checking to see if a given set is a subspace, always check the **0** first. Of all the conditions a subspace must have, containing the zero vector is the easiest condition to verify.

2. Theorem 4.2 is an extremely useful tool for creating your own subspaces as well as verifying if a given set is a subspace.

3. If you are still struggling with subspaces, Holt has a nice step-by-step method described on page 169.

4. To impress on you the importance of the Big Theorem, we restate version 4 here:
   Let $\mathcal{A} = \{\mathbf{a}_1, \ldots, \mathbf{a}_n\}$ be a set of n vectors in $\mathbf{R}^n$, let $A = [\mathbf{a}_1 \cdots \mathbf{a}_n]$, and let $T : \mathbf{R}^n \to \mathbf{R}^n$ be given by $T(\mathbf{x}) = A\mathbf{x}$. Then the following are equivalent:

   (a) $\mathcal{A}$ spans $\mathbf{R}^n$.
   (b) $\mathcal{A}$ is linearly independent.
   (c) $A\mathbf{x} = \mathbf{b}$ has an unique solution for all $\mathbf{b}$ in $\mathbf{R}^n$.
   (d) T is onto.
   (e) T is one-to-one.
   (f) $\mathcal{A}$ is invertible.
   (g) $\ker(T) = \{\mathbf{0}\}$.

5. We can extend our list from section 3.3 study hints to the following:

   - $\mathbf{v}$ is an element of $\text{span}\{\mathbf{u}_1, \mathbf{u}_1, \ldots, \mathbf{u}_m\}$.
   - $\mathbf{v}$ is in $\text{span}\{\mathbf{u}_1, \mathbf{u}_1, \ldots, \mathbf{u}_m\}$.
   - $\mathbf{v}$ can be reached by VecMobileM.
   - $\mathbf{v}$ lies on the "pizza box" with strings $\mathbf{u}_1, \mathbf{u}_2, \ldots, \mathbf{u}_m$.
   - $\mathbf{v}$ is a linear combination of $\mathbf{u}_1, \mathbf{u}_2, \ldots, \mathbf{u}_m$.
   - The linear system represented by the augmented matrix $[\mathbf{u}_1 \mathbf{u}_2 \cdots \mathbf{u}_m \mathbf{v}]$ has at least one solution.
   - The linear system represented by the augmented matrix $[\mathbf{u}_1 \mathbf{u}_2 \cdots \mathbf{u}_m \mathbf{v}]$ is consistent.
   - The vector equation $x_1 \mathbf{u}_1 + x_2 \mathbf{u}_2 \cdots x_m \mathbf{u}_m = \mathbf{v}$ has at least one solution.
   - $\text{span}\{\mathbf{v}, \mathbf{u}_1, \mathbf{u}_2, \ldots, \mathbf{u}_m\} = \text{span}\{\mathbf{u}_1, \mathbf{u}_2, \ldots, \mathbf{u}_m\}$.

- If $U = [\mathbf{u}_1\mathbf{u}_2\cdots\mathbf{u}_m]$, then the equation $U\mathbf{x} = \mathbf{v}$ has at least one solution.
- The set $\{\mathbf{v}, \mathbf{u}_1, \mathbf{u}_2, \ldots, \mathbf{u}_m\}$ is linearly dependent.
- If $W = [\mathbf{v}, \mathbf{u}_1\mathbf{u}_2\cdots\mathbf{u}_m]$, then the equation $W\mathbf{x} = \mathbf{0}$ has more than the trivial solution.
- The vector equation $x_1\mathbf{u}_1 + x_2\mathbf{u}_2 \cdots x_m\mathbf{u}_m + x_{m+1}\mathbf{v} = \mathbf{0}$ has more than the trivial solution.
- The linear system represented by the augmented matrix $[\mathbf{u}_1\mathbf{u}_2\cdots\mathbf{u}_m\mathbf{v}\ \mathbf{0}]$ has more than the trivial solution.
- If $T(\mathbf{x}) = U\mathbf{x}$, then $\mathbf{v}$ is in the range of $T(\mathbf{x})$.
- If $S(\mathbf{x}) = W\mathbf{x}$, then S is not one-to-one.
- S is not invertible.
- null(W) contains more than the zero vector.
- ker(S) contains more than the zero vector.

4.2 Basis and Dimension

Section Goals

1. Define basis for a subspace in terms of linear independent sets and spanning sets.

2. Develop two methods for finding a basis of a given subspace.

3. Define $\mathbf{e_1}, \mathbf{e_2}, \ldots, \mathbf{e_m}$ as the standard basis for $\mathbf{R}^m$.

4. Describe a method for adding vectors to a given set to form a basis for a given subspace.

5. Expand the Big Theorem.

6. Introduce Vandermonde matrices.

7. Linear algebra vocabulary: basis, dimension, standard basis, and nullity.

Study Hints

1. Finding a basis for a given subspace can be difficult. The methods shown in examples one and two will work every time, so use your favorite. After each of these examples, Holt provides step-by-step instructions.

2. Remember the dimension of $\{\mathbf{0}\}$ is 0! By definition, $\{\mathbf{0}\}$ has no basis and so has no basis vectors. Since dimension is defined to be the number of basis vectors, the dimension of $\{\mathbf{0}\}$ is 0.

3. Theorem 4.17 provides a quick method for determining if a given set is linear dependent or does not span a given subspace. You are encouraged to study the results of this theorem.

4. To impress on you the importance of the Big Theorem, we restate version 5 here:
   Let $\mathcal{A} = \{\mathbf{a}_1, \ldots, \mathbf{a}_n\}$ be a set of n vectors in $\mathbf{R}^n$, let $A = [\mathbf{a}_1 \cdots \mathbf{a}_n]$, and let $T : \mathbf{R}^n \to \mathbf{R}^n$ be given by $T(\mathbf{x}) = A\mathbf{x}$. Then the following are equivalent:

   (a) *$\mathcal{A}$ spans $\mathbf{R}^n$.*

   (b) *$\mathcal{A}$ is linearly independent.*

   (c) *$A\mathbf{x} = \mathbf{b}$ has an unique solution for all $\mathbf{b}$ in $\mathbf{R}^n$.*

   (d) *T is onto.*

   (e) *T is one-to-one.*

   (f) *$\mathcal{A}$ is invertible.*

   (g) ker$(T) = \{\mathbf{0}\}$.

(h) $\mathcal{A}$ *is a basis for* $\mathbf{R}^n$.

5. We can extend our list from section 4.1 study hints to the following:

- $\mathbf{v}$ is an element of $\mathrm{span}\{\mathbf{u}_1, \mathbf{u}_1, \ldots, \mathbf{u}_m\}$.
- $\mathbf{v}$ is in $\mathrm{span}\{\mathbf{u}_1, \mathbf{u}_1, \ldots, \mathbf{u}_m\}$.
- $\mathbf{v}$ can be reached by VecMobileM.
- $\mathbf{v}$ lies on the "pizza box" with strings $\mathbf{u}_1, \mathbf{u}_2, \ldots, \mathbf{u}_m$.
- $\mathbf{v}$ is a linear combination of $\mathbf{u}_1, \mathbf{u}_2, \ldots, \mathbf{u}_m$.
- The linear system represented by the augmented matrix $[\mathbf{u}_1 \mathbf{u}_2 \cdots \mathbf{u}_m \mathbf{v}]$ has at least one solution.
- The linear system represented by the augmented matrix $[\mathbf{u}_1 \mathbf{u}_2 \cdots \mathbf{u}_m \mathbf{v}]$ is consistent.
- The vector equation $x_1 \mathbf{u}_1 + x_2 \mathbf{u}_2 \cdots x_m \mathbf{u}_m = \mathbf{v}$ has at least one solution.
- $\mathrm{span}\{\mathbf{v}, \mathbf{u}_1, \mathbf{u}_2, \ldots, \mathbf{u}_m\} = \mathrm{span}\{\mathbf{u}_1, \mathbf{u}_2, \ldots, \mathbf{u}_m\}$.
- If $U = [\mathbf{u}_1 \mathbf{u}_2 \cdots \mathbf{u}_m]$, then the equation $U\mathbf{x} = \mathbf{v}$ has at least one solution.
- The set $\{\mathbf{v}, \mathbf{u}_1, \mathbf{u}_2, \ldots, \mathbf{u}_m\}$ is linearly dependent.
- If $W = [\mathbf{v}, \mathbf{u}_1 \mathbf{u}_2 \cdots \mathbf{u}_m]$, then the equation $W\mathbf{x} = \mathbf{0}$ has more than the trivial solution.
- The vector equation $x_1 \mathbf{u}_1 + x_2 \mathbf{u}_2 \cdots x_m \mathbf{u}_m + x_{m+1} \mathbf{v} = \mathbf{0}$ has more than the trivial solution.
- The linear system represented by the augmented matrix $[\mathbf{u}_1 \mathbf{u}_2 \cdots \mathbf{u}_m \mathbf{v}\ \mathbf{0}]$ has more than the trivial solution.
- If $T(\mathbf{x}) = U\mathbf{x}$, then $\mathbf{v}$ is in the range of $T(\mathbf{x})$.
- If $S(\mathbf{x}) = W\mathbf{x}$, then S is not one-to-one.
- S is not invertible.
- $\mathrm{null}(W)$ contains more than the zero vector.
- $\ker(S)$ contains more than the zero vector.
- $\dim(\mathrm{range}(T)) = \dim(\mathrm{range}(S))$.
- $\dim(\ker(S)) > 0$.
- $\dim(\mathrm{null}(W)) > 0$.

4.3 Row and Column Spaces

Section Goals

1. Introduce row and column spaces and their connection to the echelon form of a given matrix.

2. Discuss the rank of a given matrix and its relationship to row and column space.

3. Expand the Big Theorem.

4. Linear algebra vocabulary: row vectors, column vectors, row space, column space, and rank of a matrix.

Study Hints

1. It is important that you have mastered the techniques for various computations developed in sections 4.1 and 4.2. A thoughtful study of the results in section 4.3 is essential to a deeper understanding of subspaces. You do not want to be bogged down with unfamiliar computations.

2. Theorem 4.23, the Rank-Nullity Theorem, is an extremely useful tool in the study of Linear algebra. The reader is encouraged to become very familiar with this result.

3. To impress on you the importance of the Big Theorem, we restate version 6 here:

   Let $\mathcal{A} = \{\mathbf{a}_1, \ldots, \mathbf{a}_n\}$ be a set of n vectors in $\mathbf{R}^n$, let $A = [\mathbf{a}_1 \cdots \mathbf{a}_n]$, and let $T : \mathbf{R}^n \to \mathbf{R}^n$ be given by $T(\mathbf{x}) = A\mathbf{x}$. Then the following are equivalent:

   (a) $\mathcal{A}$ spans $\mathbf{R}^n$.

   (b) $\mathcal{A}$ is linearly independent.

   (c) $A\mathbf{x} = \mathbf{b}$ has an unique solution for all $\mathbf{b}$ in $\mathbf{R}^n$.

   (d) T is onto.

   (e) T is one-to-one.

   (f) $\mathcal{A}$ is invertible.

   (g) $\ker(T) = \{\mathbf{0}\}$.

   (h) $\mathcal{A}$ is a basis for $\mathbf{R}^n$.

   (i) $\text{col}(A) = \mathbf{R}^n$.

   (j) $\text{row}(A) = \mathbf{R}^n$.

   (k) $\text{rank}(A) = n$.

4. We can extend our list from section 4.2 study hints to the following:

   - $\mathbf{v}$ is an element of $\text{span}\{\mathbf{u}_1, \mathbf{u}_1, \ldots, \mathbf{u}_m\}$.
   - $\mathbf{v}$ is in $\text{span}\{\mathbf{u}_1, \mathbf{u}_1, \ldots, \mathbf{u}_m\}$.
   - $\mathbf{v}$ can be reached by VecMobileM.
   - $\mathbf{v}$ lies on the "pizza box" with strings $\mathbf{u}_1, \mathbf{u}_2, \ldots, \mathbf{u}_m$.
   - $\mathbf{v}$ is a linear combination of $\mathbf{u}_1, \mathbf{u}_2, \ldots, \mathbf{u}_m$.
   - The linear system represented by the augmented matrix $[\mathbf{u}_1 \mathbf{u}_2 \cdots \mathbf{u}_m \mathbf{v}]$ has at least one solution.
   - The linear system represented by the augmented matrix $[\mathbf{u}_1 \mathbf{u}_2 \cdots \mathbf{u}_m \mathbf{v}]$ is consistent.
   - The vector equation $x_1 \mathbf{u}_1 + x_2 \mathbf{u}_2 \cdots x_m \mathbf{u}_m = \mathbf{v}$ has at least one solution.
   - $\text{span}\{\mathbf{v}, \mathbf{u}_1, \mathbf{u}_2, \ldots, \mathbf{u}_m\} = \text{span}\{\mathbf{u}_1, \mathbf{u}_2, \ldots, \mathbf{u}_m\}$.
   - If $U = [\mathbf{u}_1 \mathbf{u}_2 \cdots \mathbf{u}_m]$, then the equation $U\mathbf{x} = \mathbf{v}$ has at least one solution.
   - The set $\{\mathbf{v}, \mathbf{u}_1, \mathbf{u}_2, \ldots, \mathbf{u}_m\}$ is linearly dependent.
   - If $W = [\mathbf{v}, \mathbf{u}_1 \mathbf{u}_2 \cdots \mathbf{u}_m]$, then the equation $W\mathbf{x} = \mathbf{0}$ has more than the trivial solution.
   - The vector equation $x_1 \mathbf{u}_1 + x_2 \mathbf{u}_2 \cdots x_m \mathbf{u}_m + x_{m+1} \mathbf{v} = \mathbf{0}$ has more than the trivial solution.
   - The linear system represented by the augmented matrix $[\mathbf{u}_1 \mathbf{u}_2 \cdots \mathbf{u}_m \mathbf{v} \ \mathbf{0}]$ has more than the trivial solution.
   - If $T(\mathbf{x}) = U\mathbf{x}$, then $\mathbf{v}$ is in the range of $T(\mathbf{x})$.
   - If $S(\mathbf{x}) = W\mathbf{x}$, then S is not one-to-one.
   - S is not invertible.
   - $\text{null}(W)$ contains more than the zero vector.
   - $\ker(S)$ contains more than the zero vector.
   - $\dim(\text{range}(T)) = \dim(\text{range}(S))$.
   - $\dim(\ker(S)) > 0$.
   - $\dim(\text{null}(W)) > 0$.
   - $\mathbf{v}$ is in the column space of U.
   - $\dim(\text{col}(U)) = \dim(\text{col}(W))$.
   - $\text{rank}(U) = \text{rank}(W)$.
   - $\dim(\text{row}(U)) = \dim(\text{row}(W))$.

5 Determinants

5.1 The Determinant Function

Section Goals

1. Introduce determinants and their connection to the inverse of a given matrix.

2. Provide the formula for the determinant of a 2×2 matrix.

3. Describe the method for finding the determinant of larger matrices.

4. Expand the Big Theorem.

5. Establish techniques for determinants of triangular matrices, diagonal matrices, transpose of matrices and products of matrices.

6. Linear algebra vocabulary: determinant, minor, cofactor, and cofactor expansion.

Study Hints

1. This chapter contains theorems and exercises requiring proof via induction. If you are unfamiliar with this method of proof, you are encouraged to consult the appendix "Reading and Writing Proofs."

2. To impress on you the importance of the Big Theorem, we restate version 7 here:
   Let $\mathcal{A} = \{\mathbf{a}_1, \ldots, \mathbf{a}_n\}$ be a set of n vectors in $\mathbf{R}^n$, let $A = [\mathbf{a}_1 \cdots \mathbf{a}_n]$, and let $T : \mathbf{R}^n \to \mathbf{R}^n$ be given by $T(\mathbf{x}) = A\mathbf{x}$. Then the following are equivalent:

   (a) $\mathcal{A}$ spans $\mathbf{R}^n$.

   (b) $\mathcal{A}$ is linearly independent.

   (c) $A\mathbf{x} = \mathbf{b}$ has an unique solution for all $\mathbf{b}$ in $\mathbf{R}^n$.

   (d) T is onto.

   (e) T is one-to-one.

   (f) $\mathcal{A}$ is invertible.

   (g) $\ker(T) = \{\mathbf{0}\}$.

   (h) $\mathcal{A}$ is a basis for $\mathbf{R}^n$.

   (i) $\mathrm{col}(A) = \mathbf{R}^n$.

   (j) $\mathrm{row}(A) = \mathbf{R}^n$.

   (k) $\mathrm{rank}(A) = n$.

   (l) $\det(A) \neq 0$.

3. When applying cofactor expansion, remember to always choose the row or column with the most number of 0 entries. This will reduce the number of computations required.

4. Some may find it helpful to use the following sign pattern to aid in implementing cofactor expansion:
$$\begin{bmatrix} + & - & + & \cdots \\ - & + & - & \cdots \\ + & - & + & \cdots \\ \vdots & \vdots & \vdots & \ddots \end{bmatrix}$$
   The following example displays how to use this sign pattern:
   Determine if the following matrix is invertible: $A = \begin{bmatrix} 1 & -1 & 0 \\ 0 & 2 & 1 \\ -1 & -2 & 1 \end{bmatrix}$. We will use the new edition to
   the Big Theorem to answer this question. That is, we will see if $\det(A) \neq 0$. Since A is a 3×3 matrix

we will use the sign pattern that is a 3×3 as well: $\begin{bmatrix} + & - & + \\ - & + & - \\ + & - & + \end{bmatrix}$.

Observing that row 1 has a zero entry in it, let's expand by row 1. Beginning with the first entry $\begin{bmatrix} \textcircled{1} & -1 & 0 \\ 0 & 2 & 1 \\ -1 & -2 & 1 \end{bmatrix}$, which corresponds to the $+$ symbol in our sign pattern $\begin{bmatrix} \oplus & - & + \\ - & + & - \\ + & - & + \end{bmatrix}$, thus starting

out the computations with $+1\det\begin{bmatrix} 2 & 1 \\ -2 & 1 \end{bmatrix}$. Next, we continue with the second entry in our row 1

expansion $\begin{bmatrix} 1 & \textcircled{-1} & 0 \\ 0 & 2 & 1 \\ -1 & -2 & 1 \end{bmatrix}$, which corresponds to the $-$ symbol in our sign pattern $\begin{bmatrix} + & \ominus & + \\ - & + & - \\ + & - & + \end{bmatrix}$,

thus adding to our computations $+1\det\begin{bmatrix} 2 & 1 \\ -2 & 1 \end{bmatrix} - (-1)\det\begin{bmatrix} 0 & 1 \\ -1 & 1 \end{bmatrix}$. Lastly, we continue with the

third entry in our row 1 expansion $\begin{bmatrix} 1 & -1 & \textcircled{0} \\ 0 & 2 & 1 \\ -1 & -2 & 1 \end{bmatrix}$, which corresponds to the $+$ symbol in our sign

pattern $\begin{bmatrix} + & - & \oplus \\ - & + & - \\ + & - & + \end{bmatrix}$, thus adding to our computations $+1\det\begin{bmatrix} 2 & 1 \\ -2 & 1 \end{bmatrix} - (-1)\det\begin{bmatrix} 0 & 1 \\ -1 & 1 \end{bmatrix} +$

$0\det\begin{bmatrix} 0 & 2 \\ -1 & -2 \end{bmatrix}$. Using the formula $a_{11}a_{22} - a_{12}a_{21}$ for the determinant of a 2×2, we get the following computations:

$+1\det\begin{bmatrix} 2 & 1 \\ -2 & 1 \end{bmatrix} - (-1)\det\begin{bmatrix} 0 & 1 \\ -1 & 1 \end{bmatrix} + 0\det\begin{bmatrix} 0 & 2 \\ -1 & -2 \end{bmatrix} =$
$1(2(1) - 1(-2)) + (0(1) - 1(-1)) + 0 = (2 + 2) + (0 + 1) = 4 + 1 = 5$
Thus, $\det(A) = 5 \neq 0$ and so A is invertible.

5. Pay attention to theorems 5.9, 5.10, 5.11, and 5.12; they can be quite the time savers!

6. Watch out! If the matrix is not square, then the determinant is not defined.

7. We can extend our list from section 4.3 study hints to the following:

   - $\mathbf{v}$ is an element of span$\{\mathbf{u}_1, \mathbf{u}_1, \ldots, \mathbf{u}_m\}$.
   - $\mathbf{v}$ is in span$\{\mathbf{u}_1, \mathbf{u}_1, \ldots, \mathbf{u}_m\}$.
   - $\mathbf{v}$ can be reached by VecMobileM.
   - $\mathbf{v}$ lies on the "pizza box" with strings $\mathbf{u}_1, \mathbf{u}_2, \ldots, \mathbf{u}_m$.
   - $\mathbf{v}$ is a linear combination of $\mathbf{u}_1, \mathbf{u}_2, \ldots, \mathbf{u}_m$.
   - The linear system represented by the augmented matrix $[\mathbf{u}_1\mathbf{u}_2 \cdots \mathbf{u}_m\mathbf{v}]$ has at least one solution.
   - The linear system represented by the augmented matrix $[\mathbf{u}_1\mathbf{u}_2 \cdots \mathbf{u}_m\mathbf{v}]$ is consistent.
   - The vector equation $x_1\mathbf{u}_1 + x_2\mathbf{u}_2 \cdots x_m\mathbf{u}_m = \mathbf{v}$ has at least one solution.
   - span$\{\mathbf{v}, \mathbf{u}_1, \mathbf{u}_2, \ldots, \mathbf{u}_m\}$ = span$\{\mathbf{u}_1, \mathbf{u}_2, \ldots, \mathbf{u}_m\}$.
   - If $U = [\mathbf{u}_1\mathbf{u}_2 \cdots \mathbf{u}_m]$, then the equation $U\mathbf{x} = \mathbf{v}$ has at least one solution.
   - The set $\{\mathbf{v}, \mathbf{u}_1, \mathbf{u}_2, \ldots, \mathbf{u}_m\}$ is linearly dependent.
   - If $W = [\mathbf{v}, \mathbf{u}_1\mathbf{u}_2 \cdots \mathbf{u}_m]$, then the equation $W\mathbf{x} = \mathbf{0}$ has more than the trivial solution.
   - The vector equation $x_1\mathbf{u}_1 + x_2\mathbf{u}_2 \cdots x_m\mathbf{u}_m + x_{m+1}\mathbf{v} = \mathbf{0}$ has more than the trivial solution.
   - The linear system represented by the augmented matrix $[\mathbf{u}_1\mathbf{u}_2 \cdots \mathbf{u}_m\mathbf{v}\ \mathbf{0}]$ has more than the trivial solution.
   - If $T(\mathbf{x}) = U\mathbf{x}$, then $\mathbf{v}$ is in the range of $T(\mathbf{x})$.
   - If $S(\mathbf{x}) = W\mathbf{x}$, then S is not one-to-one.

- S is not invertible.
- null(W) contains more than the zero vector.
- ker(S) contains more than the zero vector.
- dim(range(T)) = dim(range(S)).
- dim(ker(S)) > 0.
- dim(null(W)) > 0.
- $\mathbf{v}$ is in the column space of U.
- dim(col(U)) = dim(col(W)).
- rank(U) = rank(W).
- dim(row(U)) = dim(row(W)).
- det(W) = 0.

5.2 Properties of the Determinant

Section Goals

1. Establish the effects of row operations on the determinant of a given matrix.

2. Formulate the determinant of partitioned matrix in terms of the determinants of its square block submatrices.

3. Prove the Big Theorem part l, theorem 5.12, and theorem 5.13.

4. Provide a formula for the determinant of the inverse of a matrix.

Study Hints

1. Theorem 5.13 provides a convenient way to simplify finding a determinant via row operations. You are encouraged to familiarize yourself with these results.

2. Finding the determinant of a matrix is usually the quickest way to discover if a matrix is invertible.

3. Exercises in this section reveal some useful shortcuts for finding determinants. You are encouraged to practice the techniques needed to complete these exercises.

5.3 Applications of the Determinant

Section Goals

1. Discuss a few applications of determinants.

2. Develop and prove Cramer's Rule, a method for finding components of the solution vector to a given system.

3. Introduce the adjoint of a given matrix and its connection to the cofactor matrix.

4. Establish the relationship between the determinant and the area of a parallelogram.

5. Linear algebra vocabulary: cofactor matrix and adjoint matrix.

Study Hints

1. You may wish to review transformations studied earlier in section 3.1 including:

   - Reflection.

- Rotation.
- Shear.

2. Theorem 5.18 provides an alternative formula for the inverse of a given matrix using its cofactors.

3. The connection between determinants and mapping regions under a given transformation is a concept that appears in other areas of mathematics. Understanding this connection may aid you in other Math courses.

6 Eigenvalues and Eigenvectors

6.1 Eigenvalues and Eigenvectors

Section Goals

1. Introduce the concepts of eigenvalues and eigenvectors.

2. Develop an algebraic method for finding eigenvalues and eigenvectors.

3. Provide a geometric representation of eigenvectors behavior under matrix multiplication.

4. Prove the set of eigenvectors associated with a given eigenvalue along with the zero vector forms a vector space called the eigenspace.

5. Introduce the characteristic polynomial as a tool for finding eigenvalues.

6. Provide many examples of finding characteristic polynomial, eigenvalues, and eigenvectors.

7. Expand the Big Theorem.

8. Linear algebra vocabulary: eigenvalue, eigenvector, eigenspace, characteristic polynomial, characteristic equation, and multiplicity.

Study Hints

1. As you will see later, the concepts of eigenvalues and eigenvectors arise in many other areas of Mathematics. A thorough investigation of the definitions and computations presented in this section will prove helpful.

2. Notice that an eigenspace is just a type of null space. Review of section 4.1 may be beneficial.

3. You may want to review properties of polynomials. In particular, focus on properties of their roots and how to find them.

4. In order to find the characteristic polynomial, you need to be able to comfortably take determinants a review of section 5.1 may be beneficial.

5. To impress on you the importance of the Big Theorem, we restate version 8 here:
   Let $\mathcal{A} = \{\mathbf{a}_1, \ldots, \mathbf{a}_n\}$ be a set of n vectors in $\mathbf{R}^n$, let $A = [\mathbf{a}_1 \cdots \mathbf{a}_n]$, and let $T : \mathbf{R}^n \to \mathbf{R}^n$ be given by $T(\mathbf{x}) = A\mathbf{x}$. Then the following are equivalent:

   (a) $\mathcal{A}$ spans $\mathbf{R}^n$.

   (b) $\mathcal{A}$ is linearly independent.

   (c) $A\mathbf{x} = \mathbf{b}$ has an unique solution for all $\mathbf{b}$ in $\mathbf{R}^n$.

   (d) T is onto.

   (e) T is one-to-one.

   (f) A is invertible.

   (g) $\ker(T) = \{\mathbf{0}\}$.

   (h) $\mathcal{A}$ is a basis for $\mathbf{R}^n$.

   (i) $\mathrm{col}(A) = \mathbf{R}^n$.

   (j) $\mathrm{row}(A) = \mathbf{R}^n$.

   (k) $\mathrm{rank}(A) = n$.

   (l) $\det(A) \neq 0$.

   (m) $\lambda = 0$ is not an eigenvalue of A.

6. Remember, an eigenvalue can be equal to 0, but an eigenvector cannot be equal to **0**. This fact is essential to some of the proofs we will encounter later.

7. We can extend our list from section 5.1 study hints to the following:

   - **v** is an element of span$\{\mathbf{u}_1, \mathbf{u}_1, \ldots, \mathbf{u}_m\}$.
   - **v** is in span$\{\mathbf{u}_1, \mathbf{u}_1, \ldots, \mathbf{u}_m\}$.
   - **v** can be reached by VecMobileM.
   - **v** lies on the "pizza box" with strings $\mathbf{u}_1, \mathbf{u}_2, \ldots, \mathbf{u}_m$.
   - **v** is a linear combination of $\mathbf{u}_1, \mathbf{u}_2, \ldots, \mathbf{u}_m$.
   - The linear system represented by the augmented matrix $[\mathbf{u}_1\mathbf{u}_2 \cdots \mathbf{u}_m\mathbf{v}]$ has at least one solution.
   - The linear system represented by the augmented matrix $[\mathbf{u}_1\mathbf{u}_2 \cdots \mathbf{u}_m\mathbf{v}]$ is consistent.
   - The vector equation $x_1\mathbf{u}_1 + x_2\mathbf{u}_2 \cdots x_m\mathbf{u}_m = \mathbf{v}$ has at least one solution.
   - span$\{\mathbf{v}, \mathbf{u}_1, \mathbf{u}_2, \ldots, \mathbf{u}_m\}$ = span$\{\mathbf{u}_1, \mathbf{u}_2, \ldots, \mathbf{u}_m\}$.
   - If $U = [\mathbf{u}_1\mathbf{u}_2 \cdots \mathbf{u}_m]$, then the equation $U\mathbf{x} = \mathbf{v}$ has at least one solution.
   - The set $\{\mathbf{v}, \mathbf{u}_1, \mathbf{u}_2, \ldots, \mathbf{u}_m\}$ is linearly dependent.
   - If $W = [\mathbf{v}, \mathbf{u}_1\mathbf{u}_2 \cdots \mathbf{u}_m]$, then the equation $W\mathbf{x} = \mathbf{0}$ has more than the trivial solution.
   - The vector equation $x_1\mathbf{u}_1 + x_2\mathbf{u}_2 \cdots x_m\mathbf{u}_m + x_{m+1}\mathbf{v} = \mathbf{0}$ has more than the trivial solution.
   - The linear system represented by the augmented matrix $[\mathbf{u}_1\mathbf{u}_2 \cdots \mathbf{u}_m\mathbf{v} \ \mathbf{0}]$ has more than the trivial solution.
   - If $T(\mathbf{x}) = U\mathbf{x}$, then **v** is in the range of $T(\mathbf{x})$.
   - If $S(\mathbf{x}) = W\mathbf{x}$, then S is not one-to-one.
   - S is not invertible.
   - null(W) contains more than the zero vector.
   - ker(S) contains more than the zero vector.
   - dim(range(T)) = dim(range(S)).
   - dim(ker(S)) > 0.
   - dim(null(W)) > 0.
   - **v** is in the column space of U.
   - dim(col(U)) = dim(col(W)).
   - rank(U) = rank(W).
   - dim(row(U)) = dim(row(W)).
   - det(W) = 0.
   - 0 is an eigenvalue of W.
   - 0 is a root of the characteristic polynomial for W.

6.2 Approximation Methods

Section Goals

1. Introduce more practical approximation methods for finding eigenvalues and eigenvectors of larger matrices.

2. Develop the power method, the shifted power method, the inverse power method, and the shifted inverse power method.

3. Discuss a few computational considerations for these methods.

4. Linear algebra vocabulary: scaling factor, dominant eigenvalue, and shifted power method.

Study Hints

1. Keep in mind all the method thats discussed in this section have their basis in the power method. You should make sure you are comfortable with the power method before continuing to the other methods discussed in this section.

2. Make sure you are dividing by the largest component when implementing the power method.

3. The power method usually converges to the dominant eigenvalue (λ_1) and is not overly sensitive to roundoff errors.

4. The shifted power method usually converges to the eigenvalue (λ_2) furthest from (λ_1).

5. The inverse power method usually converges to the eigenvalue (λ_3) closest to 0.

6. The shifted inverse power method can help turn a rough estimate into an accurate approximation.

7. Note the use of the word "usually." Under computational comments, Holt provides a nice example whose initial guess does not converge to the largest eigenvalue.

6.3 Change of Basis

Section Goals

1. Develop a general procedure for changing from one basis to another.

2. Observe the connection between change of basis and a linear transformation.

3. Display a variety of basis, both standard and nonstandard.

4. Linear algebra vocabulary: coordinate vector and change of basis matrix.

Study Hints

1. If $\mathbf{y}_\mathcal{B}$ is the coordinate vector of $\mathbf{y}$ with respect to the basis $\mathcal{B} = \{\mathbf{b}_1, \ldots, \mathbf{b}_k\}$ then the following are also true:

   (a) If $B = [\mathbf{b}_1 \ldots \mathbf{b}_k]$, then $B\mathbf{y}_\mathcal{B} = \mathbf{y}$.
   (b) $\mathbf{y}$ is a linear combination of the vectors $\mathbf{b}_1, \ldots, \mathbf{b}_k$ with coefficients from the entries of $\mathbf{y}_\mathcal{B}$.
   (c) $\mathbf{y}_\mathcal{B}$ is a solution to the linear system corresponding to the augmented matrix $[B\mathbf{y}]$.

2. Caution! If no subscript is given on the vector, the basis is assumed to be the standard basis.

3. You may want to review inverses. This prerequisite knowledge is vital to your success in this section.

4. Example 6 is a great example that connects eigenvectors, linear transformations, and change of basis.

6.4 Diagonalization

Section Goals

1. Develop a procedure for expressing a square matrix A as a product of three matrices, $A = PDP^{-1}$, where D is a diagonal matrix.

2. Observe how P and D are formed using linearly independent eigenvectors and eigenvalues.

3. Using diagonalization techniques, create a matrix with prescribed eigenvectors and eigenvalues.

4. Illustrate the convenience of using diagonalization for computing matrix powers.

5. Linear algebra vocabulary: diagonalizable matrix.

Study Hints

1. To check your diagonalization computations, we can save time by not computing P^{-1} and check that $AP = PD$ instead.

2. Do not worry if you are lacking linearly independent eigenvectors; this just means that your matrix A is not diagonalizable.

3. Pay attention to theorem 6.16. It can provide a quick way to determine if a given matrix is diagonalizable.

6.5 Complex Eigenvalues

Section Goals

1. Expand our study of eigenvalues to complex eigenvalues.

2. Review complex arithmetic, algebraic properties of complex numbers, graphing, and basic vocabulary.

3. Introduce rotation-dilation matrices and hidden rotation-dilation matrices.

4. Linear algebra vocabulary: real part, imaginary part, modulus, rectangular coordinates, polar coordinates, argument, complex conjugate, rotation-dilation matrices, and hidden rotation-dilation matrices.

Study Hints

1. If it has been a long time since you have studied complex numbers, it is recommended that you review the basics before studying this section. You can usually find this material in a College algebra or Pre-Calculus textbook.

6.6 Systems of Differential Equations

Section Goals

1. Introduce a variety of applications where systems of equations arise involving one or more functions and the derivatives of those functions including:

    (a) Concentrations of insulin and glucose.
    (b) Arms races.
    (c) Systems with complex solutions.

2. Linear algebra vocabulary: differential equation and general solution.

Study Hints

1. Although not necessary, a basic understanding of differential equations may provide some insight into the topics discussed in this section.

7 Vector Spaces

7.1 Vector Spaces and Subspaces

Section Goals

1. Adopt a broader notion of vectors and the spaces containing them.

2. Develop the general definition for a vector and vector space.

3. Introduce and verify the following vector spaces:

    (a) $\mathbf{P}^n$, the vector space of polynomials of degree n or less.

    (b) $\mathbf{R}^{n \times n}$, the vector space of all real $n \times n$ matrices.

    (c) $\mathbf{R}^\infty$, the vector space of all infinite real sequences.

    (d) $C[a, b]$, the vector space of all real-valued continuous functions on $[a, b]$.

4. Discuss a variety of subspaces.

5. Linear algebra vocabulary: vector space, vector, additive inverse, zero vector, subspace, trivial subspaces, and trace.

Study Hints

1. It is recommended you review the properties of the Euclidean Space ($\mathbf{R}^n$) covered in section 4.1.

2. Sometimes, the transition from the Euclidean space to general vector spaces can be difficult. Holt provides three important points to guide you, we summarize them now:

    - Specify both the set of vectors and the arithmetic operations.
    - Vectors are not always a column of numbers.
    - A vector space refers to any set satisfying Definition 7.1.

3. You are encouraged to study the comparison between $\mathbf{R}^3$ and $\mathbf{P}^2$ provided at the beginning of this section.

4. Not every set of vectors will create a vector space! Check out example 2 for an example of a set of vectors that do not form a vector space.

5. Keep in mind that all of $(a) - (f)$ in theorem 7.2 must hold in a vector space. Make sure to verify that the set you are testing satisfies each one.

6. As with the Euclidean space, when verifying a given set is a subspace always check the zero vector first.

7. Holt provides a few examples in this section which require a working knowledge of Calculus. Understanding these examples is not essential to your success but provides a nice connection between Linear algebra and other areas of Mathematics.

7.2 Span and Linear Independence

Section Goals

1. Extend the concepts or span and linear independence from the Euclidean space to general vector spaces.

2. Provide a variety of examples of linearly independent/dependent sets and spanning sets.

3. Linear algebra vocabulary: linear combination, span, linear independence, and linear dependence.

Study Hints

1. It is recommended you review linearly independent/dependent sets and spanning sets covered in section 4.1.

2. You may find it helpful to review basic trigonometric identities before studying this section. A working knowledge of such identities will be essential to determining if a vector is a linear combination of a given set of vectors.

3. Keep in mind there are a few quick checks for linear dependence that carry over from Chapter 4:

   - If the zero vector is in the set, the set is linearly dependent.
   - A set of two vectors in linearly dependent if and only if one of the vectors is a multiple of the other.
   - A set with just one vector is linearly dependent if and only if it is the zero vector.
   - A set of vectors is linearly dependent if and only if one vector in the set is in the span of the others.
   - A set of vectors is linearly dependent if and only if one vector in the set is a linear combination of the others.

7.3 Basis and Dimension

Section Goals

1. Extend the concepts of basis and dimension from the Euclidean space to general vector spaces.

2. Introduce the standard basis for:

   (a) $\mathbf{P}^n$, the vector space of polynomials of degree n or less.

   (b) $\mathbf{R}^{2 \times 2}$, the vector space of all real 2×2 matrices.

3. Provide examples of how to extend a set to form a basis or remove vectors from a set to form a basis for non-Euclidean vector spaces.

4. Linear algebra vocabulary: basis, standard basis, and dimension.

Study Hints

1. It is recommended that you have a solid understanding of the material in sections 4.2, 7.1, and 7.2 before beginning this section.

2. As in the Euclidean space, a basis must both be linearly independent and a spanning set. The process is the same as with the Euclidean space. Suppose we have a set of vectors, S, from a vector space, V, with dimension k.

   (a) Suppose there is a non-trivial solution to the homogeneous equation.

      - Then remove vectors from S until a nontrivial solution exist and go to option b.

   (b) Suppose the only solution to the homogeneous equation is the trivial solution.

      - If the size of S is less than k, then S can be extended to a basis for V.
      - If the size of S is equal to k, then S forms a basis for V.

8 Orthogonality

8.1 Dot Products and Orthogonal Sets

Section Goals

1. Introduce the dot product and orthogonal sets.

2. Extend the notions of perpendicular, angle, and length in $\mathbf{R}^2$ and $\mathbf{R}^3$ to higher dimensions.

3. Develop a formula for using the dot product to find the coefficients needed to write a vector as a linear combination of orthogonal vectors.

4. Linear algebra vocabulary: dot product, norm of a vector, length of a vector, distance between vectors, orthogonal, orthogonal vectors, orthogonal complement, orthogonal set, and orthogonal basis.

Study Hints

1. There are many names for $\|\mathbf{x}\|$ including:

   (a) The length of $\mathbf{x}$.

   (b) The magnitude of $\mathbf{x}$.

   (c) The norm of $\mathbf{x}$.

2. Keep in mind that orthogonal is the same concept as perpendicular. Refer to example 6 for a connection between orthogonal vectors and the pythagorean theorem.

3. Why study orthogonal vectors? They are guaranteed to be linearly independent, so they make good candidates for a basis.

4. Orthogonal complement $S^\perp$ is sometimes referred to as "S perp."

5. A subspace and its complement can be a hard concept to grasp, so we provide a few more examples using the standard basis for $\mathbf{R}^3$:

   (a) If $S = \{\mathbf{i}\}$, then $S^\perp = \{\mathbf{j}, \mathbf{k}\}$.

   (b) If $S = \{\mathbf{j}\}$, then $S^\perp = \{\mathbf{i}, \mathbf{k}\}$.

   (c) If $S = \{\mathbf{k}, \mathbf{i}\}$, then $S^\perp = \{\mathbf{j}\}$.

   (d) If $S = \{\mathbf{i}, \mathbf{j}, \mathbf{k}\}$, then $S^\perp = \{\mathbf{0}\}$.

8.2 Projections and the Gram-Schmidt Process

Section Goals

1. Discuss creating orthogonal basis using projections and the Gram-Schmidt process.

2. Provide a variety of step-by-step examples implementing the Gram-Schmidt process.

3. Linear algebra vocabulary: projection onto a vector, projection onto a subspace, Gram-Schmidt process, orthonormal set, and normalizing.

Study Hints

1. Careful with the notation:

   - $proj_{\mathbf{v}}\,\mathbf{u}$ is the projection of $\mathbf{u}$ onto $\mathbf{v}$.

   - $proj_{\mathbf{u}}\,\mathbf{v}$ is the projection of $\mathbf{v}$ onto $\mathbf{u}$.

2. Theorem 8.14 provides some useful properties of $proj_{\mathbf{v}}\mathbf{u}$, and you are encouraged to studied its results.

3. You can think of projecting a vector onto a subspace like shining a flashlight on an stick (the vector) and viewing its shadow on a flat surface (the subspace).

4. The Gram-Schmidt process takes practice to master, computational errors can occur easily.

8.3 Diagonalizing Symmetric Matrices and QR Factorization

Section Goals

1. Revisit the problem of diagonalizing matrices.

2. Discuss symmetric matrices and their diagonalization properties.

3. Introduce the spectral theorem.

4. Develop QR factorization.

5. Linear algebra vocabulary: orthogonal matrix, orthogonally diagonalizable, and QR factorization.

Study Hints

1. You are encouraged to review section 6.4 before studying this section.

2. It is easy to make a computational mistake when attempting to diagonalize a matrix. You should always double check your answers.

3. Careful! An orthogonal matrix has columns that are orthonormal (orthogonal and have norm one).

4. There are some very useful properties of symmetric matrices:

    (a) All eigenvalues are real.

    (b) Each eigenspace has dimension equal to the multiplicity of the associated eigenvalue.

    (c) The matrix is orthogonally diagonalizable.

5. When implementing QR factorization keep in mind that once Q has been found, we can use matrix multiplication to find R. $R = Q^T A$.

6. The order of the columns of Q will effect the corresponding R. So be careful when checking your answer with another student.

8.4 The Singular Value Decomposition

Section Goals

1. Develop singular value decomposition (SVD).

2. Apply SVD to image compression.

3. Introduce numerical rank and the use of SVD to find it.

4. Linear algebra vocabulary: singular value decomposition, singular values, and machine ϵ.

Study Hints

1. The use of the singular value decomposition is convenient since it can be used on any matrix, including non-square matrices.

2. Holt provides extremely detailed step-by-step examples in example 1 and 2. If you are struggling, you are encouraged to review these.

3. Keep in mind when implementing SVD that the order of V and Σ must match. If $\mathbf{x}$ is an unit eigenvector of $A^T A$ corresponding to λ, then the column that $\mathbf{x}$ is placed in V must be the same column that λ is placed in Σ.

8.5 Least Squares Regression

Section Goals

1. Discuss the problem of fitting functions to data.

2. Develop an approximation method that gives us a way to change a linear system that has no solutions into a new system that has a solutions.

3. Introduce least squares regression.

4. Provide conditions for a unique least squares solution.

5. Display a variety of applications including:

   (a) The boiling point of water and barometric pressure.
   (b) Quarterly surveys to measure consumer confidence.
   (c) The length of a day as a function of the numbers of years that have passed.
   (d) Predicting planetary orbits.

6. Linear algebra vocabulary: least squares solution and normal equations.

Study Hints

1. You are encouraged to review projections covered in section 8.2.

2. Observe that this section highlights how a small perturbation in your system can cause significantly different results.

3. Keep in mind that when using least squares regression, the least squares solution is only an approximation.

9   Linear Transformations

9.1   Definitions and Properties

Section Goals

1. Introduce definitions and basic properties of a linear transformation in the context of a vector space. Extending the definition of a linear transformation to allow domains and codomains that are general vector spaces.

2. Provide detailed examples of transformation that are both linear and non-linear.

3. Restate previous theorems in terms of a linear transformation over general vector spaces.

4. Linear algebra vocabulary: linear transformation, domain, codomain, image, range, kernel, one-to-one, and onto.

Study Hints

1. You are strongly encouraged to review Chapters 3 and 4 before studying this chapter. Most of the definitions, such as one-to-one, onto, kernel, inverses, and coordinate vector carry over almost word-for-word from $\mathbf{R}^n$.

2. The use of the notation $\mathbf{0}_V$ is very important. Considering the "zero vector" could be $\begin{bmatrix} 0 \\ 0 \\ 0 \end{bmatrix}$, $0x^5 + 0x^4 + 0x^3 + 0x^2 + 0x + 0$, or $\begin{bmatrix} 0 & 0 \\ 0 & 0 \end{bmatrix}$.

9.2   Isomorphisms

Section Goals

1. Focus on a special type of linear transformation called an isomorphism.

2. Develop methods for determining when two different vector spaces have the same essential structure.

3. Redefine inverse linear transformations in terms of general vector spaces.

4. Linear algebra vocabulary: isomorphism, isomorphic, inverse, and invertible.

Study Hints

1. You have seen isomorphisms already, we just didn't called them that. In the beginning of section 7.1, we displayed the isomorphism between $\mathbf{R}^3$ and $\mathbf{P}^2$. We prove that this is indeed an isomorphism in example one of this section.

2. Some general points about isomorphisms:

   (a) There does not always exist an isomorphism between any two vector spaces.

   (b) For every isomorphism $T : V \to W$, there is also an isomorphism $S : W \to V$ (usually denoted T^{-1}).

   (c) V is isomorphic to W if and only if $\dim(V) = \dim(W)$.

   (d) If a linear transformation is an isomorphism, then it is invertible.

9.3   The Matrix of a Linear Transformation

Section Goals

1. Establish matrix representations for linear transformations that are similar to those for linear transformations in a Euclidean space.

2. Discuss coordinate vectors in terms of general vector spaces.

3. Linear algebra vocabulary: coordinate vector, matrix of linear transformation, and transformation matrix.

Study Hints

1. You are encouraged to review coordinate vectors in section 6.3.

2. Some important reminders about $\mathbf{v}_{\mathcal{G}} = \begin{bmatrix} c_1 \\ \vdots \\ c_m \end{bmatrix}_{\mathcal{G}}$

   (a) Just because the $\mathbf{v}$ vector "looks like" a vector in Euclidean space, don't forget that it represents a vector in V the space with basis $\mathcal{G}$. For example, $\mathbf{v}_{\mathcal{G}} = \begin{bmatrix} -2 \\ 0 \\ 1 \end{bmatrix}_{\mathcal{G}}$, where $\mathcal{G} = \{1, t^2, t^4\}$ represents $-2 + t^4$.

   (b) Choice of basis matters! For instance, if we changed the basis above to $\mathcal{G} = \{1, t, t^2\}$, we get $-2 + t^2$.

   (c) The order the basis vectors are displayed matters! For instance, if we change the basis above to $\mathcal{G} = \{t, t^2, 1\}$, we get $-2t + 1$.

   (d) Each vector has a unique coordinate vector with respect to $\mathcal{G}$.

3. The matrix of linear transformation is sometimes referred to as the transformation matrix.

9.4 Similarity

Section Goals

1. Consider similar matrices as a way to group matrices based on their relationship to a linear transformation.

2. Continue our exploration of matrix representatives of linear transformations, now focusing on the special cases when $T : V \to V$.

3. Provide a variety of examples demonstrating how to find the change of basis matrix.

4. Linear algebra vocabulary: change of basis matrix, similar matrices, and similarity transformation.

Study Hints

1. You are encouraged to review coordinate vectors in section 6.3.

2. The change of basis matrix is sometimes called the change of coordinates matrix.

3. Very often, we denote the fact that two matrices A and B are similar with $A \sim B$.

4. Theorem 9.21 is a very useful theorem, but use it carefully. Having the same determinant does not imply similarity. Consider: $\det\left(\begin{bmatrix} 0 & 0 \\ 0 & 0 \end{bmatrix} \right) = 0$ and $\det\left(\begin{bmatrix} 1 & 1 \\ 1 & 1 \end{bmatrix} \right) = 0$, yet these matrices are not similar. So $\det(A) = \det(B)$ doesn't tell us much about how A and B relate. Indeed, A and B can be different sizes and still have equal determinants!

10 Inner Product Spaces

10.1 Inner Products

Section Goals

1. Extend the dot product to a similar product in a general vector space.

2. Provide examples of inner products.

3. Generalize results and definitions covered in Chapter 8 in terms of inner products.

4. Linear algebra vocabulary: inner product, inner product space, orthogonal vectors, norm of a vector, and projection onto a vector.

Study Hints

1. This chapter generalizes many results and definitions covered in Chapter 8. You are encouraged to review the material covered in Chapter 8 before beginning this section. Observe how the dot product is replaced by the inner product in most every instance. This is not a coincidence; the dot product is just one example of an inner product.

2. This chapter has many examples and applications that involve Calculus. Although not a requirement, your enjoyment of this chapter would benefit from an understanding of Calculus.

3. You may have seen the $<, >$ notation before in a physics or calculus class as the notation for a vector. In this text, this notation is strictly used for the inner product.

4. Always check how the inner product is defined before making any computations. As with linear transformations, you must know what the rule is first before you can begin work.

10.2 The Gram-Schmidt Process Revisited

Section Goals

1. Develop the Gram-Schmidt process in context of a general inner product space.

2. Define orthogonal sets in terms of general inner products.

3. Provide examples displaying the Gram-Schmidt process using a variety of inner products.

4. Linear algebra vocabulary: orthogonal set, orthogonal basis, normal vector, orthonormal basis, and projection onto a subspace.

Study Hints

1. Holt provides ten detailed examples in this section! A thorough study of these, as well as a review of examples in section 8.2, is encouraged.

10.3 Applications of Inner Products

Section Goals

1. Introduce a few applications of inner products involving:

    (a) Least suares regression.

    (b) Fourier approximations.

    (c) Discrete fourier transformations.

2. Linear algebra vocabulary: least squares regression, weighted least squares regression, weights, n^{th}-order Fourier approximation, Fourier coefficients, discrete Fourier coefficients, and n^{th}-order discrete Fourier approximation.

Study Hints

1. The applications presented in this section are commonly seen in other areas of mathematics. You are encouraged to study these examples thoroughly.

2. In this section, a vector $\vec{s}$ is closest to $\vec{v}$ when the norm of their difference $||\vec{v} - \vec{s}||$ is as small as possible.

11 Additional Topics and Applications

11.1 Quadratic Forms

Section Goals

1. Study the quadratic form for varying Q.

2. Discuss how to find both the quadratic form, Q, and the matrix of the quadratic form, A.

3. Linear algebra vocabulary: quadratic form, matrix of quadratic form, cross-product, principal axes, standard position, positive definite, positive semidefinite, negative definite, negative semidefinite, and indefinite.

Study Hints

1. Finding the matrix of the quadratic form can be difficult, but a quick observation can ease your search. As discussed in example 2, the diagonal entries of A are the coefficients of x_i^2 in Q.

2. This section requires you to orthogonally diagonalize a matrix a review of section 8.3 is recommended.

3. Steps for graphing the set of solutions to a given quadratic form:

    (a) Find the matrix of the quadratic form (example 2).

    (b) Find D and P, essentially orthogonally diagonalizing A (example 3).

    (c) Graph $\mathbf{y}^T D \mathbf{y} = c$ (standard form, examples 4 and 5).

    (d) Rotate the graph so that the axes of symmetry align with the principal axes of A to graph $\mathbf{x}^T A \mathbf{x} = c$ (examples 4 and 5).

11.2 Positive Definite Matrices

Section Goals

1. Extend definition 11.3 to the matrix of a quadratic form.

2. Discuss a class of matrices which are guaranteed to have an LU-factorization.

3. Introduce Cholesky decomposition.

4. Linear algebra vocabulary: positive definite, negative definite, indefinite, leading principal submatrix, and Cholesky decomposition.

Study Hints

1. You are encouraged to review LU-factorization in section 3.4 before studying this section.

2. Holt provides a step-by-step demonstration of LU-factorization in example 4. If you are struggling, you are encouraged to follow the steps carefully laid out in this example.

11.3 Constrained Optimization

Section Goals

1. Consider the problem of finding the maximum or minimum value of a quadratic form $Q(\mathbf{x})$, where $\mathbf{x}$ ranges over a set of vectors that satisfies one of the following types of constraints:

    (a) $\|\mathbf{x}\| = 1$.

(b) $\|\mathbf{x}\| = c$.

(c) Polynomial constraints.

(d) Orthogonal constraints.

Study Hints

1. Techniques presented in this section require finding eigenvalues, quadratic forms, and matrix of a quadratic form. You are encouraged to review these concepts before studying this section.

11.4 Complex Vector Spaces

Section Goals

1. Introduce complex vector spaces.

2. Extend concepts such as linear combination, linear independence, span, inner product, basis, and subspaces from real to complex vector spaces.

3. Linear algebra vocabulary: complex vector space, complex vector, inner product, complex inner product space, unitary space, complex dot product, norm, length, Cauchy-Schwarz inequality, triangle inequality, orthogonal, orthonormal, orthogonal set, and orthonormal set.

Study Hints

1. Review section 7.1 and concepts such as linear combination, linear independence, span, inner product, basis, and subspaces over real vector spaces.

2. Notice that the only real difference between this section and section 7.1 is that we allow for complex scalars.

11.5 Hermitian Matrices

Section Goals

1. Discuss orthogonally diagonalizable A when A is complex.

2. Introduce unitary and Hermitian matrices.

3. Develop a procedure for diagonalizing complex A.

4. Establish the correct analog of the Spectral Theorem for complex matrices.

5. Linear algebra vocabulary: complex conjugate, conjugate transpose, unitary matrix, unitarily diagonalizable, Hermitian, and normal.

Study Hints

1. Review section 6.4 over diagonalization, 6.5 over basic properties of complex numbers, and 8.3 over the spectral theorem.

2. Steps for diagonalizing a normal matrix A:

    (a) Find the eigenvalues and corresponding eigenvectors of A.

    (b) For each distinct eigenvalue, apply Gram-Schmidt as needed to find an orthonormal basis for the associated eigenspace.

    (c) Define D to be the diagonal matrix with the eigenvalues of A along the diagonal and P to be the unitary matrix with the corresponding eigenvectors for columns.

Part 2: Solutions to Odd Numbered Exercises

Chapter 1

Systems of Linear Equations

1.1 Lines and Linear Equations

1. $2\,(1) - 5(-2) = 12 \neq 9$, so $(1, -2)$ does not lie on the line $2x_1 - 5x_2 = 9$.
   $2\,(-3) - 5(-3) = 9$, so $(-3, -3)$ lies on the line $2x_1 - 5x_2 = 9$.
   $2\,(-2) - 5(-3) = 11 \neq 9$, so $(-2, -3)$ does not lie on the line $2x_1 - 5x_2 = 9$.

3. $3(-1) + (2) = -1$ and $(-5)(-1) + 2(2) = 9 \neq 20$, so $(-1, 2)$ does not lie on both lines $3x_1 + x_2 = -1$ and $-5x_1 + 2x_2 = 20$.
   $3(-2) + (5) = -1$ and $(-5)(-2) + 2(5) = 20 = 20$, so $(-2, 5)$ lies on both lines $3x_1 + x_2 = -1$ and $-5x_1 + 2x_2 = 20$.
   $3(1) + (-5) = -2 \neq -1$ and $(-5)(1) + 2(-5) = -15 \neq 20$, so $(1, -5)$ does not lie on both lines $3x_1 + x_2 = -1$ and $-5x_1 + 2x_2 = 20$.

5. $-2(1) + 9(2) - (3) = 13 \neq -10$, so $(1, 2, 3)$ does not satisfy the first equation of the linear system.
   $-2(1) + 9(-1) - (1) = -12 \neq -10$, so $(1, -1, 1)$ does not satisfy the first equation of the linear system.
   $(-1) - 5(-2) + 2(-6) = -3 \neq 4$ so $(-1, -2, -6)$ does not satisfy the second equation of linear system.

7. (a) Not a solution, since $-2(-3 + s_1 + s_2) + 3(s_1) + 2(s_2) = s_1 + 6 \neq 6$ for every s_1.

   (b) A solution, since $-2(-3 + 3s_1 + s_2) + 3(2s_1) + 2(s_2) = 6$.

   (c) A solution, since $-2(3s_1 + s_2) + 3(2s_1 + 2) + 2(s_2) = 6$.

   (d) A solution, since $-2(s_1) + 3(s_2) + 2(3 - 3s_2/2 + s_1) = 6$.

9. $3x_1 + 5x_2 = 4 \;\Rightarrow\; x_2 = \frac{4}{5} - \frac{3}{5}x_1$. Substitute into the second equation to obtain $2x_1 - 7\left(\frac{4}{5} - \frac{3}{5}x_1\right) = 13 \;\Rightarrow\; x_1 = 3$. Thus $x_2 = (4 - 3(3))/5 = -1$.

11. $-10x_1 + 4x_2 = 2 \;\Rightarrow\; x_2 = \frac{5}{2}x_1 + \frac{1}{2}$. Substitute into the second equation to obtain $15x_1 - 6\left(\frac{5}{2}x_1 + \frac{1}{2}\right) = -3 \;\Rightarrow\; -3 = -3$, which is true for all x_1. Hence we may set x_1 as a free variable, $x_1 = s_1$ and then $x_2 = \frac{5}{2}s_1 + \frac{1}{2}$.

13. $7x_1 - 3x_2 = -1 \;\Rightarrow\; x_2 = \frac{7}{3}x_1 + \frac{1}{3}$. Substitute into the second equation to obtain $-5x_1 + 8\left(\frac{7}{3}x_1 + \frac{1}{3}\right) = 0 \;\Rightarrow\; x_1 = -\frac{8}{41}$. Thus $x_2 = \frac{7}{3}\left(-\frac{8}{41}\right) + \frac{1}{3} = -\frac{5}{41}$.

15. Echelon form. Leading variables: x_1 and x_2. No free variables.

17. Echelon form. Leading variables: x_1 and x_3. Free variable: x_2.

19. Not in echelon form since x_2 is a leading variable in both equations 2 and 3.

21. Echelon form. Leading variables: x_1 and x_3. Free variables: x_2 and x_4.

23. Equation 2 $\;\Rightarrow\; x_2 = 5$. Substitute into equation 1, $-5x_1 - 3(5) = 4 \;\Rightarrow\; x_1 = -\frac{19}{5}$.

25. x_2 is a free variable, so let $x_2 = s_1$. Substitute, $-3x_1 + 4s_1 = 2 \Rightarrow x_1 = \frac{4}{3}s_1 - \frac{2}{3}$.

27. x_3 is a free variable, so let $x_3 = s_1$. Equation 3 $\Rightarrow$ $x_4 = 5$. Substitute into equation 2, $-2x_2 + s_1 - 5 = -1 \Rightarrow x_2 = \frac{1}{2}s_1 - 2$. Substitute into equation 1, $x_1 + 5\left(\frac{1}{2}s_1 - 2\right) - 2s_1 = 0 \Rightarrow x_1 = 10 - \frac{1}{2}s_1$.

29. x_2 and x_4 are free variables, so let $x_2 = s_1$ and $x_4 = s_2$. Equation 2 $\Rightarrow$ $-3x_3 + s_2 = -4 \Rightarrow x_3 = \frac{1}{3}s_2 + \frac{4}{3}$. Substitute into equation 1, $-2x_1 + s_1 + 2\left(\frac{1}{3}s_2 + \frac{4}{3}\right) = 1 \Rightarrow x_1 = \frac{1}{2}s_1 + \frac{1}{3}s_2 + \frac{5}{6}$.

31. Interchange equations 1 and 2, to obtain:

$$3x_1 + 2x_2 = 1$$
$$-5x_2 = 4$$

Equation 2 $\Rightarrow$ $x_2 = -4/5$, and substituting into equation 1, $3x_1 + 2(-4/5) = 1 \Rightarrow x_1 = \frac{13}{15}$.

33. Interchange equations 1 and 2 to obtain:

$$x_1 + 3x_2 - 2x_3 + 2x_4 = -1$$
$$2x_2 + x_3 - 5x_4 = 0$$

x_3 and x_4 are free variables, so let $x_3 = s_1$ and $x_4 = s_2$. Substitute into equation 2, $2x_2 + s_1 - 5s_2 = 0 \Rightarrow x_2 = \frac{5}{2}s_2 - \frac{1}{2}s_1$. Substitute into equation 1, $x_1 + 3\left(\frac{5}{2}s_2 - \frac{1}{2}s_1\right) - 2s_1 + 2s_2 = -1 \Rightarrow x_1 = \frac{7}{2}s_1 - \frac{19}{2}s_2 - 1$.

35. From the first equation, $6x_1 - 5x_2 = 4$, we obtain $x_1 = \frac{5}{6}x_2 + \frac{2}{3}$. Substitute into equation 2, $9\left(\frac{5}{6}x_2 + \frac{2}{3}\right) + kx_2 = 1 \Rightarrow \left(\frac{15}{2} + k\right)x_2 = -5$. Hence the system is consistent provided $\frac{15}{2} + k \neq 0$, which means $k \neq -\frac{15}{2}$.

37. Subtract the equations to obtain $(2-h)x_1 = -1 - k$. If $h \neq 2$, the system will be consistent. If $h = 2$, then $0 = -1 - k$, and the system has no solutions if $-1 - k \neq 0$, i.e. $k \neq -1$. Hence, the system has no solutions if and only if $h = 2$ and $k \neq -1$.
    Alternatively, there will be no solution if and only if the two lines are parallel and distinct. Thus we conclude that $h = 2$ and $k \neq -1$.

39. There are 9 variables, as every variable is either a leading variable or free variable.

41. There are 7 leading variables, since the number of leading variables of a system in echelon form is equal to the number of equations.

43. For example,

$$x_1 = 0$$
$$x_2 = 0$$
$$x_3 = 0$$

45. For example,

$$
\begin{array}{ccccccc}
x_1 & + & x_2 &   &   & = & 0 \\
x_1 & + & x_2 & - & x_3 & = & 0 \\
    &   &   &   & x_3 & = & 0 \\
x_1 & + & x_2 & + & x_3 & = & 0 \\
\end{array}
$$

47. On Monday, I bought 3 apples and 4 oranges and spent $0.55. On Tuesday I bought 6 oranges and spent $0.60. How much does each apple and orange cost?
    Solution: let x_1 be the price of an apple, and x_2 the price of an orange, then we have the following system in echelon form:

$$3x_1 + 4x_2 = 0.55$$
$$6x_2 = 0.60$$

From equation 2, $x_2 = 0.10$. Substitute into equation 1, $3x_1 + 4(0.10) = 0.55$, $\Rightarrow x_1 = 0.05$. Hence apples cost 5 cents each and oranges cost 10 cents each.

49. For example,

$$
\begin{array}{rcrcrcr}
x_1 & - & x_2 & & & = & -3 \\
3x_1 & & & - & x_3 & = & 4
\end{array}
$$

51. False. Example:

$$
\begin{array}{rcrcr}
x_1 & & & = & 0 \\
& & x_2 & = & 0 \\
x_1 & + & x_2 & = & 0
\end{array}
$$

53. False. Consider the equation $x_1 + x_2 = 1$. One can set $x_1 = s_1$ and then $x_2 = 1 - s_1$. Or one can set $x_2 = s_1$ and then $x_1 = 1 - s_1$.

55. True. The leading variable moves one column to the right each time you descend one row.

57. False. Each equation in an echelon system has a unique leading variable, so back substitution is always possible. Hence a solution always exists.

59. True. The last equation would be $c_n x_n = b_n$, so $x_n = \frac{b_n}{c_n}$ is rational. And then using back substitution, each proceeding variable would be rational, as it is determined from a sum of rational numbers, divided by an integer. In this manner, we see that each variable, x_i in the solution is a rational number.

61. The total amount of glycol needed is now $0.29(300) = 87.0$ liters. Thus the system of equations becomes

$$
\begin{array}{rcrcr}
x & + & y & = & 300 \\
0.18x & + & 0.50y & = & 87
\end{array}
$$

Solving the first equation for x, we obtain $x = 300 - y$. Substitute into the second equation to get $0.18(300 - y) + 0.50y = 87$, $\Rightarrow$ $y = 103.125$ liters. Hence $x = 300 - 103.125 = 196.875$ liters.

63. Let x be the number of adults who attended, and y the number of children who attended. Since the total number of people who attended is 385, we have $x + y = 385$. The total revenue from the sale of tickets will be the revenue due to the adult tickets purchased and the children's tickets purchased. We obtain a second equation, $11x + 8y = 3974$. Solving the first equation for x, we have $x = 385 - y$. Substitute into the second equation, we have $11(385 - y) + 8y = 3974 \Rightarrow 4235 - 3y = 3974$, $\Rightarrow$ $y = 87$. Solving now for x, we determine $x = 385 - 87 = 298$. So 298 adults and 87 children attended.

65. Using $f(0) = 5$, we have $5 = a_1 e^{2(0)} + a_2 e^{-3(0)} = a_1 + a_2$. Using $f'(0) = -1$, we have $-1 = 2a_1 e^{2(0)} - 3a_2 e^{-3(0)} = 2a_1 - 3a_2$. Solving the first equation for a_1, we have $a_1 = 5 - a_2$. Substitute into the second equation, $-1 = 2(5 - a_2) - 3a_2 \Rightarrow a_2 = \frac{11}{5}$. Therefore $a_1 = 5 - \frac{11}{5} = \frac{14}{5}$.

67. Using the freezing point of water, we have $0 = a(32) + b$. From the boiling point of water, $100 = a(212) + b$. From the first equation, $b = -32a$. Substitute into the second equation, $100 = a(212) + (-32a) = 180a \Rightarrow a = \frac{5}{9}$. Hence $b = -32\left(\frac{5}{9}\right) = -\frac{160}{9}$.

69. After experimenting a bit, we get that 4 nickels and 8 quarters just about cover the long side. The short side is covered by either 9 nickels and 1 quarter, or 1 nickel and 8 quarters. Let n be the diameter of a nickel, and q the diameter of a quarter. The first equation becomes $4n + 8q = 11$, so $n = \frac{11}{4} - 2q$. With the choice of 9 nickels and 1 quarter, the second equation is $9n + q = 8.5$. Substituting for n, $9\left(\frac{11}{4} - 2q\right) + q = 8.5$, and hence $q = 0.95588$ in. And thus $n = \frac{11}{4} - 2(0.95588) = 0.83824$ in.
Using instead 1 nickel and 8 quarters for the second equation, we have $n + 8q = 8.5$. Substituting, $\left(\frac{11}{4} - 2q\right) + 8q = 8.5$, and we get $q = 0.95833$ in. Thus $n = \frac{11}{4} - 2(0.95833) = 0.83334$ in. The published values from the United States Mint are $q = 0.955$ in and $n = 0.835$ in.

71. $x_1 = 12$, $x_2 = 5$.

73. $x_1 = \frac{33}{8}s_1$, $x_2 = \frac{9}{4}s_1 - \frac{23}{11}$, $x_3 = s_1 - \frac{5}{33}$

75. $x_1 = \frac{47}{8}s_1$, $x_2 = -2s_1 + \frac{69}{47}$, $x_3 = \frac{7}{4}s_1 + \frac{565}{141}$, $x_4 = s_1 + \frac{202}{141}$

1.2   Linear Systems and Matrices

1. $\begin{array}{rrrrr} 4x_1 & + & 2x_2 & - & x_3 & = & 2 \\ -x_1 & & & + & 5x_3 & = & 7 \end{array}$

3. $\begin{array}{rrrrrrr} & & 12x_2 & - & 3x_3 & - & 9x_4 & = & 17 \\ -12x_1 & + & 5x_2 & - & 3x_3 & + & 11x_4 & = & 0 \\ 6x_1 & + & 8x_2 & + & 2x_3 & + & 10x_4 & = & -8 \\ 17x_1 & & & & & + & 13x_4 & = & -1 \end{array}$

5. Echelon form.

7. Not echelon form.

9. Echelon form.

11. $-2R_1 \Rightarrow R_1$

13. $-2R_2 + R_3 \Rightarrow R_3$

15. $R_1 \Leftrightarrow R_2$, $\begin{bmatrix} -1 & 4 & \mathbf{3} \\ 3 & 7 & -2 \\ 5 & 0 & -3 \end{bmatrix}$

17. $2R_1 \Rightarrow R_1$, $\begin{bmatrix} \mathbf{0} & 6 & -2 & 4 \\ -1 & -9 & 4 & 1 \\ 5 & 0 & 7 & 2 \end{bmatrix}$

19. $\begin{bmatrix} 2 & 1 & 1 \\ -4 & -1 & 3 \end{bmatrix} \underset{\sim}{\overset{2R_1+R_2 \Rightarrow R_2}{}} \begin{bmatrix} 2 & 1 & 1 \\ 0 & 1 & 5 \end{bmatrix}$

   Row 2 $\Rightarrow x_2 = 5$. Row 1 $\Rightarrow 2x_1 + (5) = 1 \Rightarrow x_1 = -2$.

21. $\begin{bmatrix} -2 & 5 & -10 & 4 \\ 1 & -2 & 3 & -1 \\ 7 & -17 & 34 & -16 \end{bmatrix} \underset{\sim}{\overset{R_1 \Leftrightarrow R_2}{}} \begin{bmatrix} 1 & -2 & 3 & -1 \\ -2 & 5 & -10 & 4 \\ 7 & -17 & 34 & -16 \end{bmatrix}$

   $\underset{\sim}{\overset{2R_1+R_2 \Rightarrow R_2}{\scriptstyle -7R_1+R_3 \Rightarrow R_3}} \begin{bmatrix} 1 & -2 & 3 & -1 \\ 0 & 1 & -4 & 2 \\ 0 & -3 & 13 & -9 \end{bmatrix}$

   $\underset{\sim}{\overset{3R_2+R_3 \Rightarrow R_3}{}} \begin{bmatrix} 1 & -2 & 3 & -1 \\ 0 & 1 & -4 & 2 \\ 0 & 0 & 1 & -3 \end{bmatrix}$

   Row 3 $\Rightarrow x_3 = -3$. Row 2 $\Rightarrow x_2 - 4(-3) = 2 \Rightarrow x_2 = -10$. Row 1 $\Rightarrow x_1 - 2(-10) + 3(-3) = -1 \Rightarrow x_1 = -12$.

23. $\begin{bmatrix} 2 & 2 & -1 & 8 \\ -1 & -1 & 0 & -3 \\ 3 & 3 & 1 & 7 \end{bmatrix} \underset{\sim}{\overset{R_1 \Leftrightarrow R_2}{}} \begin{bmatrix} -1 & -1 & 0 & -3 \\ 2 & 2 & -1 & 8 \\ 3 & 3 & 1 & 7 \end{bmatrix}$

   $\underset{\sim}{\overset{2R_1+R_2 \Rightarrow R_2}{\scriptstyle 3R_1+R_3 \Rightarrow R_3}} \begin{bmatrix} -1 & -1 & 0 & -3 \\ 0 & 0 & -1 & 2 \\ 0 & 0 & 1 & -2 \end{bmatrix}$

   $\underset{\sim}{\overset{R_2+R_3 \Rightarrow R_3}{}} \begin{bmatrix} -1 & -1 & 0 & -3 \\ 0 & 0 & -1 & 2 \\ 0 & 0 & 0 & 0 \end{bmatrix}$

   Free variable, $x_2 = s_1$. Row 2 $\Rightarrow x_3 = -2$. Row 1 $\Rightarrow -x_1 - (s_1) = -3 \Rightarrow x_1 = 3 - s_1$.

25. $\begin{bmatrix} 2 & 6 & -9 & -4 & 0 \\ -3 & -11 & 9 & -1 & 0 \\ 1 & 4 & -2 & 1 & 0 \end{bmatrix}$ $\underset{\sim}{R_1 \Leftrightarrow R_3}$ $\begin{bmatrix} 1 & 4 & -2 & 1 & 0 \\ -3 & -11 & 9 & -1 & 0 \\ 2 & 6 & -9 & -4 & 0 \end{bmatrix}$

$\underset{\sim}{\overset{3R_1+R_2 \Rightarrow R_2}{-2R_1+R_3 \Rightarrow R_3}}$ $\begin{bmatrix} 1 & 4 & -2 & 1 & 0 \\ 0 & 1 & 3 & 2 & 0 \\ 0 & -2 & -5 & -6 & 0 \end{bmatrix}$

$\underset{\sim}{2R_2+R_3 \Rightarrow R_3}$ $\begin{bmatrix} 1 & 4 & -2 & 1 & 0 \\ 0 & 1 & 3 & 2 & 0 \\ 0 & 0 & 1 & -2 & 0 \end{bmatrix}$

Free variable, $x_4 = s_1$. Row 3 $\Rightarrow$ $x_3 - 2s_1 = 0 \Rightarrow x_3 = 2s_1$. Row 2 $\Rightarrow$ $x_2 + 3(2s_1) + 2s_1 = 0 \Rightarrow$ $x_2 = -8s_1$. Row 1 $\Rightarrow$ $x_1 + 4(-8s_1) - 2(2s_1) + s_1 = 0 \Rightarrow x_1 = 35s_1$.

27. $\begin{bmatrix} -2 & -5 & 0 \\ 1 & 3 & 1 \end{bmatrix}$ $\underset{\sim}{R_1 \Leftrightarrow R_2}$ $\begin{bmatrix} 1 & 3 & 1 \\ -2 & -5 & 0 \end{bmatrix}$

$\underset{\sim}{2R_1+R_2 \Rightarrow R_2}$ $\begin{bmatrix} 1 & 3 & 1 \\ 0 & 1 & 2 \end{bmatrix}$

$\underset{\sim}{-3R_2+R_1 \Rightarrow R_1}$ $\begin{bmatrix} 1 & 0 & -5 \\ 0 & 1 & 2 \end{bmatrix}$

Thus $x_1 = -5$ and $x_2 = 2$.

29. $\begin{bmatrix} 2 & 1 & 0 & 2 \\ -1 & -1 & -1 & 1 \end{bmatrix}$ $\underset{\sim}{R_1 \Leftrightarrow R_2}$ $\begin{bmatrix} -1 & -1 & -1 & 1 \\ 2 & 1 & 0 & 2 \end{bmatrix}$

$\underset{\sim}{2R_1+R_2 \Rightarrow R_2}$ $\begin{bmatrix} -1 & -1 & -1 & 1 \\ 0 & -1 & -2 & 4 \end{bmatrix}$

$\underset{\sim}{-R_2+R_1 \Rightarrow R_1}$ $\begin{bmatrix} -1 & 0 & 1 & -3 \\ 0 & -1 & -2 & 4 \end{bmatrix}$

$\underset{\sim}{\overset{-R_1 \Rightarrow R_1}{-R_2 \Rightarrow R_2}}$ $\begin{bmatrix} 1 & 0 & -1 & 3 \\ 0 & 1 & 2 & -4 \end{bmatrix}$

Free variable, $x_3 = s_1$. Row 1 $\Rightarrow$ $x_1 = 3 + s_1$. Row 2 $\Rightarrow$ $x_2 = -4 - 2s_1$.

31. $(1/5)R_1 \Rightarrow R_1$

33. $R_1 \Leftrightarrow R_3$

35. $5R_2 + R_6 \Rightarrow R_6$

37. $\begin{bmatrix} 1 & 1 & 1 & 1 & 1 \\ 0 & 1 & 1 & 1 & 1 \\ 0 & 0 & 1 & 1 & 1 \end{bmatrix}$

39. $\begin{bmatrix} 1 & 0 & 0 & 4 \\ 0 & 1 & 0 & 3 \\ 0 & 0 & 1 & 2 \\ 0 & 0 & 0 & 1 \end{bmatrix}$

41. $\begin{aligned} x_1 & & & = 0 \\ & x_2 & & = 0 \\ & & x_3 + x_4 & = 0 \end{aligned}$

43. True, by definition of equivalent matrices.

45. False, by Theorem 1.5.

47. False. For example, all seven equations in the system could be $x_1 + x_2 + x_3 + x_4 = 0$, which is consistent, making the system consistent.

49. True. If it is consistent, there will be at least one free variable, and hence infinitely many solutions.

51. Exactly one solution. The last row produces a unique value for the last variable, and then back substitution produces a unique value for each preceding variable.

53. Either the system has free variables or not. If there are no free variables and the system is consistent, then every variable is a leading variable, and there will be exactly one solution. If there exists a free variable, then there will be infinitely many solutions. Thus, if there are two distinct solutions then it follows that there must be infinitely many solutions.

55. Every homogeneous system is consistent. There will be free variables, as the number of leading variables is no greater than the number of equations, and there are more variables than equations. Since there are free variables, there must be infinitely many solutions.

57. Apply $f(1) = 4$ to obtain $a(1)^2 + b(1) + c = 4 \Rightarrow a + b + c = 4$. From $f(2) = 7$, we have $a(2)^2 + b(2) + c = 7 \Rightarrow 4a + 2b + c = 7$. And $f(3) = 14 \Rightarrow a(3)^2 + b(3) + c = 14 \Rightarrow 9a + 3b + c = 14$. Write these equations as an augmented matrix and solve.

$$\begin{bmatrix} 1 & 1 & 1 & 4 \\ 4 & 2 & 1 & 7 \\ 9 & 3 & 1 & 14 \end{bmatrix} \begin{array}{c} -4R_1 + R_2 \Rightarrow R_2 \\ -9R_1 + R_3 \Rightarrow R_3 \\ \sim \end{array} \begin{bmatrix} 1 & 1 & 1 & 4 \\ 0 & -2 & -3 & -9 \\ 0 & -6 & -8 & -22 \end{bmatrix}$$

$$\begin{array}{c} -3R_2 + R_3 \Rightarrow R_3 \\ \sim \end{array} \begin{bmatrix} 1 & 1 & 1 & 4 \\ 0 & -2 & -3 & -9 \\ 0 & 0 & 1 & 5 \end{bmatrix}$$

Row 3 $\Rightarrow c = 5$. Row 2 $\Rightarrow -2b - 3(5) = -9 \Rightarrow b = -3$. Row 1 $\Rightarrow a + (-3) + (5) = 4 \Rightarrow a = 2$. Thus $f(x) = 2x^2 - 3x + 5$.

59. From a plot, the points do not appear linear, so we use a quadratic to model the data. Let $E(x) = ax^2 + bx + c$. Then $E(20) = 288$, $E(40) = 364$, and $E(60) = 360$. We obtain the three equations

$$\begin{array}{ccccccc} 400a & + & 20b & + & c & = & 288 \\ 1600a & + & 40b & + & c & = & 364 \\ 3600a & + & 60b & + & c & = & 360 \end{array}$$

and solve using the corresponding augmented matrix using a computer algebra system. We obtain $a = -\frac{1}{10}$, $b = \frac{49}{5}$, and $c = 132$. Thus $E(x) = -\frac{1}{10}x^2 + \frac{49}{5}x + 132$.

61. Using a computer algebra system, the row echelon form is $\begin{bmatrix} 1 & 0 & 0 & -\frac{157}{181} \\ 0 & 1 & 0 & \frac{20}{181} \\ 0 & 0 & 1 & -\frac{58}{181} \end{bmatrix}$. Hence $x_1 = -\frac{157}{181}$, $x_2 = \frac{20}{181}$, $x_3 = -\frac{58}{181}$. (Or, as a decimal, we obtain $\begin{bmatrix} 1 & 0 & 0 & -0.8674 \\ 0 & 1 & 0 & 0.1105 \\ 0 & 0 & 1 & -0.3204 \end{bmatrix}$, so $x_1 = -0.8674$, $x_2 = 0.1105$, and $x_3 = -0.3204$.)

63. Using a computer algebra system, the row echelon form is $\begin{bmatrix} 1 & 0 & 0 & 1 & \frac{7}{9} \\ 0 & 1 & 0 & 1 & -\frac{23}{9} \\ 0 & 0 & 1 & -1 & -\frac{22}{27} \end{bmatrix}$. We have a free variable, $x_4 = s_1$. Thus $x_1 = \frac{7}{9} - s_1$, $x_2 = -\frac{23}{9} - s_1$, and $x_3 = -\frac{22}{27} + s_1$. (Or, as a decimal, we obtain $\begin{bmatrix} 1 & 0 & 0 & 1 & 0.77778 \\ 0 & 1 & 0 & 1 & -2.5556 \\ 0 & 0 & 1 & -1 & -0.81481 \end{bmatrix}$, so $x_1 = 0.77778 - s_1$, $x_2 = -2.5556 - s_1$, and $x_3 = -0.81\,81 + s_1$.)

65. Using a computer algebra system, the row echelon form is $\begin{bmatrix} 1 & 0 & 0 & 0 \\ 0 & 1 & 0 & 0 \\ 0 & 0 & 1 & 0 \\ 0 & 0 & 0 & 1 \end{bmatrix}$. Since the last row corresponds to $0 = 1$, the linear system is inconsistent, and there are no solutions.

67. Using a computer algebra system, the row echelon form is $\begin{bmatrix} 1 & 0 & 0 & 0 & -\frac{46}{579} & 0 \\ 0 & 1 & 0 & 0 & -\frac{745}{579} & 0 \\ 0 & 0 & 1 & 0 & \frac{2264}{579} & 0 \\ 0 & 0 & 0 & 1 & \frac{655}{386} & 0 \end{bmatrix}$. We have a free

variable, $x_5 = s_1$.

Thus $x_1 = \frac{46}{579}s_1$, $x_2 = -\frac{745}{579}s_1$, $x_3 = -\frac{2264}{579}s_1$, and $x_4 = -\frac{655}{386}s_1$.

(*Or, as a decimal, we obtain* $\begin{bmatrix} 1 & 0 & 0 & 0 & -0.07947 & 0 \\ 0 & 1 & 0 & 0 & -1.2867 & 0 \\ 0 & 0 & 1 & 0 & 3.9102 & 0 \\ 0 & 0 & 0 & 1 & 1.6969 & 0 \end{bmatrix}$, *so* $x_1 = 0.07947s_1$, $x_2 = 1.2867s_1$,

$x_3 = -3.9102s_1$ *and* $x_4 = -1.6969s_1$.)

1.3   Numerical Solutions

1. $\begin{bmatrix} -2 & 3 & 4 \\ 5 & -2 & 1 \end{bmatrix}$ $\underset{\sim}{\overset{R_1 \Leftrightarrow R_2}{}}$ $\begin{bmatrix} 5 & -2 & 1 \\ -2 & 3 & 4 \end{bmatrix}$

$\underset{\sim}{\overset{(2/5)R_1 + R_2 \Rightarrow R_2}{}}$ $\begin{bmatrix} 5 & -2 & 1 \\ 0 & \frac{11}{5} & \frac{22}{5} \end{bmatrix}$

Row 2 $\Rightarrow$ $\frac{11}{5}x_2 = \frac{22}{5}$ $\Rightarrow$ $x_2 = 2$. Row 1 $\Rightarrow$ $5x_1 - 2(2) = 1$ $\Rightarrow$ $x_1 = 1$.

3. $\begin{bmatrix} 1 & 1 & -2 & -3 \\ 3 & -2 & 2 & 9 \\ 6 & -7 & -1 & 4 \end{bmatrix}$ $\underset{\sim}{\overset{R_1 \Leftrightarrow R_3}{}}$ $\begin{bmatrix} 6 & -7 & -1 & 4 \\ 3 & -2 & 2 & 9 \\ 1 & 1 & -2 & -3 \end{bmatrix}$

$\underset{\sim}{\overset{(-1/2)R_1 + R_2 \Rightarrow R_2}{\overset{(-1/6)R_1 + R_3 \Rightarrow R_3}{}}}$ $\begin{bmatrix} 6 & -7 & -1 & 4 \\ 0 & \frac{3}{2} & \frac{5}{2} & 7 \\ 0 & \frac{13}{6} & -\frac{11}{6} & -\frac{11}{3} \end{bmatrix}$

$\underset{\sim}{\overset{R_2 \Leftrightarrow R_3}{}}$ $\begin{bmatrix} 6 & -7 & -1 & 4 \\ 0 & \frac{13}{6} & -\frac{11}{6} & -\frac{11}{3} \\ 0 & \frac{3}{2} & \frac{5}{2} & 7 \end{bmatrix}$

$\underset{\sim}{\overset{(-9/13)R_2 + R_3 \Rightarrow R_3}{}}$ $\begin{bmatrix} 6 & -7 & -1 & 4 \\ 0 & \frac{13}{6} & -\frac{11}{6} & -\frac{11}{3} \\ 0 & 0 & \frac{49}{13} & \frac{124}{13} \end{bmatrix}$

Row 3 $\Rightarrow$ $\frac{49}{13}x_3 = \frac{124}{13}$ $\Rightarrow$ $x_3 = \frac{124}{49}$. Row 2 $\Rightarrow$ $\frac{13}{6}x_2 - \frac{11}{6}\left(\frac{124}{49}\right) = -\frac{11}{3}$ $\Rightarrow$ $x_2 = \frac{22}{49}$. Row
1 $\Rightarrow$ $6x_1 - 7\left(\frac{22}{49}\right) - \left(\frac{124}{49}\right) = 4$ $\Rightarrow$ $x_1 = \frac{79}{49}$.

5. Using Gaussian elimination with 3 significant digits of accuracy:

$\begin{bmatrix} 2 & 975 & 41 \\ 53 & -82 & -13 \end{bmatrix}$ $\underset{\sim}{\overset{(-53/2)R_1 + R_2 \Rightarrow R_2}{}}$ $\begin{bmatrix} 2 & 975 & 41 \\ 0 & -2.59 \times 10^4 & -1.10 \times 10^3 \end{bmatrix}$

Row 2 $\Rightarrow$ $x_2 = \frac{-1.10 \times 10^3}{-2.59 \times 10^4} = 4.25 \times 10^{-2}$. Row 1 $\Rightarrow$ $2x_1 + 975\left(4.25 \times 10^{-2}\right) = 41$ $\Rightarrow$ $x_1 = -0.219$.

Using partial pivoting:

$\begin{bmatrix} 2 & 975 & 41 \\ 53 & -82 & -13 \end{bmatrix}$ $\underset{\sim}{\overset{R_1 \Leftrightarrow R_2}{}}$ $\begin{bmatrix} 53 & -82 & -13 \\ 2 & 975 & 41 \end{bmatrix}$

$\underset{\sim}{\overset{(-2/53)R_1 + R_2 \Rightarrow R_2}{}}$ $\begin{bmatrix} 53 & -82 & -13 \\ 0 & 9.78 \times 10^2 & 4.15 \times 10^1 \end{bmatrix}$

Row 2 $\Rightarrow$ $x_2 = \frac{4.15 \times 10^1}{9.78 \times 10^2} = 4.24 \times 10^{-2}$. Row 1 $\Rightarrow$ $53x_1 - 82\left(4.24 \times 10^{-2}\right) = -13 \Rightarrow x_1 = -0.180$.

7. Using Gaussian elimination with 3 significant digits of accuracy:

$$\begin{bmatrix} 3 & -7 & 639 & 12 \\ -2 & 5 & 803 & 7 \\ 56 & -41 & 79 & 10 \end{bmatrix} \quad \begin{matrix} (2/3)R_1 + R_2 \Rightarrow R_2 \\ (-56/3)R_1 + R_3 \Rightarrow R_3 \\ \sim \end{matrix}$$

$$\begin{bmatrix} 3 & -7 & 639 & 12 \\ 0 & 0.333 & 1.23 \times 10^3 & 1.5 \times 10^1 \\ 0 & 8.97 \times 10^1 & -1.18 \times 10^4 & -2.14 \times 10^2 \end{bmatrix} \quad \begin{matrix} \left(-8.97 \times 10^1 / 0.333\right)R_2 + R_3 \Rightarrow R_3 \\ \sim \end{matrix}$$

$$\begin{bmatrix} 3 & -7 & 639 & 12 \\ 0 & 0.333 & 1.23 \times 10^3 & 1.5 \times 10^1 \\ 0 & 0 & -3.43 \times 10^5 & -4.25 \times 10^3 \end{bmatrix}$$

Row 3 $\Rightarrow$ $x_3 = \frac{-4.25 \times 10^3}{-3.43 \times 10^5} = 1.24 \times 10^{-2}$. Row 2 $\Rightarrow$ $0.333\,x_2 + 1.23 \times 10^3\left(1.24 \times 10^{-2}\right) = 1.5 \times 10 \Rightarrow$ $x_2 = -0.757$. Row 1 $\Rightarrow$ $3x_1 - 7\left(-0.757\right) + 639\left(1.24 \times 10^{-2}\right) = 12 \Rightarrow x_1 = -0.407$.
Using partial pivoting.

$$\begin{bmatrix} 3 & -7 & 639 & 12 \\ -2 & 5 & 803 & 7 \\ 56 & -41 & 79 & 10 \end{bmatrix} \quad \begin{matrix} R_1 \Leftrightarrow R_3 \\ \sim \end{matrix} \quad \begin{bmatrix} 56 & -41 & 79 & 10 \\ -2 & 5 & 803 & 7 \\ 3 & -7 & 639 & 12 \end{bmatrix}$$

$$\begin{matrix} (1/28)R_1 + R_2 \Rightarrow R_2 \\ (-3/56)R_1 + R_3 \Rightarrow R_3 \\ \sim \end{matrix} \quad \begin{bmatrix} 56 & -41 & 79 & 10 \\ 0 & 3.54 & 8.06 \times 10^2 & 7.36 \\ 0 & -4.8 & 6.35 \times 10^2 & 1.15 \times 10^1 \end{bmatrix}$$

$$\begin{matrix} R_2 \Leftrightarrow R_3 \\ \sim \end{matrix} \quad \begin{bmatrix} 56 & -41 & 79 & 10 \\ 0 & -4.8 & 6.35 \times 10^2 & 1.15 \times 10^1 \\ 0 & 3.54 & 8.06 \times 10^2 & 7.36 \end{bmatrix}$$

$$\begin{matrix} (3.54/4.8)R_2 + R_3 \Rightarrow R_3 \\ \sim \end{matrix} \quad \begin{bmatrix} 56 & -41 & 79 & 10 \\ 0 & -4.8 & 6.35 \times 10^2 & 1.15 \times 10^1 \\ 0 & 0 & 1.27 \times 10^3 & 1.58 \times 10^1 \end{bmatrix}$$

Row 3 $\Rightarrow$ $x_3 = \frac{1.58 \times 10^1}{1.27 \times 10^3} = 1.24 \times 10^{-2}$. Row 2 $\Rightarrow$ $-4.8x_2 + 6.35 \times 10^2\left(1.24 \times 10^{-2}\right) = 1.15 \times 10^1 \Rightarrow$ $x_2 = -0.755$. Row 1 $\Rightarrow$ $56x_1 - 41\left(-0.755\right) + 79\left(1.24 \times 10^{-2}\right) = 10 \Rightarrow x_1 = -0.392$

9.

n	x_1	x_2
0	0	0
1	-1.2	0.2
2	-1.12	0.56
3	-0.976	0.536

Exact solution: $x_1 = -1$, $x_2 = 0.5$.

11.

n	x_1	x_2	x_3
0	0	0	0
1	-1.3	2.3	2.6
2	-2.295	3.34	1.42
3	-2.156	3.185	0.805

Exact solution: $x_1 = -2$, $x_2 = 3$, $x_3 = 1$.

13.

n	x_1	x_2
0	0	0
1	-1.2	0.56
2	-0.976	0.4928
3	-1.0029	0.5009

Exact solution: $x_1 = -1$, $x_2 = 0.5$.

15.

n	x_1	x_2	x_3
0	0	0	0
1	-1.3	2.56	1.316
2	-2.013	3.0974	0.9584
3	-2.0042	2.9884	1.0038

Exact solution: $x_1 = -2$, $x_2 = 3$, $x_3 = 1$.

17. Not diagonally dominant. Not possible to reorder to obtain diagonal dominance.

19. Not diagonally dominant. Not possible to reorder to obtain diagonal dominance, since none of the coefficients in equation three has absolute value greater than the sum of the absolute values of the other coefficients.

21. Jacobi iteration of given linear system:

n	x_1	x_2
0	0	0
1	-1	-1
2	-3	-3
3	-7	-7
4	-15	-15

Diagonally dominant system:
$$\begin{aligned} 2x_1 &- x_2 &= 1 \\ x_1 &- 2x_2 &= -1 \end{aligned}$$

Jacobi iteration of diagonally dominant system:

n	x_1	x_2
0	0	0
1	0.5	0.5
2	0.75	0.75
3	0.875	0.875
4	0.9375	0.9375

23. Jacobi iteration of given linear system:

n	x_1	x_2	x_3
0	0	0	0
1	-1	8	-0.3333
2	16.67	12.33	27
3	-111.3	-21.33	29.67
4	-192	624	2.778

Diagonally dominant system:
$$\begin{aligned} 5x_1 &+ x_2 &- 2x_3 &= 8 \\ 2x_1 &- 10x_2 &+ 3x_3 &= -1 \\ x_1 &- 2x_2 &+ 5x_3 &= -1 \end{aligned}$$

Jacobi iteration of diagonally dominant system:

n	x_1	x_2	x_3
0	0	0	0
1	1.6	0.1	-0.2
2	1.5	0.36	-0.48
3	1.336	0.256	-0.356
4	1.406	0.2604	-0.3648

25. Gauss-Seidel iteration of given linear system:

n	x_1	x_2
0	0	0
1	-1	-3
2	-7	-15
3	-31	-63
4	-127	-255

Diagonally dominant system:
$$\begin{aligned} 2x_1 &- x_2 &= 1 \\ x_1 &- 2x_2 &= -1 \end{aligned}$$

Gauss-Seidel iteration of diagonally dominant system:

n	x_1	x_2
0	0	0
1	0.5	0.75
2	0.875	0.9375
3	0.9688	0.9844
4	0.9922	0.9961

27. Gauss-Seidel iteration of given linear system:

n	x_1	x_2	x_3
0	0	0	0
1	-1	13	43.67
2	-193.3	1062	3669
3	-1.622×10^4	8.844×10^4	3.056×10^5
4	-1.351×10^6	7.367×10^6	2.546×10^7

Diagonally dominant system:
$$\begin{aligned} 5x_1 &+ x_2 &- 2x_3 &= 8 \\ 2x_1 &- 10x_2 &+ 3x_3 &= -1 \\ x_1 &- 2x_2 &+ 5x_3 &= -1 \end{aligned}$$

Gauss-Seidel iteration of diagonally dominant system:

n	x_1	x_2	x_3
0	0	0	0
1	1.6	0.42	-0.352
2	1.375	0.2694	-0.3673
3	1.399	0.2697	-0.3712
4	1.397	0.2679	-0.3723

29. Let $x_i(n)$ be the value of the nth iteration of x_i. Then we have $x_1(n+1) = b_1 - a_1 x_2(n)$. Applying this with $n = 0$ and $n = 1$, we obtain the 2 equations

$$\begin{aligned} 1 &= b_1 - a_1(0) \\ 5 &= b_1 - a_1(-2) \end{aligned}$$

Solve this system for the quantities b_1 and a_1 to obtain $b_1 = 1$ and $a_1 = 2$. Thus $x_1(n+1) = 1 - 2x_2(n)$, and hence $x_1(3) = 1 - 2(2) = -3$. Similarly we have $x_2(n+1) = b_2 - a_2 x_1(n)$, and so with $n = 0$ and $n = 1$, we obtain the 2 equations

$$\begin{aligned} -2 &= b_2 - a_2(0) \\ 2 &= b_2 - a_2(1) \end{aligned}$$

Solve this system for the quantities b_2 and a_2 to obtain $b_2 = -2$ and $a_2 = -4$. Thus $x_2(n+1) = -2 + 4x_1(n)$, and hence $x_2(3) = -2 + 4(5) = 18$.

31. Let $x_i(n)$ be the value of the nth iteration of x_i. Then we have $x_1(n+1) = b_1 - a_1 x_2(n)$. Applying this with $n = 0$ and $n = 1$, we obtain the 2 equations

$$\begin{aligned} 3 &= b_1 - a_1(0) \\ -5 &= b_1 - a_1(4) \end{aligned}$$

Solve this system for the quantities b_1 and a_1 to obtain $b_1 = 3$ and $a_1 = 2$. Thus $x_1(n+1) = 3 - 2x_2(n)$, and hence $x_1(3) = 3 - 2(-12) = 27$. With Gauss-Seidel iteration we have $x_2(n) = b_2 - a_2 x_1(n)$, and so using $n = 1$ and $n = 2$, we obtain the 2 equations

$$\begin{aligned} 4 &= b_2 - a_2(3) \\ -12 &= b_2 - a_2(-5) \end{aligned}$$

Solve this system for the quantities b_2 and a_2 to obtain $b_2 = -2$ and $a_2 = -2$. Thus $x_2(n) = -2 + 2x_1(n)$, and hence $x_2(3) = -2 + 2(27) = 52$.

1.4   Applications of Linear Systems

1. The number of cars entering and leaving each intersection must be the same, resulting in the three equations

$$\begin{aligned} \text{A:} \quad && x_2 &= x_3 + 20 \\ \text{B:} \quad && x_3 + 35 + 50 &= x_1 + 10 \\ \text{C:} \quad && x_1 + 40 &= x_2 + 45 + 50 \end{aligned}$$

which is equivalent to

$$
\begin{array}{rcrcl}
x_2 & - & x_3 & = & 20 \\
-x_1 & + & x_3 & = & -75 \\
x_1 & - & x_2 & = & 55
\end{array}
$$

The solution of this system, with $x_3 = s_1$ is $x_1 = 75 + s_1$ and $x_2 = 20 + s_1$. Restricting each $x_i \geq 0$ implies that $s_1 \geq 0$. Therefore the minimum volume of traffic from C to A is $x_2 = 20 + 0 = 20$ vehicles.

3. We obtain the following system of equations

$$
\begin{array}{rrcl}
\text{A:} & x_4 + 30 + 40 & = & x_1 + 50 \\
\text{B:} & x_1 + 25 + 40 & = & x_2 + x_3 + 55 \\
\text{C:} & x_2 + 50 & = & x_4 + 25
\end{array}
$$

which is equivalent to

$$
\begin{array}{rcrcrcrcl}
-x_1 & & & & & + & x_4 & = & -20 \\
x_1 & - & x_2 & - & x_3 & & & = & -10 \\
& & x_2 & & & - & x_4 & = & -25
\end{array}
$$

Using a computer algebra system, with free variable $x_4 = s_1$, we obtain $x_1 = 20 + 4s_1$, $x_2 = -25 + s_1$ and $x_3 = 55 + 3s_1$. The minimum traffic from C to A is determined by the restrictions $x_i \geq 0$, which implies $s_1 \geq 25$. Therefore the minimum volume of traffic from C to A is $x_4 = s_1 = 25$ vehicles.

5. We consider $x_1 H_2 + x_2 O \longrightarrow x_3 H_2 O$, which implies

$$
\begin{array}{rcrcl}
2x_1 & & - & 2x_3 & = & 0 \\
& x_2 & - & x_3 & = & 0
\end{array}
$$

From the augmented matrix
$\begin{bmatrix} 2 & 0 & -2 & 0 \\ 0 & 1 & -1 & 0 \end{bmatrix}$ we set $x_3 = s_1$ as a free variable, and thus $x_2 = s_1$ and $x_1 = s_1$. We set $s_1 = 1$ to obtain $x_1 = 1$, $x_2 = 1$, $x_3 = 1$, and the balanced equation

$$H_2 + O \longrightarrow H_2 O$$

7. We consider $x_1 Fe + x_2 O_2 \longrightarrow x_3 Fe_2 O_3$, which implies

$$
\begin{array}{rcrcl}
x_1 & & - & 2x_3 & = & 0 \\
& 2x_2 & - & 3x_3 & = & 0
\end{array}
$$

From the augmented matrix
$\begin{bmatrix} 1 & 0 & -2 & 0 \\ 0 & 2 & -3 & 0 \end{bmatrix}$ we set $x_3 = s_1$ as a free variable, and thus $x_2 = \frac{3}{2} s_1$ and $x_1 = 2s_1$. We set $s_1 = 2$ to obtain $x_1 = 4$, $x_2 = 3$, $x_3 = 2$, and the balanced equation

$$4Fe + 3O_2 \longrightarrow 2Fe_2 O_3$$

9. We consider $x_1 C_3 H_8 + x_2 O_2 \longrightarrow x_3 CO_2 + x_4 H_2 O$, which implies

$$
\begin{array}{rcrcrcrcl}
3x_1 & & & - & x_3 & & & = & 0 \\
8x_1 & & & & & - & 2x_4 & = & 0 \\
& 2x_2 & & - & 2x_3 & - & x_4 & = & 0
\end{array}
$$

Row-reduce the augmented matrix

$$
\begin{bmatrix}
3 & 0 & -1 & 0 & 0 \\
8 & 0 & 0 & -2 & 0 \\
0 & 2 & -2 & -1 & 0
\end{bmatrix}
\underset{\sim}{\overset{(-8/3)R_1 + R_2 \Rightarrow R_2}{}}
\begin{bmatrix}
3 & 0 & -1 & 0 & 0 \\
0 & 0 & \frac{8}{3} & -2 & 0 \\
0 & 2 & -2 & -1 & 0
\end{bmatrix}
$$

$$
\underset{\sim}{\overset{R_2 \Leftrightarrow R_3}{}}
\begin{bmatrix}
3 & 0 & -1 & 0 & 0 \\
0 & 2 & -2 & -1 & 0 \\
0 & 0 & \frac{8}{3} & -2 & 0
\end{bmatrix}
$$

We set $x_4 = s_1$ as a free variable. From row 3, $\frac{8}{3}x_3 - 2s_1 = 0 \Rightarrow x_3 = \frac{3}{4}s_1$. From row 2, $2x_2 - 2x_3 - x_4 = 0 \Rightarrow 2x_2 - 2\left(\frac{3}{4}s_1\right) - s_1 = 0 \Rightarrow x_2 = \frac{5}{4}s_1$ From row 1, $3x_1 - x_3 = 0 \Rightarrow 3x_1 - \left(\frac{3}{4}s_1\right) = 0 \Rightarrow x_1 = \frac{1}{4}s_1$. We set $s_1 = 4$ to obtain $x_1 = 1$, $x_2 = 5$, $x_3 = 3$, $x_4 = 4$, and the balanced equation

$$C_3H_8 + 5O_2 \longrightarrow 3CO_2 + 4H_2O$$

11. We consider $x_1 KO_2 + x_2 CO_2 \longrightarrow x_3 K_2CO_3 + x_4 O_2$, which implies

$$
\begin{array}{rcrcrcrcl}
x_1 & & & - & 2x_3 & & & = & 0 \\
2x_1 & + & 2x_2 & - & 3x_3 & - & 2x_4 & = & 0 \\
& & x_2 & - & x_3 & & & = & 0
\end{array}
$$

Row-reduce the augmented matrix

$$
\begin{bmatrix}
1 & 0 & -2 & 0 & 0 \\
2 & 2 & -3 & -2 & 0 \\
0 & 1 & -1 & 0 & 0
\end{bmatrix}
\underset{\sim}{\overset{-2R_1 + R_2 \Rightarrow R_2}{}}
\begin{bmatrix}
1 & 0 & -2 & 0 & 0 \\
0 & 2 & 1 & -2 & 0 \\
0 & 1 & -1 & 0 & 0
\end{bmatrix}
$$

$$
\underset{\sim}{\overset{(-1/2)R_2 + R_3 \Rightarrow R_3}{}}
\begin{bmatrix}
1 & 0 & -2 & 0 & 0 \\
0 & 2 & 1 & -2 & 0 \\
0 & 0 & -\frac{3}{2} & 1 & 0
\end{bmatrix}
$$

We set $x_4 = s_1$ as a free variable. From row 3, $-\frac{3}{2}x_3 + s_1 = 0 \Rightarrow x_3 = \frac{2}{3}s_1$. From row 2, $2x_2 + x_3 - 2x_4 = 0 \Rightarrow 2x_2 + \left(\frac{2}{3}s_1\right) - 2s_1 = 0 \Rightarrow x_2 = \frac{2}{3}s_1$. From row 1, $x_1 - 2x_3 = 0 \Rightarrow x_1 - 2\left(\frac{2}{3}s_1\right) = 0 \Rightarrow x_1 = \frac{4}{3}s_1$. We set $s_1 = 3$ to obtain $x_1 = 4$, $x_2 = 2$, $x_3 = 2$, $x_4 = 3$, and the balanced equation

$$4KO_2 + 2CO_2 \longrightarrow 2K_2CO_3 + 3O_2$$

13. Assuming $p = ad^b$, so that $\ln(p) = \ln(a) + b\ln(d)$, and letting $a_1 = \ln(a)$, we obtain the following equations using the data for Earth and Mars

$$a_1 + b\ln(149.6) = \ln(365.2)$$
$$a_1 + b\ln(227.9) = \ln(687)$$

The solution to this system is $a_1 = -1.6171$ and $b = 1.5011$. Thus $a = e^{a_1} = e^{-1.6171} = 0.19847$. Hence $p = (0.19847)d^{1.5011}$.

15. Assuming $p = ad^b$, so that $\ln(p) = \ln(a) + b\ln(d)$, and letting $a_1 = \ln(a)$, we obtain the following equations using the data for Venus and Neptune

$$a_1 + b\ln(108.2) = \ln(224.7)$$
$$a_1 + b\ln(4495.1) = \ln(59800)$$

The solution to this system is $a_1 = -1.6035$ and $b = 1.49835$. Thus $a = e^{-1.6035} = 0.20120$. Hence $p = (0.20120)d^{1.49835}$.

17. Assuming $d = as^k$, so that $\ln(d) = \ln(a) + k\ln(s)$, and letting $a_1 = \ln(a)$, we obtain the following equations using the data for $s = 10$ and $s = 20$

$$a_1 + k\ln(10) = \ln(4.5)$$
$$a_1 + k\ln(20) = \ln(18)$$

The solution to this system is $a_1 = -3.1010$ and $k = 2$. Thus $a = e^{-3.1010} = 0.04500$. Hence $d = (0.04500)s^2$. The predicted distance for each speed is as follows:

Speed (MPH)	10	20	30	40
Distance (Feet)	4.50	18.0	40.5	72.0

Chapter 2

Euclidean Space

2.1 Vectors

1. $\mathbf{u} - \mathbf{v} = \begin{bmatrix} 3 \\ -2 \\ 0 \end{bmatrix} - \begin{bmatrix} -4 \\ 1 \\ 5 \end{bmatrix} = \begin{bmatrix} 3-(-4) \\ -2-1 \\ 0-5 \end{bmatrix} = \begin{bmatrix} 7 \\ -3 \\ -5 \end{bmatrix}$

3. $\mathbf{w} + 3\mathbf{v} = \begin{bmatrix} 2 \\ -7 \\ -1 \end{bmatrix} + 3\begin{bmatrix} -4 \\ 1 \\ 5 \end{bmatrix} = \begin{bmatrix} 2+3(-4) \\ -7+3(1) \\ -1+3(5) \end{bmatrix} = \begin{bmatrix} -10 \\ -4 \\ 14 \end{bmatrix}$

5. $-\mathbf{u} + \mathbf{v} + \mathbf{w} = -\begin{bmatrix} 3 \\ -2 \\ 0 \end{bmatrix} + \begin{bmatrix} -4 \\ 1 \\ 5 \end{bmatrix} + \begin{bmatrix} 2 \\ -7 \\ -1 \end{bmatrix} = \begin{bmatrix} -3-4+2 \\ -(-2)+1-7 \\ -(0)+5-1 \end{bmatrix} = \begin{bmatrix} -5 \\ -4 \\ 4 \end{bmatrix}$

7. $\begin{aligned} 3x_1 &- x_2 &= 8 \\ 2x_1 &+ 5x_2 &= 13 \end{aligned}$

9. $\begin{aligned} -6x_1 &+ 5x_2 & &= 4 \\ 5x_1 &- 3x_2 &+ 2x_3 &= 16 \end{aligned}$

11. $x_1\begin{bmatrix} 2 \\ -1 \end{bmatrix} + x_2\begin{bmatrix} 8 \\ -3 \end{bmatrix} + x_3\begin{bmatrix} -4 \\ 5 \end{bmatrix} = \begin{bmatrix} -10 \\ 4 \end{bmatrix}$

13. $x_1\begin{bmatrix} 1 \\ -2 \\ -3 \end{bmatrix} + x_2\begin{bmatrix} -1 \\ 2 \\ -3 \end{bmatrix} + x_3\begin{bmatrix} -3 \\ 6 \\ 10 \end{bmatrix} + x_4\begin{bmatrix} -1 \\ 2 \\ 0 \end{bmatrix} = \begin{bmatrix} -1 \\ -1 \\ 5 \end{bmatrix}$

15. $\begin{bmatrix} x_1 \\ x_2 \end{bmatrix} = \begin{bmatrix} -4 \\ 0 \end{bmatrix} + s_1\begin{bmatrix} 3 \\ 1 \end{bmatrix}$

17. $\begin{bmatrix} x_1 \\ x_2 \\ x_3 \\ x_4 \end{bmatrix} = \begin{bmatrix} 4 \\ 0 \\ -9 \\ 0 \end{bmatrix} + s_1\begin{bmatrix} 6 \\ 0 \\ 3 \\ 1 \end{bmatrix} + s_2\begin{bmatrix} -5 \\ 1 \\ 0 \\ 0 \end{bmatrix}$

19. $1\mathbf{u} + 0\mathbf{v} = \mathbf{u} = \begin{bmatrix} 3 \\ -2 \end{bmatrix}$, $0\mathbf{u} + 1\mathbf{v} = \mathbf{v} = \begin{bmatrix} -1 \\ -4 \end{bmatrix}$, $1\mathbf{u} + 1\mathbf{v} = \begin{bmatrix} 3 \\ -2 \end{bmatrix} + \begin{bmatrix} -1 \\ -4 \end{bmatrix} = \begin{bmatrix} 2 \\ -6 \end{bmatrix}$

21. $1\mathbf{u} + 0\mathbf{v} + 0\mathbf{w} = \mathbf{u} = \begin{bmatrix} -4 \\ 0 \\ -3 \end{bmatrix}$, $0\mathbf{u} + 1\mathbf{v} + 0\mathbf{w} = \mathbf{v} = \begin{bmatrix} -2 \\ -1 \\ 5 \end{bmatrix}$, $0\mathbf{u} + 0\mathbf{v} + 1\mathbf{w} = \mathbf{w} = \begin{bmatrix} 9 \\ 6 \\ 11 \end{bmatrix}$.

23. $-3 \begin{bmatrix} a \\ 3 \end{bmatrix} + 4 \begin{bmatrix} -1 \\ b \end{bmatrix} = \begin{bmatrix} -10 \\ 19 \end{bmatrix} \Rightarrow \begin{bmatrix} -3a - 4 \\ -9 + 4b \end{bmatrix} = \begin{bmatrix} -10 \\ 19 \end{bmatrix} \Rightarrow -3a - 4 = -10$ and $-9 + 4b = 19$.
Solving these equations, we obtain $a = 2$ and $b = 7$.

25. $- \begin{bmatrix} -1 \\ a \\ 2 \end{bmatrix} + 2 \begin{bmatrix} 3 \\ -2 \\ b \end{bmatrix} = \begin{bmatrix} c \\ -7 \\ 8 \end{bmatrix} \Rightarrow \begin{bmatrix} 1 + 6 \\ -a - 4 \\ -2 + 2b \end{bmatrix} = \begin{bmatrix} c \\ -7 \\ 8 \end{bmatrix} \Rightarrow$

$7 = c$, $-a - 4 = -7$, and $-2 + 2b = 8$. Solving these equations, we obtain $a = 3$, $b = 5$, and $c = 7$.

27. $x_1 \mathbf{a}_1 + x_2 \mathbf{a}_2 = \mathbf{b} \quad \Leftrightarrow \quad x_1 \begin{bmatrix} -2 \\ 5 \end{bmatrix} + x_2 \begin{bmatrix} 7 \\ -3 \end{bmatrix} = \begin{bmatrix} 8 \\ 9 \end{bmatrix} \quad \Leftrightarrow \quad \begin{bmatrix} -2x_1 + 7x_2 \\ 5x_1 - 3x_2 \end{bmatrix} = \begin{bmatrix} 8 \\ 9 \end{bmatrix} \quad \Leftrightarrow$ the

augmented matrix $\begin{bmatrix} -2 & 7 & 8 \\ 5 & -3 & 9 \end{bmatrix}$ has a solution:

$$\begin{bmatrix} -2 & 7 & 8 \\ 5 & -3 & 9 \end{bmatrix} \overset{(5/2)R_1 + R_2 \Rightarrow R_2}{\sim} \begin{bmatrix} -2 & 7 & 8 \\ 0 & \frac{29}{2} & 29 \end{bmatrix}$$

From row 2, $\frac{29}{2} x_2 = 29 \Rightarrow x_2 = 2$. From row 1, $-2x_1 + 7(2) = 8 \Rightarrow x_1 = 3$. Hence $\mathbf{b}$ is a linear combination of $\mathbf{a}_1$ and $\mathbf{a}_2$, with $\mathbf{b} = 3\mathbf{a}_1 + 2\mathbf{a}_2$.

29. $x_1 \mathbf{a}_1 + x_2 \mathbf{a}_2 = \mathbf{b} \quad \Leftrightarrow \quad x_1 \begin{bmatrix} 2 \\ -3 \\ 1 \end{bmatrix} + x_2 \begin{bmatrix} 0 \\ 3 \\ -3 \end{bmatrix} = \begin{bmatrix} 6 \\ 3 \\ -9 \end{bmatrix} \quad \Leftrightarrow \quad \begin{bmatrix} 2x_1 \\ -3x_1 + 3x_2 \\ x_1 - 3x_2 \end{bmatrix} = \begin{bmatrix} 6 \\ 3 \\ -9 \end{bmatrix}$. The
first equation $2x_1 = 6 \Rightarrow x_1 = 3$. Then the second equation $-3(3) + 3x_2 = 3 \Rightarrow x_2 = 4$. We check the third equation, $3 - 3(4) = -9$. Hence $\mathbf{b}$ is a linear combination of $\mathbf{a}_1$ and $\mathbf{a}_2$, with $\mathbf{b} = 3\mathbf{a}_1 + 4\mathbf{a}_2$.

31. Using vectors, we calculate

$$(2) \begin{bmatrix} 29 \\ 3 \\ 4 \end{bmatrix} + (1) \begin{bmatrix} 18 \\ 25 \\ 6 \end{bmatrix} = \begin{bmatrix} 76 \\ 31 \\ 14 \end{bmatrix}$$

Hence we have 76 pounds of nitrogen, 31 pounds of phosphoric acid, and 14 pounds of potash.

33. Let x_1 be the amount of Vigoro, x_2 the amount of Parker's, and then we need

$$x_1 \begin{bmatrix} 29 \\ 3 \\ 4 \end{bmatrix} + x_2 \begin{bmatrix} 18 \\ 25 \\ 6 \end{bmatrix} = \begin{bmatrix} 112 \\ 81 \\ 26 \end{bmatrix}$$

Solve using the corresponding augmented matrix:

$$\begin{bmatrix} 29 & 18 & 112 \\ 3 & 25 & 81 \\ 4 & 6 & 26 \end{bmatrix} \overset{(-3/29)R_1 + R_2 \Rightarrow R_2}{\underset{(-4/29)R_1 + R_3 \Rightarrow R_3}{\sim}} \begin{bmatrix} 29 & 18 & 112 \\ 0 & \frac{671}{29} & \frac{2013}{29} \\ 0 & \frac{102}{29} & \frac{306}{29} \end{bmatrix}$$

$$\overset{(-102/671)R_2 + R_3 \Rightarrow R_3}{\sim} \begin{bmatrix} 29 & 18 & 112 \\ 0 & \frac{671}{29} & \frac{2013}{29} \\ 0 & 0 & 0 \end{bmatrix}$$

From row 2, we have $\frac{671}{29} x_2 = \frac{2013}{29} \Rightarrow x_2 = 3$. Form row 1, we have $29x_1 + 18(3) = 112 \Rightarrow x_1 = 2$.
Thus we need 2 bags of Vigoro and 3 bags of Parker's.

35. Let x_1 be the amount of Vigoro, x_2 the amount of Parker's, and then we need

$$x_1 \begin{bmatrix} 29 \\ 3 \\ 4 \end{bmatrix} + x_2 \begin{bmatrix} 18 \\ 25 \\ 6 \end{bmatrix} = \begin{bmatrix} 123 \\ 59 \\ 24 \end{bmatrix}$$

Solve using the corresponding augmented matrix:

$$\begin{bmatrix} 29 & 18 & 123 \\ 3 & 25 & 59 \\ 4 & 6 & 24 \end{bmatrix} \begin{array}{c} (-3/29)R_1+R_2\Rightarrow R_2 \\ (-4/29)R_1+R_3\Rightarrow R_3 \\ \sim \end{array} \begin{bmatrix} 29 & 18 & 123 \\ 0 & \frac{671}{29} & \frac{1342}{29} \\ 0 & \frac{102}{29} & \frac{204}{29} \end{bmatrix}$$

$$\begin{array}{c} (29/671)R_2\Rightarrow R_2 \\ (-102/29)R_2+R_3\Rightarrow R_3 \\ \sim \end{array} \begin{bmatrix} 29 & 18 & 123 \\ 0 & 1 & 2 \\ 0 & 0 & 0 \end{bmatrix}$$

Back substituting gives $x_2 = 2$ and $x_1 = 3$. Hence we need 3 bags of Vigoro and 2 bags of Parker's.

37. Let x_1 be the amount of Vigoro, x_2 the amount of Parker's, and then we need

$$x_1 \begin{bmatrix} 29 \\ 3 \\ 4 \end{bmatrix} + x_2 \begin{bmatrix} 18 \\ 25 \\ 6 \end{bmatrix} = \begin{bmatrix} 148 \\ 131 \\ 40 \end{bmatrix}$$

Solve using the corresponding augmented matrix:

$$\begin{bmatrix} 29 & 18 & 148 \\ 3 & 25 & 131 \\ 4 & 6 & 40 \end{bmatrix} \begin{array}{c} (-3/29)R_1+R_2\Rightarrow R_2 \\ (-4/29)R_1+R_3\Rightarrow R_3 \\ \sim \end{array} \begin{bmatrix} 29 & 18 & 148 \\ 0 & \frac{671}{29} & \frac{3355}{29} \\ 0 & \frac{102}{29} & \frac{568}{29} \end{bmatrix}$$

$$\begin{array}{c} (-102/671)R_2+R_3\Rightarrow R_3 \\ \sim \end{array} \begin{bmatrix} 29 & 18 & 148 \\ 0 & \frac{671}{29} & \frac{3355}{29} \\ 0 & 0 & 2 \end{bmatrix}$$

Since row 3 corresponds to the equation $0 = 2$, the system has no solutions.

39. Let x_1 be the amount of Vigoro, x_2 the amount of Parker's, and then we need

$$x_1 \begin{bmatrix} 29 \\ 3 \\ 4 \end{bmatrix} + x_2 \begin{bmatrix} 18 \\ 25 \\ 6 \end{bmatrix} = \begin{bmatrix} 25 \\ 72 \\ 14 \end{bmatrix}$$

Solve using the corresponding augmented matrix:

$$\begin{bmatrix} 29 & 18 & 25 \\ 3 & 25 & 72 \\ 4 & 6 & 14 \end{bmatrix} \begin{array}{c} (-3/29)R_1+R_2\Rightarrow R_2 \\ (-4/29)R_1+R_3\Rightarrow R_3 \\ \sim \end{array} \begin{bmatrix} 29 & 18 & 25 \\ 0 & \frac{671}{29} & \frac{2013}{29} \\ 0 & \frac{102}{29} & \frac{306}{29} \end{bmatrix}$$

$$\begin{array}{c} (-102/671)R_2+R_3\Rightarrow R_3 \\ \sim \end{array} \begin{bmatrix} 29 & 18 & 25 \\ 0 & \frac{671}{29} & \frac{2013}{29} \\ 0 & 0 & 0 \end{bmatrix}$$

From row 2, we have $\frac{671}{29}x_2 = \frac{2013}{29} \Rightarrow x_2 = 3$. From row 1, we have $29x_1 + 18(3) = 25 \Rightarrow x_1 = -1$. Since we can not use a negative amount, we conclude that there is no solution.

41. Let x_1 be the number of cans of Red Bull, and x_2 the number of cans of Jolt Cola, and then we need

$$x_1 \begin{bmatrix} 27 \\ 80 \end{bmatrix} + x_2 \begin{bmatrix} 94 \\ 280 \end{bmatrix} = \begin{bmatrix} 148 \\ 440 \end{bmatrix}$$

Solve using the corresponding augmented matrix:

$$\begin{bmatrix} 27 & 94 & 148 \\ 80 & 280 & 440 \end{bmatrix} \begin{array}{c} (-80/27)R_1+R_2\Rightarrow R_2 \\ \sim \end{array} \begin{bmatrix} 27 & 94 & 148 \\ 0 & \frac{40}{27} & \frac{40}{27} \end{bmatrix}$$

From row 2, we have $\frac{40}{27}x_2 = \frac{40}{27} \Rightarrow x_2 = 1$. From row 1, $27x_1 + 94(1) = 148 \Rightarrow x_1 = 2$. Thus we need to drink 2 cans of Red Bull and 1 can of Jolt Cola.

43. Let x_1 be the number of cans of Red Bull, and x_2 the number of cans of Jolt Cola, and then we need

$$x_1 \begin{bmatrix} 27 \\ 80 \end{bmatrix} + x_2 \begin{bmatrix} 94 \\ 280 \end{bmatrix} = \begin{bmatrix} 242 \\ 720 \end{bmatrix}$$

Solve using the corresponding augmented matrix:

$$\begin{bmatrix} 27 & 94 & 242 \\ 80 & 280 & 720 \end{bmatrix} \underset{\sim}{\overset{(-80/27)R_1+R_2 \Rightarrow R_2}{}} \begin{bmatrix} 27 & 94 & 242 \\ 0 & \frac{40}{27} & \frac{80}{27} \end{bmatrix}$$

From row 2, we have $\frac{40}{27}x_2 = \frac{80}{27} \Rightarrow x_2 = 2$. From row 1, $27x_1 + 94(2) = 242 \Rightarrow x_1 = 2$. Thus we need to drink 2 cans of Red Bull and 2 cans of Jolt Cola.

45. Let x_1 be the number of servings of Lucky Charms and x_2 the number of servings of Raisin Bran, and then we need

$$x_1 \begin{bmatrix} 10 \\ 25 \\ 25 \end{bmatrix} + x_2 \begin{bmatrix} 2 \\ 25 \\ 10 \end{bmatrix} = \begin{bmatrix} 40 \\ 200 \\ 125 \end{bmatrix}$$

Solve using the corresponding augmented matrix:

$$\begin{bmatrix} 10 & 2 & 40 \\ 25 & 25 & 200 \\ 25 & 10 & 125 \end{bmatrix} \underset{\sim}{\overset{(-5/2)R_1+R_2 \Rightarrow R_2}{\overset{(-5/2)R_1+R_3 \Rightarrow R_3}{}}} \begin{bmatrix} 10 & 2 & 40 \\ 0 & 20 & 100 \\ 0 & 5 & 25 \end{bmatrix}$$

$$\underset{\sim}{\overset{(-1/4)R_2+R_3 \Rightarrow R_3}{}} \begin{bmatrix} 10 & 2 & 40 \\ 0 & 20 & 100 \\ 0 & 0 & 0 \end{bmatrix}$$

From row 2, we have $20x_2 = 100 \Rightarrow x_2 = 5$. From row 1, $10x_1 + 2(5) = 40 \Rightarrow x_1 = 3$. Thus we need 3 servings of Lucky Charms and 5 servings of Raisin Bran.

47. Let x_1 be the number of servings of Lucky Charms and x_2 the number of servings of Raisin Bran, and then we need

$$x_1 \begin{bmatrix} 10 \\ 25 \\ 25 \end{bmatrix} + x_2 \begin{bmatrix} 2 \\ 25 \\ 10 \end{bmatrix} = \begin{bmatrix} 26 \\ 125 \\ 80 \end{bmatrix}$$

Solve using the corresponding augmented matrix:

$$\begin{bmatrix} 10 & 2 & 26 \\ 25 & 25 & 125 \\ 25 & 10 & 80 \end{bmatrix} \underset{\sim}{\overset{(-5/2)R_1+R_2 \Rightarrow R_2}{\overset{(-5/2)R_1+R_3 \Rightarrow R_3}{}}} \begin{bmatrix} 10 & 2 & 26 \\ 0 & 20 & 60 \\ 0 & 5 & 15 \end{bmatrix}$$

$$\underset{\sim}{\overset{(-1/4)R_2+R_3 \Rightarrow R_3}{}} \begin{bmatrix} 10 & 2 & 26 \\ 0 & 20 & 60 \\ 0 & 0 & 0 \end{bmatrix}$$

From row 2, we have $20x_2 = 60 \Rightarrow x_2 = 3$. From row 1, $10x_1 + 2(3) = 26 \Rightarrow x_1 = 2$. Thus we need 2 servings of Lucky Charms and 3 servings of Raisin Bran.

49. (a) $\mathbf{a} = \begin{bmatrix} 2000 \\ 8000 \end{bmatrix}$, $\mathbf{b} = \begin{bmatrix} 3000 \\ 10000 \end{bmatrix}$

(b) $8\mathbf{b} = (8) \begin{bmatrix} 3000 \\ 10000 \end{bmatrix} = \begin{bmatrix} 24000 \\ 80000 \end{bmatrix}$. The company produces 24000 computer monitors and 80000 flat panel televisions at facility B in 8 weeks.

(c) $6\mathbf{a} + 6\mathbf{b} = 6 \begin{bmatrix} 2000 \\ 8000 \end{bmatrix} + 6 \begin{bmatrix} 3000 \\ 10000 \end{bmatrix} = \begin{bmatrix} 30000 \\ 108000 \end{bmatrix}$. The company produces 30000 computer monitors and 108000 flat panel televisions at facilities A and B in 6 weeks.

(d) Let x_1 be the number of weeks of production at facility A, and x_2 the number of weeks of production at facility B, and then we need

$$x_1 \begin{bmatrix} 2000 \\ 8000 \end{bmatrix} + x_2 \begin{bmatrix} 3000 \\ 10000 \end{bmatrix} = \begin{bmatrix} 24000 \\ 92000 \end{bmatrix}$$

Solve using the corresponding augmented matrix:

$$\begin{bmatrix} 2000 & 3000 & 24000 \\ 8000 & 10000 & 92000 \end{bmatrix} \overset{(-4)R_1+R_2 \Rightarrow R_2}{\sim} \begin{bmatrix} 2000 & 3000 & 24000 \\ 0 & -2000 & -4000 \end{bmatrix}$$

From row 2, we have $-2000x_2 = -4000 \Rightarrow x_2 = 2$. From row 1, $2000x_1 + 3000(2) = 24000 \Rightarrow x_1 = 9$. Thus we need 9 weeks of production at facility A and 2 weeks of production at facility B.

51.

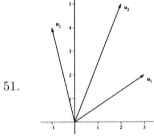

$$\overline{\mathbf{v}} = \frac{5\mathbf{u}_1+3\mathbf{u}_2+2\mathbf{u}_3}{5+3+2} = \frac{1}{10}\left(5\begin{bmatrix} 3 \\ 2 \end{bmatrix} + 3\begin{bmatrix} -1 \\ 4 \end{bmatrix} + 2\begin{bmatrix} 2 \\ 5 \end{bmatrix}\right) = \frac{1}{10}\begin{bmatrix} 16 \\ 32 \end{bmatrix} = \begin{bmatrix} \frac{8}{5} \\ \frac{16}{5} \end{bmatrix}$$

53. Let $x_1, x_2,$ and x_3 be the mass of $\mathbf{u}_1$, $\mathbf{u}_2$, and $\mathbf{u}_3$ respectively. Then

$$\overline{\mathbf{v}} = \frac{x_1\mathbf{u}_1 + x_2\mathbf{u}_2 + x_3\mathbf{u}_3}{11} = \frac{1}{11}\left(x_1\begin{bmatrix} -1 \\ 3 \end{bmatrix} + x_2\begin{bmatrix} 3 \\ -2 \end{bmatrix} + x_3\begin{bmatrix} 5 \\ 2 \end{bmatrix}\right)$$

$$= \begin{bmatrix} -\frac{1}{11}x_1 + \frac{3}{11}x_2 + \frac{5}{11}x_3 \\ \frac{3}{11}x_1 - \frac{2}{11}x_2 + \frac{2}{11}x_3 \end{bmatrix} = \begin{bmatrix} \frac{13}{11} \\ \frac{16}{11} \end{bmatrix}$$

We obtain the 2 equations, $-x_1 + 3x_2 + 5x_3 = 13$ and $3x_1 - 2x_2 + 2x_3 = 16$. Together with the equation $x_1 + x_2 + x_3 = 11$, we have 3 equations and solve the corresponding augmented matrix:

$$\begin{bmatrix} -1 & 3 & 5 & 13 \\ 3 & -2 & 2 & 16 \\ 1 & 1 & 1 & 11 \end{bmatrix} \overset{\substack{3R_1+R_2 \Rightarrow R_2 \\ R_1+R_3 \Rightarrow R_3}}{\sim} \begin{bmatrix} -1 & 3 & 5 & 13 \\ 0 & 7 & 17 & 55 \\ 0 & 4 & 6 & 24 \end{bmatrix}$$

$$\overset{(-4/7)R_2+R_3 \Rightarrow R_3}{\sim} \begin{bmatrix} -1 & 3 & 5 & 13 \\ 0 & 7 & 17 & 55 \\ 0 & 0 & -\frac{26}{7} & -\frac{52}{7} \end{bmatrix}$$

From row 3, $-\frac{26}{7}x_3 = -\frac{52}{7} \Rightarrow x_3 = 2$. From row 2, $7x_2 + 17(2) = 55 \Rightarrow x_2 = 3$. From row 1, $-x_1 + 3(3) + 5(2) = 13 \Rightarrow x_1 = 6$.

55. For example, $\mathbf{u} = (0, 0, -1)$ and $\mathbf{v} = (3, 2, 0)$.

57. For example, $\mathbf{u} = (1, 0, 0)$, $\mathbf{v} = (1, 0, 0)$, and $\mathbf{w} = (-2, 0, 0)$.

59. For example, $\mathbf{u} = (1, 0)$ and $\mathbf{v} = (2, 0)$.

61. Simply, $x_1 = 3$ and $x_2 = -2$.

63. True, since $-2 \begin{bmatrix} -3 \\ 5 \end{bmatrix} = \begin{bmatrix} (-2)(-3) \\ (-2)(5) \end{bmatrix} = \begin{bmatrix} 6 \\ -10 \end{bmatrix}$.

65. True, by Theorem 2.3(b).

67. False. The sum $c_1 + \mathbf{u}_1$ of a scalar and a vector is undefined.

69. True, by Definition 2.1, where it is stated that vectors can be expressed in column or row form.

71. False. It works regardless of the quadrant, and can be established algebraically for vectors positioned anywhere.

73. (a) Let $\mathbf{u} = \begin{bmatrix} u_1 \\ u_2 \\ \vdots \\ u_n \end{bmatrix}$. Then $(a+b)\mathbf{u} = (a+b) \begin{bmatrix} u_1 \\ u_2 \\ \vdots \\ u_n \end{bmatrix} = \begin{bmatrix} (a+b)\,u_1 \\ (a+b)\,u_2 \\ \vdots \\ (a+b)\,u_n \end{bmatrix}$

$$= \begin{bmatrix} au_1 + bu_1 \\ au_2 + bu_2 \\ \vdots \\ au_n + bu_n \end{bmatrix} = \begin{bmatrix} au_1 \\ au_2 \\ \vdots \\ au_n \end{bmatrix} + \begin{bmatrix} bu_1 \\ bu_2 \\ \vdots \\ bu_n \end{bmatrix} = a \begin{bmatrix} u_1 \\ u_2 \\ \vdots \\ u_n \end{bmatrix} + b \begin{bmatrix} u_1 \\ u_2 \\ \vdots \\ u_n \end{bmatrix} = a\mathbf{u} + b\mathbf{u}.$$

(b) Let $\mathbf{u} = \begin{bmatrix} u_1 \\ u_2 \\ \vdots \\ u_n \end{bmatrix}$, $\mathbf{v} = \begin{bmatrix} v_1 \\ v_2 \\ \vdots \\ v_n \end{bmatrix}$, and $\mathbf{w} = \begin{bmatrix} w_1 \\ w_2 \\ \vdots \\ w_n \end{bmatrix}$. Then

$$(\mathbf{u} + \mathbf{v}) + \mathbf{w} = \left(\begin{bmatrix} u_1 \\ u_2 \\ \vdots \\ u_n \end{bmatrix} + \begin{bmatrix} v_1 \\ v_2 \\ \vdots \\ v_n \end{bmatrix} \right) + \begin{bmatrix} w_1 \\ w_2 \\ \vdots \\ w_n \end{bmatrix} = \begin{bmatrix} u_1 + v_1 \\ u_2 + v_2 \\ \vdots \\ u_n + v_n \end{bmatrix} + \begin{bmatrix} w_1 \\ w_2 \\ \vdots \\ w_n \end{bmatrix}$$

$$= \begin{bmatrix} (u_1 + v_1) + w_1 \\ (u_2 + v_2) + w_2 \\ \vdots \\ (u_n + v_n) + w_n \end{bmatrix} = \begin{bmatrix} u_1 + (v_1 + w_1) \\ u_2 + (v_2 + w_2) \\ \vdots \\ u_n + (v_n + w_n) \end{bmatrix} = \begin{bmatrix} u_1 \\ u_2 \\ \vdots \\ u_n \end{bmatrix} + \begin{bmatrix} v_1 + w_1 \\ v_2 + w_2 \\ \vdots \\ v_n + w_n \end{bmatrix}$$

$$= \begin{bmatrix} u_1 \\ u_2 \\ \vdots \\ u_n \end{bmatrix} + \left(\begin{bmatrix} v_1 \\ v_2 \\ \vdots \\ v_n \end{bmatrix} + \begin{bmatrix} w_1 \\ w_2 \\ \vdots \\ w_n \end{bmatrix} \right) = \mathbf{u} + (\mathbf{v} + \mathbf{w}).$$

(c) Let $\mathbf{u} = \begin{bmatrix} u_1 \\ u_2 \\ \vdots \\ u_n \end{bmatrix}$. Then $a(b\mathbf{u}) = a \left(b \begin{bmatrix} u_1 \\ u_2 \\ \vdots \\ u_n \end{bmatrix} \right) = a \left(\begin{bmatrix} bu_1 \\ bu_2 \\ \vdots \\ bu_n \end{bmatrix} \right)$

$$= \begin{bmatrix} a\,(bu_1) \\ a\,(bu_2) \\ \vdots \\ a\,(bu_n) \end{bmatrix} = \begin{bmatrix} (ab)\,u_1 \\ (ab)\,u_2 \\ \vdots \\ (ab)\,u_n \end{bmatrix} = (ab) \begin{bmatrix} u_1 \\ u_2 \\ \vdots \\ u_n \end{bmatrix} = (ab)\mathbf{u}.$$

(d) Let $\mathbf{u} = \begin{bmatrix} u_1 \\ u_2 \\ \vdots \\ u_n \end{bmatrix}$. Then $\mathbf{u} + (-\mathbf{u}) = \begin{bmatrix} u_1 \\ u_2 \\ \vdots \\ u_n \end{bmatrix} + \left(- \begin{bmatrix} u_1 \\ u_2 \\ \vdots \\ u_n \end{bmatrix} \right)$

$$= \begin{bmatrix} u_1 \\ u_2 \\ \vdots \\ u_n \end{bmatrix} + \begin{bmatrix} -u_1 \\ -u_2 \\ \vdots \\ -u_n \end{bmatrix} = \begin{bmatrix} u_1 - u_1 \\ u_2 - u_2 \\ \vdots \\ u_n - u_n \end{bmatrix} = \begin{bmatrix} 0 \\ 0 \\ \vdots \\ 0 \end{bmatrix} = \mathbf{0}.$$

(e) Let $\mathbf{u} = \begin{bmatrix} u_1 \\ u_2 \\ \vdots \\ u_n \end{bmatrix}$. Then $\mathbf{u} + \mathbf{0} = \begin{bmatrix} u_1 \\ u_2 \\ \vdots \\ u_n \end{bmatrix} + \begin{bmatrix} 0 \\ 0 \\ \vdots \\ 0 \end{bmatrix} = \begin{bmatrix} u_1 + 0 \\ u_2 + 0 \\ \vdots \\ u_n + 0 \end{bmatrix} = \begin{bmatrix} u_1 \\ u_2 \\ \vdots \\ u_n \end{bmatrix} = \mathbf{u}.$ Likewise,

$$\mathbf{0} + \mathbf{u} = \begin{bmatrix} 0 \\ 0 \\ \vdots \\ 0 \end{bmatrix} + \begin{bmatrix} u_1 \\ u_2 \\ \vdots \\ u_n \end{bmatrix} = \begin{bmatrix} 0 + u_1 \\ 0 + u_2 \\ \vdots \\ 0 + u_n \end{bmatrix} = \begin{bmatrix} u_1 \\ u_2 \\ \vdots \\ u_n \end{bmatrix} = \mathbf{u}.$$

(f) Let $\mathbf{u} = \begin{bmatrix} u_1 \\ u_2 \\ \vdots \\ u_n \end{bmatrix}$. Then $1\mathbf{u} = (1) \begin{bmatrix} u_1 \\ u_2 \\ \vdots \\ u_n \end{bmatrix} = \begin{bmatrix} (1)\,u_1 \\ (1)\,u_2 \\ \vdots \\ (1)\,u_n \end{bmatrix} = \begin{bmatrix} u_1 \\ u_2 \\ \vdots \\ u_n \end{bmatrix} = \mathbf{u}.$

75.

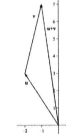

77.

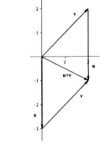

79.

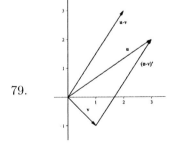

81. We obtain the three equations $2x_1 + 2x_2 + 5x_3 = 0$, $7x_1 + 4x_2 + x_3 = 3$, and $3x_1 + 2x_2 + 6x_3 = 5$. Using a computer algebra system to solve this system, we get $x_1 = 4$, $x_2 = -6.5$, and $x_3 = 1$.

2.2 Span

1. $0\mathbf{u}_1 + 0\mathbf{u}_2 = 0\begin{bmatrix} 2 \\ 6 \end{bmatrix} + 0\begin{bmatrix} 9 \\ 15 \end{bmatrix} = \begin{bmatrix} 0 \\ 0 \end{bmatrix}$, $1\mathbf{u}_1 + 0\mathbf{u}_2 = 1\begin{bmatrix} 2 \\ 6 \end{bmatrix} + 0\begin{bmatrix} 9 \\ 15 \end{bmatrix} = \begin{bmatrix} 2 \\ 6 \end{bmatrix}$, $0\mathbf{u}_1 + 1\mathbf{u}_2 =$
$0\begin{bmatrix} 2 \\ 6 \end{bmatrix} + 1\begin{bmatrix} 9 \\ 15 \end{bmatrix} = \begin{bmatrix} 9 \\ 15 \end{bmatrix}$

3. $0\mathbf{u}_1 + 0\mathbf{u}_2 = 0\begin{bmatrix} 2 \\ 5 \\ -3 \end{bmatrix} + 0\begin{bmatrix} 1 \\ 0 \\ 4 \end{bmatrix} = \begin{bmatrix} 0 \\ 0 \\ 0 \end{bmatrix}$, $1\mathbf{u}_1 + 0\mathbf{u}_2 = 1\begin{bmatrix} 2 \\ 5 \\ -3 \end{bmatrix} + 0\begin{bmatrix} 1 \\ 0 \\ 4 \end{bmatrix} = \begin{bmatrix} 2 \\ 5 \\ -3 \end{bmatrix}$, $0\mathbf{u}_1 + 1\mathbf{u}_2 =$
$0\begin{bmatrix} 2 \\ 5 \\ -3 \end{bmatrix} + 1\begin{bmatrix} 1 \\ 0 \\ 4 \end{bmatrix} = \begin{bmatrix} 1 \\ 0 \\ 4 \end{bmatrix}$

5. $0\mathbf{u}_1 + 0\mathbf{u}_2 + 0\mathbf{u}_3 = 0\begin{bmatrix} 2 \\ 0 \\ 0 \end{bmatrix} + 0\begin{bmatrix} 4 \\ 1 \\ 6 \end{bmatrix} + 0\begin{bmatrix} -4 \\ 0 \\ 7 \end{bmatrix} = \begin{bmatrix} 0 \\ 0 \\ 0 \end{bmatrix}$, $1\mathbf{u}_1 + 0\mathbf{u}_2 + 0\mathbf{u}_3 = 1\begin{bmatrix} 2 \\ 0 \\ 0 \end{bmatrix} + 0\begin{bmatrix} 4 \\ 1 \\ 6 \end{bmatrix} +$
$0\begin{bmatrix} -4 \\ 0 \\ 7 \end{bmatrix} = \begin{bmatrix} 2 \\ 0 \\ 0 \end{bmatrix}$, $0\mathbf{u}_1 + 1\mathbf{u}_2 + 0\mathbf{u}_3 = 0\begin{bmatrix} 2 \\ 0 \\ 0 \end{bmatrix} + 1\begin{bmatrix} 4 \\ 1 \\ 6 \end{bmatrix} + 0\begin{bmatrix} -4 \\ 0 \\ 7 \end{bmatrix} = \begin{bmatrix} 4 \\ 1 \\ 6 \end{bmatrix}$

7. Set $x_1\mathbf{a}_1 = \mathbf{b} \Rightarrow x_1\begin{bmatrix} 3 \\ 5 \end{bmatrix} = \begin{bmatrix} 9 \\ -15 \end{bmatrix} \Rightarrow \begin{bmatrix} 3x_1 \\ 5x_1 \end{bmatrix} = \begin{bmatrix} 9 \\ -15 \end{bmatrix}$.
From the first component, $x_1 = 3$, but from the second component $x_1 = -3$. Thus $\mathbf{b}$ is not in the span of $\mathbf{a}_1$.

9. Set $x_1\mathbf{a}_1 = \mathbf{b} \Rightarrow x_1\begin{bmatrix} 4 \\ -2 \\ 10 \end{bmatrix} = \begin{bmatrix} 2 \\ -1 \\ -5 \end{bmatrix} \Rightarrow \begin{bmatrix} 4x_1 \\ -2x_1 \\ 10x_1 \end{bmatrix} = \begin{bmatrix} 2 \\ -1 \\ -5 \end{bmatrix}$.
From the first and second components, $x_1 = \frac{1}{2}$, but from the third component $x_1 = -\frac{1}{2}$. Thus $\mathbf{b}$ is not in the span of $\mathbf{a}_1$.

11. Set $x_1\mathbf{a}_1 + x_2\mathbf{a}_2 = \mathbf{b} \Rightarrow x_1\begin{bmatrix} -1 \\ 4 \\ -3 \end{bmatrix} + x_2\begin{bmatrix} 2 \\ 8 \\ -7 \end{bmatrix} = \begin{bmatrix} -10 \\ -8 \\ 7 \end{bmatrix} \Rightarrow$

$\begin{bmatrix} -x_1 + 2x_2 \\ 4x_1 + 8x_2 \\ -3x_1 - 7x_2 \end{bmatrix} = \begin{bmatrix} -10 \\ -8 \\ 7 \end{bmatrix}$. We obtain 3 equations and row-reduce the associated augmented matrix
to determine if there are solutions.

$$\begin{bmatrix} -1 & 2 & -10 \\ 4 & 8 & -8 \\ -3 & -7 & 7 \end{bmatrix} \underset{\sim}{\overset{\substack{4R_1+R_2\Rightarrow R_2 \\ -3R_1+R_3\Rightarrow R_3}}{}} \begin{bmatrix} -1 & 2 & -10 \\ 0 & 16 & -48 \\ 0 & -13 & 37 \end{bmatrix}$$

$$\underset{\sim}{\overset{(13/16)R_2+R_3\Rightarrow R_3}{}} \begin{bmatrix} -1 & 2 & -10 \\ 0 & 16 & -48 \\ 0 & 0 & -2 \end{bmatrix}$$

From the third row, $0 = -2$, and hence there are no solutions. We conclude that there do not exist x_1 and x_2 such that $x_1\mathbf{a}_1 + x_2\mathbf{a}_2 = \mathbf{b}$, and therefore $\mathbf{b}$ is not in the span of $\mathbf{a}_1$ and $\mathbf{a}_2$.

13. $A = \begin{bmatrix} 2 & 8 & -4 \\ -1 & -3 & 5 \end{bmatrix}$, $\mathbf{x} = \begin{bmatrix} x_1 \\ x_2 \\ x_3 \end{bmatrix}$, $\mathbf{b} = \begin{bmatrix} -10 \\ 4 \end{bmatrix}$

15. $A = \begin{bmatrix} 1 & -1 & -3 & -1 \\ -2 & 2 & 6 & 2 \\ -3 & -3 & 10 & 0 \end{bmatrix}$, $\mathbf{x} = \begin{bmatrix} x_1 \\ x_2 \\ x_3 \\ x_4 \end{bmatrix}$, $\mathbf{b} = \begin{bmatrix} -1 \\ -1 \\ 5 \end{bmatrix}$

17. $x_1 \begin{bmatrix} 5 \\ 1 \end{bmatrix} + x_2 \begin{bmatrix} 7 \\ -5 \end{bmatrix} + x_3 \begin{bmatrix} -2 \\ -4 \end{bmatrix} = \begin{bmatrix} 9 \\ 2 \end{bmatrix}$

19. $x_1 \begin{bmatrix} 4 \\ 0 \\ 3 \end{bmatrix} + x_2 \begin{bmatrix} -2 \\ -5 \\ 8 \end{bmatrix} + x_3 \begin{bmatrix} -3 \\ 7 \\ 2 \end{bmatrix} + x_4 \begin{bmatrix} 5 \\ 3 \\ -1 \end{bmatrix} = \begin{bmatrix} 12 \\ 6 \\ 2 \end{bmatrix}$

21. Row-reduce to echelon form:

$$\begin{bmatrix} 15 & -6 \\ -5 & 2 \end{bmatrix} \overset{(1/3)R_1+R_2 \Rightarrow R_2}{\sim} \begin{bmatrix} 15 & -6 \\ 0 & 0 \end{bmatrix}$$

Since there is a row of zeros, there exists a vector $\mathbf{b}$ which is not in the span of the columns of the matrix, and therefore the columns of the matrix do not span $\mathbf{R}^2$.

23. Row-reduce to echelon form:

$$\begin{bmatrix} 2 & 1 & 0 \\ 6 & -3 & -1 \end{bmatrix} \overset{-3R_1+R_2 \Rightarrow R_2}{\sim} \begin{bmatrix} 2 & 1 & 0 \\ 0 & -6 & -1 \end{bmatrix}$$

Since there is not a row of zeros, every choice of $\mathbf{b}$ is in the span of the columns of the given matrix, and therefore the columns of the matrix span $\mathbf{R}^2$.

25. Row-reduce to echelon form:

$$\begin{bmatrix} 3 & 1 & 0 \\ 5 & -2 & -1 \\ 4 & -4 & -3 \end{bmatrix} \overset{(-5/3)R_1+R_2 \Rightarrow R_2}{\underset{(-4/3)R_1+R_3 \Rightarrow R_3}{\sim}} \begin{bmatrix} 3 & 1 & 0 \\ 0 & -\frac{11}{3} & -1 \\ 0 & -\frac{16}{3} & -3 \end{bmatrix}$$

$$\overset{(-16/11)R_2+R_3 \Rightarrow R_3}{\sim} \begin{bmatrix} 3 & 1 & 0 \\ 0 & -\frac{11}{3} & -1 \\ 0 & 0 & -\frac{17}{11} \end{bmatrix}$$

Since there is not a row of zeros, every choice of $\mathbf{b}$ is in the span of the columns of the given matrix, and therefore the columns of the matrix span $\mathbf{R}^3$.

27. Row-reduce to echelon form:

$$\begin{bmatrix} 2 & 1 & -3 & 5 \\ 1 & 4 & 2 & 6 \\ 0 & 3 & 3 & 3 \end{bmatrix} \overset{(-1/2)R_1+R_2 \Rightarrow R_2}{\sim} \begin{bmatrix} 2 & 1 & -3 & 5 \\ 0 & \frac{7}{2} & \frac{7}{2} & \frac{7}{2} \\ 0 & 3 & 3 & 3 \end{bmatrix}$$

$$\overset{(-6/7)R_2+R_3 \Rightarrow R_3}{\sim} \begin{bmatrix} 2 & 1 & -3 & 5 \\ 0 & \frac{7}{2} & \frac{7}{2} & \frac{7}{2} \\ 0 & 0 & 0 & 0 \end{bmatrix}$$

Since there is a row of zeros, there exists a vector $\mathbf{b}$ which is not in the span of the columns of the matrix, and therefore the columns of the matrix do not span $\mathbf{R}^3$.

29. Row-reduce A to echelon form:

$$\begin{bmatrix} 3 & -4 \\ 4 & 2 \end{bmatrix} \overset{(-4/3)R_1+R_2 \Rightarrow R_2}{\sim} \begin{bmatrix} 3 & -4 \\ 0 & \frac{22}{3} \end{bmatrix}$$

Since there is not a row of zeros, for every choice of $\mathbf{b}$ there is a solution of $A\mathbf{x} = \mathbf{b}$.

31. Since the number of columns, $m = 2$, is less than $n = 3$, the columns of A do not span $\mathbf{R}^3$, and by Theorem 2.10, there is a choice of $\mathbf{b}$ for which $A\mathbf{x} = \mathbf{b}$ has no solution.

33. Row-reduce A to echelon form:

$$\begin{bmatrix} -3 & 2 & 1 \\ 1 & -1 & -1 \\ 5 & -4 & -3 \end{bmatrix} \begin{matrix} (1/3)R_1+R_2 \Rightarrow R_2 \\ (5/3)R_1+R_3 \Rightarrow R_3 \\ \sim \end{matrix} \begin{bmatrix} -3 & 2 & 1 \\ 0 & -\frac{1}{3} & -\frac{2}{3} \\ 0 & -\frac{2}{3} & -\frac{4}{3} \end{bmatrix}$$

$$\begin{matrix} (-2)R_2+R_3 \Rightarrow R_3 \\ \sim \end{matrix} \begin{bmatrix} -3 & 2 & 1 \\ 0 & -\frac{1}{3} & -\frac{2}{3} \\ 0 & 0 & 0 \end{bmatrix}$$

Since there is a row of zeros, there is a choice of $\mathbf{b}$ for which $A\mathbf{x} = \mathbf{b}$ has no solution.

35. $\mathbf{b} = \begin{bmatrix} 0 \\ 1 \end{bmatrix}$ is not in span $\left\{ \begin{bmatrix} 1 \\ -2 \end{bmatrix}, \begin{bmatrix} -3 \\ 6 \end{bmatrix} \right\}$, since span $\left\{ \begin{bmatrix} 1 \\ -2 \end{bmatrix}, \begin{bmatrix} -3 \\ 6 \end{bmatrix} \right\} = $ span $\left\{ \begin{bmatrix} 1 \\ -2 \end{bmatrix} \right\}$ and $\mathbf{b} \neq c \begin{bmatrix} 1 \\ -2 \end{bmatrix}$ for any scalar c.

37. $\mathbf{b} = \begin{bmatrix} 0 \\ 0 \\ 1 \end{bmatrix}$ is not in span $\left\{ \begin{bmatrix} 1 \\ 3 \\ -2 \end{bmatrix}, \begin{bmatrix} 2 \\ -1 \\ 1 \end{bmatrix} \right\}$, since $c_1 \begin{bmatrix} 1 \\ 3 \\ -2 \end{bmatrix} + c_2 \begin{bmatrix} 2 \\ -1 \\ 1 \end{bmatrix} = \begin{bmatrix} 0 \\ 0 \\ 1 \end{bmatrix}$ has no solutions.

39. $h \neq 3$, since when $h = 3$ the vectors $\begin{bmatrix} 2 \\ 4 \end{bmatrix}$ and $\begin{bmatrix} 3 \\ 6 \end{bmatrix}$ are parallel and do not span $\mathbf{R}^2$.

41. $h \neq 4$. This value for h was determined by row-reducing

$$\begin{bmatrix} 2 & h & 1 \\ 4 & 8 & 2 \\ 5 & 10 & 6 \end{bmatrix} \sim \begin{bmatrix} 2 & h & 1 \\ 0 & 8-2h & 0 \\ 0 & 0 & \frac{7}{2} \end{bmatrix}$$

Then $c_1 \begin{bmatrix} 2 \\ 4 \\ 5 \end{bmatrix} + c_2 \begin{bmatrix} h \\ 8 \\ 10 \end{bmatrix} + c_3 \begin{bmatrix} 1 \\ 2 \\ 6 \end{bmatrix} = \begin{bmatrix} x \\ y \\ z \end{bmatrix}$ has a solution provided $h \neq 4$.

43. $\mathbf{u}_1 = (1,0,0)$, $\mathbf{u}_2 = (0,1,0)$, $\mathbf{u}_3 = (0,0,1)$, $\mathbf{u}_4 = (1,1,1)$

45. $\mathbf{u}_1 = (1,0,0)$, $\mathbf{u}_2 = (2,0,0)$, $\mathbf{u}_3 = (3,0,0)$, $\mathbf{u}_4 = (4,0,0)$

47. $\mathbf{u}_1 = (1,0,0)$, $\mathbf{u}_2 = (0,1,0)$

49. $\mathbf{u}_1 = (1,-1,0)$, $\mathbf{u}_2 = (1,0,-1)$

51. True, by Example 5.

53. False, since every column of A may be a zero column.

55. False. Consider $A = [1]$.

57. True, the span of a set of vectors can only increase (with respect to set containment) when adding a vector to the set.

59. False. Consider $\mathbf{u}_1 = (0,0,0)$, $\mathbf{u}_2 = (1,0,0)$, $\mathbf{u}_3 = (0,1,0)$, and $\mathbf{u}_4 = (0,0,1)$.

61. True. span $\{\mathbf{u}_1, \mathbf{u}_2, \mathbf{u}_3\} \subseteq $ span $\{\mathbf{u}_1, \mathbf{u}_2, \mathbf{u}_3, \mathbf{u}_4\}$ is always true. If a vector $\mathbf{w} \in $ span $\{\mathbf{u}_1, \mathbf{u}_2, \mathbf{u}_3, \mathbf{u}_4\}$, then since $\mathbf{u}_4$ is a linear combination of $\{\mathbf{u}_1, \mathbf{u}_2, \mathbf{u}_3\}$, we can express $\mathbf{w}$ as a linear combination of just the vectors $\mathbf{u}_1, \mathbf{u}_2$, and $\mathbf{u}_3$. Hence $\mathbf{w}$ is in span $\{\mathbf{u}_1, \mathbf{u}_2, \mathbf{u}_3\}$, and we have span $\{\mathbf{u}_1, \mathbf{u}_2, \mathbf{u}_3, \mathbf{u}_4\} \subseteq $ span $\{\mathbf{u}_1, \mathbf{u}_2, \mathbf{u}_3\}$.

63. False. Consider $\mathbf{u}_1 = (1,0,0,0)$, $\mathbf{u}_2 = (0,1,0,0)$, $\mathbf{u}_3 = (0,0,1,0)$, and $\mathbf{u}_4 = (0,0,0,1)$.

65. (a) Cannot possibly span $\mathbf{R}^3$, since $m = 1 < n = 3$.

   (b) Cannot possibly span $\mathbf{R}^3$, since $m = 2 < n = 3$.

   (c) Can possibly span $\mathbf{R}^3$. For example, $\mathbf{u}_1 = (1, 0, 0)$, $\mathbf{u}_2 = (0, 1, 0)$, $\mathbf{u}_3 = (0, 0, 1)$.

   (d) Can possibly span $\mathbf{R}^3$. For example, $\mathbf{u}_1 = (1, 0, 0)$, $\mathbf{u}_2 = (0, 1, 0)$, $\mathbf{u}_3 = (0, 0, 1)$, $\mathbf{u}_4 = (0, 0, 0)$.

67. Let $\mathbf{w} \in$ span $\{\mathbf{u}\}$, then $\mathbf{w} = x_1\mathbf{u} = \left(\frac{x_1}{c}\right)(c\mathbf{u})$, so $\mathbf{w} \in$ span $\{c\mathbf{u}\}$ and thus span $\{\mathbf{u}\} \subseteq$ span $\{c\mathbf{u}\}$. Now let $\mathbf{w} \in$ span $\{c\mathbf{u}\}$, then $\mathbf{w} = x_1(c\mathbf{u}) = (x_1c)(\mathbf{u})$, so $\mathbf{w} \in$ span $\{\mathbf{u}\}$ and thus span $\{c\mathbf{u}\} \subseteq$ span $\{\mathbf{u}\}$. Together, we conclude span $\{\mathbf{u}\} =$ span $\{c\mathbf{u}\}$.

69. We may let $S_1 = \{\mathbf{u}_1, \mathbf{u}_2, \dots, \mathbf{u}_m\}$ and $S_2 = \{\mathbf{u}_1, \mathbf{u}_2, \dots, \mathbf{u}_m, \mathbf{u}_{m+1}, \dots \mathbf{u}_n\}$ where $m \leq n$. Let $\mathbf{w} \in$ span(S_1), then

$$\mathbf{w} = x_1\mathbf{u}_1 + x_2\mathbf{u}_2 + \cdots + x_m\mathbf{u}_m$$
$$= x_1\mathbf{u}_1 + x_2\mathbf{u}_2 + \cdots + x_m\mathbf{u}_m + 0\mathbf{u}_{m+1} + \cdots + 0\mathbf{u}_n$$

and thus $\mathbf{w} \in$ span(S_2). We conclude that span $(S_1) \subseteq$ span (S_2).

71. Let $\mathbf{b} \in \mathbf{R}^3$, then $\mathbf{b} = x_1\mathbf{u}_1 + x_2\mathbf{u}_2 + x_3\mathbf{u}_3$ for some scalars x_1, x_2, and x_3 because span $\{\mathbf{u}_1, \mathbf{u}_2, \mathbf{u}_3\} = \mathbf{R}^3$. We can rewrite $\mathbf{b} = \frac{x_1+x_2-x_3}{2}(\mathbf{u}_1 + \mathbf{u}_2) + \frac{x_1-x_2+x_3}{2}(\mathbf{u}_1 + \mathbf{u}_3) + \frac{-x_1+x_2+x_3}{2}(\mathbf{u}_2 + \mathbf{u}_3)$, thus $\mathbf{b} \in$ span $\{\mathbf{u}_1 + \mathbf{u}_2, \mathbf{u}_1 + \mathbf{u}_3, \mathbf{u}_2 + \mathbf{u}_3\}$. Since $\mathbf{b}$ was arbitrary, span $\{\mathbf{u}_1 + \mathbf{u}_2, \mathbf{u}_1 + \mathbf{u}_3, \mathbf{u}_2 + \mathbf{u}_3\} = \mathbf{R}^3$.

73. Let $A = [\mathbf{u}_1 \cdots \mathbf{u}_m]$ and suppose $A \sim B$, where B is in echelon form. Since $m < n$, the last row of B must consist of zeros. Form B_1 by appending to B the vector $\mathbf{e} = \begin{bmatrix} 0 \\ \vdots \\ 1 \end{bmatrix}$, so that $B_1 = [B \quad \mathbf{e}]$. If B_1 is viewed as an augmented matrix, then the bottom row corresponds to the equation $0 = 1$, so the corresponding linear system is inconsistent. Now reverse the row operations used to transform A to B, and apply these to B_1. Then the resulting matrix will have the form $[A \quad \mathbf{e}']$. This implies that $\mathbf{e}'$ is not in the span of the columns of A, as required.

75. True. Using a computer algebra system, the row-reduced echelon form of the matrix with the given vectors as columns does not have any zero rows. Hence the vectors span $\mathbf{R}^3$.

77. False. Using a computer algebra system, the row-reduced echelon form of the matrix with the given vectors as columns does have a zero row. Hence the vectors do not span $\mathbf{R}^4$.

2.3 Linear Independence

1. Consider $x_1\mathbf{u} + x_2\mathbf{v} = \mathbf{0}$, and solve using the corresponding augmented matrix:

$$\begin{bmatrix} 3 & -1 & 0 \\ -2 & -4 & 0 \end{bmatrix} \overset{(2/3)R_1 + R_2 \Rightarrow R_2}{\sim} \begin{bmatrix} 3 & -1 & 0 \\ 0 & -\frac{14}{3} & 0 \end{bmatrix}$$

Since the only solution is the trivial solution, the vectors are linearly independent.

3. Consider $x_1\mathbf{u} + x_2\mathbf{v} = \mathbf{0}$, and solve using the corresponding augmented matrix:

$$\begin{bmatrix} 7 & 5 & 0 \\ 1 & -3 & 0 \\ -13 & 2 & 0 \end{bmatrix} \overset{\substack{(-1/7)R_1 + R_2 \Rightarrow R_2 \\ (13/7)R_1 + R_3 \Rightarrow R_3}}{\sim} \begin{bmatrix} 7 & 5 & 0 \\ 0 & -\frac{26}{7} & 0 \\ 0 & \frac{79}{7} & 0 \end{bmatrix}$$
$$\overset{(79/26)R_2 + R_3 \Rightarrow R_3}{\sim} \begin{bmatrix} 7 & 5 & 0 \\ 0 & -\frac{26}{7} & 0 \\ 0 & 0 & 0 \end{bmatrix}$$

Since the only solution is the trivial solution, the vectors are linearly independent.

5. Consider $x_1\mathbf{u} + x_2\mathbf{v} + x_3\mathbf{w} = \mathbf{0}$, and solve using the corresponding augmented matrix:

$$
\begin{bmatrix} 3 & 0 & 2 & 0 \\ -1 & 4 & 4 & 0 \\ 2 & 1 & 7 & 0 \end{bmatrix}
\begin{array}{c} (1/3)R_1+R_2\Rightarrow R_2 \\ (-2/3)R_1+R_3\Rightarrow R_3 \\ \sim \end{array}
\begin{bmatrix} 3 & 0 & 2 & 0 \\ 0 & 4 & \frac{14}{3} & 0 \\ 0 & 1 & \frac{17}{3} & 0 \end{bmatrix}
$$

$$
\begin{array}{c} (-1/4)R_2+R_3\Rightarrow R_3 \\ \sim \end{array}
\begin{bmatrix} 3 & 0 & 2 & 0 \\ 0 & 4 & \frac{14}{3} & 0 \\ 0 & 0 & \frac{9}{2} & 0 \end{bmatrix}
$$

Since the only solution is the trivial solution, the vectors are linearly independent.

7. We solve the homogeneous system of equations using the corresponding augmented matrix:

$$
\begin{bmatrix} 15 & -6 & 0 \\ -5 & 2 & 0 \end{bmatrix}
\begin{array}{c} (2/3)R_1+R_2\Rightarrow R_2 \\ \sim \end{array}
\begin{bmatrix} 15 & -6 & 0 \\ 0 & 0 & 0 \end{bmatrix}
$$

Since there exist nontrivial solutions, the columns of the matrix are not linearly independent.

9. We solve the homogeneous system of equations using the corresponding augmented matrix:

$$
\begin{bmatrix} 1 & 0 & 0 \\ -2 & 2 & 0 \\ 5 & -7 & 0 \end{bmatrix}
\begin{array}{c} 2R_1+R_2\Rightarrow R_2 \\ -5R_1+R_3\Rightarrow R_3 \\ \sim \end{array}
\begin{bmatrix} 1 & 0 & 0 \\ 0 & 2 & 0 \\ 0 & -7 & 0 \end{bmatrix}
$$

$$
\begin{array}{c} (7/2)R_2+R_3\Rightarrow R_3 \\ \sim \end{array}
\begin{bmatrix} 1 & 0 & 0 \\ 0 & 2 & 0 \\ 0 & 0 & 0 \end{bmatrix}
$$

There is only the trivial solution, the columns of the matrix are linearly independent.

11. We solve the homogeneous system of equations using the corresponding augmented matrix:

$$
\begin{bmatrix} 3 & 1 & 0 & 0 \\ 5 & -2 & -1 & 0 \\ 4 & -4 & -3 & 0 \end{bmatrix}
\begin{array}{c} (-5/3)R_1+R_2\Rightarrow R_2 \\ (-4/3)R_1+R_3\Rightarrow R_3 \\ \sim \end{array}
\begin{bmatrix} 3 & 1 & 0 & 0 \\ 0 & -\frac{11}{3} & -1 & 0 \\ 0 & -\frac{16}{3} & -3 & 0 \end{bmatrix}
$$

$$
\begin{array}{c} (-16/11)R_2+R_3\Rightarrow R_3 \\ \sim \end{array}
\begin{bmatrix} 3 & 1 & 0 & 0 \\ 0 & -\frac{11}{3} & -1 & 0 \\ 0 & 0 & -\frac{17}{11} & 0 \end{bmatrix}
$$

Since the only solution is the trivial solution, the columns of the matrix are linearly independent.

13. We solve the homogeneous equation using the corresponding augmented matrix:

$$
\begin{bmatrix} -3 & 5 & 0 \\ 4 & 1 & 0 \end{bmatrix}
\begin{array}{c} (4/3)R_1+R_2\Rightarrow R_2 \\ \sim \end{array}
\begin{bmatrix} -3 & 5 & 0 \\ 0 & \frac{23}{3} & 0 \end{bmatrix}
$$

Since the only solution is the trivial solution, the homogeneous equation $A\mathbf{x} = \mathbf{0}$ has only the trivial solution.

15. We solve the homogeneous equation using the corresponding augmented matrix:

$$
\begin{bmatrix} 8 & 1 & 0 \\ 0 & -1 & 0 \\ -3 & 2 & 0 \end{bmatrix}
\begin{array}{c} (3/8)R_1+R_3\Rightarrow R_3 \\ \sim \end{array}
\begin{bmatrix} 8 & 1 & 0 \\ 0 & -1 & 0 \\ 0 & \frac{19}{8} & 0 \end{bmatrix}
$$

$$
\begin{array}{c} (19/8)R_2+R_3\Rightarrow R_3 \\ \sim \end{array}
\begin{bmatrix} 8 & 1 & 0 \\ 0 & -1 & 0 \\ 0 & 0 & 0 \end{bmatrix}
$$

Since the only solution is the trivial solution, the homogeneous equation $A\mathbf{x} = \mathbf{0}$ has only the trivial solution.

17. We solve the homogeneous equation using the corresponding augmented matrix:

$$\begin{bmatrix} -1 & 3 & 1 & 0 \\ 4 & -3 & -1 & 0 \\ 3 & 0 & 5 & 0 \end{bmatrix} \begin{array}{c} 4R_1+R_2\Rightarrow R_2 \\ 3R_1+R_3\Rightarrow R_3 \\ \sim \end{array} \begin{bmatrix} -1 & 3 & 1 & 0 \\ 0 & 9 & 3 & 0 \\ 0 & 9 & 8 & 0 \end{bmatrix}$$

$$\begin{array}{c} -R_2+R_3\Rightarrow R_3 \\ \sim \end{array} \begin{bmatrix} -1 & 3 & 1 & 0 \\ 0 & 9 & 3 & 0 \\ 0 & 0 & 5 & 0 \end{bmatrix}$$

The homogeneous equation $A\mathbf{x} = \mathbf{0}$ has only the trivial solution.

19. Linearly dependent. Notice that $\mathbf{u} = 2\mathbf{v}$, so $\mathbf{u} - 2\mathbf{v} = \mathbf{0}$.

21. Linearly dependent. Apply Theorem 2.13.

23. Linearly dependent. Any collection of vectors containing the zero vector must be linearly dependent.

25. We solve the homogeneous system of equations using the corresponding augmented matrix:

$$\begin{bmatrix} 6 & 1 & 0 \\ 2 & 7 & 0 \\ -5 & 0 & 0 \end{bmatrix} \begin{array}{c} (-1/3)R_1+R_2\Rightarrow R_2 \\ (5/6)R_1+R_3\Rightarrow R_3 \\ \sim \end{array} \begin{bmatrix} 6 & 1 & 0 \\ 0 & \frac{20}{3} & 0 \\ 0 & \frac{5}{6} & 0 \end{bmatrix}$$

$$\begin{array}{c} (-1/8)R_2+R_3\Rightarrow R_3 \\ \sim \end{array} \begin{bmatrix} 6 & 1 & 0 \\ 0 & \frac{20}{3} & 0 \\ 0 & 0 & 0 \end{bmatrix}$$

Since the only solution is the trivial solution, the columns of the matrix are linearly independent. By Theorem 2.14, none of the vectors is in the span of the other vectors.

27. We solve the homogeneous system of equations using the corresponding augmented matrix:

$$\begin{bmatrix} 4 & 3 & -5 & 0 \\ -1 & 5 & 7 & 0 \\ 3 & -2 & -7 & 0 \end{bmatrix} \begin{array}{c} (1/4)R_1+R_2\Rightarrow R_2 \\ (-3/4)R_1+R_3\Rightarrow R_3 \\ \sim \end{array} \begin{bmatrix} 4 & 3 & -5 & 0 \\ 0 & \frac{23}{4} & \frac{23}{4} & 0 \\ 0 & -\frac{17}{4} & -\frac{13}{4} & 0 \end{bmatrix}$$

$$\begin{array}{c} (17/23)R_2+R_3\Rightarrow R_3 \\ \sim \end{array} \begin{bmatrix} 4 & 3 & -5 & 0 \\ 0 & \frac{23}{4} & \frac{23}{4} & 0 \\ 0 & 0 & 1 & 0 \end{bmatrix}$$

Since the only solution is the trivial solution, the columns of the matrix are linearly independent. By Theorem 2.14, none of the vectors is in the span of the other vectors.

29. We solve the homogeneous system of equations using the corresponding augmented matrix:

$$\begin{bmatrix} 2 & -1 & 0 & 0 \\ 1 & 0 & 1 & 0 \\ -3 & 4 & 5 & 0 \end{bmatrix} \begin{array}{c} (-1/2)R_1+R_2\Rightarrow R_2 \\ (3/2)R_1+R_3\Rightarrow R_3 \\ \sim \end{array} \begin{bmatrix} 2 & -1 & 0 & 0 \\ 0 & \frac{1}{2} & 1 & 0 \\ 0 & \frac{5}{2} & 5 & 0 \end{bmatrix}$$

$$\begin{array}{c} -5R_2+R_3\Rightarrow R_3 \\ \sim \end{array} \begin{bmatrix} 2 & -1 & 0 & 0 \\ 0 & \frac{1}{2} & 1 & 0 \\ 0 & 0 & 0 & 0 \end{bmatrix}$$

Since there exist nontrivial solutions, the columns of the matrix are linearly dependent. By The Big Theorem, $A\mathbf{x} = \mathbf{b}$ does not have a unique solution for all $\mathbf{b}$ in $\mathbf{R}^3$.

31. We solve the homogeneous system of equations using the corresponding augmented matrix:

$$\begin{bmatrix} 3 & -2 & 1 & 0 \\ -4 & 1 & 0 & 0 \\ -5 & 0 & 1 & 0 \end{bmatrix} \begin{array}{c} (4/3)R_1+R_2 \Rightarrow R_2 \\ (5/3)R_1+R_3 \Rightarrow R_3 \\ \sim \end{array} \begin{bmatrix} 3 & -2 & 1 & 0 \\ 0 & -\frac{5}{3} & \frac{4}{3} & 0 \\ 0 & -\frac{10}{3} & \frac{8}{3} & 0 \end{bmatrix}$$

$$\begin{array}{c} -2R_2+R_3 \Rightarrow R_3 \\ \sim \end{array} \begin{bmatrix} 3 & -2 & 1 & 0 \\ 0 & -\frac{5}{3} & \frac{4}{3} & 0 \\ 0 & 0 & 0 & 0 \end{bmatrix}$$

   Since there exist nontrivial solutions, the columns of the matrix are linearly dependent. By The Big Theorem, $A\mathbf{x} = \mathbf{b}$ does not have a unique solution for all $\mathbf{b}$ in $\mathbf{R}^3$.

33. $\mathbf{u} = (1,0,0,0)$, $\mathbf{v} = (0,1,0,0)$, $\mathbf{w} = (1,1,0,0)$

35. $\mathbf{u} = (1,0)$, $\mathbf{v} = (2,0)$, $\mathbf{w} = (3,0)$

37. $\mathbf{u} = (1,0,0)$, $\mathbf{v} = (0,1,0)$, $\mathbf{w} = (1,1,0)$

39. False. For example, $\mathbf{u} = (1,0)$ and $\mathbf{v} = (2,0)$ are linearly dependent but do not span $\mathbf{R}^2$.

41. False. For example, $A = \begin{bmatrix} 1 & -1 \\ 2 & -2 \\ 0 & 0 \end{bmatrix}$ has more rows than columns but the columns are linearly dependent.

43. False. $A\mathbf{x} = \mathbf{0}$ corresponds to $x_1\mathbf{a}_1 + \cdots + x_n\mathbf{a}_n = \mathbf{0}$, and by linear independence, each $x_i = 0$.

45. False. Consider for example $\mathbf{u}_4 = \mathbf{0}$.

47. True. Consider $x_1\mathbf{u}_1 + x_2\mathbf{u}_2 + x_3\mathbf{u}_3 = \mathbf{0}$. If one of the $x_i \neq 0$, then $x_1\mathbf{u}_1 + x_2\mathbf{u}_2 + x_3\mathbf{u}_3 + 0\mathbf{u}_4 = \mathbf{0}$ would imply that $\{\mathbf{u}_1, \mathbf{u}_2, \mathbf{u}_3, \mathbf{u}_4\}$ is linearly dependent, a contradiction. Hence each $x_i = 0$, and $\{\mathbf{u}_1, \mathbf{u}_2, \mathbf{u}_3\}$ is linearly independent.

49. False. If $\mathbf{u}_4 = x_1\mathbf{u}_1 + x_2\mathbf{u}_2 + x_3\mathbf{u}_3$, then $x_1\mathbf{u}_1 + x_2\mathbf{u}_2 + x_3\mathbf{u}_3 - \mathbf{u}_4 = \mathbf{0}$, and since the coefficient of $\mathbf{u}_4$ is -1, $\{\mathbf{u}_1, \mathbf{u}_2, \mathbf{u}_3, \mathbf{u}_4\}$ is linearly dependent.

51. False. Consider $\mathbf{u}_1 = (1,0,0)$, $\mathbf{u}_2 = (1,0,0)$, $\mathbf{u}_3 = (1,0,0)$, $\mathbf{u}_4 = (0,1,0)$.

53. (a), (b), and (c). For example, consider $\mathbf{u}_1 = (1,0,0)$, $\mathbf{u}_2 = (1,0,0)$, and $\mathbf{u}_3 = (1,0,0)$. (d) cannot be linearly independent, by Theorem 2.13.

55. Consider $x_1(c_1\mathbf{u}_1) + x_2(c_2\mathbf{u}_2) + x_3(c_3\mathbf{u}_3) = \mathbf{0}$. Then $(x_1c_1)\mathbf{u}_1 + (x_2c_2)\mathbf{u}_2 + (x_3c_3)\mathbf{u}_3 = \mathbf{0}$, and since $\{\mathbf{u}_1, \mathbf{u}_2, \mathbf{u}_3\}$ is linearly independent, $x_1c_1 = 0$, $x_2c_2 = 0$, and $x_3c_3 = 0$. Since each $c_i \neq 0$, we must have each $x_i = 0$. Hence, $\{c_1\mathbf{u}_1, c_2\mathbf{u}_2, c_3\mathbf{u}_3\}$ is linearly independent.

57. Consider $x_1(\mathbf{u}_1 + \mathbf{u}_2) + x_2(\mathbf{u}_1 + \mathbf{u}_3) + x_3(\mathbf{u}_2 + \mathbf{u}_3) = \mathbf{0}$. This implies $(x_1 + x_2)\mathbf{u}_1 + (x_1 + x_3)\mathbf{u}_2 + (x_2 + x_3)\mathbf{u}_3 = \mathbf{0}$. Since $\{\mathbf{u}_1, \mathbf{u}_2, \mathbf{u}_3\}$ is linearly independent, $x_1 + x_2 = 0$, $x_1 + x_3 = 0$, and $x_2 + x_3 = 0$. Solving this system, we obtain $x_1 = 0$, $x_2 = 0$, and $x_3 = 0$. Thus $\{\mathbf{u}_1 + \mathbf{u}_2, \mathbf{u}_1 + \mathbf{u}_3, \mathbf{u}_2 + \mathbf{u}_3\}$ is linearly independent.

59. Suppose $\{\mathbf{u}_1, \mathbf{u}_2, \ldots, \mathbf{u}_n\}$ is linearly dependent set, and we add vectors to form a new set $\{\mathbf{u}_1, \mathbf{u}_2, \ldots, \mathbf{u}_n, \ldots \mathbf{u}_m\}$. There exist x_i with a least one $x_i \neq 0$ such that $x_1\mathbf{u}_1 + x_2\mathbf{u}_2 + \cdots + x_n\mathbf{u}_n = \mathbf{0}$. Thus $x_1\mathbf{u}_1 + x_2\mathbf{u}_2 + \cdots + x_n\mathbf{u}_n + 0\mathbf{u}_{n+1} + \cdots + 0\mathbf{u}_m = \mathbf{0}$, and so $\{\mathbf{u}_1, \mathbf{u}_2, \ldots, \mathbf{u}_n, \ldots \mathbf{u}_m\}$ is linearly dependent.

61. $\mathbf{u}$ and $\mathbf{v}$ are linearly dependent if and only if there exist scalars x_1 and x_2, not both zero, such that $x_1\mathbf{u} + x_2\mathbf{v} = \mathbf{0}$. If $x_1 \neq 0$, then $\mathbf{u} = (-x_2/x_1)\mathbf{v} = c\mathbf{v}$. If $x_2 \neq 0$, then $\mathbf{v} = (-x_1/x_2)\mathbf{u} = c\mathbf{u}$.

63. Suppose $A = \begin{bmatrix} \mathbf{a}_1 & \mathbf{a}_2 & \cdots & \mathbf{a}_m \end{bmatrix}$, $\mathbf{x} = (x_1, x_2, \ldots, x_m)$ and $\mathbf{y} = (y_1, y_2, \ldots, y_m)$. Then we have $\mathbf{x} - \mathbf{y} = (x_1 - y_1, x_2 - y_2, \ldots, x_m - y_m)$, and thus

$$
\begin{aligned}
A(\mathbf{x} - \mathbf{y}) &= (x_1 - y_1)\,\mathbf{a}_1 + (x_2 - y_2)\,\mathbf{a}_2 + \cdots + (x_m - y_m)\,\mathbf{a}_m \\
&= (x_1\mathbf{a}_1 + x_2\mathbf{a}_2 + \cdots + x_m\mathbf{a}_m) - (y_1\mathbf{a}_1 + y_2\mathbf{a}_2 + \cdots + y_m\mathbf{a}_m) \\
&= A\mathbf{x} - A\mathbf{y}
\end{aligned}
$$

65. We row-reduce the corresponding augmented matrix:

$$
\begin{bmatrix} 1 & 4 & 2 & 0 \\ -2 & 2 & 6 & 0 \\ 5 & 0 & 3 & 0 \end{bmatrix}
\begin{array}{c} 2R_1 + R_2 \Rightarrow R_2 \\ -5R_1 + R_3 \Rightarrow R_3 \\ \sim \end{array}
\begin{bmatrix} 1 & 4 & 2 & 0 \\ 0 & 10 & 10 & 0 \\ 0 & -20 & -7 & 0 \end{bmatrix}
$$

$$
\begin{array}{c} 2R_2 + R_3 \Rightarrow R_3 \\ \sim \end{array}
\begin{bmatrix} 1 & 4 & 2 & 0 \\ 0 & 10 & 10 & 0 \\ 0 & 0 & 13 & 0 \end{bmatrix}
$$

Since the only solution is the trivial solution, the columns of the matrix are linearly independent. Thus, no vector is in the span of the others, there is no redundancy in the vectors, and they span $\mathbf{R}^3$.

67. Using a computer algebra system, the vectors are linearly independent.

69. Using a computer algebra system, the vectors are linearly independent.

71. Using a computer algebra system, $A\mathbf{x} = \mathbf{b}$ has a unique solution for all $\mathbf{b}$ in $\mathbf{R}^3$.

73. Using a computer algebra system, $A\mathbf{x} = \mathbf{b}$ does not have a unique solution for all $\mathbf{b}$ in $\mathbf{R}^4$.

Chapter 3

Matrices

3.1 Linear Transformations

1. $T(\mathbf{u}_1) = A\mathbf{u}_1 = \begin{bmatrix} 2 & 1 \\ -3 & 5 \end{bmatrix} \begin{bmatrix} -4 \\ -2 \end{bmatrix} = \begin{bmatrix} -10 \\ 2 \end{bmatrix}, T(\mathbf{u}_2) = A\mathbf{u}_2 = \begin{bmatrix} 2 & 1 \\ -3 & 5 \end{bmatrix} \begin{bmatrix} 1 \\ -6 \end{bmatrix} = \begin{bmatrix} -4 \\ -33 \end{bmatrix}$

3. $T(\mathbf{u}_1) = A\mathbf{u}_1 = \begin{bmatrix} 0 & -4 & 2 \\ 3 & 1 & -2 \end{bmatrix} \begin{bmatrix} 3 \\ 2 \\ 1 \end{bmatrix} = \begin{bmatrix} -6 \\ 9 \end{bmatrix}, T(\mathbf{u}_2) = A\mathbf{u}_2 = \begin{bmatrix} 0 & -4 & 2 \\ 3 & 1 & -2 \end{bmatrix} \begin{bmatrix} 4 \\ -5 \\ -2 \end{bmatrix}$
   $= \begin{bmatrix} 16 \\ 11 \end{bmatrix}$

5. We consider $T(\mathbf{x}) = A\mathbf{x} = \mathbf{y}$, and row-reduce the corresponding augmented matrix:

   $$\begin{bmatrix} 1 & -2 & 0 & -3 \\ 3 & 2 & 1 & 6 \end{bmatrix} \overset{-3R_1 + R_2 \Rightarrow R_2}{\sim} \begin{bmatrix} 1 & -2 & 0 & -3 \\ 0 & 8 & 1 & 15 \end{bmatrix}$$

   Since there exists a solution $\mathbf{x}$ to $A\mathbf{x} = \mathbf{y}$, $\mathbf{y}$ is in the range of T.

7. We consider $T(\mathbf{x}) = A\mathbf{x} = \mathbf{y}$, and row-reduce the corresponding augmented matrix:

   $$\begin{bmatrix} 1 & -2 & 0 & 2 \\ 3 & 2 & 1 & 7 \end{bmatrix} \overset{-3R_1 + R_2 \Rightarrow R_2}{\sim} \begin{bmatrix} 1 & -2 & 0 & 2 \\ 0 & 8 & 1 & 1 \end{bmatrix}$$

   Since there exists a solution $\mathbf{x}$ to $A\mathbf{x} = \mathbf{y}$, $\mathbf{y}$ is in the range of T.

9. $T(-2\mathbf{u}_1 + 3\mathbf{u}_2) = -2T(\mathbf{u}_1) + 3T(\mathbf{u}_2) = -2 \begin{bmatrix} 2 \\ 1 \end{bmatrix} + 3 \begin{bmatrix} -3 \\ 2 \end{bmatrix} = \begin{bmatrix} -13 \\ 4 \end{bmatrix}$

11. $T(-\mathbf{u}_1 + 4\mathbf{u}_2 - 3\mathbf{u}_3) = -T(\mathbf{u}_1) + 4T(\mathbf{u}_2) - 3T(\mathbf{u}_3)$
    $= -\begin{bmatrix} -3 \\ 0 \end{bmatrix} + 4 \begin{bmatrix} 2 \\ -1 \end{bmatrix} - 3 \begin{bmatrix} 0 \\ 5 \end{bmatrix} = \begin{bmatrix} 11 \\ -19 \end{bmatrix}$

13. Linear transformation, with $A = \begin{bmatrix} 3 & 1 \\ -2 & 4 \end{bmatrix}$.

15. Not a linear transformation, since $T(0(0,0,0)) = T(0,0,0) = (2\cos 0, 3\sin 0, 0) = (2,0,0)$, but $0(T(0,0,0)) = (0,0,0)$.

17. Linear transformation, with $A = \begin{bmatrix} -4 & 0 & 1 \\ 6 & 5 & 0 \end{bmatrix}$.

67

19. Linear transformation, with $A = \begin{bmatrix} 0 & \sin\frac{\pi}{4} \\ \ln 2 & 0 \end{bmatrix}$.

21. We consider $T(\mathbf{x}) = A\mathbf{x} = \mathbf{b}$, and row-reduce the corresponding augmented matrix:

$$\begin{bmatrix} 1 & -3 & b_1 \\ -2 & 5 & b_2 \end{bmatrix} \overset{2R_1 + R_2 \Rightarrow R_2}{\sim} \begin{bmatrix} 1 & -3 & b_1 \\ 0 & -1 & 2b_1 + b_2 \end{bmatrix}$$

Since there exists a unique solution $\mathbf{x}$ to $A\mathbf{x} = \mathbf{b}$, by The Big Theorem - Version 2, T is both one-to-one and onto.

23. Since $n = 2 < m = 3$, by Theorem 3.6 T is not one-to-one. To determine if T is onto, we row-reduce the corresponding augmented matrix:

$$\begin{bmatrix} 5 & 4 & -2 & b_1 \\ 3 & -1 & 0 & b_2 \end{bmatrix} \overset{(-3/5)R_1 + R_2 \Rightarrow R_2}{\sim} \begin{bmatrix} 5 & 4 & -2 & b_1 \\ 0 & -\frac{17}{5} & \frac{6}{5} & -\frac{3}{5}b_1 + b_2 \end{bmatrix}$$

Since there exists a solution $\mathbf{x}$ to $A\mathbf{x} = \mathbf{b}$ for all $\mathbf{b}$, the columns of A span $\mathbb{R}^n$, and by Theorem 3.7 T is onto.

25. Since $n = 3 > m = 2$, by Theorem 3.7 T is not onto. To determine if T is one-to-one, we row-reduce the corresponding augmented matrix:

$$\begin{bmatrix} 1 & -2 & 0 \\ -3 & 5 & 0 \\ 2 & -7 & 0 \end{bmatrix} \overset{\substack{3R_1 + R_2 \Rightarrow R_2 \\ -2R_1 + R_3 \Rightarrow R_3}}{\sim} \begin{bmatrix} 1 & -2 & 0 \\ 0 & -1 & 0 \\ 0 & -3 & 0 \end{bmatrix}$$

$$\overset{-3R_2 + R_3 \Rightarrow R_3}{\sim} \begin{bmatrix} 1 & -2 & 0 \\ 0 & -1 & 0 \\ 0 & 0 & 0 \end{bmatrix}$$

Since $T(\mathbf{x}) = A\mathbf{x} = \mathbf{0}$ has only the trivial solution, by Theorem 3.5 T is one-to-one.

27. We consider $T(\mathbf{x}) = A\mathbf{x} = \mathbf{b}$, and row-reduce the corresponding augmented matrix:

$$\begin{bmatrix} 2 & 8 & 4 & b_1 \\ 3 & 2 & 3 & b_2 \\ 1 & 14 & 5 & b_3 \end{bmatrix} \overset{\substack{(-3/2)R_1 + R_2 \Rightarrow R_2 \\ (-1/2)R_1 + R_3 \Rightarrow R_3}}{\sim} \begin{bmatrix} 2 & 8 & 4 & b_1 \\ 0 & -10 & -3 & (-3/2)b_1 + b_2 \\ 0 & 10 & 3 & (-1/2)b_1 + b_3 \end{bmatrix}$$

$$\overset{R_2 + R_3 \Rightarrow R_3}{\sim} \begin{bmatrix} 2 & 8 & 4 & b_1 \\ 0 & -10 & -3 & (-3/2)b_1 + b_2 \\ 0 & 0 & 0 & -2b_1 + b_2 + b_3 \end{bmatrix}$$

If $-2b_1 + b_2 + b_3 \neq 0$, there does not exist a unique solution $\mathbf{x}$ to $A\mathbf{x} = \mathbf{b}$. By The Big Theorem - Version 2, T is neither one-to-one nor onto.

29.

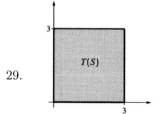

31.

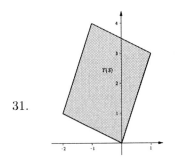

33. $T(\mathbf{x}) = \begin{bmatrix} 2 & 0 \\ 3 & 0 \end{bmatrix} \mathbf{x}$

35. $T(\mathbf{x}) = \begin{bmatrix} 7/3 & 0 \\ 0 & 0 \end{bmatrix} \mathbf{x}$

37. $T(\mathbf{x}) = \begin{bmatrix} 1 & -2 \\ 3 & 1 \end{bmatrix} \mathbf{x}$

39. False. For instance $T : \mathbf{R}^2 \to \mathbf{R}^2$ defined by $T(\mathbf{x}) = \mathbf{0}$ for all $\mathbf{x}$ has range$(T) = \{\mathbf{0}\}$ but codomain equal to $\mathbf{R}^2$.

41. True, by definition of the range of T.

43. True. If T is linear, then $T(\mathbf{0}) = \mathbf{0}$ and so $\mathbf{b} = \mathbf{0}$. If $\mathbf{b} = \mathbf{0}$, then T is linear by Theorem 3.2.

45. False. W will be linear, but not necessarily one-to-one. Consider $T_2(\mathbf{x}) = -T_1(\mathbf{x})$ where T_1 is one-to-one.

47. False, by Theorem 3.5.

49. (a) $A = \begin{bmatrix} r & 0 \\ 0 & r \end{bmatrix}$

(b)
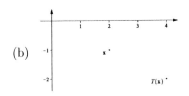

51. Let $\mathbf{u} = (u_1, \ldots, u_n)$. Then

$$
\begin{aligned}
T(\mathbf{x} + \mathbf{y}) &= \mathbf{u} \cdot (\mathbf{x} + \mathbf{y}) \\
&= (u_1, \ldots, u_n) \cdot ((x_1, \ldots, x_n) + (y_1, \ldots, y_n)) \\
&= (u_1, \ldots, u_n) \cdot (x_1 + y_1, \ldots, x_n + y_n) \\
&= u_1(x_1 + y_1) + \cdots + u_n(x_n + y_n) \\
&= (u_1 x_1 + u_1 y_1) + \cdots + (u_n x_n + u_n y_n) \\
&= (u_1 x_1 + \cdots + u_n x_n) + (u_1 y_1 + \cdots + u_n y_n) \\
&= \mathbf{u} \cdot \mathbf{x} + \mathbf{u} \cdot \mathbf{y} \\
&= T(\mathbf{x}) + T(\mathbf{y}),
\end{aligned}
$$

and

$$T(r\mathbf{x}) = \mathbf{u} \cdot (r\mathbf{x})$$
$$= (u_1, \ldots, u_n) \cdot (r(x_1, \ldots, x_n))$$
$$= (u_1, \ldots, u_n) \cdot (rx_1, \ldots, rx_n)$$
$$= u_1(rx_1) + \cdots + u_n(rx_n)$$
$$= r(u_1 x_1 + \cdots + u_n x_n)$$
$$= r\mathbf{u} \cdot \mathbf{x}$$
$$= rT(\mathbf{x}).$$

Thus T is a linear transformation.

53. Let $\mathbf{u}_1 = (1,0,0)$, $\mathbf{u}_2 = (0,1,0)$, and $\mathbf{u}_3 = (0,0,1)$. Since $T(\mathbf{u}_1)$, $T(\mathbf{u}_2)$, $T(\mathbf{u}_3)$ are three vectors in $\mathbf{R}^2$, they must be linearly dependent, and therefore there exist scalars c_1, c_2, and c_3 with at least one $c_i \neq 0$ and $c_1 T(\mathbf{u}_1) + c_2 T(\mathbf{u}_2) + c_3 T(\mathbf{u}_3) = \mathbf{0}$. Since T is a linear transformation, $T(c_1\mathbf{u}_1 + c_2\mathbf{u}_2 + c_3\mathbf{u}_3) = \mathbf{0}$. Also, since $\mathbf{u}_1$, $\mathbf{u}_2$, and $\mathbf{u}_3$ are linearly independent and one of the $c_i \neq 0$, it follows that $c_1\mathbf{u}_1 + c_2\mathbf{u}_2 + c_3\mathbf{u}_3 \neq \mathbf{0}$. Noting that $T(\mathbf{0}) = \mathbf{0}$, $T(c_1\mathbf{u}_1 + c_2\mathbf{u}_2 + c_3\mathbf{u}_3) = \mathbf{0}$, and $c_1\mathbf{u}_1 + c_2\mathbf{u}_2 + c_3\mathbf{u}_3 \neq \mathbf{0}$, we conclude that T is not one-to-one.

55. $T(\mathbf{0}) = T(\mathbf{0} + \mathbf{0}) = T(\mathbf{0}) + T(\mathbf{0}) \Rightarrow T(\mathbf{0}) = \mathbf{0}$, upon subtracting $T(\mathbf{0})$ from both sides. (*Another proof, using the scalar property:* $T(\mathbf{0}) = T(2(\mathbf{0})) = 2T(\mathbf{0}) \Rightarrow T(\mathbf{0}) = \mathbf{0}$, *upon subtracting* $T(\mathbf{0})$ *from both sides.*)

57. $T(r\mathbf{u}) = A(r\mathbf{u}) = r(A\mathbf{u}) = rT(\mathbf{u})$ for all scalars r and all vectors $\mathbf{u}$.

59. Suppose $T : \mathbf{R}^m \to \mathbf{R}^n$ is one-to-one, and let $T(\mathbf{u}) = T(\mathbf{v}) = \mathbf{w}$. Since there exists at most one vector whose image under T is $\mathbf{w}$, it follows that $\mathbf{u} = \mathbf{v}$. Now suppose $T(\mathbf{u}) = T(\mathbf{v})$ implies $\mathbf{u} = \mathbf{v}$. Let $\mathbf{w} \in \mathbf{R}^n$, and suppose $T(\mathbf{u}) = T(\mathbf{v}) = \mathbf{w}$. Then we must have $\mathbf{u} = \mathbf{v}$, and therefore there is at most one vector whose image under T is $\mathbf{w}$. Hence T is one-to-one.

61. Consider $c_1\mathbf{u}_1 + c_2\mathbf{u}_2 = \mathbf{0}$, and apply the linear transformation T to obtain $c_1 T(\mathbf{u}_1) + c_2 T(\mathbf{u}_2) = T(\mathbf{0}) = \mathbf{0}$. Since $T(\mathbf{u}_1)$ and $T(\mathbf{u}_2)$ are linearly independent, $c_1 = c_2 = 0$. This shows that $\mathbf{u}_1$ and $\mathbf{u}_2$ are linearly independent.

63. Since $\mathbf{y}$ is in the range of T, there exists a vector $\mathbf{w}$ such that $T(\mathbf{w}) = \mathbf{y}$. For each $r \in \mathbf{R}$ define the vector $\mathbf{x}_r = \mathbf{w} + r\mathbf{u}$. Then $T(\mathbf{x}_r) = T(\mathbf{w} + r\mathbf{u}) = T(\mathbf{w}) + rT(\mathbf{u}) = \mathbf{y} + r\mathbf{0} = \mathbf{y}$ for every r. Moreover, each $\mathbf{x}_r$ is distinct, for if $\mathbf{x}_r = \mathbf{x}_s$, then $\mathbf{w} + r\mathbf{u} = \mathbf{w} + s\mathbf{u} \Rightarrow (r - s)\mathbf{u} = \mathbf{0} \Rightarrow r = s$, since $\mathbf{u} \neq \mathbf{0}$. Therefore there are infinitely many vectors $\mathbf{x}$ such that $T(\mathbf{x}) = \mathbf{0}$.

65. Let $\mathbf{u} = (1,0)$ and $\mathbf{v} = (0,1)$, and $A = [\ \mathbf{a}_1\ \ \mathbf{a}_2\ ]$. The unit square consists of all vectors $\mathbf{x} = s\mathbf{u} + t\mathbf{v}$ where $0 \leq s \leq 1$ and $0 \leq t \leq 1$. The image consists of all vectors $T(\mathbf{x}) = T(s\mathbf{u}+t\mathbf{v}) = s(A\mathbf{u}) + t(A\mathbf{v}) = s\mathbf{a}_1 + t\mathbf{a}_2$. If the columns of A are linearly independent, then $\{\mathbf{a}_1, \mathbf{a}_2\}$ is linearly independent, and neither vector is a multiple of the other. Thus $\{s\mathbf{a}_1 + t\mathbf{a}_2 : 0 \leq s \leq 1, 0 \leq t \leq 1\}$ is a parallelogram. If the columns of A are linearly dependent, then $\{\mathbf{a}_1, \mathbf{a}_2\}$ is linearly dependent, so one vector is a multiple of the other, and $\{s\mathbf{a}_1 + t\mathbf{a}_2 : 0 \leq s \leq 1, 0 \leq t \leq 1\}$ is a segment. In the linearly dependent case, if $\mathbf{a}_1 = \mathbf{a}_2 = \mathbf{0}$, then $\{s\mathbf{a}_1 + t\mathbf{a}_2 : 0 \leq s \leq 1, 0 \leq t \leq 1\} = \{\mathbf{0}\}$, a point.

67. (a) Let $p_1(x) = a_1 x^2 + b_1 x + c_1$ and $p_2(x) = a_2 x^2 + b_2 x + c_2$ be two polynomials of degree 2 or less, and r a constant. Then

$$(p_1 + p_2)(x) = (a_1 x^2 + b_1 x + c_1) + (a_2 x^2 + b_2 x + c_2)$$
$$= (a_1 + a_2)x^2 + (b_1 + b_2)x + (c_1 + c_2)$$
$$\leftrightarrow \begin{bmatrix} a_1 + a_2 \\ b_1 + b_2 \\ c_1 + c_2 \end{bmatrix} = \begin{bmatrix} a_1 \\ b_1 \\ c_1 \end{bmatrix} + \begin{bmatrix} a_2 \\ b_2 \\ c_2 \end{bmatrix}$$
$$\leftrightarrow p_1(x) + p_2(x)$$

and

$$(rp_1)(x) = r\left(a_1 x^2 + b_1 x + c_1\right)$$
$$= (ra_1)\, x^2 + (rb_1)\, x + (rc_1)$$
$$\leftrightarrow \begin{bmatrix} ra_1 \\ rb_1 \\ rc_1 \end{bmatrix} = r \begin{bmatrix} a_1 \\ b_1 \\ c_1 \end{bmatrix}$$
$$\leftrightarrow r\left(p_1(x)\right)$$

Therefore addition of polynomials corresponds to addition of vectors, and scalar multiplication of polynomials corresponds to scalar multiplication of vectors.

(b) With p_1 and p_2 as above, we have

$$T(p_1(x) + p_2(x)) = \left(\left(a_1 x^2 + b_1 x + c_1\right) + \left(a_2 x^2 + b_2 x + c_2\right)\right)'$$
$$= \left((a_1 + a_2)x^2 + (b_1 + b_2)\, x + (c_1 + c_2)\right)'$$
$$= (2(a_1 + a_2))x + (b_1 + b_2)$$
$$\leftrightarrow \begin{bmatrix} 0 \\ 2(a_1 + a_2) \\ b_1 + b_2 \end{bmatrix} = \begin{bmatrix} 0 \\ 2a_1 \\ b_1 \end{bmatrix} + \begin{bmatrix} 0 \\ 2a_2 \\ b_2 \end{bmatrix}$$
$$\leftrightarrow (2a_1 x + b_1) + (2a_2 x + b_2)$$
$$= \left(a_1 x^2 + b_1 x + c_1\right)' + \left(a_2 x^2 + b_2 x + c_2\right)'$$
$$= T(p_1(x)) + T(p_2(x))$$

and

$$T(rp_1(x)) = (rp_1(x))'$$
$$= \left(ra_1 x^2 + rb_1 x + rc_1\right)'$$
$$= (2ra_1 x + rb_1)$$
$$\leftrightarrow \begin{bmatrix} 0 \\ 2ra_1 \\ rb_1 \end{bmatrix} = r \begin{bmatrix} 0 \\ 2a_1 \\ b_1 \end{bmatrix}$$
$$\leftrightarrow r\,(2a_1 x + b_1)$$
$$= r\left(a_1 x^2 + b_1 x + c_1\right)'$$
$$= rT(p_1(x))$$

Thus T is a linear transformation.

(c) Since $p(x) = ax^2 + bx + c$ has derivative $p'(x) = (2a)x + b$, we can represent T by

$$T(\mathbf{x}) = A\mathbf{x} = \begin{bmatrix} 0 & 0 & 0 \\ 2 & 0 & 0 \\ 0 & 1 & 0 \end{bmatrix} \mathbf{x}$$

(d) T is not onto, as there is no polynomial $p(x)$ of degree 2 or less with $p'(x) = x^2$. T is not one-to-one, as $T(x^2) = T(x^2 + 1)$. One can also use the Big Theorem - Version 2, and the observation that the columns of the matrix A are linearly dependent to conclude that T is neither onto nor one-to-one.

69. (a) $T(x^2 + \sin x) = \left(x^2 + \sin x\right)' = 2x + \cos x$

(b)   i. $T\left(f(x) + g(x)\right) = (f(x) + g(x))' = f'(x) + g'(x) = T\left(f(x)\right) + T\left(g(x)\right)$

ii. $T\left(rf(x)\right) = \left(rf(x)\right)' = rf'(x) = rT\left(f(x)\right)$

71. $T\left(\begin{bmatrix} 5 \\ 3 \\ 6 \end{bmatrix}\right) = \begin{bmatrix} 15 & 46 & 65 \\ 14 & 43 & 61 \\ 20 & 60 & 81 \end{bmatrix}\begin{bmatrix} 5 \\ 3 \\ 6 \end{bmatrix} = \begin{bmatrix} 603 \\ 565 \\ 766 \end{bmatrix}$

73. $T\left(\begin{bmatrix} 14 \\ 10 \\ 9 \end{bmatrix}\right) = \begin{bmatrix} 15 & 46 & 65 \\ 14 & 43 & 61 \\ 20 & 60 & 81 \end{bmatrix}\begin{bmatrix} 14 \\ 10 \\ 9 \end{bmatrix} = \begin{bmatrix} 1255 \\ 1175 \\ 1609 \end{bmatrix}$

75. Using a computer algebra system, the matrix has row-reduced echelon form

$$\begin{bmatrix} 4 & 2 & -5 & 2 & 6 \\ 7 & -2 & 0 & -4 & 1 \\ 0 & 3 & -5 & 7 & -1 \end{bmatrix} \sim \begin{bmatrix} 1 & 0 & 0 & -6 & 13 \\ 0 & 1 & 0 & -19 & 45 \\ 0 & 0 & 1 & -\frac{64}{5} & \frac{136}{5} \end{bmatrix}$$

Hence T is onto, since $A\mathbf{x} = \mathbf{b}$ has solutions for all $\mathbf{b}$. T is not one-to-one, since $A\mathbf{x} = \mathbf{0}$ has nontrivial solutions.

77. Using a computer algebra system, the matrix has row-reduced echelon form

$$\begin{bmatrix} 2 & -1 & 4 & 0 \\ 3 & -3 & 1 & 1 \\ 1 & -1 & 8 & 3 \\ 0 & -2 & 1 & 4 \end{bmatrix} \sim \begin{bmatrix} 1 & 0 & 0 & -\frac{37}{23} \\ 0 & 1 & 0 & -\frac{42}{23} \\ 0 & 0 & 1 & \frac{8}{23} \\ 0 & 0 & 0 & 0 \end{bmatrix}$$

Hence T is not onto, since $A\mathbf{x} = \mathbf{b}$ does not have solutions for all $\mathbf{b}$. T is not one-to-one, since $A\mathbf{x} = \mathbf{0}$ has nontrivial solutions.

79. Using a computer algebra system, the matrix has row-reduced echelon form

$$\begin{bmatrix} 2 & -3 & 5 & 1 \\ 6 & 0 & 3 & -2 \\ -4 & 2 & 1 & 1 \\ 8 & 2 & 3 & -4 \\ -1 & 2 & 5 & -3 \end{bmatrix} \sim \begin{bmatrix} 1 & 0 & 0 & 0 \\ 0 & 1 & 0 & 0 \\ 0 & 0 & 1 & 0 \\ 0 & 0 & 0 & 1 \\ 0 & 0 & 0 & 0 \end{bmatrix}$$

Hence T is not onto, since $A\mathbf{x} = \mathbf{b}$ does not have solutions for all $\mathbf{b}$. T is one-to-one, since $A\mathbf{x} = \mathbf{0}$ has only the trivial solution.

3.2 Matrix Algebra

1. (a) $A + B = \begin{bmatrix} -3 & 1 \\ 2 & -1 \end{bmatrix} + \begin{bmatrix} 0 & 4 \\ -2 & 5 \end{bmatrix} = \begin{bmatrix} -3 & 5 \\ 0 & 4 \end{bmatrix}$

   (b) $AB + I_2 = \begin{bmatrix} -3 & 1 \\ 2 & -1 \end{bmatrix}\begin{bmatrix} 0 & 4 \\ -2 & 5 \end{bmatrix} + \begin{bmatrix} 1 & 0 \\ 0 & 1 \end{bmatrix} = \begin{bmatrix} -2 & -7 \\ 2 & 3 \end{bmatrix} + \begin{bmatrix} 1 & 0 \\ 0 & 1 \end{bmatrix} = \begin{bmatrix} -1 & -7 \\ 2 & 4 \end{bmatrix}$

   (c) $A + C$ is not possible, since A and C are different sizes.

3. (a) $(AB)^T = \left(\begin{bmatrix} -3 & 1 \\ 2 & -1 \end{bmatrix}\begin{bmatrix} 0 & 4 \\ -2 & 5 \end{bmatrix}\right)^T = \begin{bmatrix} -2 & -7 \\ 2 & 3 \end{bmatrix}^T = \begin{bmatrix} -2 & 2 \\ -7 & 3 \end{bmatrix}$

   (b) CE is not defined, since C has 2 columns, and E has 3 rows.

   (c) $(A - B)D = \left(\begin{bmatrix} -3 & 1 \\ 2 & -1 \end{bmatrix} - \begin{bmatrix} 0 & 4 \\ -2 & 5 \end{bmatrix}\right)\begin{bmatrix} 1 & 0 & -3 \\ -2 & 5 & -1 \end{bmatrix}$

   $= \begin{bmatrix} -3 & -3 \\ 4 & -6 \end{bmatrix}\begin{bmatrix} 1 & 0 & -3 \\ -2 & 5 & -1 \end{bmatrix} = \begin{bmatrix} 3 & -15 & 12 \\ 16 & -30 & -6 \end{bmatrix}$

5. (a) $(C + E)B$ is not possible, since C and E are different sizes.

(b) $B\left(C^T + D\right) = \begin{bmatrix} 0 & 4 \\ -2 & 5 \end{bmatrix} \left(\begin{bmatrix} 5 & 0 \\ -1 & 4 \\ 3 & 3 \end{bmatrix}^T + \begin{bmatrix} 1 & 0 & -3 \\ -2 & 5 & -1 \end{bmatrix} \right)$

$= \begin{bmatrix} 0 & 4 \\ -2 & 5 \end{bmatrix} \left(\begin{bmatrix} 5 & -1 & 3 \\ 0 & 4 & 3 \end{bmatrix} + \begin{bmatrix} 1 & 0 & -3 \\ -2 & 5 & -1 \end{bmatrix} \right)$

$= \begin{bmatrix} 0 & 4 \\ -2 & 5 \end{bmatrix} \begin{bmatrix} 6 & -1 & 0 \\ -2 & 9 & 2 \end{bmatrix} = \begin{bmatrix} -8 & 36 & 8 \\ -22 & 47 & 10 \end{bmatrix}$

(c) $E + CD = \begin{bmatrix} 1 & 4 & -5 \\ -2 & 1 & -3 \\ 0 & 2 & 6 \end{bmatrix} + \begin{bmatrix} 5 & 0 \\ -1 & 4 \\ 3 & 3 \end{bmatrix} \begin{bmatrix} 1 & 0 & -3 \\ -2 & 5 & -1 \end{bmatrix}$

$= \begin{bmatrix} 1 & 4 & -5 \\ -2 & 1 & -3 \\ 0 & 2 & 6 \end{bmatrix} + \begin{bmatrix} 5 & 0 & -15 \\ -9 & 20 & -1 \\ -3 & 15 & -12 \end{bmatrix} = \begin{bmatrix} 6 & 4 & -20 \\ -11 & 21 & -4 \\ -3 & 17 & -6 \end{bmatrix}$

7. $\begin{bmatrix} 2 & a \\ 3 & -2 \end{bmatrix} \begin{bmatrix} b & -3 \\ -1 & 2 \end{bmatrix} = \begin{bmatrix} 2b - a & 2a - 6 \\ 3b + 2 & -13 \end{bmatrix} = \begin{bmatrix} 3 & -8 \\ 5 & c \end{bmatrix} \Rightarrow$
$c = -13,\ 3b + 2 = 5 \Rightarrow b = 1,\ 2a - 6 = -8 \Rightarrow a = -1.$

9. $\begin{bmatrix} a & 3 & -2 \\ 3 & -2 & 4 \end{bmatrix} \begin{bmatrix} 2 & -1 \\ 0 & b \\ c & 1 \end{bmatrix} = \begin{bmatrix} 2a - 2c & 3b - a - 2 \\ 4c + 6 & 1 - 2b \end{bmatrix} = \begin{bmatrix} 4 & d \\ -6 & -5 \end{bmatrix} \Rightarrow$
$4c + 6 = -6 \Rightarrow c = -3,\ 1 - 2b = -5 \Rightarrow b = 3,\ 2a - 2c = 4 \Rightarrow 2a - 2(-3) = 4 \Rightarrow a = -1,$
$3b - a - 2 = d \Rightarrow 3(3) - (-1) - 2 = d \Rightarrow d = 8$

11. $A^2 = \begin{bmatrix} 5 & -10 \\ a & -4 \end{bmatrix} \begin{bmatrix} 5 & -10 \\ a & -4 \end{bmatrix} = \begin{bmatrix} 25 - 10a & -10 \\ a & 16 - 10a \end{bmatrix}$. Setting this equal to A, we obtain $25 - 10a = 5 \Rightarrow a = 2$. We check that all entries of A^2 and A are equal when $a = 2$.

13. We first determine that $T_1(\mathbf{x}) = A_1\mathbf{x} = \begin{bmatrix} 3 & 5 \\ -2 & 7 \end{bmatrix}$ and $T_2(\mathbf{x}) = A_2\mathbf{x} = \begin{bmatrix} -2 & 9 \\ 0 & 5 \end{bmatrix}$.

(a) $T_1(T_2(\mathbf{x})) = T_1(A_2\mathbf{x}) = A_1(A_2\mathbf{x}) = (A_1A_2)\mathbf{x}$.
So $A = A_1A_2 = \begin{bmatrix} 3 & 5 \\ -2 & 7 \end{bmatrix} \begin{bmatrix} -2 & 9 \\ 0 & 5 \end{bmatrix} = \begin{bmatrix} -6 & 52 \\ 4 & 17 \end{bmatrix}$

(b) $T_2(T_1(\mathbf{x})) = T_2(A_1\mathbf{x}) = A_2(A_1\mathbf{x}) = (A_2A_1)\mathbf{x}$.
So $A = A_2A_1 = \begin{bmatrix} -2 & 9 \\ 0 & 5 \end{bmatrix} \begin{bmatrix} 3 & 5 \\ -2 & 7 \end{bmatrix} = \begin{bmatrix} -24 & 53 \\ -10 & 35 \end{bmatrix}$

(c) $T_1(T_1(\mathbf{x})) = T_1(A_1\mathbf{x}) = A_1(A_1\mathbf{x}) = (A_1A_1)\mathbf{x}$.
So $A = A_1A_1 = \begin{bmatrix} 3 & 5 \\ -2 & 7 \end{bmatrix} \begin{bmatrix} 3 & 5 \\ -2 & 7 \end{bmatrix} = \begin{bmatrix} -1 & 50 \\ -20 & 39 \end{bmatrix}$

(d) $T_2(T_2(\mathbf{x})) = T_2(A_2\mathbf{x}) = A_2(A_2\mathbf{x}) = (A_2A_2)\mathbf{x}$.
So $A = A_2A_2 = \begin{bmatrix} -2 & 9 \\ 0 & 5 \end{bmatrix} \begin{bmatrix} -2 & 9 \\ 0 & 5 \end{bmatrix} = \begin{bmatrix} 4 & 27 \\ 0 & 25 \end{bmatrix}$

15. $(A + I)(A - I) = A(A - I) + I(A - I) = A(A) - A(I) + (A - I) = A^2 - A + A - I = A^2 - I$

17. $\left(A + B^2\right)(BA - A) = A(BA - A) + B^2(BA - A) = A(BA) - A(A) + B^2(BA) - B^2A$
$= ABA - A^2 + B^3A - B^2A$

19. $(A + B)^2 = (A + B)(A + B) = A(A + B) + B(A + B) = A^2 + AB + BA + B^2$. This only is equal to $A^2 + 2AB + B^2$ when $AB + BA = 2AB \Leftrightarrow AB = BA$, which in general is not true.

21. $(A - B)(A + B) = A(A + B) - B(A + B) = A^2 + AB - BA - B^2 = (A^2 - B^2) + AB - BA$, so $A^2 - B^2 = (A - B)(A + B) - AB + BA$. This only is equal to $(A - B)(A + B)$ when $-AB + BA = 0_{n \times n} \Leftrightarrow AB = BA$, which in general is not true.

23. AB is 4×5, 4 rows and 5 columns.

25. $A = \begin{bmatrix} 1 & -2 & -1 & 3 \\ -2 & 0 & 1 & 4 \\ -1 & 2 & -2 & 0 \\ 0 & 1 & 2 & 1 \end{bmatrix} = \left[\begin{array}{cc|cc} 1 & -2 & -1 & 3 \\ -2 & 0 & 1 & 4 \\ \hline -1 & 2 & -2 & 0 \\ 0 & 1 & 2 & 1 \end{array} \right] = \left[\begin{array}{c|c} A_{11} & A_{12} \\ \hline A_{21} & A_{22} \end{array} \right]$

$B = \begin{bmatrix} 2 & 0 & -1 & 1 \\ -3 & 1 & 2 & 1 \\ 0 & -1 & -2 & 3 \\ 2 & 2 & -1 & -2 \end{bmatrix} = \left[\begin{array}{cc|cc} 2 & 0 & -1 & 1 \\ -3 & 1 & 2 & 1 \\ \hline 0 & -1 & -2 & 3 \\ 2 & 2 & -1 & -2 \end{array} \right] = \left[\begin{array}{c|c} B_{11} & B_{12} \\ \hline B_{21} & B_{22} \end{array} \right]$

(a) $A - B = \left[\begin{array}{c|c} A_{11} - B_{11} & A_{12} - B_{12} \\ \hline A_{21} - B_{21} & A_{22} - B_{22} \end{array} \right] = \left[\begin{array}{cc|cc} -1 & -2 & 0 & 2 \\ 1 & -1 & -1 & 3 \\ \hline -1 & 3 & 0 & -3 \\ -2 & -1 & 3 & 3 \end{array} \right]$

(b) $AB = \left[\begin{array}{c|c} A_{11}B_{11} + A_{12}B_{21} & A_{11}B_{12} + A_{12}B_{22} \\ \hline A_{21}B_{11} + A_{22}B_{21} & A_{21}B_{12} + A_{22}B_{22} \end{array} \right] = \left[\begin{array}{cc|cc} 14 & 5 & -6 & -10 \\ 4 & 7 & -4 & -7 \\ \hline -8 & 4 & 9 & -5 \\ -1 & 1 & -3 & 5 \end{array} \right]$

(c) $BA = \left[\begin{array}{c|c} B_{11}A_{11} + B_{12}A_{21} & B_{11}A_{12} + B_{12}A_{22} \\ \hline B_{21}A_{11} + B_{22}A_{21} & B_{21}A_{12} + B_{22}A_{22} \end{array} \right] = \left[\begin{array}{cc|cc} 3 & -5 & 2 & 7 \\ -7 & 11 & 2 & -4 \\ \hline 4 & -1 & 9 & -1 \\ -1 & -8 & -2 & 12 \end{array} \right]$

27. $A = \begin{bmatrix} 1 & -2 & -1 & 3 \\ -2 & 0 & 1 & 4 \\ -1 & 2 & -2 & 0 \\ 0 & 1 & 2 & 1 \end{bmatrix} = \left[\begin{array}{ccc|c} 1 & -2 & -1 & 3 \\ \hline -2 & 0 & 1 & 4 \\ -1 & 2 & -2 & 0 \\ 0 & 1 & 2 & 1 \end{array} \right] = \left[\begin{array}{c|c} A_{11} & A_{12} \\ \hline A_{21} & A_{22} \end{array} \right]$

$B = \begin{bmatrix} 2 & 0 & -1 & 1 \\ -3 & 1 & 2 & 1 \\ 0 & -1 & -2 & 3 \\ 2 & 2 & -1 & -2 \end{bmatrix} = \left[\begin{array}{ccc|c} 2 & 0 & -1 & 1 \\ \hline -3 & 1 & 2 & 1 \\ 0 & -1 & -2 & 3 \\ 2 & 2 & -1 & -2 \end{array} \right] = \left[\begin{array}{c|c} B_{11} & B_{12} \\ \hline B_{21} & B_{22} \end{array} \right]$

(a) $B - A = \left[\begin{array}{c|c} A_{11} - B_{11} & A_{12} - B_{12} \\ \hline A_{21} - B_{21} & A_{22} - B_{22} \end{array} \right] = \left[\begin{array}{ccc|c} 1 & 2 & 0 & -2 \\ \hline -1 & 1 & 1 & -3 \\ 1 & -3 & 0 & 3 \\ 2 & 1 & -3 & -3 \end{array} \right]$

(b) $AB = \left[\begin{array}{c|c} A_{11}B_{11} + A_{12}B_{21} & A_{11}B_{12} + A_{12}B_{22} \\ \hline A_{21}B_{11} + A_{22}B_{21} & A_{21}B_{12} + A_{22}B_{22} \end{array} \right] = \left[\begin{array}{ccc|c} 14 & 5 & -6 & -10 \\ \hline 4 & 7 & -4 & -7 \\ -8 & 4 & 9 & -5 \\ -1 & 1 & -3 & 5 \end{array} \right]$

(c) $BA + A = \left[\begin{array}{c|c} B_{11}A_{11} + B_{12}A_{21} & B_{11}A_{12} + B_{12}A_{22} \\ \hline B_{21}A_{11} + B_{22}A_{21} & B_{21}A_{12} + B_{22}A_{22} \end{array} \right] + \left[\begin{array}{c|c} A_{11} & A_{12} \\ \hline A_{21} & A_{22} \end{array} \right] =$

$\left[\begin{array}{ccc|c} 3 & -5 & 2 & 7 \\ \hline -7 & 11 & 2 & -4 \\ 4 & -1 & 9 & -1 \\ -1 & -8 & -2 & 12 \end{array} \right] + \left[\begin{array}{ccc|c} 1 & -2 & -1 & 3 \\ \hline -2 & 0 & 1 & 4 \\ -1 & 2 & -2 & 0 \\ 0 & 1 & 2 & 1 \end{array} \right] = \left[\begin{array}{ccc|c} 4 & -7 & 1 & 10 \\ \hline -9 & 11 & 3 & 0 \\ 3 & 1 & 7 & -1 \\ -1 & -7 & 0 & 13 \end{array} \right]$

29. (a) $E = \begin{bmatrix} 0 & 1 & 0 \\ 1 & 0 & 0 \\ 0 & 0 & 1 \end{bmatrix}$

   (b) $E = \begin{bmatrix} 0 & 0 & 1 \\ 0 & 1 & 0 \\ 1 & 0 & 0 \end{bmatrix}$

   (c) $E = \begin{bmatrix} 1 & 0 & 0 \\ 0 & -2 & 0 \\ 0 & 0 & 1 \end{bmatrix}$

31. For example, $A = \begin{bmatrix} 0 & 1 & 0 \\ 0 & 0 & 0 \\ 0 & 0 & 0 \end{bmatrix}$, $B = \begin{bmatrix} 1 & 0 & 0 \\ 0 & 0 & 0 \\ 0 & 0 & 0 \end{bmatrix}$.

Then $AB = \begin{bmatrix} 0 & 0 & 0 \\ 0 & 0 & 0 \\ 0 & 0 & 0 \end{bmatrix}$, and $BA = \begin{bmatrix} 0 & 1 & 0 \\ 0 & 0 & 0 \\ 0 & 0 & 0 \end{bmatrix}$.

33. For example, $A = \begin{bmatrix} 0 & 1 \\ 0 & 0 \end{bmatrix}$, $B = \begin{bmatrix} 1 & 0 \\ 0 & 0 \end{bmatrix}$. Then $AB = \begin{bmatrix} 0 & 0 \\ 0 & 0 \end{bmatrix}$.

35. For example, $A = \begin{bmatrix} 1 & 1 \\ 1 & 1 \end{bmatrix}$, $B = \begin{bmatrix} 1 & 1 \\ -1 & -1 \end{bmatrix}$. Then $AB = \begin{bmatrix} 0 & 0 \\ 0 & 0 \end{bmatrix}$.

37. For example, $A = \begin{bmatrix} 1 & 2 \\ 2 & 1 \end{bmatrix}$, $B = \begin{bmatrix} 2 & 1 \\ 1 & 2 \end{bmatrix}$, $C = \begin{bmatrix} 1 & 1 \\ 1 & 1 \end{bmatrix}$. Then $AC = BC = \begin{bmatrix} 3 & 3 \\ 3 & 3 \end{bmatrix}$.

39. False. Consider $A = [1]$, $B = [-1]$.

41. True. If $i < j$, then $(A^T)_{ij} = A_{ji} = 0$, since A is upper triangular.

43. False. $C = [0]$, $I_1 = [1]$, but $C + I_1 = [1] \neq [0] = C$.

45. True. Using Theorem 3.15(c), we have $(ABC)^T = ((AB)C)^T = C^T(AB)^T = C^T(B^TA^T) = C^TB^TA^T$.

47. True. Using Theorem 3.15(a,c) and Theorem 3.11(a), we have $(AB + C)^T = (C + AB)^T = C^T + (AB)^T = C^T + B^TA^T$.

49. Let $A = \begin{bmatrix} a_{11} & \cdots & a_{1m} \\ \vdots & & \vdots \\ a_{n1} & \cdots & a_{nm} \end{bmatrix}$, $B = \begin{bmatrix} b_{11} & \cdots & b_{1m} \\ \vdots & & \vdots \\ b_{n1} & \cdots & b_{nm} \end{bmatrix}$, $C = \begin{bmatrix} c_{11} & \cdots & c_{1m} \\ \vdots & & \vdots \\ c_{n1} & \cdots & c_{nm} \end{bmatrix}$ and s and t scalars.

(a) $\begin{aligned} A + B &= \begin{bmatrix} a_{11} & \cdots & a_{1m} \\ \vdots & & \vdots \\ a_{n1} & \cdots & a_{nm} \end{bmatrix} + \begin{bmatrix} b_{11} & \cdots & b_{1m} \\ \vdots & & \vdots \\ b_{n1} & \cdots & b_{nm} \end{bmatrix} \\ &= \begin{bmatrix} a_{11} + b_{11} & \cdots & a_{1m} + b_{1m} \\ \vdots & & \vdots \\ a_{n1} + b_{n1} & \cdots & a_{nm} + b_{nm} \end{bmatrix} = \begin{bmatrix} b_{11} + a_{11} & \cdots & b_{1m} + a_{1m} \\ \vdots & & \vdots \\ b_{n1} + a_{n1} & \cdots & b_{nm} + a_{nm} \end{bmatrix} \\ &= \begin{bmatrix} b_{11} & \cdots & b_{1m} \\ \vdots & & \vdots \\ b_{n1} & \cdots & b_{nm} \end{bmatrix} + \begin{bmatrix} a_{11} & \cdots & a_{1m} \\ \vdots & & \vdots \\ a_{n1} & \cdots & a_{nm} \end{bmatrix} = B + A \end{aligned}$

(b) $\quad s\left(A+B\right) = s\left(\begin{bmatrix} a_{11} & a_{12} & \cdots & a_{1m} \\ a_{21} & a_{22} & \cdots & a_{2m} \\ \vdots & \vdots & & \vdots \\ a_{n1} & a_{n2} & \cdots & a_{nm} \end{bmatrix} + \begin{bmatrix} b_{11} & b_{12} & \cdots & b_{1m} \\ b_{21} & b_{22} & \cdots & b_{2m} \\ \vdots & \vdots & & \vdots \\ b_{n1} & b_{n2} & \cdots & b_{nm} \end{bmatrix}\right)$

$$= \begin{bmatrix} sa_{11}+sb_{11} & sa_{12}+sb_{12} & \cdots & sa_{1m}+sb_{1m} \\ sa_{21}+sb_{21} & sa_{22}+sb_{22} & \cdots & sa_{2m}+sb_{2m} \\ \vdots & & \vdots & & \vdots \\ sa_{n1}+sb_{n1} & sa_{n2}+sb_{n2} & \cdots & sa_{nm}+sb_{nm} \end{bmatrix}$$

$$= \begin{bmatrix} sa_{11} & sa_{12} & \cdots & sa_{1m} \\ sa_{21} & sa_{22} & \cdots & sa_{2m} \\ \vdots & \vdots & & \vdots \\ sa_{n1} & sa_{n2} & \cdots & sa_{nm} \end{bmatrix} + \begin{bmatrix} sb_{11} & sb_{12} & \cdots & sb_{1m} \\ sb_{21} & sb_{22} & \cdots & sb_{2m} \\ \vdots & \vdots & & \vdots \\ sb_{n1} & sb_{n2} & \cdots & sb_{nm} \end{bmatrix}$$

$$= s\begin{bmatrix} a_{11} & a_{12} & \cdots & a_{1m} \\ a_{21} & a_{22} & \cdots & a_{2m} \\ \vdots & \vdots & & \vdots \\ a_{n1} & a_{n2} & \cdots & a_{nm} \end{bmatrix} + s\begin{bmatrix} b_{11} & b_{12} & \cdots & b_{1m} \\ b_{21} & b_{22} & \cdots & b_{2m} \\ \vdots & \vdots & & \vdots \\ b_{n1} & b_{n2} & \cdots & b_{nm} \end{bmatrix} = sA+sB$$

(c) $\quad (s+t)\,A = (s+t)\begin{bmatrix} a_{11} & \cdots & a_{1m} \\ \vdots & & \vdots \\ a_{n1} & \cdots & a_{nm} \end{bmatrix} = \begin{bmatrix} (s+t)\,a_{11} & \cdots & (s+t)\,a_{1m} \\ \vdots & & \vdots \\ (s+t)\,a_{n1} & \cdots & (s+t)\,a_{nm} \end{bmatrix}$

$$= \begin{bmatrix} sa_{11}+ta_{11} & \cdots & sa_{1m}+ta_{1m} \\ \vdots & & \vdots \\ sa_{n1}+ta_{n1} & \cdots & sa_{nm}+ta_{nm} \end{bmatrix} = \begin{bmatrix} sa_{11} & \cdots & sa_{1m} \\ \vdots & & \vdots \\ sa_{n1} & \cdots & sa_{nm} \end{bmatrix} + \begin{bmatrix} ta_{11} & \cdots & ta_{1m} \\ \vdots & & \vdots \\ ta_{n1} & \cdots & ta_{nm} \end{bmatrix}$$

$$= s\begin{bmatrix} a_{11} & \cdots & a_{1m} \\ \vdots & & \vdots \\ a_{n1} & \cdots & a_{nm} \end{bmatrix} + t\begin{bmatrix} a_{11} & \cdots & a_{1m} \\ \vdots & & \vdots \\ a_{n1} & \cdots & a_{nm} \end{bmatrix} = sA+tA$$

(d) $\quad (A+B)+C = \left(\begin{bmatrix} a_{11} & \cdots & a_{1m} \\ \vdots & & \vdots \\ a_{n1} & \cdots & a_{nm} \end{bmatrix} + \begin{bmatrix} b_{11} & \cdots & b_{1m} \\ \vdots & & \vdots \\ b_{n1} & \cdots & b_{nm} \end{bmatrix}\right) + \begin{bmatrix} c_{11} & \cdots & c_{1m} \\ \vdots & & \vdots \\ c_{n1} & \cdots & c_{nm} \end{bmatrix}$

$$= \begin{bmatrix} a_{11}+b_{11} & \cdots & a_{1m}+b_{1m} \\ \vdots & & \vdots \\ a_{n1}+b_{n1} & \cdots & a_{nm}+b_{nm} \end{bmatrix} + \begin{bmatrix} c_{11} & \cdots & c_{1m} \\ \vdots & & \vdots \\ c_{n1} & \cdots & c_{nm} \end{bmatrix}$$

$$= \begin{bmatrix} (a_{11}+b_{11})+c_{11} & \cdots & (a_{1m}+b_{1m})+c_{1m} \\ \vdots & & \vdots \\ (a_{n1}+b_{n1})+c_{n1} & \cdots & (a_{nm}+b_{nm})+c_{nm} \end{bmatrix}$$

$$= \begin{bmatrix} a_{11}+(b_{11}+c_{11}) & \cdots & a_{1m}+(b_{1m}+c_{1m}) \\ \vdots & & \vdots \\ a_{n1}+(b_{n1}+c_{n1}) & \cdots & a_{nm}+(b_{nm}+c_{nm}) \end{bmatrix}$$

$$= \begin{bmatrix} a_{11} & \cdots & a_{1m} \\ \vdots & & \vdots \\ a_{n1} & \cdots & a_{nm} \end{bmatrix} + \begin{bmatrix} b_{11}+c_{11} & \cdots & b_{1m}+c_{1m} \\ \vdots & & \vdots \\ b_{n1}+c_{n1} & \cdots & b_{nm}+c_{nm} \end{bmatrix}$$

$$= \begin{bmatrix} a_{11} & \cdots & a_{1m} \\ \vdots & & \vdots \\ a_{n1} & \cdots & a_{nm} \end{bmatrix} + \left(\begin{bmatrix} b_{11} & \cdots & b_{1m} \\ \vdots & & \vdots \\ b_{n1} & \cdots & b_{nm} \end{bmatrix} + \begin{bmatrix} c_{11} & \cdots & c_{1m} \\ \vdots & & \vdots \\ c_{n1} & \cdots & c_{nm} \end{bmatrix}\right)$$

$$= A+(B+C)$$

(f) $\quad A + 0_{nm} \;=\; \begin{bmatrix} a_{11} & \cdots & a_{1m} \\ \vdots & & \vdots \\ a_{n1} & \cdots & a_{nm} \end{bmatrix} + \begin{bmatrix} 0 & \cdots & 0 \\ \vdots & & \vdots \\ 0 & \cdots & 0 \end{bmatrix}$

$\qquad\qquad\quad = \begin{bmatrix} a_{11}+0 & \cdots & a_{1m}+0 \\ \vdots & & \vdots \\ a_{n1}+0 & \cdots & a_{nm}+0 \end{bmatrix} = \begin{bmatrix} a_{11} & \cdots & a_{1m} \\ \vdots & & \vdots \\ a_{n1} & \cdots & a_{nm} \end{bmatrix} = A$

51. (a) Let $A = [a_{ij}]$, $B = [b_{ij}]$, $C = [c_{ij}] = A+B$, $D = [d_{ij}] = A^T$, $E = [e_{ij}] = B^T$, $F = [f_{ij}] = D+E = A^T + B^T$, and $G = [g_{ij}] = C^T = (A+B)^T$. Then

$$\begin{aligned} g_{ij} &= c_{ji} \\ &= a_{ji} + b_{ji} \\ &= d_{ij} + e_{ij} \\ &= f_{ij}, \end{aligned}$$

hence $G = F$, and so $(A+B)^T = A^T + B^T$.

(b) Let $A = [a_{ij}]$, $B = [b_{ij}] = sA$, $C = [c_{ij}] = A^T$, $D = [d_{ij}] = sC = sA^T$, and $E = [e_{ij}] = B^T = (sA)^T$. Then

$$\begin{aligned} e_{ij} &= b_{ji} \\ &= sa_{ji} \\ &= sc_{ij} \\ &= d_{ij}, \end{aligned}$$

hence $E = D$, and so $(sA)^T = sA^T$.

53. $(AB)^T = B^T A^T$ (by Theorem 3.15c) $= BA$ (since A and B are symmetric) $= AB$. Hence AB is symmetric.

55. (a) If A is $n \times m$, then A^T is $m \times n$, and so $A^T A$ is $m \times m$.

(b) $(A^T A)^T = (A^T)(A^T)^T = A^T A$, hence $A^T A$ is symmetric.

57. Let $A = [a_{ij}]$, $B = [b_{ij}]$, and $C = [c_{ij}]$. Then if $i > j$,

$$\begin{aligned} c_{ij} &= a_{i1}b_{1j} + a_{i2}b_{2j} + \cdots + a_{in}b_{nj} \\ &= a_{ii}b_{ij} + \cdots + a_{in}b_{nj} \text{ (since } a_{ik} = 0 \text{ if } i > k) \\ &= 0 \text{ (since } b_{kj} = 0 \text{ when } k \geq i > j). \end{aligned}$$

Therefore $C = AB$ is upper triangular.

59. Proof by induction. Assume A^n is upper(lower) triangular, and note that when $n = 1$, A is upper(lower) triangular. Since $A^{n+1} = A^n A$, by exercise 57(58), since both A^n and A are upper(lower) triangular, A^{n+1} is upper(lower) triangular.

61. (a) For example, $A = \begin{bmatrix} 0 & 1 & 2 \\ -1 & 0 & 3 \\ -2 & -3 & 0 \end{bmatrix}$.

(b) Since $A^T = -A \Rightarrow A + A^T = 0_n$, and since the diagonal entry a_{ii} of A and A^T are the same, we have $a_{ii} + a_{ii} = 0$, and hence $a_{ii} = 0$.

63. Let $A = [a_{ij}]$, $B = [b_{ij}] = A^T$, and $C = [c_{ij}] = B^T = (A^T)^T$. Then $c_{ij} = b_{ji} = a_{ij}$, hence $C = A$. Therefore $A = (A^T)^T$.

65. Let $A = \begin{bmatrix} .80 & .05 & .05 \\ .10 & .90 & .10 \\ .10 & .05 & .85 \end{bmatrix}$ and $\mathbf{x} = \begin{bmatrix} 8000 \\ 1500 \\ 500 \end{bmatrix}$. Using a computer algebra system, the distribution

after one year is $A\mathbf{x} = \begin{bmatrix} .80 & .05 & .05 \\ .10 & .90 & .10 \\ .10 & .05 & .85 \end{bmatrix} \begin{bmatrix} 8000 \\ 1500 \\ 500 \end{bmatrix} = \begin{bmatrix} 6500 \\ 2200 \\ 1300 \end{bmatrix}$;

after two years $A(A\mathbf{x}) = \begin{bmatrix} .80 & .05 & .05 \\ .10 & .90 & .10 \\ .10 & .05 & .85 \end{bmatrix} \begin{bmatrix} 6500 \\ 2200 \\ 1300 \end{bmatrix} = \begin{bmatrix} 5375 \\ 2760 \\ 1865 \end{bmatrix}$;

after three years $A(A^2\mathbf{x}) = \begin{bmatrix} .80 & .05 & .05 \\ .10 & .90 & .10 \\ .10 & .05 & .85 \end{bmatrix} \begin{bmatrix} 5375 \\ 2760 \\ 1865 \end{bmatrix} \approx \begin{bmatrix} 4531 \\ 3208 \\ 2261 \end{bmatrix}$;

and after four years $A(A^3\mathbf{x}) \approx \begin{bmatrix} .80 & .05 & .05 \\ .10 & .90 & .10 \\ .10 & .05 & .85 \end{bmatrix} \begin{bmatrix} 4531 \\ 3208 \\ 2261 \end{bmatrix} \approx \begin{bmatrix} 3898 \\ 3566 \\ 2535 \end{bmatrix}$.

67. The transition matrix is $A = \begin{bmatrix} .85 & .40 \\ .15 & .60 \end{bmatrix}$ and the initial distribution is $\mathbf{x} = \begin{bmatrix} 760 \\ 240 \end{bmatrix}$.

Using a computer algebra system, the distribution tomorrow is $A\mathbf{x} = \begin{bmatrix} .85 & .40 \\ .15 & .60 \end{bmatrix} \begin{bmatrix} 760 \\ 240 \end{bmatrix} = \begin{bmatrix} 742 \\ 258 \end{bmatrix}$,

the next day $A(A\mathbf{x}) = \begin{bmatrix} .85 & .40 \\ .15 & .60 \end{bmatrix} \begin{bmatrix} 742 \\ 258 \end{bmatrix} \approx \begin{bmatrix} 734 \\ 266 \end{bmatrix}$,

and the day after that $A(A^2\mathbf{x}) \approx \begin{bmatrix} .85 & .40 \\ .15 & .60 \end{bmatrix} \begin{bmatrix} 734 \\ 266 \end{bmatrix} \approx \begin{bmatrix} 730 \\ 270 \end{bmatrix}$.

69. (a) $A + B = \begin{bmatrix} 2 & -1 & 0 & 4 \\ 0 & 3 & 3 & -1 \\ 6 & 8 & 1 & 1 \\ 5 & -3 & 1 & -2 \end{bmatrix} + \begin{bmatrix} -6 & 2 & -3 & 1 \\ -5 & 2 & 0 & 3 \\ 0 & 3 & -1 & 4 \\ 8 & 5 & -2 & 0 \end{bmatrix} = \begin{bmatrix} -4 & 1 & -3 & 5 \\ -5 & 5 & 3 & 2 \\ 6 & 11 & 0 & 5 \\ 13 & 2 & -1 & -2 \end{bmatrix}$

(b) $BA - I_4 = \begin{bmatrix} -6 & 2 & -3 & 1 \\ -5 & 2 & 0 & 3 \\ 0 & 3 & -1 & 4 \\ 8 & 5 & -2 & 0 \end{bmatrix} \begin{bmatrix} 2 & -1 & 0 & 4 \\ 0 & 3 & 3 & -1 \\ 6 & 8 & 1 & 1 \\ 5 & -3 & 1 & -2 \end{bmatrix} - \begin{bmatrix} 1 & 0 & 0 & 0 \\ 0 & 1 & 0 & 0 \\ 0 & 0 & 1 & 0 \\ 0 & 0 & 0 & 1 \end{bmatrix}$

$= \begin{bmatrix} -26 & -15 & 4 & -31 \\ 5 & 1 & 9 & -28 \\ 14 & -11 & 11 & -12 \\ 4 & -9 & 13 & 24 \end{bmatrix}$

(c) $D + C$ is not possible, since they are not the same size.

71. (a) $AB = \begin{bmatrix} 2 & -1 & 0 & 4 \\ 0 & 3 & 3 & -1 \\ 6 & 8 & 1 & 1 \\ 5 & -3 & 1 & -2 \end{bmatrix} \begin{bmatrix} -6 & 2 & -3 & 1 \\ -5 & 2 & 0 & 3 \\ 0 & 3 & -1 & 4 \\ 8 & 5 & -2 & 0 \end{bmatrix} = \begin{bmatrix} 25 & 22 & -14 & -1 \\ -23 & 10 & -1 & 21 \\ -68 & 36 & -21 & 34 \\ -31 & -3 & -12 & 0 \end{bmatrix}$

(b) $CD = \begin{bmatrix} 2 & 0 & 1 & 1 & 1 \\ 5 & 1 & 2 & 4 & 3 \\ 6 & 2 & 4 & 0 & 8 \\ 7 & 3 & 3 & 3 & 2 \end{bmatrix} \begin{bmatrix} 5 & 2 & 0 & 0 \\ 2 & 5 & 1 & 3 \\ 0 & 7 & 1 & 4 \\ 3 & 6 & 9 & 2 \\ 1 & 4 & 7 & 1 \end{bmatrix} = \begin{bmatrix} 14 & 21 & 17 & 7 \\ 42 & 65 & 60 & 22 \\ 42 & 82 & 62 & 30 \\ 52 & 76 & 47 & 29 \end{bmatrix}$

(c) $(A - B)C^T$ is not possible, as $A - B$ has 4 columns and C^T has 5 rows.

73. (a) $(C + A)B$ is not possible since C and A are different sizes.

(b) $C\left(C^T + D\right) = \begin{bmatrix} 2 & 0 & 1 & 1 & 1 \\ 5 & 1 & 2 & 4 & 3 \\ 6 & 2 & 4 & 0 & 8 \\ 7 & 3 & 3 & 3 & 2 \end{bmatrix} \left(\begin{bmatrix} 2 & 0 & 1 & 1 & 1 \\ 5 & 1 & 2 & 4 & 3 \\ 6 & 2 & 4 & 0 & 8 \\ 7 & 3 & 3 & 3 & 2 \end{bmatrix}^T + \begin{bmatrix} 5 & 2 & 0 & 0 \\ 2 & 5 & 1 & 3 \\ 0 & 7 & 1 & 4 \\ 3 & 6 & 9 & 2 \\ 1 & 4 & 7 & 1 \end{bmatrix} \right)$

$= \begin{bmatrix} 21 & 40 & 41 & 29 \\ 61 & 120 & 124 & 84 \\ 66 & 146 & 182 & 106 \\ 74 & 138 & 123 & 109 \end{bmatrix}$

(c) $A + CD = \begin{bmatrix} 2 & -1 & 0 & 4 \\ 0 & 3 & 3 & -1 \\ 6 & 8 & 1 & 1 \\ 5 & -3 & 1 & -2 \end{bmatrix} + \begin{bmatrix} 2 & 0 & 1 & 1 & 1 \\ 5 & 1 & 2 & 4 & 3 \\ 6 & 2 & 4 & 0 & 8 \\ 7 & 3 & 3 & 3 & 2 \end{bmatrix} \begin{bmatrix} 5 & 2 & 0 & 0 \\ 2 & 5 & 1 & 3 \\ 0 & 7 & 1 & 4 \\ 3 & 6 & 9 & 2 \\ 1 & 4 & 7 & 1 \end{bmatrix}$

$= \begin{bmatrix} 16 & 20 & 17 & 11 \\ 42 & 68 & 63 & 21 \\ 48 & 90 & 63 & 31 \\ 57 & 73 & 48 & 27 \end{bmatrix}$

3.3 Inverses

1. $\begin{bmatrix} 7 & 3 \\ 2 & 1 \end{bmatrix}^{-1} = \dfrac{1}{7(1) - 3(2)} \begin{bmatrix} 1 & -3 \\ -2 & 7 \end{bmatrix} = \dfrac{1}{1} \begin{bmatrix} 1 & -3 \\ -2 & 7 \end{bmatrix} = \begin{bmatrix} 1 & -3 \\ -2 & 7 \end{bmatrix}$

3. $\begin{bmatrix} 2 & -5 \\ -4 & 10 \end{bmatrix}^{-1}$ does not exist, since $2(10) - (-5)(-4) = 0$.

5. $\begin{bmatrix} 1 & 4 & 1 & 0 \\ 2 & 9 & 0 & 1 \end{bmatrix} \underset{\sim}{^{-2R_1+R_2 \Rightarrow R_2}} \begin{bmatrix} 1 & 4 & 1 & 0 \\ 0 & 1 & -2 & 1 \end{bmatrix}$

$\underset{\sim}{^{-4R_2+R_1 \Rightarrow R_1}} \begin{bmatrix} 1 & 0 & 9 & -4 \\ 0 & 1 & -1 & 1 \end{bmatrix}$,

so $\begin{bmatrix} 1 & 4 \\ 2 & 9 \end{bmatrix}^{-1} = \begin{bmatrix} 9 & -4 \\ -2 & 1 \end{bmatrix}$

7. $\begin{bmatrix} 1 & 0 & 1 & 1 & 0 & 0 \\ 0 & 1 & 0 & 0 & 1 & 0 \\ 1 & 1 & 1 & 0 & 0 & 1 \end{bmatrix} \underset{\sim}{^{-R_1+R_3 \Rightarrow R_3}} \begin{bmatrix} 1 & 0 & 1 & 1 & 0 & 0 \\ 0 & 1 & 0 & 0 & 1 & 0 \\ 0 & 1 & 0 & -1 & 0 & 1 \end{bmatrix}$

$\underset{\sim}{^{-R_2+R_3 \Rightarrow R_3}} \begin{bmatrix} 1 & 0 & 1 & 1 & 0 & 0 \\ 0 & 1 & 0 & 0 & 1 & 0 \\ 0 & 0 & 0 & -1 & -1 & 1 \end{bmatrix}$,

and we conclude that the inverse does not exist, since the left part of the augmented matrix cannot be reduced to the identity matrix.

9. $\begin{bmatrix} 1 & 2 & -1 & 1 & 0 & 0 \\ 0 & 1 & 3 & 0 & 1 & 0 \\ 0 & 0 & 1 & 0 & 0 & 1 \end{bmatrix} \underset{\sim}{\overset{-3R_3+R_2 \Rightarrow R_2}{R_3+R_1 \Rightarrow R_1}} \begin{bmatrix} 1 & 2 & 0 & 1 & 0 & 1 \\ 0 & 1 & 0 & 0 & 1 & -3 \\ 0 & 0 & 1 & 0 & 0 & 1 \end{bmatrix}$

$\underset{\sim}{^{-2R_2+R_1 \Rightarrow R_1}} \begin{bmatrix} 1 & 0 & 0 & 1 & -2 & 7 \\ 0 & 1 & 0 & 0 & 1 & -3 \\ 0 & 0 & 1 & 0 & 0 & 1 \end{bmatrix}$,

so $\begin{bmatrix} 1 & 2 & -1 \\ 0 & 1 & 3 \\ 0 & 0 & 1 \end{bmatrix}^{-1} = \begin{bmatrix} 1 & -2 & 7 \\ 0 & 1 & -3 \\ 0 & 0 & 1 \end{bmatrix}$.

11. $\begin{bmatrix} 1 & -3 & 1 & 1 & 0 & 0 \\ 2 & -5 & 4 & 0 & 1 & 0 \\ -2 & 3 & -8 & 0 & 0 & 1 \end{bmatrix}$ $\underset{\sim}{\overset{-2R_1+R_2\Rightarrow R_2}{2R_1+R_3\Rightarrow R_3}}$ $\begin{bmatrix} 1 & -3 & 1 & 1 & 0 & 0 \\ 0 & 1 & 2 & -2 & 1 & 0 \\ 0 & -3 & -6 & 2 & 0 & 1 \end{bmatrix}$

$\underset{\sim}{3R_2+R_3\Rightarrow R_3}$ $\begin{bmatrix} 1 & -3 & 1 & 1 & 0 & 0 \\ 0 & 1 & 2 & -2 & 1 & 0 \\ 0 & 0 & 0 & -4 & 3 & 1 \end{bmatrix}$,

and we conclude that the inverse does not exist, since the left part of the augmented matrix cannot be reduced to the identity matrix.

13. $\begin{bmatrix} 0 & 0 & 1 & 0 & 1 & 0 & 0 & 0 \\ 1 & 0 & 0 & 0 & 0 & 1 & 0 & 0 \\ 0 & 0 & 0 & 1 & 0 & 0 & 1 & 0 \\ 0 & 1 & 0 & 0 & 0 & 0 & 0 & 1 \end{bmatrix}$ $\underset{\sim}{\overset{R_1\Leftrightarrow R_2}{R_3\Leftrightarrow R_4}}$ $\begin{bmatrix} 1 & 0 & 0 & 0 & 0 & 1 & 0 & 0 \\ 0 & 0 & 1 & 0 & 1 & 0 & 0 & 0 \\ 0 & 1 & 0 & 0 & 0 & 0 & 0 & 1 \\ 0 & 0 & 0 & 1 & 0 & 0 & 1 & 0 \end{bmatrix}$

$\underset{\sim}{R_2\Leftrightarrow R_3}$ $\begin{bmatrix} 1 & 0 & 0 & 0 & 0 & 1 & 0 & 0 \\ 0 & 1 & 0 & 0 & 0 & 0 & 0 & 1 \\ 0 & 0 & 1 & 0 & 1 & 0 & 0 & 0 \\ 0 & 0 & 0 & 1 & 0 & 0 & 1 & 0 \end{bmatrix}$,

so $\begin{bmatrix} 0 & 0 & 1 & 0 \\ 1 & 0 & 0 & 0 \\ 0 & 0 & 0 & 1 \\ 0 & 1 & 0 & 0 \end{bmatrix}^{-1} = \begin{bmatrix} 0 & 1 & 0 & 0 \\ 0 & 0 & 0 & 1 \\ 1 & 0 & 0 & 0 \\ 0 & 0 & 1 & 0 \end{bmatrix}$.

15. $\begin{bmatrix} 1 & 3 & 1 & -4 & 1 & 0 & 0 & 0 \\ 0 & 1 & -2 & 2 & 0 & 1 & 0 & 0 \\ 0 & 0 & 1 & 1 & 0 & 0 & 1 & 0 \\ 0 & 0 & 0 & 1 & 0 & 0 & 0 & 1 \end{bmatrix}$ $\underset{\sim}{\overset{-R_4+R_3\Rightarrow R_3}{\overset{-2R_4+R_2\Rightarrow R_2}{4R_4+R_1\Rightarrow R_1}}}$ $\begin{bmatrix} 1 & 3 & 1 & 0 & 1 & 0 & 0 & 4 \\ 0 & 1 & -2 & 0 & 0 & 1 & 0 & -2 \\ 0 & 0 & 1 & 0 & 0 & 0 & 1 & -1 \\ 0 & 0 & 0 & 1 & 0 & 0 & 0 & 1 \end{bmatrix}$

$\underset{\sim}{\overset{2R_3+R_2\Rightarrow R_2}{-R_3+R_1\Rightarrow R_1}}$ $\begin{bmatrix} 1 & 3 & 0 & 0 & 1 & 0 & -1 & 5 \\ 0 & 1 & 0 & 0 & 0 & 1 & 2 & -4 \\ 0 & 0 & 1 & 0 & 0 & 0 & 1 & -1 \\ 0 & 0 & 0 & 1 & 0 & 0 & 0 & 1 \end{bmatrix}$

$\underset{\sim}{-3R_2+R_1\Rightarrow R_1}$ $\begin{bmatrix} 1 & 0 & 0 & 0 & 1 & -3 & -7 & 17 \\ 0 & 1 & 0 & 0 & 0 & 1 & 2 & -4 \\ 0 & 0 & 1 & 0 & 0 & 0 & 1 & -1 \\ 0 & 0 & 0 & 1 & 0 & 0 & 0 & 1 \end{bmatrix}$,

so $\begin{bmatrix} 1 & 3 & 1 & -4 \\ 0 & 1 & -2 & 2 \\ 0 & 0 & 1 & 1 \\ 0 & 0 & 0 & 1 \end{bmatrix}^{-1} = \begin{bmatrix} 1 & -3 & -7 & 17 \\ 0 & 1 & 2 & -4 \\ 0 & 0 & 1 & -1 \\ 0 & 0 & 0 & 1 \end{bmatrix}$.

17. The linear system is equivalent to $A\mathbf{x} = \mathbf{b}$, with $A = \begin{bmatrix} 4 & 13 \\ 1 & 3 \end{bmatrix}$, and $\mathbf{b} = \begin{bmatrix} -3 \\ 2 \end{bmatrix}$. Thus $\mathbf{x} = A^{-1}\mathbf{b} =$

$\begin{bmatrix} 4 & 13 \\ 1 & 3 \end{bmatrix}^{-1} \begin{bmatrix} -3 \\ 2 \end{bmatrix} = \begin{bmatrix} -3 & 13 \\ 1 & -4 \end{bmatrix} \begin{bmatrix} -3 \\ 2 \end{bmatrix} = \begin{bmatrix} 35 \\ -11 \end{bmatrix}$. Hence $x_1 = 35$ and $x_2 = -11$.

19. The linear system is equivalent to $A\mathbf{x} = \mathbf{b}$, with $A = \begin{bmatrix} 3 & -1 & 9 \\ 1 & -1 & 4 \\ 2 & -2 & 10 \end{bmatrix}$, and $\mathbf{b} = \begin{bmatrix} 4 \\ -1 \\ 3 \end{bmatrix}$. Thus $\mathbf{x} =$

$A^{-1}\mathbf{b} = \begin{bmatrix} 3 & -1 & 9 \\ 1 & -1 & 4 \\ 2 & -2 & 10 \end{bmatrix}^{-1} \begin{bmatrix} 4 \\ -1 \\ 3 \end{bmatrix} = \begin{bmatrix} \frac{1}{2} & 2 & -\frac{5}{4} \\ \frac{1}{2} & -3 & \frac{3}{4} \\ 0 & -1 & \frac{1}{2} \end{bmatrix} \begin{bmatrix} 4 \\ -1 \\ 3 \end{bmatrix} = \begin{bmatrix} -\frac{15}{4} \\ \frac{29}{4} \\ \frac{5}{2} \end{bmatrix}$. Hence $x_1 = -\frac{15}{4}$, $x_2 =$

$\frac{29}{4}$ and $x_3 = \frac{5}{2}$.

21. $T\left(\begin{bmatrix} x_1 \\ x_2 \end{bmatrix}\right) = T(\mathbf{x}) = A\mathbf{x}$, where $A = \begin{bmatrix} 4 & 3 \\ 3 & 2 \end{bmatrix}$. We determine A^{-1}:

$$\begin{bmatrix} 4 & 3 & 1 & 0 \\ 3 & 2 & 0 & 1 \end{bmatrix} \quad \underset{\sim}{(-3/4)R_1+R_2\Rightarrow R_2} \quad \begin{bmatrix} 4 & 3 & 1 & 0 \\ 0 & -\frac{1}{4} & -\frac{3}{4} & 1 \end{bmatrix}$$

$$\underset{\sim}{12R_2+R_1\Rightarrow R_1} \quad \begin{bmatrix} 4 & 0 & -8 & 12 \\ 0 & -\frac{1}{4} & -\frac{3}{4} & 1 \end{bmatrix}$$

$$\underset{\sim}{\substack{(1/4)R_1\Rightarrow R_1 \\ -4R_2\Rightarrow R_2}} \quad \begin{bmatrix} 1 & 0 & -2 & 3 \\ 0 & 1 & 3 & -4 \end{bmatrix},$$

hence $A^{-1} = \begin{bmatrix} 4 & 3 \\ 3 & 2 \end{bmatrix}^{-1} = \begin{bmatrix} -2 & 3 \\ 3 & -4 \end{bmatrix}$. Consequently, $T^{-1}\left(\begin{bmatrix} x_1 \\ x_2 \end{bmatrix}\right) = T^{-1}(\mathbf{x}) = A^{-1}\mathbf{x} = \left(\begin{bmatrix} -2x_1 + 3x_2 \\ 3x_1 - 4x_2 \end{bmatrix}\right).$

23. $T\left(\begin{bmatrix} x_1 \\ x_2 \end{bmatrix}\right) = T(\mathbf{x}) = A\mathbf{x}$, where $A = \begin{bmatrix} 1 & -5 \\ -2 & 10 \end{bmatrix}$. We seek to determine A^{-1}:

$$\begin{bmatrix} 1 & -5 & 1 & 0 \\ -2 & 10 & 0 & 1 \end{bmatrix} \quad \underset{\sim}{2R_1+R_2\Rightarrow R_2} \quad \begin{bmatrix} 1 & -5 & 1 & 0 \\ 0 & 0 & 2 & 1 \end{bmatrix}.$$

Thus A^{-1} does not exist, and so T^{-1} does not exist.

25. $T\left(\begin{bmatrix} x_1 \\ x_2 \\ x_3 \end{bmatrix}\right) = T(\mathbf{x}) = A\mathbf{x}$, where $A = \begin{bmatrix} 1 & 2 & -1 \\ 1 & 1 & -1 \end{bmatrix}$. Since A is not square, A^{-1} does not exist, and hence T^{-1} does not exist.

27. $T_1\left(\begin{bmatrix} x_1 \\ x_2 \end{bmatrix}\right) = T_1(\mathbf{x}) = A_1\mathbf{x}$, where $A_1 = \begin{bmatrix} 2 & 1 \\ 1 & 1 \end{bmatrix}$, and $T_2\left(\begin{bmatrix} x_1 \\ x_2 \end{bmatrix}\right) = T_2(\mathbf{x}) = A_2\mathbf{x}$, where $A_2 = \begin{bmatrix} 3 & 2 \\ 1 & 1 \end{bmatrix}$.

(a) $T_1^{-1}(T_2(\mathbf{x})) = A\mathbf{x} = A_1^{-1}A_2\mathbf{x}$, so $A = A_1^{-1}A_2$. Now $A_1^{-1} = \begin{bmatrix} 2 & 1 \\ 1 & 1 \end{bmatrix}^{-1} = \begin{bmatrix} 1 & -1 \\ -1 & 2 \end{bmatrix}$,

so $A = \begin{bmatrix} 1 & -1 \\ -1 & 2 \end{bmatrix}\begin{bmatrix} 3 & 2 \\ 1 & 1 \end{bmatrix} = \begin{bmatrix} 2 & 1 \\ -1 & 0 \end{bmatrix}.$

(b) $T_1(T_2^{-1}(\mathbf{x})) = A\mathbf{x} = A_1 A_2^{-1}\mathbf{x}$, so $A = A_1 A_2^{-1}$. Now $A_2^{-1} = \begin{bmatrix} 3 & 2 \\ 1 & 1 \end{bmatrix}^{-1} = \begin{bmatrix} 1 & -2 \\ -1 & 3 \end{bmatrix}$,

so $A = \begin{bmatrix} 2 & 1 \\ 1 & 1 \end{bmatrix}\begin{bmatrix} 1 & -2 \\ -1 & 3 \end{bmatrix} = \begin{bmatrix} 1 & -1 \\ 0 & 1 \end{bmatrix}.$

(c) $T_2^{-1}(T_1(\mathbf{x})) = A\mathbf{x} = A_2^{-1}A_1\mathbf{x}$, so $A = A_2^{-1}A_1$. Now $A_2^{-1} = \begin{bmatrix} 3 & 2 \\ 1 & 1 \end{bmatrix}^{-1} = \begin{bmatrix} 1 & -2 \\ -1 & 3 \end{bmatrix}$,

so $A = \begin{bmatrix} 1 & -2 \\ -1 & 3 \end{bmatrix}\begin{bmatrix} 2 & 1 \\ 1 & 1 \end{bmatrix} = \begin{bmatrix} 0 & -1 \\ 1 & 2 \end{bmatrix}.$

(d) $T_2(T_1^{-1}(\mathbf{x})) = A\mathbf{x} = A_2 A_1^{-1}\mathbf{x}$, so $A = A_2 A_1^{-1}$. Now $A_1^{-1} = \begin{bmatrix} 2 & 1 \\ 1 & 1 \end{bmatrix}^{-1} = \begin{bmatrix} 1 & -1 \\ -1 & 2 \end{bmatrix}$,

so $A = \begin{bmatrix} 3 & 2 \\ 1 & 1 \end{bmatrix}\begin{bmatrix} 1 & -1 \\ -1 & 2 \end{bmatrix} = \begin{bmatrix} 1 & 1 \\ 0 & 1 \end{bmatrix}.$

29. $A = \begin{bmatrix} A_{11} & A_{12} \\ \hline A_{21} & A_{22} \end{bmatrix} = \begin{bmatrix} 1 & 0 & 0 \\ \hline 0 & 2 & 7 \\ 0 & 1 & 4 \end{bmatrix}$, so

$$A^{-1} = \begin{bmatrix} A_{11}^{-1} & 0_{12} \\ \hline 0_{21} & A_{22}^{-1} \end{bmatrix} = \begin{bmatrix} 1^{-1} & 0 & 0 \\ \hline 0 & \begin{bmatrix} 2 & 7 \\ 1 & 4 \end{bmatrix}^{-1} \end{bmatrix} = \begin{bmatrix} 1 & 0 & 0 \\ \hline 0 & 4 & -7 \\ 0 & -1 & 2 \end{bmatrix}.$$

31. $A = \begin{bmatrix} A_{11} & A_{12} \\ \hline A_{21} & A_{22} \end{bmatrix} = \begin{bmatrix} 2 & 5 & 0 & 0 \\ 3 & 8 & 0 & 0 \\ \hline 0 & 0 & 1 & 4 \\ 0 & 0 & 1 & 3 \end{bmatrix}$, so

$$A^{-1} = \begin{bmatrix} A_{11}^{-1} & 0_{12} \\ \hline 0_{21} & A_{22}^{-1} \end{bmatrix} = \begin{bmatrix} \begin{bmatrix} 2 & 5 \\ 3 & 8 \end{bmatrix}^{-1} & 0 & 0 \\ & 0 & 0 \\ \hline 0 & 0 & \begin{bmatrix} 1 & 4 \\ 1 & 3 \end{bmatrix}^{-1} \end{bmatrix} = \begin{bmatrix} 8 & -5 & 0 & 0 \\ -3 & 2 & 0 & 0 \\ \hline 0 & 0 & -3 & 4 \\ 0 & 0 & 1 & -1 \end{bmatrix}.$$

33. $A = \begin{bmatrix} A_{11} & A_{12} \\ \hline A_{21} & A_{22} \end{bmatrix} = \begin{bmatrix} 1 & 3 & 0 & 0 & 0 \\ 3 & 8 & 0 & 0 & 0 \\ \hline -1 & 2 & 1 & 2 & -2 \\ 4 & 3 & 0 & 1 & 0 \\ 1 & -2 & 0 & 0 & 1 \end{bmatrix}$, so $A^{-1} = \begin{bmatrix} A_{11}^{-1} & 0_{12} \\ \hline -A_{22}^{-1} A_{21} A_{11}^{-1} & A_{22}^{-1} \end{bmatrix}$

$$= \begin{bmatrix} \begin{bmatrix} 1 & 3 \\ 3 & 8 \end{bmatrix}^{-1} & 0 & 0 & 0 \\ & 0 & 0 & 0 \\ \hline -\begin{bmatrix} 1 & 2 & -2 \\ 0 & 1 & 0 \\ 0 & 0 & 1 \end{bmatrix}^{-1} \begin{bmatrix} -1 & 2 \\ 4 & 3 \\ 1 & -2 \end{bmatrix} \begin{bmatrix} 1 & 3 \\ 3 & 8 \end{bmatrix}^{-1} & \begin{bmatrix} 1 & 2 & -2 \\ 0 & 1 & 0 \\ 0 & 0 & 1 \end{bmatrix}^{-1} \end{bmatrix}$$

$$= \begin{bmatrix} -8 & 3 & 0 & 0 & 0 \\ 3 & -1 & 0 & 0 & 0 \\ \hline -32 & 13 & 1 & -2 & 2 \\ 23 & -9 & 0 & 1 & 0 \\ 14 & -5 & 0 & 0 & 1 \end{bmatrix}.$$

35. $A = \begin{bmatrix} 1 & 0 & 0 \\ 0 & 1 & 0 \\ 0 & 0 & 1 \end{bmatrix}$

37. $A = \begin{bmatrix} 1 & 0 \\ 0 & 1 \end{bmatrix}, B = \begin{bmatrix} 3 & 0 \\ 0 & 3 \end{bmatrix}$

39. $A = \begin{bmatrix} 1 & 0 & 0 \\ 0 & 1 & 0 \end{bmatrix}, B = \begin{bmatrix} 1 & 0 \\ 0 & 1 \\ 0 & 0 \end{bmatrix}$. Then $AB = \begin{bmatrix} 1 & 0 & 0 \\ 0 & 1 & 0 \end{bmatrix} \begin{bmatrix} 1 & 0 \\ 0 & 1 \\ 0 & 0 \end{bmatrix} = \begin{bmatrix} 1 & 0 \\ 0 & 1 \end{bmatrix} = I_2$, but $BA =$
$\begin{bmatrix} 1 & 0 \\ 0 & 1 \\ 0 & 0 \end{bmatrix} \begin{bmatrix} 1 & 0 & 0 \\ 0 & 1 & 0 \end{bmatrix} = \begin{bmatrix} 1 & 0 & 0 \\ 0 & 1 & 0 \\ 0 & 0 & 0 \end{bmatrix} \neq I_3$.

41. False. If A is invertible, then $A\mathbf{x} = \mathbf{b}$ will have one solution for all vectors $\mathbf{b}$.

43. True. A matrix is equivalent to I_n if and only if it is invertible, and since A^{-1} is invertible, A^{-1} is equivalent to I_n.

45. True. As shown in the example, the Caesar cipher corresponds to an invertible matrix.

47. True, since $\left(B^{-1} A^{-1} \right) (AB) = B^{-1} \left(A^{-1} A \right) B = B^{-1} I_n B = B^{-1} B = I_n$.

49. True, since $A(A^{-1}) = I_n$ we conclude that the inverse of A^{-1} is A, so $\left(A^{-1}\right)^{-1} = A$.

51. $AX = B \;\Rightarrow\; A^{-1}(AX) = A^{-1}B \;\Rightarrow\; \left(A^{-1}A\right)X = A^{-1}B \;\Rightarrow\; I_nX = A^{-1}B \;\Rightarrow\; X = A^{-1}B.$

53. $B(X + A)^{-1} = C \;\Rightarrow\; (X + A)^{-1} = B^{-1}C \;\Rightarrow\; X + A = \left(B^{-1}C\right)^{-1} = C^{-1}B \;\Rightarrow\; X = C^{-1}B - A.$

55. If $A^{-1} = A$, then $A\left(A^{-1}\right) = A^2 = I_2$. If $A = \begin{bmatrix} a & b \\ c & d \end{bmatrix}$, then $A^2 = \begin{bmatrix} a & b \\ c & d \end{bmatrix}\begin{bmatrix} a & b \\ c & d \end{bmatrix} = \begin{bmatrix} a^2 + bc & ab + bd \\ ac + cd & d^2 + bc \end{bmatrix} = \begin{bmatrix} 1 & 0 \\ 0 & 1 \end{bmatrix}$. From $ab + bd = 0$ we determine that either $b = 0$ or $d = -a$. From $ac + cd = 0$, we have that $c = 0$ or $d = -a$. *Case 1: $d \neq -a$.* Then $b = 0$ and $c = 0$, and so $a^2 + bc = 1 \;\Rightarrow\; a = \pm 1$. Since $d \neq -a$, we have from $d^2 + bc = 1$ that $d = \pm 1$, and hence either $a = d = 1$ or $a = d = -1$. We now have two matrices $\begin{bmatrix} 1 & 0 \\ 0 & 1 \end{bmatrix}$ and $\begin{bmatrix} -1 & 0 \\ 0 & -1 \end{bmatrix}$. *Case 2: $d = -a$.* Then we need to satisfy $a^2 + bc = 1$. If $b = 0$, then this implies $a = \pm 1$, so $d = \mp 1$, and we get the matrices $\begin{bmatrix} 1 & 0 \\ 0 & -1 \end{bmatrix}$ and $\begin{bmatrix} -1 & 0 \\ 0 & 1 \end{bmatrix}$. If $b \neq 0$, then we may freely choose a and b, and obtain $c = \frac{1-a^2}{b}$, and the family of matrices $\begin{bmatrix} a & b \\ \frac{1-a^2}{b} & -a \end{bmatrix}$.

57. With two equal columns, the columns of A are not linearly independent. By The Big Theorem A is not invertible.

59. $A = \begin{bmatrix} 1 & 1 \\ c & c^2 \end{bmatrix}$ is invertible provided the columns of A are linearly independent, which will be the case if $c \neq c^2$. Thus we require that $c \neq 0$ and $c \neq 1$.

61. Suppose that $A\mathbf{x} = \mathbf{b}$ has solution $\mathbf{x}_b$. If A is $n \times n$ and not invertible, then it follows that the system $A\mathbf{x} = \mathbf{0}$ has a nontrivial solution $\mathbf{x}_0$. Then $A(\mathbf{x}_b + \mathbf{x}_0) = A\mathbf{x}_b + A\mathbf{x}_0 = \mathbf{b} + \mathbf{0} = \mathbf{b}$. Therefore $A\mathbf{x} = \mathbf{b}$ does not have a unique solution, a contradiction.

63. $AC = CB \;\Rightarrow\; C^{-1}(AC) = C^{-1}(CB) \;\Rightarrow\; C^{-1}AC = \left(C^{-1}C\right)B \;\Rightarrow\; C^{-1}AC = B.$

65. $(B - C)A = 0_{nm} \;\Rightarrow\; ((B - C)A)A^{-1} = 0_{nm}A^{-1} \;\Rightarrow\; (B - C)\left(AA^{-1}\right) = 0_{nm} \;\Rightarrow\; (B - C)I_m = 0_{nm} \;\Rightarrow\; B - C = 0_{nm} \;\Rightarrow\; B = C.$

67. Since B is singular, there exists $\mathbf{x} \neq \mathbf{0}$ such that $B\mathbf{x} = \mathbf{0}$. Thus, $(AB)\mathbf{x} = A(B\mathbf{x}) = A(\mathbf{0}) = \mathbf{0}$, and hence AB is singular.

69. Let $\mathbf{x} = T(\mathbf{y})$, then $T^{-1}(r\mathbf{x}) = T^{-1}(rT(\mathbf{y})) = T^{-1}(T(r\mathbf{y})) = r\mathbf{y} = rT^{-1}(\mathbf{x}).$

71. $T^{-1}(\mathbf{x}) = A^{-1}\mathbf{x} = \begin{bmatrix} 15 & 46 & 65 \\ 14 & 43 & 61 \\ 20 & 60 & 81 \end{bmatrix}^{-1}\begin{bmatrix} 1070 \\ 1002 \\ 1368 \end{bmatrix} = \begin{bmatrix} 6 \\ 10 \\ 8 \end{bmatrix}.$

73. $T^{-1}(\mathbf{x}) = A^{-1}\mathbf{x} = \begin{bmatrix} 15 & 46 & 65 \\ 14 & 43 & 61 \\ 20 & 60 & 81 \end{bmatrix}^{-1}\begin{bmatrix} 2045 \\ 1965 \\ 2615 \end{bmatrix} = \begin{bmatrix} 8710 \\ -4230 \\ 1015 \end{bmatrix}$, which is not reasonable, as it means a negative production of J40 type MP3 players.

75. $T(\mathbf{x}) = A\mathbf{x} = \begin{bmatrix} 29 & 18 & 50 \\ 3 & 25 & 19 \\ 4 & 6 & 9 \end{bmatrix}\mathbf{x}$, $T^{-1}(\mathbf{y}) = A^{-1}\mathbf{y} = \begin{bmatrix} 29 & 18 & 50 \\ 3 & 25 & 19 \\ 4 & 6 & 9 \end{bmatrix}^{-1}\mathbf{y}$

$= \begin{bmatrix} 111 & 138 & -908 \\ 49 & 61 & -401 \\ -82 & -102 & 671 \end{bmatrix}\mathbf{y}$, thus $T^{-1}\left(\begin{bmatrix} 409 \\ 204 \\ 81 \end{bmatrix}\right) = \begin{bmatrix} 111 & 138 & -908 \\ 49 & 61 & -401 \\ -82 & -102 & 671 \end{bmatrix}\begin{bmatrix} 409 \\ 204 \\ 81 \end{bmatrix} = \begin{bmatrix} 3 \\ 4 \\ 5 \end{bmatrix}$

77. $T\left(\mathbf{x}\right)=A\mathbf{x}=\begin{bmatrix} 29 & 18 & 50 \\ 3 & 25 & 19 \\ 4 & 6 & 9 \end{bmatrix}\mathbf{x}$, $T^{-1}\left(\mathbf{y}\right)=A^{-1}\mathbf{y}=\begin{bmatrix} 29 & 18 & 50 \\ 3 & 25 & 19 \\ 4 & 6 & 9 \end{bmatrix}^{-1}\mathbf{y}$

$=\begin{bmatrix} 111 & 138 & -908 \\ 49 & 61 & -401 \\ -82 & -102 & 671 \end{bmatrix}\mathbf{y}$, thus $T^{-1}\left(\begin{bmatrix} 1092 \\ 589 \\ 223 \end{bmatrix}\right)=\begin{bmatrix} 111 & 138 & -908 \\ 49 & 61 & -401 \\ -82 & -102 & 671 \end{bmatrix}\begin{bmatrix} 1092 \\ 589 \\ 223 \end{bmatrix}=\begin{bmatrix} 10 \\ 14 \\ 11 \end{bmatrix}.$

79. $A^{-1}\begin{bmatrix} 41 & 7 \\ 161 & 79 \\ -306 & -142 \end{bmatrix}=\begin{bmatrix} 1 & -3 & 2 \\ 2 & -7 & 9 \\ -4 & 14 & -17 \end{bmatrix}^{-1}\begin{bmatrix} 41 & 7 \\ 161 & 79 \\ -306 & -142 \end{bmatrix}=\begin{bmatrix} 12 & 20 \\ 1 & 15 \\ 16 & 16 \end{bmatrix}$. Hence the decoded

message is $\left\{\begin{bmatrix} 12 \\ 1 \\ 16 \end{bmatrix},\begin{bmatrix} 20 \\ 15 \\ 16 \end{bmatrix}\right\}\rightarrow\left\{\begin{bmatrix} l \\ a \\ p \end{bmatrix},\begin{bmatrix} t \\ o \\ p \end{bmatrix}\right\}$, *i.e.* "laptop".

81. $A^{-1}\begin{bmatrix} 7 & -25 & 47 \\ 75 & -37 & 158 \\ -136 & 79 & -303 \end{bmatrix}=\begin{bmatrix} 1 & -3 & 2 \\ 2 & -7 & 9 \\ -4 & 14 & -17 \end{bmatrix}^{-1}\begin{bmatrix} 7 & -25 & 47 \\ 75 & -37 & 158 \\ -136 & 79 & -303 \end{bmatrix}=\begin{bmatrix} 6 & 1 & 24 \\ 9 & 12 & 1 \\ 14 & 5 & 13 \end{bmatrix}$. Hence

the decoded message is $\left\{\begin{bmatrix} 6 \\ 9 \\ 14 \end{bmatrix},\begin{bmatrix} 1 \\ 12 \\ 5 \end{bmatrix},\begin{bmatrix} 24 \\ 1 \\ 13 \end{bmatrix}\right\}\rightarrow\left\{\begin{bmatrix} f \\ i \\ n \end{bmatrix},\begin{bmatrix} a \\ l \\ e \end{bmatrix},\begin{bmatrix} x \\ a \\ m \end{bmatrix}\right\}$, *i.e.* "final exam".

83. $A^{-1}=\begin{bmatrix} 3 & 1 & -2 & 0 \\ 2 & 2 & 5 & 1 \\ -3 & 0 & -2 & 2 \\ 4 & 1 & 2 & 3 \end{bmatrix}^{-1}=\begin{bmatrix} \frac{8}{145} & -\frac{14}{145} & -\frac{23}{145} & \frac{4}{29} \\ \frac{67}{145} & \frac{64}{145} & \frac{43}{145} & -\frac{10}{29} \\ -\frac{27}{145} & \frac{11}{145} & -\frac{13}{145} & \frac{1}{29} \\ -\frac{3}{29} & -\frac{2}{29} & \frac{5}{29} & \frac{7}{29} \end{bmatrix}.$

85. $A^{-1}=\begin{bmatrix} 5 & 1 & 2 & 1 & 2 \\ -3 & 2 & 2 & 1 & 0 \\ 2 & 3 & 1 & 0 & 1 \\ 5 & -1 & -1 & -1 & 3 \\ 0 & 0 & 3 & 2 & 1 \end{bmatrix}^{-1}$ does not exist.

3.4   LU Factorization

1. $\begin{bmatrix} 1 & 0 \\ -7 & 1 \end{bmatrix}\begin{bmatrix} 2 & 2 & -3 \\ 0 & 1 & -4 \end{bmatrix}=\begin{bmatrix} 2 & 2 & -3 \\ -14 & -13 & 17 \end{bmatrix}$, so $a=2$ and $b=-14$.

3. $\begin{bmatrix} 1 & 0 & 0 \\ 3 & 1 & 0 \\ a & 2 & 1 \end{bmatrix}\begin{bmatrix} 5 & 2 \\ 0 & b \\ 0 & 0 \end{bmatrix}=\begin{bmatrix} 5 & 2 \\ 15 & b+6 \\ 5a & 2a+2b \end{bmatrix}$. Equating this to $\begin{bmatrix} 5 & c \\ 15 & 9 \\ 20 & 14 \end{bmatrix}$ we obtain that $c=2$,

   $9=b+6 \Rightarrow b=3$, and $20=5a \Rightarrow a=4$. We check that $2a+2b=2(4)+2(3)=14$.

5. Solve $L\mathbf{y}=\mathbf{b}$, $\begin{bmatrix} 1 & 0 \\ -2 & 1 \end{bmatrix}\mathbf{y}=\begin{bmatrix} 2 \\ 2 \end{bmatrix}$, using back substitution, to obtain $\mathbf{y}=\begin{bmatrix} 2 \\ 6 \end{bmatrix}$. Now solve $U\mathbf{x}=\mathbf{y}$,

   $\begin{bmatrix} 2 & -2 \\ 0 & 3 \end{bmatrix}\mathbf{x}=\begin{bmatrix} 2 \\ 6 \end{bmatrix}$, using back substitution, to obtain $\mathbf{x}=\begin{bmatrix} 3 \\ 2 \end{bmatrix}$.

7. Solve $L\mathbf{y}=\mathbf{b}$, $\begin{bmatrix} 1 & 0 & 0 \\ -1 & 1 & 0 \\ 2 & -2 & 1 \end{bmatrix}\mathbf{y}=\begin{bmatrix} 4 \\ 0 \\ -4 \end{bmatrix}$, using back substitution, to obtain $\mathbf{y}=\begin{bmatrix} 4 \\ 4 \\ -4 \end{bmatrix}$. Now

   solve $U\mathbf{x}=\mathbf{y}$, $\begin{bmatrix} 2 & -1 & 3 \\ 0 & 1 & 2 \\ 0 & 0 & -2 \end{bmatrix}\mathbf{x}=\begin{bmatrix} 4 \\ 4 \\ -4 \end{bmatrix}$, using back substitution, to obtain $\mathbf{x}=\begin{bmatrix} -1 \\ 0 \\ 2 \end{bmatrix}$.

9. Solve $L\mathbf{y} = \mathbf{b}$, $\begin{bmatrix} 1 & 0 & 0 \\ 2 & 1 & 0 \\ -3 & 4 & 1 \end{bmatrix} \mathbf{y} = \begin{bmatrix} 0 \\ 1 \\ 4 \end{bmatrix}$, using back substitution, to obtain $\mathbf{y} = \begin{bmatrix} 0 \\ 1 \\ 0 \end{bmatrix}$. Now solve

$U\mathbf{x} = \mathbf{y}$, $\begin{bmatrix} 1 & -2 \\ 0 & 1 \\ 0 & 0 \end{bmatrix} \mathbf{x} = \begin{bmatrix} 0 \\ 1 \\ 0 \end{bmatrix}$, using back substitution, to obtain $\mathbf{x} = \begin{bmatrix} 2 \\ 1 \end{bmatrix}$.

11. Solve $L\mathbf{y} = \mathbf{b}$, $\begin{bmatrix} 1 & 0 & 0 & 0 \\ -2 & 1 & 0 & 0 \\ 0 & 3 & 1 & 0 \\ 2 & -1 & 0 & 1 \end{bmatrix} \mathbf{y} = \begin{bmatrix} 0 \\ 0 \\ -1 \\ 0 \end{bmatrix}$, using back substitution, to obtain $\mathbf{y} = \begin{bmatrix} 0 \\ 0 \\ -1 \\ 0 \end{bmatrix}$.

Now solve $U\mathbf{x} = \mathbf{y}$, $\begin{bmatrix} 1 & -2 & 0 & -1 \\ 0 & 1 & 1 & 3 \\ 0 & 0 & 1 & -1 \\ 0 & 0 & 0 & 1 \end{bmatrix} \mathbf{x} = \begin{bmatrix} 0 \\ 0 \\ -1 \\ 0 \end{bmatrix}$, using back substitution, to obtain $\mathbf{x} = \begin{bmatrix} 2 \\ 1 \\ -1 \\ 0 \end{bmatrix}$.

13. $\begin{bmatrix} 1 & -4 \\ -2 & 9 \end{bmatrix} \overset{2R_1 + R_2 \Rightarrow R_2}{\sim} \begin{bmatrix} 1 & -4 \\ 0 & 1 \end{bmatrix} \Rightarrow L = \begin{bmatrix} 1 & \bullet \\ -2 & 1 \end{bmatrix}$

Thus $L = \begin{bmatrix} 1 & 0 \\ -2 & 1 \end{bmatrix}$ and $U = \begin{bmatrix} 1 & -4 \\ 0 & 1 \end{bmatrix}$.

15. $\begin{bmatrix} -2 & -1 & 1 \\ -6 & 0 & 4 \\ 2 & -2 & -1 \end{bmatrix} \overset{-3R_1 + R_2 \Rightarrow R_2}{\underset{R_1 + R_3 \Rightarrow R_3}{\sim}} \begin{bmatrix} -2 & -1 & 1 \\ 0 & 3 & 1 \\ 0 & -3 & 0 \end{bmatrix} \Rightarrow L = \begin{bmatrix} 1 & \bullet & \bullet \\ 3 & 1 & \bullet \\ -1 & -1 & \bullet \end{bmatrix}$

$\begin{bmatrix} -2 & -1 & 1 \\ 0 & 3 & 1 \\ 0 & -3 & 0 \end{bmatrix} \overset{R_2 + R_3 \Rightarrow R_3}{\sim} \begin{bmatrix} -2 & -1 & 1 \\ 0 & 3 & 1 \\ 0 & 0 & 1 \end{bmatrix} \Rightarrow L = \begin{bmatrix} 1 & \bullet & \bullet \\ 3 & 1 & \bullet \\ -1 & -1 & 1 \end{bmatrix}$

Thus $L = \begin{bmatrix} 1 & 0 & 0 \\ 3 & 1 & 0 \\ -1 & -1 & 1 \end{bmatrix}$ and $U = \begin{bmatrix} -2 & -1 & 1 \\ 0 & 3 & 1 \\ 0 & 0 & 1 \end{bmatrix}$.

17. $\begin{bmatrix} -1 & 0 & -1 & 2 \\ 1 & 3 & 2 & -2 \\ -2 & -9 & -3 & 3 \\ -1 & 9 & -2 & 5 \end{bmatrix} \overset{R_1 + R_2 \Rightarrow R_2}{\underset{\sim}{\overset{-2R_1 + R_3 \Rightarrow R_3}{-R_1 + R_4 \Rightarrow R_4}}} \begin{bmatrix} -1 & 0 & -1 & 2 \\ 0 & 3 & 1 & 0 \\ 0 & -9 & -1 & -1 \\ 0 & 9 & -1 & 3 \end{bmatrix} \Rightarrow L = \begin{bmatrix} 1 & \bullet & \bullet & \bullet \\ -1 & 1 & \bullet & \bullet \\ 2 & -3 & \bullet & \bullet \\ 1 & 3 & \bullet & \bullet \end{bmatrix}$

$\begin{bmatrix} -1 & 0 & -1 & 2 \\ 0 & 3 & 1 & 0 \\ 0 & -9 & -1 & -1 \\ 0 & 9 & -1 & 3 \end{bmatrix} \overset{3R_2 + R_3 \Rightarrow R_3}{\underset{-3R_2 + R_4 \Rightarrow R_4}{\sim}} \begin{bmatrix} -1 & 0 & -1 & 2 \\ 0 & 3 & 1 & 0 \\ 0 & 0 & 2 & -1 \\ 0 & 0 & -4 & 3 \end{bmatrix} \Rightarrow L = \begin{bmatrix} 1 & \bullet & \bullet & \bullet \\ -1 & 1 & \bullet & \bullet \\ 2 & -3 & 1 & \bullet \\ 1 & 3 & -2 & \bullet \end{bmatrix}$

$\begin{bmatrix} -1 & 0 & -1 & 2 \\ 0 & 3 & 1 & 0 \\ 0 & 0 & 2 & -1 \\ 0 & 0 & -4 & 3 \end{bmatrix} \overset{2R_3 + R_4 \Rightarrow R_4}{\sim} \begin{bmatrix} -1 & 0 & -1 & 2 \\ 0 & 3 & 1 & 0 \\ 0 & 0 & 2 & -1 \\ 0 & 0 & 0 & 1 \end{bmatrix} \Rightarrow L = \begin{bmatrix} 1 & \bullet & \bullet & \bullet \\ -1 & 1 & \bullet & \bullet \\ 2 & -3 & 1 & \bullet \\ 1 & 3 & -2 & 1 \end{bmatrix}$

Thus $L = \begin{bmatrix} 1 & 0 & 0 & 0 \\ -1 & 1 & 0 & 0 \\ 2 & -3 & 1 & 0 \\ 1 & 3 & -2 & 1 \end{bmatrix}$ and $U = \begin{bmatrix} -1 & 0 & -1 & 2 \\ 0 & 3 & 1 & 0 \\ 0 & 0 & 2 & -1 \\ 0 & 0 & 0 & 1 \end{bmatrix}$.

19. $\begin{bmatrix} -1 & 2 & 1 & 3 \\ 4 & -7 & -7 & -17 \\ -2 & 6 & -3 & -2 \end{bmatrix} \overset{4R_1 + R_2 \Rightarrow R_2}{\underset{-2R_1 + R_3 \Rightarrow R_3}{\sim}} \begin{bmatrix} -1 & 2 & 1 & 3 \\ 0 & 1 & -3 & -5 \\ 0 & 2 & -5 & -8 \end{bmatrix} \Rightarrow L = \begin{bmatrix} 1 & \bullet & \bullet \\ -4 & 1 & \bullet \\ 2 & 2 & \bullet \end{bmatrix}$

$\begin{bmatrix} -1 & 2 & 1 & 3 \\ 0 & 1 & -3 & -5 \\ 0 & 2 & -5 & -8 \end{bmatrix} \overset{-2R_2 + R_3 \Rightarrow R_3}{\sim} \begin{bmatrix} -1 & 2 & 1 & 3 \\ 0 & 1 & -3 & -5 \\ 0 & 0 & 1 & 2 \end{bmatrix} \Rightarrow L = \begin{bmatrix} 1 & \bullet & \bullet \\ -4 & 1 & \bullet \\ 2 & 2 & 1 \end{bmatrix}$

Thus $L = \begin{bmatrix} 1 & 0 & 0 \\ -4 & 1 & 0 \\ 2 & 2 & 1 \end{bmatrix}$ and $U = \begin{bmatrix} -1 & 2 & 1 & 3 \\ 0 & 1 & -3 & -5 \\ 0 & 0 & 1 & 2 \end{bmatrix}$.

21. $\begin{bmatrix} 1 & 1 & 0 \\ 1 & 0 & -1 \\ 1 & -1 & 0 \\ 0 & 1 & -1 \end{bmatrix} \begin{array}{c} -R_1+R_2 \Rightarrow R_2 \\ -R_1+R_3 \Rightarrow R_3 \\ \sim \end{array} \begin{bmatrix} 1 & 1 & 0 \\ 0 & -1 & -1 \\ 0 & -2 & 0 \\ 0 & 1 & -1 \end{bmatrix} \Rightarrow L = \begin{bmatrix} 1 & \bullet & \bullet & \bullet \\ 1 & 1 & \bullet & \bullet \\ 1 & 2 & \bullet & \bullet \\ 0 & -1 & \bullet & \bullet \end{bmatrix}$

$\begin{bmatrix} 1 & 1 & 0 \\ 0 & -1 & -1 \\ 0 & -2 & 0 \\ 0 & 1 & -1 \end{bmatrix} \begin{array}{c} -2R_2+R_3 \Rightarrow R_3 \\ R_2+R_4 \Rightarrow R_4 \\ \sim \end{array} \begin{bmatrix} 1 & 1 & 0 \\ 0 & -1 & -1 \\ 0 & 0 & 2 \\ 0 & 0 & -2 \end{bmatrix} \Rightarrow L = \begin{bmatrix} 1 & \bullet & \bullet & \bullet \\ 1 & 1 & \bullet & \bullet \\ 1 & 2 & 1 & \bullet \\ 0 & -1 & -1 & \bullet \end{bmatrix}$

$\begin{bmatrix} 1 & 1 & 0 \\ 0 & -1 & -1 \\ 0 & 0 & 2 \\ 0 & 0 & -2 \end{bmatrix} \begin{array}{c} R_3+R_4 \Rightarrow R_4 \\ \sim \end{array} \begin{bmatrix} 1 & 1 & 0 \\ 0 & -1 & -1 \\ 0 & 0 & 2 \\ 0 & 0 & 0 \end{bmatrix} \Rightarrow L = \begin{bmatrix} 1 & \bullet & \bullet & \bullet \\ 1 & 1 & \bullet & \bullet \\ 1 & 2 & 1 & \bullet \\ 0 & -1 & -1 & 1 \end{bmatrix}$

Thus $L = \begin{bmatrix} 1 & 0 & 0 & 0 \\ 1 & 1 & 0 & 0 \\ 1 & 2 & 1 & 0 \\ 0 & -1 & -1 & 1 \end{bmatrix}$ and $U = \begin{bmatrix} 1 & 1 & 0 \\ 0 & -1 & -1 \\ 0 & 0 & 2 \\ 0 & 0 & 0 \end{bmatrix}$.

23. $\begin{bmatrix} -2 & 1 & 3 \\ 2 & 0 & 8 \\ -4 & 1 & 12 \\ 2 & 0 & -10 \\ -4 & 2 & 7 \end{bmatrix} \begin{array}{c} R_1+R_2 \Rightarrow R_2 \\ -2R_1+R_3 \Rightarrow R_3 \\ R_1+R_4 \Rightarrow R_4 \\ -2R_1+R_5 \Rightarrow R_5 \\ \sim \end{array} \begin{bmatrix} -2 & 1 & 3 \\ 0 & 1 & 11 \\ 0 & -1 & 6 \\ 0 & 1 & -7 \\ 0 & 0 & 1 \end{bmatrix} \Rightarrow L = \begin{bmatrix} 1 & \bullet & \bullet & \bullet & \bullet \\ -1 & 1 & \bullet & \bullet & \bullet \\ 2 & -1 & \bullet & \bullet & \bullet \\ -1 & 1 & \bullet & \bullet & \bullet \\ 2 & 0 & \bullet & \bullet & \bullet \end{bmatrix}$

$\begin{bmatrix} -2 & 1 & 3 \\ 0 & 1 & 11 \\ 0 & -1 & 6 \\ 0 & 1 & -7 \\ 0 & 0 & 1 \end{bmatrix} \begin{array}{c} R_2+R_3 \Rightarrow R_3 \\ -R_2+R_4 \Rightarrow R_4 \\ \sim \end{array} \begin{bmatrix} -2 & 1 & 3 \\ 0 & 1 & 11 \\ 0 & 0 & 17 \\ 0 & 0 & -18 \\ 0 & 0 & 1 \end{bmatrix} \Rightarrow L = \begin{bmatrix} 1 & \bullet & \bullet & \bullet & \bullet \\ -1 & 1 & \bullet & \bullet & \bullet \\ 2 & -1 & 1 & \bullet & \bullet \\ -1 & 1 & -\frac{18}{17} & \bullet & \bullet \\ 2 & 0 & \frac{1}{17} & \bullet & \bullet \end{bmatrix}$

$\begin{bmatrix} -2 & 1 & 3 \\ 0 & 1 & 11 \\ 0 & 0 & 17 \\ 0 & 0 & -18 \\ 0 & 0 & 1 \end{bmatrix} \begin{array}{c} (18/17)R_3+R_4 \Rightarrow R_4 \\ (-1/17)R_3+R_5 \Rightarrow R_5 \\ \sim \end{array} \begin{bmatrix} -2 & 1 & 3 \\ 0 & 1 & 11 \\ 0 & 0 & 17 \\ 0 & 0 & 0 \\ 0 & 0 & 0 \end{bmatrix} \Rightarrow L = \begin{bmatrix} 1 & \bullet & \bullet & \bullet & \bullet \\ -1 & 1 & \bullet & \bullet & \bullet \\ 2 & -1 & 1 & \bullet & \bullet \\ -1 & 1 & -\frac{18}{17} & 1 & 0 \\ 2 & 0 & \frac{1}{17} & 0 & 1 \end{bmatrix}$

Thus $L = \begin{bmatrix} 1 & 0 & 0 & 0 & 0 \\ -1 & 1 & 0 & 0 & 0 \\ 2 & -1 & 1 & 0 & 0 \\ -1 & 1 & -\frac{18}{17} & 1 & 0 \\ 2 & 0 & \frac{1}{17} & 0 & 1 \end{bmatrix}$ and $U = \begin{bmatrix} -2 & 1 & 3 \\ 0 & 1 & 11 \\ 0 & 0 & 17 \\ 0 & 0 & 0 \\ 0 & 0 & 0 \end{bmatrix}$.

25. $L = \begin{bmatrix} 1 & 0 \\ -2 & 1 \end{bmatrix}$. We divide the rows of $U = \begin{bmatrix} 2 & -2 \\ 0 & 3 \end{bmatrix}$ by the diagonal entries to obtain $D = \begin{bmatrix} 2 & 0 \\ 0 & 3 \end{bmatrix}$ and $U = \begin{bmatrix} 1 & -1 \\ 0 & 1 \end{bmatrix}$.

27. $L = \begin{bmatrix} 1 & 0 \\ 3 & 1 \end{bmatrix}$. We divide the rows of $U = \begin{bmatrix} 1 & -1 & 2 \\ 0 & -2 & -1 \end{bmatrix}$ by the diagonal entries to obtain $D = \begin{bmatrix} 1 & 0 \\ 0 & -2 \end{bmatrix}$ and $U = \begin{bmatrix} 1 & -1 & 2 \\ 0 & 1 & \frac{1}{2} \end{bmatrix}$.

29. $L = \begin{bmatrix} 1 & 0 & 0 \\ 3 & 1 & 0 \\ -1 & -1 & 1 \end{bmatrix}$. We divide the rows of $U = \begin{bmatrix} -2 & -1 & 1 \\ 0 & 3 & 1 \\ 0 & 0 & 1 \end{bmatrix}$ by the diagonal entries to obtain

$D = \begin{bmatrix} -2 & 0 & 0 \\ 0 & 3 & 0 \\ 0 & 0 & 1 \end{bmatrix}$ and $U = \begin{bmatrix} 1 & \frac{1}{2} & -\frac{1}{2} \\ 0 & 1 & \frac{1}{3} \\ 0 & 0 & 1 \end{bmatrix}$.

31. $E = \begin{bmatrix} 4 & 0 & 0 \\ 0 & 1 & 0 \\ 0 & 0 & 1 \end{bmatrix}$

33. $E = \begin{bmatrix} 0 & 1 & 0 \\ 1 & 0 & 0 \\ 0 & 0 & 1 \end{bmatrix}$

35. $E = \begin{bmatrix} 1 & 0 & 0 \\ 0 & 1 & 0 \\ 2 & 0 & 1 \end{bmatrix}$

37. $\{-2R_1 + R_2 \Rightarrow R_2\} \leftrightarrow E_1 = \begin{bmatrix} 1 & 0 & 0 \\ -2 & 1 & 0 \\ 0 & 0 & 1 \end{bmatrix}$ and $\{5R_3 \Rightarrow R_3\} \leftrightarrow E_2 = \begin{bmatrix} 1 & 0 & 0 \\ 0 & 1 & 0 \\ 0 & 0 & 5 \end{bmatrix}$. Thus $B =$

$E_2 E_1 = \begin{bmatrix} 1 & 0 & 0 \\ 0 & 1 & 0 \\ 0 & 0 & 5 \end{bmatrix} \begin{bmatrix} 1 & 0 & 0 \\ -2 & 1 & 0 \\ 0 & 0 & 1 \end{bmatrix} = \begin{bmatrix} 1 & 0 & 0 \\ -2 & 1 & 0 \\ 0 & 0 & 5 \end{bmatrix}$.

39. $\{R_2 \leftrightarrow R_1\} \leftrightarrow E_1 = \begin{bmatrix} 0 & 1 & 0 \\ 1 & 0 & 0 \\ 0 & 0 & 1 \end{bmatrix}$ and $\{3R_1 + R_2 \Rightarrow R_2\} \leftrightarrow E_2 = \begin{bmatrix} 1 & 0 & 0 \\ 3 & 1 & 0 \\ 0 & 0 & 1 \end{bmatrix}$. Thus $B = E_2 E_1 =$

$\begin{bmatrix} 1 & 0 & 0 \\ 3 & 1 & 0 \\ 0 & 0 & 1 \end{bmatrix} \begin{bmatrix} 0 & 1 & 0 \\ 1 & 0 & 0 \\ 0 & 0 & 1 \end{bmatrix} = \begin{bmatrix} 0 & 1 & 0 \\ 1 & 3 & 0 \\ 0 & 0 & 1 \end{bmatrix}$.

41. $\{-3R_1 \Rightarrow R_1\} \leftrightarrow E_1 = \begin{bmatrix} -3 & 0 & 0 \\ 0 & 1 & 0 \\ 0 & 0 & 1 \end{bmatrix}$. $\{R_1 \leftrightarrow R_2\} \leftrightarrow E_2 = \begin{bmatrix} 0 & 1 & 0 \\ 1 & 0 & 0 \\ 0 & 0 & 1 \end{bmatrix}$

And $\{4R_1 + R_2 \Rightarrow R_2\} \leftrightarrow E_3 = \begin{bmatrix} 1 & 0 & 0 \\ 4 & 1 & 0 \\ 0 & 0 & 1 \end{bmatrix}$. Thus $B = E_3 E_2 E_1$

$= \begin{bmatrix} 1 & 0 & 0 \\ 4 & 1 & 0 \\ 0 & 0 & 1 \end{bmatrix} \begin{bmatrix} 0 & 1 & 0 \\ 1 & 0 & 0 \\ 0 & 0 & 1 \end{bmatrix} \begin{bmatrix} -3 & 0 & 0 \\ 0 & 1 & 0 \\ 0 & 0 & 1 \end{bmatrix} = \begin{bmatrix} 0 & 1 & 0 \\ -3 & 4 & 0 \\ 0 & 0 & 1 \end{bmatrix}$.

43. $\{-6R_2 \Rightarrow R_2\} \leftrightarrow E = \begin{bmatrix} 1 & 0 & 0 & 0 \\ 0 & -6 & 0 & 0 \\ 0 & 0 & 1 & 0 \\ 0 & 0 & 0 & 1 \end{bmatrix}$. $E^{-1} = \begin{bmatrix} 1 & 0 & 0 & 0 \\ 0 & -6 & 0 & 0 \\ 0 & 0 & 1 & 0 \\ 0 & 0 & 0 & 1 \end{bmatrix}^{-1}$

$= \begin{bmatrix} 1 & 0 & 0 & 0 \\ 0 & -\frac{1}{6} & 0 & 0 \\ 0 & 0 & 1 & 0 \\ 0 & 0 & 0 & 1 \end{bmatrix} \leftrightarrow \{-\frac{1}{6}R_2 \Rightarrow R_2\}$.

45. $\{R_3 \leftrightarrow R_4\} \leftrightarrow E = \begin{bmatrix} 1 & 0 & 0 & 0 \\ 0 & 1 & 0 & 0 \\ 0 & 0 & 0 & 1 \\ 0 & 0 & 1 & 0 \end{bmatrix}$. $E^{-1} = \begin{bmatrix} 1 & 0 & 0 & 0 \\ 0 & 1 & 0 & 0 \\ 0 & 0 & 0 & 1 \\ 0 & 0 & 1 & 0 \end{bmatrix}^{-1}$

$$= \begin{bmatrix} 1 & 0 & 0 & 0 \\ 0 & 1 & 0 & 0 \\ 0 & 0 & 0 & 1 \\ 0 & 0 & 1 & 0 \end{bmatrix} \leftrightarrow \{R_3 \Leftrightarrow R_4\}.$$

47. $\{-5R_1 + R_2 \Rightarrow R_2\} \leftrightarrow E = \begin{bmatrix} 1 & 0 & 0 & 0 \\ -5 & 1 & 0 & 0 \\ 0 & 0 & 1 & 0 \\ 0 & 0 & 0 & 1 \end{bmatrix}.\; E^{-1} = \begin{bmatrix} 1 & 0 & 0 & 0 \\ -5 & 1 & 0 & 0 \\ 0 & 0 & 1 & 0 \\ 0 & 0 & 0 & 1 \end{bmatrix}^{-1}$

$$= \begin{bmatrix} 1 & 0 & 0 & 0 \\ 5 & 1 & 0 & 0 \\ 0 & 0 & 1 & 0 \\ 0 & 0 & 0 & 1 \end{bmatrix} \leftrightarrow \{5R_1 + R_2 \Rightarrow R_2\}.$$

49. $A^{-1} = (LU)^{-1} = U^{-1}L^{-1} = \begin{bmatrix} 2 & -2 \\ 0 & 3 \end{bmatrix}^{-1} \begin{bmatrix} 1 & 0 \\ -2 & 1 \end{bmatrix}^{-1} = \begin{bmatrix} \frac{1}{2} & \frac{1}{3} \\ 0 & \frac{1}{3} \end{bmatrix} \begin{bmatrix} 1 & 0 \\ 2 & 1 \end{bmatrix} = \begin{bmatrix} \frac{7}{6} & \frac{1}{3} \\ \frac{2}{3} & \frac{1}{3} \end{bmatrix}.$

51. $A^{-1} = (LU)^{-1} = U^{-1}L^{-1} = \begin{bmatrix} 1 & -2 & 0 & -1 \\ 0 & 1 & 1 & 3 \\ 0 & 0 & 1 & -1 \\ 0 & 0 & 0 & 1 \end{bmatrix}^{-1} \begin{bmatrix} 1 & 0 & 0 & 0 \\ -2 & 1 & 0 & 0 \\ 0 & 3 & 1 & 0 \\ 2 & -1 & 0 & 1 \end{bmatrix}^{-1}$

$$= \begin{bmatrix} 1 & 2 & -2 & -7 \\ 0 & 1 & -1 & -4 \\ 0 & 0 & 1 & 1 \\ 0 & 0 & 0 & 1 \end{bmatrix} \begin{bmatrix} 1 & 0 & 0 & 0 \\ 2 & 1 & 0 & 0 \\ -6 & -3 & 1 & 0 \\ 0 & 1 & 0 & 1 \end{bmatrix} = \begin{bmatrix} 17 & 1 & -2 & -7 \\ 8 & 0 & -1 & -4 \\ -6 & -2 & 1 & 1 \\ 0 & 1 & 0 & 1 \end{bmatrix}.$$

53. $A^{-1} = (LU)^{-1} = U^{-1}L^{-1} = \begin{bmatrix} -3 & 2 & 1 \\ 0 & -2 & 1 \\ 0 & 0 & 2 \end{bmatrix}^{-1} \begin{bmatrix} 1 & 0 & 0 \\ 2 & 1 & 0 \\ 0 & 4 & 1 \end{bmatrix}^{-1}$

$$= \begin{bmatrix} -\frac{1}{3} & -\frac{1}{3} & \frac{1}{3} \\ 0 & -\frac{1}{2} & \frac{1}{4} \\ 0 & 0 & \frac{1}{2} \end{bmatrix} \begin{bmatrix} 1 & 0 & 0 \\ -2 & 1 & 0 \\ 8 & -4 & 1 \end{bmatrix} = \begin{bmatrix} 3 & -\frac{5}{3} & \frac{1}{3} \\ 3 & -\frac{3}{2} & \frac{1}{4} \\ 4 & -2 & \frac{1}{2} \end{bmatrix}.$$

55. $A = \begin{bmatrix} 1 & 0 & 0 \\ 0 & 1 & 0 \\ 0 & 0 & 1 \\ 0 & 0 & 0 \end{bmatrix} = LU = \begin{bmatrix} 1 & 0 & 0 & 0 \\ 0 & 1 & 0 & 0 \\ 0 & 0 & 1 & 0 \\ 0 & 0 & 0 & 1 \end{bmatrix} \begin{bmatrix} 1 & 0 & 0 \\ 0 & 1 & 0 \\ 0 & 0 & 1 \\ 0 & 0 & 0 \end{bmatrix}$

57. $A = \begin{bmatrix} 1 & 0 \\ 0 & 1 \end{bmatrix} = LU = \begin{bmatrix} 1 & 0 \\ 0 & 1 \end{bmatrix} \begin{bmatrix} 1 & 0 \\ 0 & 1 \end{bmatrix}$

59. $A = \begin{bmatrix} 1 & 0 & 0 & 0 \\ 0 & 1 & 0 & 0 \\ 0 & 0 & 1 & 0 \\ 0 & 0 & 0 & 1 \end{bmatrix} = LU = \begin{bmatrix} 1 & 0 & 0 & 0 \\ 0 & 1 & 0 & 0 \\ 0 & 0 & 1 & 0 \\ 0 & 0 & 0 & 1 \end{bmatrix} \begin{bmatrix} 1 & 0 & 0 & 0 \\ 0 & 1 & 0 & 0 \\ 0 & 0 & 1 & 0 \\ 0 & 0 & 0 & 1 \end{bmatrix}$

61. False. If A is $m \times n$, then L is $m \times m$ and U is $m \times n$. (See answer to 19.)

63. False. We would also need that the diagonal entries of A all be non-zero, so that row exchanges are not necessary.

65. False. For example $A = \begin{bmatrix} 0 & 0 \\ 0 & 0 \end{bmatrix} = \begin{bmatrix} 1 & 0 \\ 0 & 1 \end{bmatrix} \begin{bmatrix} 0 & 0 \\ 0 & 0 \end{bmatrix}$ is one LU factorization,

$\begin{bmatrix} 0 & 0 \\ 0 & 0 \end{bmatrix} = \begin{bmatrix} 1 & 0 \\ 1 & 1 \end{bmatrix} \begin{bmatrix} 0 & 0 \\ 0 & 0 \end{bmatrix}$ is another.

67. False. If $E_1 = \begin{bmatrix} 0 & 1 & 0 \\ 1 & 0 & 0 \\ 0 & 0 & 1 \end{bmatrix}$ and $E_2 = \begin{bmatrix} 1 & 0 & 0 \\ 0 & 0 & 1 \\ 0 & 1 & 0 \end{bmatrix}$, then $E_1 E_2 = \begin{bmatrix} 0 & 1 & 0 \\ 1 & 0 & 0 \\ 0 & 0 & 1 \end{bmatrix} \begin{bmatrix} 1 & 0 & 0 \\ 0 & 0 & 1 \\ 0 & 1 & 0 \end{bmatrix}$

$= \begin{bmatrix} 0 & 0 & 1 \\ 1 & 0 & 0 \\ 0 & 1 & 0 \end{bmatrix}$, but $E_2 E_1 = \begin{bmatrix} 1 & 0 & 0 \\ 0 & 0 & 1 \\ 0 & 1 & 0 \end{bmatrix} \begin{bmatrix} 0 & 1 & 0 \\ 1 & 0 & 0 \\ 0 & 0 & 1 \end{bmatrix} = \begin{bmatrix} 0 & 1 & 0 \\ 0 & 0 & 1 \\ 1 & 0 & 0 \end{bmatrix} \neq E_1 E_2$.

69. Let $D = [a_{ij}]$ be an $n \times n$ diagonal matrix and $U = [u_{ij}]$ an $m \times n$ upper triangular matrix. Let $B = [b_{ij}] = DU$. Then

$$b_{ij} = d_{i1}u_{1j} + d_{i2}u_{2j} + \cdots + d_{in}u_{nj}$$
$$= d_{ii}u_{ij}$$

since $d_{kj} = 0$ if $k \neq j$. Thus the jth column of row i of the product DU is given by the product of the diagonal d_{ii} in row i with the entry u_{ij} in row i and column j of U. Thus the entire i^{th} row of DU is given by multiplication with the diagonal entry d_{ii} in the ith row of D.

71. Suppose the row operation is the interchange of row k and row l. Then $E = [e_{ij}]$ where $e_{ii} = 1$ if $i \notin k, l$, $e_{kl} = e_{lk} = 1$, and $e_{ij} = 0$ otherwise. Then if $B = [b_{ij}] = EA$, we have

$$b_{kj} = e_{k1}a_{1j} + e_{k2}a_{2j} + \cdots + e_{kn}a_{nj}$$
$$= (1)a_{lj},$$

hence the kth row of EA is the lth row of A. Likewise $b_{lj} = (1)a_{kj}$, so the lth row of EA is the kth row of A. For $i \notin k, l$ we have

$$b_{ij} = e_{i1}a_{1j} + e_{i2}a_{2j} + \cdots + e_{in}a_{nj}$$
$$= (1)a_{ij},$$

and the ith row of EA is the ith row of A. Hence the matrix EA is the same matrix as obtained by applying the row operation $R_i \Leftrightarrow R_j$ to A.

Now suppose the row operation is the multiplication of row k by $c \neq 0$. Then $E = [e_{ij}]$ where $e_{ii} = 1$ if $i \neq k$, $e_{kk} = c$, and $e_{ij} = 0$ otherwise. Then if $B = [b_{ij}] = EA$, we have

$$b_{kj} = e_{k1}a_{1j} + e_{k2}a_{2j} + \cdots + e_{kn}a_{nj}$$
$$= (c)a_{kj},$$

and the kth row of EA is c times the kth row of A. For $i \neq k$ we have

$$b_{ij} = e_{i1}a_{1j} + e_{i2}a_{2j} + \cdots + e_{in}a_{nj}$$
$$= (1)a_{ij},$$

and the ith row of EA is the ith row of A. Hence the matrix EA is the same matrix as obtained by applying the row operation $cR_k \Rightarrow R_k$ to A.

Finally suppose the row operation is the multiplication of row k by c added to row $l \neq k$ to produce a new row l, $cR_k + R_l \Rightarrow R_l$. Then $E = [e_{ij}]$ where $e_{ii} = 1$, $e_{lk} = c$, and $e_{ij} = 0$ otherwise. Then if $B = [b_{ij}] = EA$, we have

$$b_{lj} = e_{l1}a_{1j} + e_{l2}a_{2j} + \cdots + e_{ln}a_{nj}$$
$$= (c)a_{kj} + (1)a_{lj},$$

and the lth row of EA is c times the kth row of A added to the lth row of A. For $i \neq l$,

$$b_{ij} = e_{i1}a_{1j} + e_{i2}a_{2j} + \cdots + e_{in}a_{nj}$$
$$= (1)a_{ij},$$

and the ith row of EA is the ith row of A. Hence the matrix EA is the same matrix as obtained by applying the row operation $cR_k + R_l \Rightarrow R_l$ to A.

Thus, in each case the row operation applied to A is the same matrix as EA, where E is the matrix obtained from the row operation applied to I_n.

73. Using a computer algebra system, it is determined that A does not have a LU factorization. Also note that A can not be reduced to the a lower triangular matrix without interchanging rows.

75. Using a computer algebra system, we determine that $L = \begin{bmatrix} 1 & 0 & 0 & 0 \\ \frac{1}{2} & 1 & 0 & 0 \\ 1 & 0 & 1 & 0 \\ \frac{3}{2} & \frac{2}{7} & -\frac{1}{5} & 1 \end{bmatrix}$,

$$U = \begin{bmatrix} 10 & 2 & 0 & -4 & 2 \\ 0 & 0 & -14 & 7 & 21 \\ 0 & 0 & 0 & -5 & 0 \\ 0 & 0 & 0 & 0 & -16 \end{bmatrix}.$$

3.5   Markov Chains

1. Stochastic, since each entry is nonnegative and all column sums are one.

3. Stochastic, since each entry is nonnegative and all column sums are one.

5. Setting column sums equal to one, we obtain $a = 0.35$ and $b = 0.55$.

7. Setting column sums equal to one, we obtain $a = \frac{8}{13}$, $b = \frac{1}{7}$, and $c = \frac{1}{10}$.

9. Setting column and row sums equal to one, we obtain $a = 0.7$ and $b = 0.7$.

11. Setting column and row sums equal to one, we first obtain $a = 0.5$ (from row 1 or column 2), $c = 0.5$ (from column 3), and $d = 0.4$ (from row 3). Then we obtain $b = 0.4$ from either row 2 or column 1.

13. $\mathbf{x}_1 = A\mathbf{x}_0 = \begin{bmatrix} 0.2 & 0.6 \\ 0.8 & 0.4 \end{bmatrix} \begin{bmatrix} 0.2 \\ 0.8 \end{bmatrix} = \begin{bmatrix} 0.52 \\ 0.48 \end{bmatrix}$, $\mathbf{x}_2 = A\mathbf{x}_1 = \begin{bmatrix} 0.2 & 0.6 \\ 0.8 & 0.4 \end{bmatrix} \begin{bmatrix} 0.52 \\ 0.48 \end{bmatrix} = \begin{bmatrix} 0.392 \\ 0.608 \end{bmatrix}$,

$\mathbf{x}_3 = A\mathbf{x}_2 = \begin{bmatrix} 0.2 & 0.6 \\ 0.8 & 0.4 \end{bmatrix} \begin{bmatrix} 0.392 \\ 0.608 \end{bmatrix} = \begin{bmatrix} 0.4432 \\ 0.5568 \end{bmatrix}$.

15. $\mathbf{x}_1 = A\mathbf{x}_0 = \begin{bmatrix} \frac{1}{3} & \frac{2}{5} \\ \frac{2}{3} & \frac{3}{5} \end{bmatrix} \begin{bmatrix} \frac{1}{2} \\ \frac{1}{2} \end{bmatrix} = \begin{bmatrix} \frac{11}{30} \\ \frac{19}{30} \end{bmatrix}$, $\mathbf{x}_2 = A\mathbf{x}_1 = \begin{bmatrix} \frac{1}{3} & \frac{2}{5} \\ \frac{2}{3} & \frac{3}{5} \end{bmatrix} \begin{bmatrix} \frac{11}{30} \\ \frac{19}{30} \end{bmatrix} = \begin{bmatrix} \frac{169}{450} \\ \frac{281}{450} \end{bmatrix}$, $\mathbf{x}_3 = A\mathbf{x}_2 =$

$\begin{bmatrix} \frac{1}{3} & \frac{2}{5} \\ \frac{2}{3} & \frac{3}{5} \end{bmatrix} \begin{bmatrix} \frac{169}{450} \\ \frac{281}{450} \end{bmatrix} = \begin{bmatrix} \frac{2531}{6750} \\ \frac{4219}{6750} \end{bmatrix}$.

17. Solve $(A - I)\mathbf{x} = \mathbf{0}$ by row-reducing the augmented matrix.

$$\begin{bmatrix} -0.2 & 0.5 & 0 \\ 0.2 & -0.5 & 0 \end{bmatrix} \underset{\sim}{\overset{R_1 + R_2 \Rightarrow R_2}{}} \begin{bmatrix} -0.2 & 0.5 & 0 \\ 0 & 0 & 0 \end{bmatrix}$$

and we obtain $\mathbf{x} = s \begin{bmatrix} 2.5 \\ 1 \end{bmatrix}$. Setting the column sum of $\mathbf{x}$ equal to one, we need $s = \frac{1}{3.5}$, and so

$\mathbf{x} = \frac{1}{3.5} \begin{bmatrix} 2.5 \\ 1 \end{bmatrix} = \begin{bmatrix} 0.71429 \\ 0.28571 \end{bmatrix}$.

19. Solve $(A - I)\mathbf{x} = \mathbf{0}$ by row-reducing the augmented matrix.

$$\begin{bmatrix} -0.6 & 0.5 & 0.3 & 0 \\ 0.2 & -0.7 & 0.4 & 0 \\ 0.4 & 0.2 & -0.7 & 0 \end{bmatrix} \underset{\sim}{\overset{(1/3)R_1 + R_2 \Rightarrow R_2}{\overset{(2/3)R_1 + R_3 \Rightarrow R_3}{}}} \begin{bmatrix} -0.6 & 0.5 & 0.3 & 0 \\ 0 & -0.53333 & 0.5 & 0 \\ 0 & 0.53333 & -0.5 & 0 \end{bmatrix}$$

$$\underset{\sim}{\overset{R_2 + R_3 \Rightarrow R_3}{}} \begin{bmatrix} -0.6 & 0.5 & 0.3 & 0 \\ 0 & -0.53333 & 0.5 & 0 \\ 0 & 0 & 0 & 0 \end{bmatrix}$$

and we obtain $\mathbf{x} = s \begin{bmatrix} 1.2813 \\ 0.93751 \\ 1 \end{bmatrix}$. Setting the column sum of $\mathbf{x}$ equal to one, we need $s = \frac{1}{3.2188}$, and

so $\mathbf{x} = \frac{1}{3.2188} \begin{bmatrix} 1.2813 \\ 0.93751 \\ 1 \end{bmatrix} = \begin{bmatrix} 0.39807 \\ 0.29126 \\ 0.31067 \end{bmatrix}$.

21. A is not regular since every power A^k of an upper triangular matrix will be upper triangular, and hence will have a zero entry.

23. A is not regular since every power A^k of a block lower triangular matrix will be block lower triangular, and hence will have a zero entry.

25. $A = \begin{bmatrix} 0.1 & 0.1 & 0.1 & 0.1 \\ 0.2 & 0.2 & 0.2 & 0.2 \\ 0.3 & 0.3 & 0.3 & 0.3 \\ 0.4 & 0.4 & 0.4 & 0.4 \end{bmatrix}$.

27. $A = \begin{bmatrix} \frac{1}{3} & \frac{2}{3} \\ \frac{2}{3} & \frac{1}{3} \end{bmatrix}$.

29. $A = \begin{bmatrix} 1 & 0 & 0 \\ 0 & 0 & 1 \\ 0 & 1 & 0 \end{bmatrix}$, $\mathbf{x}_0 = \begin{bmatrix} 0 \\ 1 \\ 0 \end{bmatrix}$. Then $\mathbf{x}_1 = \begin{bmatrix} 1 & 0 & 0 \\ 0 & 0 & 1 \\ 0 & 1 & 0 \end{bmatrix} \begin{bmatrix} 0 \\ 1 \\ 0 \end{bmatrix} = \begin{bmatrix} 0 \\ 0 \\ 1 \end{bmatrix}$,

$\mathbf{x}_2 = \begin{bmatrix} 1 & 0 & 0 \\ 0 & 0 & 1 \\ 0 & 1 & 0 \end{bmatrix} \begin{bmatrix} 0 \\ 0 \\ 1 \end{bmatrix} = \begin{bmatrix} 0 \\ 1 \\ 0 \end{bmatrix} = \mathbf{x}_0$, and we obtain a Markov chain which cycles between $\mathbf{x}_1$ and

$\mathbf{x}_0$, and therefore does not converge to a steady-state vector.

31. False. For example, $A = \begin{bmatrix} 1 & 1 \\ 0 & 0 \end{bmatrix}$ is stochastic, but $A^T = \begin{bmatrix} 1 & 0 \\ 1 & 0 \end{bmatrix}$ is not.

33. False. If $A = \begin{bmatrix} 1 & 0 \\ 0 & 1 \end{bmatrix}$ and $B = \begin{bmatrix} 1 & 1 \\ 0 & 0 \end{bmatrix}$ then A and B are stochastic, but $AB^T = \begin{bmatrix} 1 & 0 \\ 1 & 0 \end{bmatrix}$ is not.

35. False. If $A = \begin{bmatrix} 0 & 1 \\ 1 & 0 \end{bmatrix}$ and $\mathbf{x}_0 = \begin{bmatrix} 1 \\ 0 \end{bmatrix}$, then the Markov chain cycles: $\begin{bmatrix} 1 \\ 0 \end{bmatrix}, \begin{bmatrix} 0 \\ 1 \end{bmatrix}, \begin{bmatrix} 1 \\ 0 \end{bmatrix}, \ldots$, and hence does not converge.

37. Let A be a stochastic matrix and $\mathbf{x}$ an initial state vector. Let $\mathbf{y} = \begin{bmatrix} 1 & 1 & \cdots & 1 \end{bmatrix}$. Since each column of A has sum one, we have $\mathbf{y}A = \mathbf{y}$. Thus $\mathbf{y}(A\mathbf{x}) = (\mathbf{y}A)\mathbf{x} = \mathbf{y}\mathbf{x} = 1$, since the sum of the entries of $\mathbf{x}$ is one. This shows that the sum of the entries of $A\mathbf{x}$ is one, and we may conclude that each state vector is a probability vector.

39. We show A^k is stochastic using induction on the power k. If $k = 1$, then $A^1 = A$ is stochastic. Assume A^{k-1} is stochastic. Then $A^k = A\left(A^{k-1}\right)$ is the product of stochastic matrices, and hence stochastic. Thus A^k is stochastic for every integer $k \geq 1$.

41. Since all column and row sums are one, we have $a + b = a + c = 1$, and thus $b = c = 1 - a$. Also, $b + d = 1$, so $d = 1 - b = 1 - (1 - a) = a$.

43. Consider $A^{k+1} = \left(A^k\right)A$, with $A^k = [b_{ij}]$. The entry in the ith row and j^{th} column of A^{k+1} will have the form

$$b_{i1}a_{1j} + b_{i2}a_{2j} + \cdots + b_{in}a_{nj}$$

None of these terms is negative, and at least one of these must be positive, since every column of A has at least one positive entry, and every entry in A^k is positive. Thus we conclude that every entry of A^{k+1} is positive. By induction we establish that A^{k+l} is regular for all $l \geq 1$, i.e. $A^{k+1}, A^{k+2}, \ldots$ all have strictly positive entries.

45. (a) Every entry of A is non-negative, and each column sum is one.

(b) We have $A = \begin{bmatrix} \alpha & 0 \\ 1-\alpha & 1 \end{bmatrix}$, $A^2 = \begin{bmatrix} \alpha & 0 \\ 1-\alpha & 1 \end{bmatrix}^2 = \begin{bmatrix} \alpha^2 & 0 \\ 1-\alpha^2 & 1 \end{bmatrix}$, A^3

$= \begin{bmatrix} \alpha & 0 \\ 1-\alpha & 1 \end{bmatrix} \begin{bmatrix} \alpha^2 & 0 \\ 1-\alpha^2 & 1 \end{bmatrix} = \begin{bmatrix} \alpha^3 & 0 \\ 1-\alpha^3 & 1 \end{bmatrix}$, and in general $A^k = \begin{bmatrix} \alpha^k & 0 \\ 1-\alpha^k & 1 \end{bmatrix}$. Since we have a zero entry for every A^k, A is not regular.

(c) $\lim_{k \to \infty} A^k = \lim_{k \to \infty} \begin{bmatrix} \alpha^k & 0 \\ 1-\alpha^k & 1 \end{bmatrix} = \begin{bmatrix} 0 & 0 \\ 1 & 1 \end{bmatrix}$, since for $0 < \alpha < 1$, $\lim_{k \to \infty} \alpha^k = 0$.

(d) $A \begin{bmatrix} 0 \\ 1 \end{bmatrix} = \begin{bmatrix} \alpha & 0 \\ 1-\alpha & 1 \end{bmatrix} \begin{bmatrix} 0 \\ 1 \end{bmatrix} = \begin{bmatrix} 0 \\ 1 \end{bmatrix}$. And $A \begin{bmatrix} x_1 \\ x_2 \end{bmatrix} = \begin{bmatrix} \alpha & 0 \\ 1-\alpha & 1 \end{bmatrix} \begin{bmatrix} x_1 \\ x_2 \end{bmatrix}$

$= \begin{bmatrix} \alpha x_1 \\ x_2 - x_1(\alpha - 1) \end{bmatrix} \neq \begin{bmatrix} x_1 \\ x_2 \end{bmatrix}$ unless $x_1 = 0$, since $\alpha \neq 0$. Thus $\begin{bmatrix} 0 \\ 1 \end{bmatrix}$ is the unique steady-state vector of A.

47. (a) $A = \begin{bmatrix} 0.9 & 0.15 \\ 0.1 & 0.85 \end{bmatrix}$

(b) $A^6 \begin{bmatrix} 1 \\ 0 \end{bmatrix} = \begin{bmatrix} 0.9 & 0.15 \\ 0.1 & 0.85 \end{bmatrix}^6 \begin{bmatrix} 1 \\ 0 \end{bmatrix} = \begin{bmatrix} 0.67119 \\ 0.32881 \end{bmatrix}$, so the probability that the sixth person in the chain hears the wrong news is 0.32881.

(c) $\mathbf{x} = \begin{bmatrix} 0.6 \\ 0.4 \end{bmatrix}$

49. (a) $A = \begin{bmatrix} 0.35 & 0.8 \\ 0.65 & 0.2 \end{bmatrix}$

(b) i. $A^2 \begin{bmatrix} 1 \\ 0 \end{bmatrix} = \begin{bmatrix} 0.35 & 0.8 \\ 0.65 & 0.2 \end{bmatrix}^2 \begin{bmatrix} 1 \\ 0 \end{bmatrix} = \begin{bmatrix} 0.6425 \\ 0.3575 \end{bmatrix}$, so the probability that she will go to McDonald's two Sundays from now is 0.3575.

   ii. $A^3 \begin{bmatrix} 1 \\ 0 \end{bmatrix} = \begin{bmatrix} 0.35 & 0.8 \\ 0.65 & 0.2 \end{bmatrix}^3 \begin{bmatrix} 1 \\ 0 \end{bmatrix} = \begin{bmatrix} 0.51088 \\ 0.48913 \end{bmatrix}$, so the probability that she will go to McDonald's two Sundays from now is 0.48913.

(c) $A^2 \begin{bmatrix} 0.4 \\ 0.6 \end{bmatrix} = \begin{bmatrix} 0.35 & 0.8 \\ 0.65 & 0.2 \end{bmatrix}^2 \begin{bmatrix} 0.4 \\ 0.6 \end{bmatrix} = \begin{bmatrix} 0.521 \\ 0.479 \end{bmatrix}$, so the probability that his third fast food experience will be at Krusty's will be 0.521.

(d) $\mathbf{x} = \begin{bmatrix} 0.55173 \\ 0.44827 \end{bmatrix}$

51. (a) The transition matrix is $A = \begin{bmatrix} 0.4 & 0.1 & 0.2 \\ 0.3 & 0.7 & 0.7 \\ 0.3 & 0.2 & 0.1 \end{bmatrix}$.

The probability that a book is at C after two more circulations is determined by

$A^2 \begin{bmatrix} 1 \\ 0 \\ 0 \end{bmatrix} = \begin{bmatrix} 0.4 & 0.1 & 0.2 \\ 0.3 & 0.7 & 0.7 \\ 0.3 & 0.2 & 0.1 \end{bmatrix}^2 \begin{bmatrix} 1 \\ 0 \\ 0 \end{bmatrix} = \begin{bmatrix} 0.25 \\ 0.54 \\ 0.21 \end{bmatrix}$; thus the probability is 0.21.

(b) $A^3 \begin{bmatrix} 0 \\ 1 \\ 0 \end{bmatrix} = \begin{bmatrix} 0.4 & 0.1 & 0.2 \\ 0.3 & 0.7 & 0.7 \\ 0.3 & 0.2 & 0.1 \end{bmatrix}^3 \begin{bmatrix} 0 \\ 1 \\ 0 \end{bmatrix} = \begin{bmatrix} 0.164 \\ 0.64 \\ 0.196 \end{bmatrix}$, so the probability that the book is at B after three more circulations is 0.64.

(c) $\mathbf{x} = \begin{bmatrix} 0.17105 \\ 0.63158 \\ 0.19737 \end{bmatrix}$

53. $A^9 \mathbf{x}_0 = \begin{bmatrix} 0.2 & 0.3 & 0.1 & 0.4 \\ 0.3 & 0.5 & 0.6 & 0.2 \\ 0.1 & 0.1 & 0.2 & 0.2 \\ 0.4 & 0.1 & 0.1 & 0.2 \end{bmatrix}^9 \begin{bmatrix} 0.25 \\ 0.25 \\ 0.25 \\ 0.25 \end{bmatrix} = \begin{bmatrix} 0.266666636 \\ 0.399999867 \\ 0.133333363 \\ 0.200000134 \end{bmatrix}$

and $A^{10} \mathbf{x}_0 = \begin{bmatrix} 0.2 & 0.3 & 0.1 & 0.4 \\ 0.3 & 0.5 & 0.6 & 0.2 \\ 0.1 & 0.1 & 0.2 & 0.2 \\ 0.4 & 0.1 & 0.1 & 0.2 \end{bmatrix}^{10} \begin{bmatrix} 0.25 \\ 0.25 \\ 0.25 \\ 0.25 \end{bmatrix} = \begin{bmatrix} 0.266666677 \\ 0.399999968 \\ 0.133333349 \\ 0.200000004 \end{bmatrix}$ agree to six decimal places, and

$k = 9$ is the smallest integer for which this is true. The steady-state vector is $\mathbf{x} = \begin{bmatrix} \frac{4}{15} \\ \frac{2}{5} \\ \frac{2}{15} \\ \frac{1}{5} \end{bmatrix}$.

55. $A \begin{bmatrix} 0 \\ 1 \\ 0 \\ 0 \end{bmatrix} = \begin{bmatrix} 0 \\ 1 \\ 0 \\ 0 \end{bmatrix}$ so $\begin{bmatrix} 0 \\ 1 \\ 0 \\ 0 \end{bmatrix}$ has itself as its steady-state vector. Also $A \begin{bmatrix} 0 \\ 0 \\ 1 \\ 0 \end{bmatrix} = \begin{bmatrix} 0 \\ 0 \\ 1 \\ 0 \end{bmatrix}$ so $\begin{bmatrix} 0 \\ 0 \\ 1 \\ 0 \end{bmatrix}$ has

itself as its steady-state vector.

Chapter 4

Subspaces

4.1 Introduction to Subspaces

1. Let S be the set of vectors of the form $\begin{bmatrix} a \\ 0 \\ b \end{bmatrix}$. Letting $a = 0$ and $b = 0$, we see that $\mathbf{0} \in S$. Suppose $\mathbf{u}$ and $\mathbf{v}$ are in S, then $\mathbf{u} = \begin{bmatrix} a_1 \\ 0 \\ b_1 \end{bmatrix}$ and $\mathbf{v} = \begin{bmatrix} a_2 \\ 0 \\ b_2 \end{bmatrix}$. Thus $\mathbf{u} + \mathbf{v} = \begin{bmatrix} a_1 \\ 0 \\ b_1 \end{bmatrix} + \begin{bmatrix} a_2 \\ 0 \\ b_2 \end{bmatrix} = \begin{bmatrix} a_1 + a_2 \\ 0 \\ b_1 + b_2 \end{bmatrix} \in S$.

Let $r \in \mathbf{R}$ and $\mathbf{u} \in S$, then $r\mathbf{u} = r\begin{bmatrix} a_1 \\ 0 \\ b_1 \end{bmatrix} = \begin{bmatrix} ra_1 \\ 0 \\ rb_1 \end{bmatrix} \in S$, and we conclude that S is a subspace.

(*Alternatively, we have* $S = \text{span} \left\{ \begin{bmatrix} 1 \\ 0 \\ 0 \end{bmatrix}, \begin{bmatrix} 0 \\ 0 \\ 1 \end{bmatrix} \right\}$, *and hence S is a subspace.*)

3. Not a subspace, because $0 + 0 \neq 1$, and thus $\mathbf{0}$ is not in this set.

5. Not a subspace, since $\mathbf{0}$ is not in this subset because the second component of $\mathbf{0}$ is $0 \neq 1$.

7. Not a subspace. Let $r = \frac{1}{2}$ and $\mathbf{u} = \begin{bmatrix} 1 \\ 0 \end{bmatrix}$. Then $\mathbf{u}$ belongs to the set, but $r\mathbf{u} = \begin{bmatrix} 1/2 \\ 1/2 \end{bmatrix}$ does not.

9. Not a subspace. Let $\mathbf{u} = \begin{bmatrix} 1 \\ 0 \\ 0 \end{bmatrix}$ and $\mathbf{v} = \begin{bmatrix} 0 \\ 1 \\ 1 \end{bmatrix}$. Then both $\mathbf{u}$ and $\mathbf{v}$ belong to the subset, but

$\mathbf{u} + \mathbf{v} = \begin{bmatrix} 1 \\ 1 \\ 1 \end{bmatrix}$ does not, since $1(1)(1) = 1 \neq 0$.

11. Not a subspace. Let $r = -1$ and $\mathbf{u} = \begin{bmatrix} 1 \\ 0 \\ 0 \end{bmatrix}$. Then $\mathbf{u}$ is in the subset, but $r\mathbf{u} = \begin{bmatrix} -1 \\ 0 \\ 0 \end{bmatrix}$ is not.

13. Not a subspace. Let $r = -1$ and $\mathbf{u} = \begin{bmatrix} 1 \\ 2 \end{bmatrix}$. Then $\mathbf{u}$ is in the subset since $1 \leq 2$, but $r\mathbf{u} = \begin{bmatrix} -1 \\ -2 \end{bmatrix}$ is not, since $-1 \nleq -2$.

15. This subset is equal to $\text{null}\left(\begin{bmatrix} 1 & \cdots & 1 \end{bmatrix}\right)$, which is a subspace.

17. It would not be closed under scalar multiplication, since there is a vector $\begin{bmatrix} a \\ 0 \end{bmatrix}$, where $0 < a$, which is

in the set, but $r \begin{bmatrix} a \\ 0 \end{bmatrix} = \begin{bmatrix} ra \\ 0 \end{bmatrix}$ would not be in the set for r sufficiently large. One might also note

that $\begin{bmatrix} b \\ b \end{bmatrix}$ and $\begin{bmatrix} b \\ -b \end{bmatrix}$ are in the set, but the sum $\begin{bmatrix} 2b \\ 0 \end{bmatrix}$ is not, for b sufficiently large.

19. It would not be closed under vector addition. For example, $\begin{bmatrix} 1 \\ 0 \end{bmatrix}$ and $\begin{bmatrix} 0 \\ -1 \end{bmatrix}$ are in the region, but

the sum $\begin{bmatrix} 1 \\ -1 \end{bmatrix}$ is not.

21. We row reduce

$$\begin{bmatrix} 1 & -3 \\ 0 & 1 \end{bmatrix} \sim \begin{bmatrix} 1 & 0 \\ 0 & 1 \end{bmatrix}.$$

Thus $A\mathbf{x} = \mathbf{0}$ has the trivial solution $\mathbf{x} = \begin{bmatrix} 0 \\ 0 \end{bmatrix}$, and thus $\text{null}(A) = \left\{ \begin{bmatrix} 0 \\ 0 \end{bmatrix} \right\}$.

23. A is row-reduced, and we see that $A\mathbf{x} = \mathbf{0}$ has solutions of the form

$$\mathbf{x} = s \begin{bmatrix} 5 \\ -2 \\ 1 \end{bmatrix},$$

and thus $\text{null}(A) = \text{span} \left\{ \begin{bmatrix} 5 \\ -2 \\ 1 \end{bmatrix} \right\}$.

25. We row reduce

$$\begin{bmatrix} 1 & -2 & 2 \\ -2 & 5 & -7 \end{bmatrix} \sim \begin{bmatrix} 1 & 0 & -4 \\ 0 & 1 & -3 \end{bmatrix}.$$

Thus $A\mathbf{x} = \mathbf{0}$ has solutions of the form

$$\mathbf{x} = s \begin{bmatrix} 4 \\ 3 \\ 1 \end{bmatrix},$$

and therefore $\text{null}(A) = \text{span} \left\{ \begin{bmatrix} 4 \\ 3 \\ 1 \end{bmatrix} \right\}$.

27. We row reduce

$$\begin{bmatrix} 1 & 3 \\ -2 & 1 \\ 3 & 2 \end{bmatrix} \sim \begin{bmatrix} 1 & 0 \\ 0 & 1 \\ 0 & 0 \end{bmatrix}.$$

Thus $A\mathbf{x} = \mathbf{0}$ has the trivial solution $\mathbf{x} = \begin{bmatrix} 0 \\ 0 \end{bmatrix}$, and thus $\text{null}(A) = \left\{ \begin{bmatrix} 0 \\ 0 \end{bmatrix} \right\}$.

29. We row reduce

$$\begin{bmatrix} 1 & -1 & 1 \\ 0 & 1 & 3 \\ 0 & 0 & 3 \end{bmatrix} \sim \begin{bmatrix} 1 & 0 & 0 \\ 0 & 1 & 0 \\ 0 & 0 & 1 \end{bmatrix}.$$

Thus $A\mathbf{x} = \mathbf{0}$ has the trivial solution $\mathbf{x} = \begin{bmatrix} 0 \\ 0 \\ 0 \end{bmatrix}$, and thus $\text{null}(A) = \left\{ \begin{bmatrix} 0 \\ 0 \\ 0 \end{bmatrix} \right\}$.

31. We row reduce

$$\begin{bmatrix} 1 & 1 & -2 & 1 \\ 0 & 1 & 1 & -1 \\ 0 & 0 & 0 & 2 \end{bmatrix} \sim \begin{bmatrix} 1 & 0 & -3 & 0 \\ 0 & 1 & 1 & 0 \\ 0 & 0 & 0 & 1 \end{bmatrix}.$$

Thus $A\mathbf{x} = \mathbf{0}$ has solutions of the form

$$\mathbf{x} = s \begin{bmatrix} 3 \\ -1 \\ 1 \\ 0 \end{bmatrix},$$

and therefore $\operatorname{null}(A) = \operatorname{span} \left\{ \begin{bmatrix} 3 \\ -1 \\ 1 \\ 0 \end{bmatrix} \right\}$.

33. $A\mathbf{b} = \begin{bmatrix} 1 & -2 \\ -3 & -1 \end{bmatrix} \begin{bmatrix} 2 \\ 1 \end{bmatrix} = \begin{bmatrix} 0 \\ -7 \end{bmatrix} \neq \mathbf{0}$, so $\mathbf{b} \notin \ker(T)$. We row-reduce to determine a solution of $A\mathbf{x} = \mathbf{c}$,

$$\begin{bmatrix} 1 & -2 & 4 \\ -3 & -1 & -7 \end{bmatrix} \sim \begin{bmatrix} 1 & 0 & \frac{18}{7} \\ 0 & 1 & -\frac{5}{7} \end{bmatrix}.$$

Thus $A \begin{bmatrix} \frac{18}{7} \\ -\frac{5}{7} \end{bmatrix} = \mathbf{c}$, so $\mathbf{c} \in \operatorname{range}(T)$.

35. $A\mathbf{b} = \begin{bmatrix} 4 & -2 \\ 1 & 3 \\ 2 & 7 \end{bmatrix} \begin{bmatrix} -5 \\ 2 \end{bmatrix} = \begin{bmatrix} -24 \\ 1 \\ 4 \end{bmatrix} \neq \mathbf{0}$, so $\mathbf{b} \notin \ker(T)$. Since the range of T is a subset of $\mathbf{R}^3$, and $\mathbf{c} \in \mathbf{R}^2$, $\mathbf{c} \notin \operatorname{range}(T)$.

37. For example, $S = \left\{ \begin{bmatrix} x \\ y \end{bmatrix} : x > 0 \right\}$.

39. For example, $S_1 = \left\{ \begin{bmatrix} x \\ 0 \\ 0 \end{bmatrix} : x \geq 0 \right\}$ and $S_2 = \left\{ \begin{bmatrix} x \\ 0 \\ 0 \end{bmatrix} : x < 0 \right\}$, then $S_1 \cup S_2 = \operatorname{span} \left\{ \begin{bmatrix} 1 \\ 0 \\ 0 \end{bmatrix} \right\}$ is a subspace.

41. Let $T(\mathbf{x}) = A\mathbf{x}$, where $A = \begin{bmatrix} 1 & 0 \\ 1 & 0 \end{bmatrix}$, then $\operatorname{range}(T) = \operatorname{span} \left\{ \begin{bmatrix} 1 \\ 1 \end{bmatrix} \right\}$.

43. Let $T(\mathbf{x}) = A\mathbf{x}$, where $A = I_3$, then $\operatorname{range}(T) = \mathbf{R}^3$.

45. True. Since $A\mathbf{0} = \mathbf{0} \neq \mathbf{b}$, $\mathbf{0}$ is not a solution to $A\mathbf{x} = \mathbf{b}$, and hence the set of solutions is not a subspace.

47. True, by Theorem 4.3.

49. False, because $\ker(T)$ is a subspace of the domain $\mathbf{R}^5$, not the codomain $\mathbf{R}^8$.

51. True, by Theorem 4.5.

53. True. Proof: Let S_1 and S_2 be two subspaces of $\mathbf{R}^n$. Since $\mathbf{0}$ must belong to both S_1 and S_2, $\mathbf{0} \in S_1 \cap S_2$. Now let $\mathbf{u}$ and $\mathbf{v}$ be two vectors in $S_1 \cap S_2$. Since $\mathbf{u}$ and $\mathbf{v}$ are both in S_1, and S_1 is a subspace, $\mathbf{u} + \mathbf{v} \in \mathbf{S}_1$. Likewise, since $\mathbf{u}$ and $\mathbf{v}$ are both in S_2, and S_2 is a subspace, $\mathbf{u} + \mathbf{v} \in \mathbf{S}_2$. Thus $\mathbf{u} + \mathbf{v} \in S_1 \cap S_2$. Now let $\mathbf{u} \in S_1 \cap S_2$ and $r \in \mathbf{R}$. Since $\mathbf{u} \in S_1$ and S_1 is a subspace, $r\mathbf{u} \in S_1$. Likewise, since $\mathbf{u} \in S_2$ and S_2 is a subspace, $r\mathbf{u} \in S_2$. Thus $r\mathbf{u} \in S_1 \cap S_2$, and we conclude that $S_1 \cap S_2$ is a subspace.

55. True. Proof: Since S_1 and S_2 are two subspaces of $\mathbf{R}^n$, $\mathbf{0}$ must belong to both S_1 and S_2, and hence $\mathbf{0} - \mathbf{0} = \mathbf{0} \in S$. Now let $\mathbf{u}$ and $\mathbf{v}$ be two vectors in S. Then there exist $\mathbf{u}_1 \in S_1$ and $\mathbf{u}_2 \in S_2$ with $\mathbf{u} = \mathbf{u}_1 - \mathbf{u}_2$. Likewise, there exist $\mathbf{v}_1 \in S_1$ and $\mathbf{v}_2 \in S_2$ with $\mathbf{v} = \mathbf{v}_1 - \mathbf{v}_2$, and thus $\mathbf{u} + \mathbf{v} = (\mathbf{u}_1 - \mathbf{u}_2) + (\mathbf{v}_1 - \mathbf{v}_2) = (\mathbf{u}_1 + \mathbf{v}_1) - (\mathbf{u}_2 + \mathbf{v}_2)$. Since S_1 is a subspace, $\mathbf{u}_1 + \mathbf{v}_1 \in S_1$, and since S_2 is a subspace $\mathbf{u}_2 + \mathbf{v}_2 \in S_2$ Thus, $\mathbf{u} + \mathbf{v} \in S$. Now let $\mathbf{u} \in S$ and $r \in \mathbf{R}$. Since $\mathbf{u} \in S$ there exist $\mathbf{u}_1 \in S_1$ and $\mathbf{u}_2 \in S_2$ with $\mathbf{u} = \mathbf{u}_1 - \mathbf{u}_2$, and so $r\mathbf{u} = r\mathbf{u}_1 - r\mathbf{u}_2$. Since S_1 is a subspace, $r\mathbf{u}_1 \in S_1$, and likewise since S_2 is a subspace, $r\mathbf{u}_2 \in S_2$. Hence, $r\mathbf{u} \in S$. We conclude that S is a subspace.

57. False. If $S \neq \{\mathbf{0}\}$, then there exists $\mathbf{v} \in S$ with $\mathbf{v} \neq \mathbf{0}$. Since S is a subspace, $r\mathbf{v} \in S$ for all scalars r. Each $r\mathbf{v}$ is a distinct vector, for if $r_1\mathbf{v} = r_2\mathbf{v}$, then $(r_1 - r_2)\mathbf{v} = \mathbf{0}$, and since $\mathbf{v} \neq \mathbf{0}$ we must have $r_1 - r_2 = 0$, and thus $r_1 = r_2$. Thus, S must contain infinitely many vectors.

59. False. For example, let $S_1 = \{x \in \mathbf{R} : x < 0\}$ and $S_2 = \{x \in \mathbf{R} : x \geq 0\}$. Then S_1 and S_2 are not subspaces, but the union $S_1 \cup S_2 = \mathbf{R}$ is a subspace.

61. Let S be a subspace of $\mathbf{R}$, with $S \neq \{\mathbf{0}\}$. Then there exists $x \in S$ with $x \neq 0$. Let $y \in R$. Set $r = \frac{y}{x}$, then $rx = \left(\frac{y}{x}\right) x = y$, and therefore since S is a subspace, $y \in S$. Since y was arbitrary, it follows that $S = \mathbf{R}$.

63. Since $A\mathbf{0} = \mathbf{0} \neq \mathbf{b}$, $\mathbf{0}$ does not belong to the set of solutions to $A\mathbf{x} = \mathbf{b}$, and therefore this set is not a subspace.

65. The subspaces of $\mathbf{R}^3$ consist of $\{\mathbf{0}\}$, $\mathbf{R}^3$, all lines which contain the origin, and all planes which contain the origin.

67. Since $A\mathbf{0} = \mathbf{0} \neq \mathbf{y}$, $\mathbf{0}$ does not belong to the set of solutions to $A\mathbf{x} = \mathbf{y}$, and therefore this set is not a subspace of $\mathbf{R}^m$.

69. Suppose $\ker(T) = \{\mathbf{0}\}$. Then $T(\mathbf{x}) = A\mathbf{x} = \mathbf{0}$ has the unique solution, $\mathbf{x} = \mathbf{0}$. If $A = [\ \mathbf{a}_1 \ \cdots \ \mathbf{a}_n\ ]$, then $c_1\mathbf{a}_1 + \cdots + c_n\mathbf{a}_n = A\mathbf{c} = \mathbf{0}$ implies $\mathbf{c} = \mathbf{0}$, and thus every $c_i = 0$. Hence the columns of A are linearly independent. Now suppose the columns of $A = [\ \mathbf{a}_1 \ \cdots \ \mathbf{a}_n\ ]$ are linearly independent. If $A\mathbf{c} = c_1\mathbf{a}_1 + \cdots + c_n\mathbf{a}_n = \mathbf{0}$, then every $c_i = 0$. Thus, $\mathbf{c} = \mathbf{0}$, which shows that $T(\mathbf{x}) = A\mathbf{x} = \mathbf{0}$ has the unique solution, $\mathbf{x} = \mathbf{0}$. Thus, $\ker(T) = \{\mathbf{0}\}$.

71. Since $-\mathbf{v} = (-1)\mathbf{v}$ and S is closed under both addition and scalar multiplication, it follows that $\mathbf{u} + (-1)\mathbf{v}$ is in S and hence $\mathbf{u} - \mathbf{v}$ is in S.

73. We solve the linear system

$$
\begin{array}{rcrcrcll}
6x_1 & - & 2x_2 & - & x_3 & = & 0 & \text{(Carbon atoms)} \\
12x_1 & - & 6x_2 & & & = & 0 & \text{(Hydrogen atoms)} \\
6x_1 & - & x_2 & - & 2x_3 & = & 0 & \text{(Oxygen atoms)}
\end{array}
$$

The general solution to this system is

$$
\mathbf{x} = s \begin{bmatrix} 1 \\ 2 \\ 2 \end{bmatrix};
$$

where s is any real number. The set of solutions is given by

$$
\text{span}\left(\begin{bmatrix} 1 \\ 2 \\ 2 \end{bmatrix} \right).
$$

75. We solve the linear system

$$
\begin{array}{rcrcrcrcll}
x_1 & & & - & x_3 & & & = & 0 & \text{(Calcium atoms)} \\
2x_1 & & & & & - & x_4 & = & 0 & \text{(Oxygen atoms)} \\
2x_1 & + & x_2 & & & - & 2x_4 & = & 0 & \text{(Hydrogen atoms)} \\
& & x_2 & - & 2x_3 & & & = & 0 & \text{(Chlorine atoms)}
\end{array}
$$

The general solution to this system is

$$\mathbf{x} = s \begin{bmatrix} 1 \\ 2 \\ 1 \\ 2 \end{bmatrix};$$

where s is any real number. The set of solutions is given by

$$\text{span}\left(\begin{bmatrix} 1 \\ 2 \\ 1 \\ 2 \end{bmatrix} \right).$$

77. Using a computer algebra system, we determine

$$\text{null}\left(\begin{bmatrix} 1 & 7 & -2 & 14 & 0 \\ 3 & 0 & 1 & -2 & 3 \\ 6 & 1 & -1 & 0 & 4 \end{bmatrix} \right) = \text{span}\left(\begin{bmatrix} \frac{3}{7} \\ -\frac{13}{7} \\ \frac{5}{7} \\ 1 \\ 0 \end{bmatrix}, \begin{bmatrix} -\frac{43}{56} \\ -\frac{5}{56} \\ -\frac{39}{56} \\ 0 \\ 1 \end{bmatrix} \right).$$

79. Using a computer algebra system, we determine

$$\text{null}\left(\begin{bmatrix} 3 & 1 & 2 & 4 \\ 5 & 0 & 2 & -1 \\ 2 & 2 & 2 & 2 \\ -1 & 0 & 3 & 1 \\ 0 & 2 & 0 & 4 \end{bmatrix} \right) = \left\{ \begin{bmatrix} 0 \\ 0 \\ 0 \\ 0 \end{bmatrix} \right\}.$$

4.2   Basis and Dimension

1. Not a basis, since $\mathbf{u}_1$ and $\mathbf{u}_2$ are not linearly independent. Also, they do not span $\mathbf{R}^2$.

3. Not a basis, since three vectors in a two-dimensional space must be linearly dependent.

5. Row-reduce the matrix with the given vectors as rows,

$$\begin{bmatrix} 1 & -4 \\ -5 & 20 \end{bmatrix} \sim \begin{bmatrix} 1 & -4 \\ 0 & 0 \end{bmatrix}.$$

Thus a basis for S is given by the non-zero row vector, $\left\{ \begin{bmatrix} 1 \\ -4 \end{bmatrix} \right\}.$

7. Row-reduce the matrix with the given vectors as rows,

$$\begin{bmatrix} 1 & 3 & -2 \\ 2 & 4 & 1 \\ -1 & 1 & -8 \end{bmatrix} \sim \begin{bmatrix} 1 & 3 & -2 \\ 0 & -2 & 5 \\ 0 & 0 & 0 \end{bmatrix}.$$

Thus a basis for S is given by the non-zero row vectors, $\left\{ \begin{bmatrix} 1 \\ 3 \\ -2 \end{bmatrix}, \begin{bmatrix} 0 \\ -2 \\ 5 \end{bmatrix} \right\}.$

9. Row-reduce the matrix with the given vectors as rows,

$$\begin{bmatrix} 1 & -2 & 3 & -2 \\ 0 & 2 & -5 & 1 \\ 2 & -2 & 1 & -3 \end{bmatrix} \sim \begin{bmatrix} 1 & -2 & 3 & -2 \\ 0 & 2 & -5 & 1 \\ 0 & 0 & 0 & 0 \end{bmatrix}.$$

Thus a basis for S is given by the non-zero row vectors, $\left\{ \begin{bmatrix} 1 \\ -2 \\ 3 \\ -2 \end{bmatrix}, \begin{bmatrix} 0 \\ 2 \\ -5 \\ 1 \end{bmatrix} \right\}$.

11. Row-reduce the matrix with the given vectors as columns,

$$\begin{bmatrix} 1 & 4 \\ 3 & -12 \end{bmatrix} \sim \begin{bmatrix} 1 & 4 \\ 0 & -24 \end{bmatrix}.$$

A basis for S is given by columns 1 and 2 of the original matrix corresponding to the pivot columns of the row-reduced matrix. Hence a basis for S is $\left\{ \begin{bmatrix} 1 \\ 3 \end{bmatrix}, \begin{bmatrix} 4 \\ -12 \end{bmatrix} \right\}$.

13. Row-reduce the matrix with the given vectors as columns,

$$\begin{bmatrix} 1 & 0 & 3 \\ 2 & 1 & -2 \\ 4 & -3 & -1 \end{bmatrix} \sim \begin{bmatrix} 1 & 0 & 3 \\ 0 & 1 & -8 \\ 0 & 0 & -37 \end{bmatrix}.$$

A basis for S is given by columns 1, 2, and 3 of the original matrix corresponding to the pivot columns of the row-reduced matrix. Hence a basis for S is $\left\{ \begin{bmatrix} 1 \\ 2 \\ 4 \end{bmatrix}, \begin{bmatrix} 0 \\ 1 \\ -3 \end{bmatrix}, \begin{bmatrix} 3 \\ -2 \\ -1 \end{bmatrix} \right\}$.

15. Row-reduce the matrix with the given vectors as columns,

$$\begin{bmatrix} 1 & 2 & 0 \\ -1 & -5 & 1 \\ 0 & 9 & -3 \\ 2 & 7 & -1 \end{bmatrix} \sim \begin{bmatrix} 1 & 2 & 0 \\ 0 & -3 & 1 \\ 0 & 0 & 0 \\ 0 & 0 & 0 \end{bmatrix}.$$

A basis for S is given by columns 1 and 2 of the original matrix corresponding to the pivot columns of the row-reduced matrix. Hence a basis for S is $\left\{ \begin{bmatrix} 1 \\ -1 \\ 0 \\ 2 \end{bmatrix}, \begin{bmatrix} 2 \\ -5 \\ 9 \\ 7 \end{bmatrix} \right\}$.

17. Since the second vector is $\frac{-3}{2}$ times the first, we eliminate the second vector, and obtain the basis $\left\{ \begin{bmatrix} 2 \\ -6 \end{bmatrix} \right\}$. The dimension is 1.

19. The third vector is 3 times the first, and is eliminated as a dependent vector. Likewise, the second vector is 2 times the first and is eliminated, leaving the basis $\left\{ \begin{bmatrix} 1 \\ 1 \\ 1 \end{bmatrix} \right\}$. The dimension is 1.

21. The first vector is eliminated, as the zero vector is always linearly dependent. The other vectors are linearly independent since the corresponding matrix

$$\begin{bmatrix} 3 & 2 & 1 \\ 0 & 1 & 2 \\ 0 & 0 & 3 \end{bmatrix}$$

is row-reduced with 3 pivots. Hence a basis is $\left\{ \begin{bmatrix} 3 \\ 0 \\ 0 \end{bmatrix}, \begin{bmatrix} 2 \\ 1 \\ 0 \end{bmatrix}, \begin{bmatrix} 1 \\ 2 \\ 3 \end{bmatrix} \right\}$. The dimension is 3.

23. One extension is $\left\{ \begin{bmatrix} 1 \\ -3 \end{bmatrix}, \begin{bmatrix} 1 \\ 0 \end{bmatrix} \right\}$, as these two vectors are linearly independent.

25. One extension is $\left\{ \begin{bmatrix} 1 \\ 0 \\ 0 \end{bmatrix}, \begin{bmatrix} 0 \\ 1 \\ 0 \end{bmatrix}, \begin{bmatrix} -1 \\ 2 \\ 1 \end{bmatrix} \right\}$, as these three vectors are linearly independent.

27. To extend this set, we row-reduce

$$\begin{bmatrix} 1 & 2 & 1 & 0 & 0 \\ 3 & -1 & 0 & 1 & 0 \\ -2 & 0 & 0 & 0 & 1 \end{bmatrix} \sim \begin{bmatrix} 1 & 2 & 1 & 0 & 0 \\ 0 & -7 & -3 & 1 & 0 \\ 0 & 0 & \frac{2}{7} & \frac{4}{7} & 1 \end{bmatrix}.$$

The extended basis is given by columns 1, 2, and 3 of the original matrix corresponding to the pivot columns of the row-reduced matrix.

Hence our basis is $\left\{ \begin{bmatrix} 1 \\ 3 \\ -2 \end{bmatrix}, \begin{bmatrix} 2 \\ -1 \\ 0 \end{bmatrix}, \begin{bmatrix} 1 \\ 0 \\ 0 \end{bmatrix} \right\}$.

29. We row reduce

$$\begin{bmatrix} -2 & -5 \\ 1 & 3 \end{bmatrix} \sim \begin{bmatrix} -2 & -5 \\ 0 & -\frac{1}{2} \end{bmatrix}.$$

Thus $A\mathbf{x} = \mathbf{0}$ has the trivial solution $\mathbf{x} = \begin{bmatrix} 0 \\ 0 \end{bmatrix}$, and thus $\mathrm{null}(A) = \left\{ \begin{bmatrix} 0 \\ 0 \end{bmatrix} \right\}$. This subspace has no basis, and $\mathrm{nullity}(A) = 0$.

31. The matrix

$$\begin{bmatrix} 1 & 1 & 2 & 1 \\ 0 & 0 & 1 & -3 \end{bmatrix}$$

is row-reduced. Thus $A\mathbf{x} = \mathbf{0}$ has solutions of the form

$$\mathbf{x} = s_1 \begin{bmatrix} -7 \\ 0 \\ 3 \\ 1 \end{bmatrix} + s_2 \begin{bmatrix} -1 \\ 1 \\ 0 \\ 0 \end{bmatrix}.$$

A basis for the null space is $\left\{ \begin{bmatrix} -7 \\ 0 \\ 3 \\ 1 \end{bmatrix}, \begin{bmatrix} -1 \\ 1 \\ 0 \\ 0 \end{bmatrix} \right\}$, and $\mathrm{nullity}(A) = 2$.

33. For example, $\left\{ \begin{bmatrix} 1 \\ 0 \end{bmatrix}, \begin{bmatrix} 1 \\ 1 \end{bmatrix}, \begin{bmatrix} 0 \\ 1 \end{bmatrix}, \begin{bmatrix} -1 \\ 1 \end{bmatrix} \right\}$.

35. For example, take the span of the first m vectors of the n standard basis vectors of $\mathbf{R}^n$.

37. For example, $S_1 = \mathrm{span} \left\{ \begin{bmatrix} 1 \\ 0 \\ 0 \\ 0 \end{bmatrix}, \begin{bmatrix} 0 \\ 1 \\ 0 \\ 0 \end{bmatrix} \right\}$ and $S_2 = \mathrm{span} \left\{ \begin{bmatrix} 0 \\ 0 \\ 1 \\ 0 \end{bmatrix}, \begin{bmatrix} 0 \\ 0 \\ 0 \\ 1 \end{bmatrix} \right\}$.

39. For example, $\mathbf{u}_1 = \begin{bmatrix} 1 \\ 0 \\ 0 \end{bmatrix}$ and $\mathbf{u}_2 = \begin{bmatrix} 0 \\ 1 \\ 0 \end{bmatrix}$.

41. False. For example, $S_1 = \mathrm{span} \left\{ \begin{bmatrix} 1 \\ 0 \end{bmatrix} \right\}$ and $S_2 = \mathrm{span} \left\{ \begin{bmatrix} 0 \\ 1 \end{bmatrix} \right\}$. Then each subspace has dimension 1, but $S_1 \neq S_2$.

43. False. For example, if $\mathcal{U} = \left\{ \begin{bmatrix} 1 \\ 0 \end{bmatrix}, \begin{bmatrix} 0 \\ 1 \end{bmatrix}, \begin{bmatrix} 1 \\ 1 \end{bmatrix} \right\}$, then $\mathcal{U}$ spans $S = \mathbf{R}^2$, but adding additional vectors will not yield a basis.

45. False. If the vectors in $\mathcal{U}$ are linearly independent, and thus a basis for S, then the removal of a vector from $\mathcal{U}$ will no longer be a basis.

47. False. If the three vectors lie in the same plane, then they must be linearly dependent, and cannot form a basis.

49. False. The set $\{\mathbf{0}\}$ is linearly dependent, and thus cannot be a basis. The subspace $\{\mathbf{0}\}$ does not have a basis.

51. False. For example, if $\mathcal{U} = \left\{ \begin{bmatrix} 1 \\ 0 \end{bmatrix}, \begin{bmatrix} 2 \\ 0 \end{bmatrix}, \begin{bmatrix} 3 \\ 0 \end{bmatrix} \right\}$, then removing vectors from $\mathcal{U}$ will not form a basis for $\mathbf{R}^2$.

53. True. The vectors $\{\mathbf{u}_1, \mathbf{u}_2\}$ are linearly independent, and hence span a two-dimensional subspace, a plane.

55. (a) The dimension of S_1 cannot exceed the dimension of S_2 since S_1 is contained in S_2. S_1 is non-zero, and thus its dimension can't be 0. Hence the possible dimensions of S_1 are 1, 2, and 3.

    (b) If $S_1 \neq S_2$, then S_1 is properly contained in S_2, and the dimension of S_1 is strictly less than the dimension of S_2. Thus the possible dimensions of S_1 are 1, and 2.

57. Let S be a subspace of $\mathbf{R}^n$ of dimension n. Then S has a basis $\{\mathbf{u}_1, \mathbf{u}_2, \ldots, \mathbf{u}_n\}$. If $S \neq \mathbf{R}^n$, then there exists a vector $\mathbf{v} \notin \mathbf{S}$ such that $\{\mathbf{u}_1, \mathbf{u}_2, \ldots, \mathbf{u}_n, \mathbf{v}\}$ is linearly independent, for otherwise one could express $\mathbf{v}$ in terms of the vectors $\mathbf{u}_i$, and then $\mathbf{v}$ would belong to S. But now the subspace span $\{\mathbf{u}_1, \mathbf{u}_2, \ldots, \mathbf{u}_n, \mathbf{v}\}$ is an $n+1$ dimensional subspace of $\mathbf{R}^n$, which is not possible. Hence we must have that $S = \mathbf{R}^n$.

59. The vectors $\{\mathbf{s}_1, \mathbf{s}_2, \ldots, \mathbf{s}_m\}$ span S, since every vector $\mathbf{s}$ in S can be written in terms of these vectors. If we consider $c_1\mathbf{s}_1 + c_2\mathbf{s}_2 + \cdots + c_m\mathbf{s}_m = \mathbf{0}$, then since also $0\mathbf{s}_1 + 0\mathbf{s}_2 + \cdots + 0\mathbf{s}_m = \mathbf{0}$, we now have the vector $\mathbf{0} \in S$ expressed in terms of the $\mathbf{s}_i$ in two ways. Since each vector in S is uniquely written in terms of the $\mathbf{s}_i$ we must have that each $c_i = 0$. Consequently, $\{\mathbf{s}_1, \mathbf{s}_2, \ldots, \mathbf{s}_m\}$ is linearly independent, and therefore a basis for S.

61. Let $\{\mathbf{u}_1, \ldots, \mathbf{u}_m\}$ be a basis for S_1. By Theorem 4.14(a) either $\{\mathbf{u}_1, \ldots, \mathbf{u}_m\}$ is a basis for S_2 or we can add vectors to form a basis of S_2. If $\{\mathbf{u}_1, \ldots, \mathbf{u}_m\}$ is a basis for S_2, then $m = \dim(S_1) = \dim(S_2)$. If we add k vectors, then $\dim(S_1) = m < m + k = \dim(S_2)$. Hence $\dim(S_1) \leq \dim(S_2)$. Now if the dimensions are equal, then the basis $\{\mathbf{u}_1, \ldots, \mathbf{u}_m\}$ for S_1 is also a basis for S_2. Thus $S_1 = \text{span}\{\mathbf{u}_1, \ldots, \mathbf{u}_m\} = S_2$.

63. Suppose the pivots occur in columns c_1, c_2, $\ldots$, c_k with $c_1 < c_2 < \cdots < c_k$. Let $\mathbf{u}_1$, $\mathbf{u}_2$, $\ldots$, $\mathbf{u}_k$ be the nonzero row vectors, and consider the equation

$$a_1\mathbf{u}_1 + a_2\mathbf{u}_2 + \cdots + a_k\mathbf{u}_k = \mathbf{0}.$$

By considering the c_1 component, we have $a_1 = 0$, since each c_1 component of $\mathbf{u}_i$ for $i > 1$ is zero. Hence we have

$$a_2\mathbf{u}_2 + a_3\mathbf{u}_3 + \cdots + a_k\mathbf{u}_k = \mathbf{0}.$$

Now consider the c_2 component, and conclude that $a_2 = 0$ as before. Continue in this way to conclude that each $a_i = 0$ for $1 \leq i \leq k$. Thus the nonzero rows are linearly independent.

65. The maximum value of $m_1 + m_2$ is n. To see why, suppose that $m_1 + m_2 > n$. Let $\{\mathbf{u}_1, \ldots, \mathbf{u}_{m_1}\}$ and $\{\mathbf{v}_1, \ldots, \mathbf{v}_{m_2}\}$ be bases for S_1 and S_2, respectively. Since $m_1 + m_2 > n$, the combined set

$$\{\mathbf{u}_1, \ldots, \mathbf{u}_{m_1}, \mathbf{v}_1, \ldots, \mathbf{v}_{m_2}\}$$

must be linearly dependent by Theorem 4.17b. Therefore there exists a nontrivial linear combination

$$a_1\mathbf{u}_1 + \cdots + a_{m_1}\mathbf{u}_{m_1} + b_1\mathbf{v}_1 + \cdots + b_{m_2}\mathbf{v}_{m_2} = \mathbf{0} \quad (**)$$

$$\implies \quad a_1\mathbf{u}_1 + \cdots + a_{m_1}\mathbf{u}_{m_1} = -b_1\mathbf{v}_1 - \cdots - b_{m_2}\mathbf{v}_{m_2}$$

Since $S_1 \cap S_2 = \{\mathbf{0}\}$, it follows that

$$a_1\mathbf{u}_1 + \cdots + a_{m_1}\mathbf{u}_{m_1} = \mathbf{0} \quad \text{and} \quad b_1\mathbf{v}_1 + \cdots + b_{m_2}\mathbf{v}_{m_2} = \mathbf{0}$$

As $\{\mathbf{u}_1, \ldots, \mathbf{u}_{m_1}\}$ and $\{\mathbf{v}_1, \ldots, \mathbf{v}_{m_2}\}$ are bases, the above equations imply that $a_1 = \cdots = a_{m_1} = 0$ and $b_1 = \cdots = b_{m_2} = 0$. But this contradicts $(**)$ being a *nontrivial* linear combination. Hence it must be that $m_1 + m_2 \leq n$, as claimed.

67. A linear dependence in the vectors $\{\mathbf{u}_1, \ldots, \mathbf{u}_m\}$ is an expression of the form

$$a_1\mathbf{u}_1 + a_2\mathbf{u}_2 + \cdots + a_k\mathbf{u}_k = \mathbf{0}.$$

This implies $U\mathbf{a} = \mathbf{0}$ where $\mathbf{a}$ is the vector with components a_i. Since U and V are equivalent, we know that $V = EU$ where E is an invertible matrix. Thus $V\mathbf{a} = (EU)\mathbf{a} = E(U\mathbf{a}) = E\mathbf{0} = \mathbf{0}$. But this implies the same linear dependence in the vectors $\{\mathbf{v}_1, \ldots, \mathbf{v}_m\}$,

$$a_1\mathbf{v}_1 + a_2\mathbf{v}_2 + \cdots + a_k\mathbf{v}_k = \mathbf{0}.$$

Similarly, an initial linear dependence in $\{\mathbf{v}_1, \ldots, \mathbf{v}_m\}$ corresponds to a linear dependence in $\{\mathbf{u}_1, \ldots, \mathbf{u}_m\}$, by writing $U = E^{-1}V$.

69. Using a computer algebra system, we determine that the span of the vectors has basis $\left\{ \begin{bmatrix} 2 \\ -1 \\ 5 \end{bmatrix}, \begin{bmatrix} -3 \\ 4 \\ -2 \end{bmatrix} \right\}$, with dimension 2. The vectors are not a basis for $\mathbf{R}^3$.

71. Using a computer algebra system, we determine that the span of the vectors has basis $\left\{ \begin{bmatrix} 3 \\ 0 \\ 1 \\ -2 \end{bmatrix}, \begin{bmatrix} 2 \\ -4 \\ 5 \\ 0 \end{bmatrix}, \begin{bmatrix} -2 \\ 7 \\ 0 \\ 4 \end{bmatrix}, \begin{bmatrix} -2 \\ 5 \\ -5 \\ 4 \end{bmatrix} \right\}$, with dimension 4. The vectors thus span $\mathbf{R}^4$.

73. Using a computer algebra system, we determine that the span of the vectors has basis $\left\{ \begin{bmatrix} 1 \\ 1 \\ -1 \\ 1 \\ 1 \end{bmatrix}, \begin{bmatrix} -1 \\ 0 \\ 1 \\ 2 \\ -1 \end{bmatrix}, \begin{bmatrix} 2 \\ 1 \\ -2 \\ 1 \\ 2 \end{bmatrix} \right\}$, with dimension 3. The vectors therefore do not span $\mathbf{R}^5$.

4.3 Row and Column Spaces

1. A basis for the column space, determined from the pivot columns 1 and 2, is $\left\{ \begin{bmatrix} 1 \\ -2 \\ -3 \end{bmatrix}, \begin{bmatrix} -3 \\ 5 \\ 8 \end{bmatrix} \right\}$. A basis for the row space is determined from the nonzero rows of the echelon form, $\left\{ \begin{bmatrix} 1 \\ 0 \\ -10 \end{bmatrix}, \begin{bmatrix} 0 \\ 1 \\ -4 \end{bmatrix} \right\}$. We solve $A\mathbf{x} = \mathbf{0}$, to obtain $\mathbf{x} = s \begin{bmatrix} 10 \\ 4 \\ 1 \end{bmatrix}$, and so our nullspace basis is $\left\{ \begin{bmatrix} 10 \\ 4 \\ 1 \end{bmatrix} \right\}$.

3. A basis for the column space, determined from the pivot columns 1 and 2, is $\left\{ \begin{bmatrix} 1 \\ -2 \\ 0 \end{bmatrix}, \begin{bmatrix} 0 \\ 1 \\ 1 \end{bmatrix} \right\}$. A basis

for the row space is determined from the nonzero rows of the echelon form, $\left\{ \begin{bmatrix} 1 \\ 0 \\ -4 \\ -3 \end{bmatrix}, \begin{bmatrix} 0 \\ 1 \\ 5 \\ -1 \end{bmatrix} \right\}$. We

solve $A\mathbf{x} = \mathbf{0}$, to obtain $\mathbf{x} = s_1 \begin{bmatrix} 4 \\ -5 \\ 1 \\ 0 \end{bmatrix} + s_2 \begin{bmatrix} 3 \\ 1 \\ 0 \\ 1 \end{bmatrix}$, and so our nullspace basis is $\left\{ \begin{bmatrix} 4 \\ -5 \\ 1 \\ 0 \end{bmatrix}, \begin{bmatrix} 3 \\ 1 \\ 0 \\ 1 \end{bmatrix} \right\}$.

5. We reduce A to echelon form:

$$\begin{bmatrix} 1 & -2 & 2 \\ 2 & -2 & 3 \\ -1 & -2 & 0 \end{bmatrix} \sim \begin{bmatrix} 1 & -2 & 2 \\ 0 & 2 & -1 \\ 0 & 0 & 0 \end{bmatrix}$$

A basis for the column space, determined from the pivot columns 1 and 2, is $\left\{ \begin{bmatrix} 1 \\ 2 \\ -1 \end{bmatrix}, \begin{bmatrix} -2 \\ -2 \\ -2 \end{bmatrix} \right\}$. A

basis for the row space is determined from the nonzero rows of the echelon form, $\left\{ \begin{bmatrix} 1 \\ -2 \\ 2 \end{bmatrix}, \begin{bmatrix} 0 \\ 2 \\ -1 \end{bmatrix} \right\}$.

We solve $A\mathbf{x} = \mathbf{0}$, to obtain $\mathbf{x} = s \begin{bmatrix} -1 \\ \frac{1}{2} \\ 1 \end{bmatrix}$, and so our nullspace basis is $\left\{ \begin{bmatrix} -1 \\ \frac{1}{2} \\ 1 \end{bmatrix} \right\}$. We have

rank $(A) = 2$, nullity $(A) = 1$, and rank (A) + nullity $(A) = 2 + 1 = 3 = m$.

7. We reduce A to echelon form:

$$\begin{bmatrix} 1 & 3 & 2 & 0 \\ 3 & 11 & 7 & 1 \\ 1 & 1 & 4 & 0 \end{bmatrix} \sim \begin{bmatrix} 1 & 3 & 2 & 0 \\ 0 & 2 & 1 & 1 \\ 0 & 0 & 3 & 1 \end{bmatrix}$$

A basis for the column space, determined from the pivot columns 1, 2, and 3,

is $\left\{ \begin{bmatrix} 1 \\ 3 \\ 1 \end{bmatrix}, \begin{bmatrix} 3 \\ 11 \\ 1 \end{bmatrix}, \begin{bmatrix} 2 \\ 7 \\ 4 \end{bmatrix} \right\}$. A basis for the row space is determined from the nonzero rows of the

echelon form, $\left\{ \begin{bmatrix} 1 \\ 3 \\ 2 \\ 0 \end{bmatrix}, \begin{bmatrix} 0 \\ 2 \\ 1 \\ 1 \end{bmatrix}, \begin{bmatrix} 0 \\ 0 \\ 3 \\ 1 \end{bmatrix} \right\}$. We solve $A\mathbf{x} = \mathbf{0}$, to obtain $\mathbf{x} = s \begin{bmatrix} \frac{5}{3} \\ -\frac{1}{3} \\ -\frac{1}{3} \\ 1 \end{bmatrix}$, and so our

nullspace basis is $\left\{ \begin{bmatrix} \frac{5}{3} \\ -\frac{1}{3} \\ -\frac{1}{3} \\ 1 \end{bmatrix} \right\}$. We have rank $(A) = 3$, nullity $(A) = 1$, and rank (A) + nullity $(A) =$

$3 + 1 = 4 = m$.

9. We require that the rows are linearly independent, so therefore $x \neq 8$.

11. We reduce A:

$$\begin{bmatrix} -1 & 2 & 1 \\ 3 & 1 & 11 \\ 4 & 3 & x \end{bmatrix} \sim \begin{bmatrix} -1 & 2 & 1 \\ 0 & 7 & 14 \\ 0 & 0 & x - 18 \end{bmatrix}$$

Thus, for rank $(A) = 2$, we need two pivots, and hence $x = 18$.

13. The dimension of the column space is 5, the same as the dimension of the row space.

15. The dimension of the row space is $4 - 1 = 3$, the number of nonzero rows in the echelon form. The dimension of the column space is also 3, and the dimension of the null space is $7 - 3 = 4$.

17. $\operatorname{rank}(A) = m - \operatorname{nullity}(A) = 5 - 3 = 2$.

19. $\operatorname{nullity}(A) = m - \operatorname{rank}(A) = 11 - 4 = 7$.

21. $\dim(\operatorname{range}(T)) = \operatorname{rank}(A) = m - \operatorname{nullity}(A) = 11 - 7 = 4$.

23. Since T is one-to-one, $\ker(T) = \mathbf{0}$, and $\dim(\ker(T)) = 0$. Hence $\operatorname{nullity}(A) = \dim(\operatorname{null}(A)) = \dim(\ker(T)) = 0$.

25. The maximum possible value for the rank of A is 5 since the echelon form can have at most 5 pivots. The minimum possible value of the nullity of A is 8, since $\operatorname{nullity}(A) = m - \operatorname{rank}(A) = 13 - \operatorname{rank}(A) \geq 13 - 5 = 8$.

27. $\operatorname{rank}(A) = 3$, the number of nonzero rows of B.

29. $\operatorname{nullity}(A) = m - \operatorname{rank}(A) = 5 - 3 = 2$, since the rank of A is the number of nonzero rows of B.

31. B has 3 nonzero rows, since the rank of A is equal to the number of nonzero rows of B.

33. A must be 7×5, since $\operatorname{col}(A)$ is a subspace of $\mathbf{R}^n$, and $\operatorname{row}(A)$ is a subspace of $\mathbf{R}^m$.

35. For example, $A = \begin{bmatrix} 1 & 0 & 0 \\ 0 & 1 & 0 \end{bmatrix}$.

37. For example, $A = \left[\begin{array}{c|c} I_{3\times 3} & 0_{3\times 1} \\ \hline 0_{6\times 3} & 0_{6\times 1} \end{array} \right]$.

39. For example, $A = \begin{bmatrix} 1 & 0 & 0 & 0 \\ 0 & 1 & 0 & 0 \\ 0 & 0 & 1 & 0 \end{bmatrix}$.

41. For example, $A = \begin{bmatrix} 1 & 0 \\ 0 & 1 \end{bmatrix}$.

43. True, by Theorem 4.21.

45. True, since the rank is the number of pivots, which can not exceed the number of rows.

47. False. For example, see A and B in Example 2.

49. False. Solutions of $A\mathbf{x} = \mathbf{b}$ are not related to $\operatorname{row}(A)$.

51. False, since $\dim(\operatorname{range}(T)) = \operatorname{rank}(A) \leq 5$, T cannot map onto $\mathbf{R}^9$.

53. True. For example, if $A = [\ I_{4\times 4} \mid 0_{4\times 9}\ ]$, then T maps onto $\mathbf{R}^4$.

55. The span of the rows of A is the same subspace as the span of the columns of A^T, since these subspaces are determined by the same vectors. Hence $\operatorname{row}(A) = \operatorname{col}(A^T)$, and thus $\operatorname{rank}(A) = \dim(\operatorname{row}(A)) = \dim(\operatorname{col}(A^T)) = \operatorname{rank}(A^T)$.

57. If $\operatorname{rank}(A) < m$, then $\operatorname{nullity}(A) = m - \operatorname{rank}(A) > m - m = 0$. Thus $\dim(\operatorname{null}(A)) > 0$, and therefore there exists nontrivial solutions to $A\mathbf{x} = \mathbf{0}$.

59. If $m > n$, then $\operatorname{rank}(A) \leq n < m$, so it must be that $\operatorname{nullity}(A) > 0$. If $m < n$, then the same reasoning applies to A^T, which is $m \times n$.

61. Using a computer algebra system, we determine

$$\text{rank}\left(\begin{bmatrix} 1 & 3 & 2 & 4 & -1 \\ 1 & 5 & -3 & 3 & -4 \\ 2 & 8 & -1 & 7 & -5 \end{bmatrix}\right) = 2, \text{ and nullity}\left(\begin{bmatrix} 1 & 3 & 2 & 4 & -1 \\ 1 & 5 & -3 & 3 & -4 \\ 2 & 8 & -1 & 7 & -5 \end{bmatrix}\right) = 3.$$

63. Using a computer algebra system, we determine

$$\text{rank}\left(\begin{bmatrix} 4 & 8 & 2 \\ 3 & 5 & 1 \\ 9 & 19 & 5 \\ 7 & 13 & 3 \\ 5 & 11 & 3 \end{bmatrix}\right) = 2, \text{ and nullity}\left(\begin{bmatrix} 4 & 8 & 2 \\ 3 & 5 & 1 \\ 9 & 19 & 5 \\ 7 & 13 & 3 \\ 5 & 11 & 3 \end{bmatrix}\right) = 1.$$

Chapter 5

Determinants

5.1 The Determinant Function

1. $M_{23} = \begin{bmatrix} 7 & 0 \\ 5 & 1 \end{bmatrix}$, $M_{31} = \begin{bmatrix} 0 & -4 \\ 6 & 2 \end{bmatrix}$

3. $M_{23} = \begin{bmatrix} 6 & 1 & 5 \\ 7 & 1 & 1 \\ 4 & 3 & 2 \end{bmatrix}$, $M_{31} = \begin{bmatrix} 1 & -1 & 5 \\ 2 & 3 & 0 \\ 3 & 1 & 2 \end{bmatrix}$

5. $M_{23} = \begin{bmatrix} 4 & 3 & 1 & 0 \\ 3 & 2 & 4 & 4 \\ 5 & 1 & 0 & 3 \\ 2 & 2 & 1 & 0 \end{bmatrix}$, $M_{31} = \begin{bmatrix} 3 & 2 & 1 & 0 \\ 1 & 2 & 0 & 5 \\ 1 & 0 & 0 & 3 \\ 2 & 4 & 1 & 0 \end{bmatrix}$

7. $C_{13} = (-1)^{1+3} |M_{13}| = \begin{vmatrix} 0 & -1 \\ 4 & 0 \end{vmatrix} = (0)(0) - (-1)(4) = 4$,

   $C_{22} = (-1)^{2+2} |M_{22}| = \begin{vmatrix} 2 & 3 \\ 4 & 1 \end{vmatrix} = (2)(1) - (3)(4) = -10$

9. $C_{13} = (-1)^{1+3} |M_{13}| = \begin{vmatrix} 4 & 3 \\ 1 & 1 \end{vmatrix} = (4)(1) - (3)(1) = 1$,

   $C_{22} = (-1)^{2+2} |M_{22}| = \begin{vmatrix} 6 & 2 \\ 1 & 1 \end{vmatrix} = (6)(1) - (2)(1) = 4$

11. (a) Using the second row,

    $\det(A) = a_{21}C_{21} + a_{22}C_{22} + a_{23}C_{23} = (0)C_{21} + (4)C_{22} + (0)C_{23} = (4)(-1)^{2+2} |M_{22}| = (4) \begin{vmatrix} 1 & -3 \\ 5 & 0 \end{vmatrix} = (4)((1)(0) - (-3)(5)) = (4)(15) = 60.$
    (b) Using the third column,

    $\det(A) = a_{13}C_{13} + a_{23}C_{23} + a_{33}C_{33} = (-3)C_{13} + (0)C_{22} + (0)C_{33} = (-3)(-1)^{1+3} |M_{13}| = (-3) \begin{vmatrix} 0 & 4 \\ 5 & -1 \end{vmatrix} = (-3)((0)(-1) - (4)(5)) = (-3)(-20) = 60$
    Since $\det(A) \neq 0$, A is invertible, and hence T is invertible.

13. (a) Using the second row,
    $\det(A) = a_{21}C_{21} + a_{22}C_{22} + a_{23}C_{23} + a_{24}C_{24} = (0)C_{21} + (3)C_{22} + (2)C_{23} + (0)C_{24} =$
    $(3)(-1)^{2+2} |M_{22}| + (2)(-1)^{2+3} |M_{23}| = (3) \begin{vmatrix} -1 & -1 & 2 \\ 1 & 0 & 1 \\ 0 & 3 & -1 \end{vmatrix} - (2) \begin{vmatrix} -1 & 1 & 2 \\ 1 & 4 & 1 \\ 0 & -1 & -1 \end{vmatrix} = 20.$

    (b) Using the first column,

$\det(A) = a_{11}C_{11} + a_{21}C_{21} + a_{31}C_{31} + a_{41}C_{41} = (-1)C_{11} + (0)C_{21} + (1)C_{31} + (0)C_{41} =$

$(-1)(-1)^{1+1}|M_{11}| + (1)(-1)^{3+1}|M_{31}| = (-1)\begin{vmatrix} 3 & 2 & 0 \\ 4 & 0 & 1 \\ -1 & 3 & -1 \end{vmatrix} + (1)\begin{vmatrix} 1 & -1 & 2 \\ 3 & 2 & 0 \\ -1 & 3 & -1 \end{vmatrix} = 20.$ Since

$\det(A) \neq 0$, A is invertible, and hence T is invertible.

15. (a) Using the first row,
$\det(A) = a_{11}C_{11} + a_{12}C_{12} + a_{13}C_{13} + a_{14}C_{14} = (-1)C_{11} + (1)C_{12} + (0)C_{13} + (0)C_{14} =$

$(-1)(-1)^{1+1}|M_{11}| + (1)(-1)^{1+2}|M_{12}| = (-1)\begin{vmatrix} 3 & 3 & 2 \\ -1 & 0 & 5 \\ 1 & 4 & -1 \end{vmatrix} - (1)\begin{vmatrix} 2 & 3 & 2 \\ 0 & 0 & 5 \\ 3 & 4 & -1 \end{vmatrix} = 51.$

(b) Using the third column,
$\det(A) = a_{13}C_{13} + a_{23}C_{23} + a_{33}C_{33} + a_{43}C_{43} = (0)C_{13} + (3)C_{23} + (0)C_{33} + (4)C_{43} =$

$(3)(-1)^{1+3}|M_{13}| + (4)(-1)^{4+3}|M_{43}| = -(3)\begin{vmatrix} -1 & 1 & 0 \\ 0 & -1 & 5 \\ 3 & 1 & -1 \end{vmatrix} - (4)\begin{vmatrix} -1 & 1 & 0 \\ 2 & 3 & 2 \\ 0 & -1 & 5 \end{vmatrix} = 51.$ Since $\det(A) \neq$

0, A is invertible, and hence T is invertible.

17. (a) Using the third row,
$\det(A) = a_{31}C_{31} + a_{32}C_{32} + a_{33}C_{33} + a_{34}C_{34} + a_{35}C_{35} = (0)C_{31} + (-1)C_{32} + (0)C_{33} + (0)C_{34} + (1)C_{35} =$

$(-1)(-1)^{3+2}|M_{32}| + (1)(-1)^{3+5}|M_{35}| = -(-1)\begin{vmatrix} 4 & 1 & 0 & 1 \\ 0 & -1 & 1 & 2 \\ 0 & 2 & 0 & 3 \\ 0 & 1 & 0 & 1 \end{vmatrix} + (1)\begin{vmatrix} 4 & 2 & 1 & 0 \\ 0 & 3 & -1 & 1 \\ 0 & 1 & 2 & 0 \\ 0 & 0 & 1 & 0 \end{vmatrix} = 8.$

(b) Using the first column,
$\det(A) = a_{11}C_{11} + a_{21}C_{21} + a_{31}C_{31} + a_{41}C_{41} + a_{51}C_{51} = (4)C_{11} + (0)C_{21} + (0)C_{31} + (0)C_{41} + (0)C_{51} =$

$(4)(-1)^{1+1}|M_{11}| = (4)\begin{vmatrix} 3 & -1 & 1 & 2 \\ -1 & 0 & 0 & 1 \\ 1 & 2 & 0 & 3 \\ 0 & 1 & 0 & 1 \end{vmatrix} = 8.$ Since $\det(A) \neq 0$, A is invertible, and hence T is

invertible.

19. $\det(A) = (4)(2) - (6)(-1) = 14.$

21. A is not a square matrix, so $\det(A)$ is not defined.

23. $\det(A) = (3)(0)(1) + (1)(4)(1) + (-1)(2)(6) - (-1)(0)(1) - (3)(4)(6) - (1)(2)(1) = -82.$

25. The shortcut method is not applicable for 4×4 matrices.

27. $\det\left(\begin{bmatrix} 2 & 3 \\ 6 & a \end{bmatrix}\right) = 2a - 18 = 0 \quad \Rightarrow \quad a = 9.$ A is not invertible if $a = 9$.

29. $\det\left(\begin{bmatrix} a & a \\ 3 & -1 \end{bmatrix}\right) = -4a = 0 \quad \Rightarrow \quad a = 0.$ A is not invertible if $a = 0$.

31. $\det\left(\begin{bmatrix} 1 & -1 & 3 \\ 0 & a & -2 \\ 2 & 4 & 3 \end{bmatrix}\right) = 12 - 3a = 0 \quad \Rightarrow \quad a = 4.$ A is not invertible if $a = 4$.

33. $\det\left(\begin{bmatrix} 1 & a & -2 \\ -1 & 0 & 1 \\ a & 3 & -4 \end{bmatrix}\right) = a^2 - 4a + 3 = (a-1)(a-3) = 0 \quad \Rightarrow \quad a = 1 \text{ or } a = 3.$ A is not invertible
if $a = 1$ or $a = 3$.

35. Since A is triangular, $\det(A) = (2)(-1)(1)(4) = -8$, the product along the diagonal.

37. $\det(A) = 0$, since the third column of A contains all zeroes.

39. $\det(A) = 0$, since the first and third rows of A are identical.

41. $\det(A) = \det\left(\begin{bmatrix} 3 & -2 \\ 4 & 1 \end{bmatrix}\right) = 11$, $\det(A^T) = \det\left(\begin{bmatrix} 3 & 4 \\ -2 & 1 \end{bmatrix}\right) = 11$.

43. $\det(A) = \det\left(\begin{bmatrix} 0 & 7 & 1 \\ 2 & 3 & 1 \\ 4 & -1 & -1 \end{bmatrix}\right) = 28$, $\det(A^T) = \det\left(\begin{bmatrix} 0 & 2 & 4 \\ 7 & 3 & -1 \\ 1 & 1 & -1 \end{bmatrix}\right) = 28$.

45. $\det(A - \lambda I_2) = \det\left(\begin{bmatrix} 2 & 4 \\ 5 & 3 \end{bmatrix} - \lambda\begin{bmatrix} 1 & 0 \\ 0 & 1 \end{bmatrix}\right) = \det\left(\begin{bmatrix} 2-\lambda & 4 \\ 5 & 3-\lambda \end{bmatrix}\right) = \lambda^2 - 5\lambda - 14 = 0 \Rightarrow$
$(\lambda + 2)(\lambda - 7) = 0 \Rightarrow \lambda = -2$ or $\lambda = 7$.

47. $\det(A - \lambda I_2) = \det\left(\begin{bmatrix} 1 & 0 \\ -5 & 1 \end{bmatrix} - \lambda\begin{bmatrix} 1 & 0 \\ 0 & 1 \end{bmatrix}\right) = \det\left(\begin{bmatrix} 1-\lambda & 0 \\ -5 & 1-\lambda \end{bmatrix}\right) = \lambda^2 - 2\lambda + 1 = 0 \Rightarrow$
$(\lambda - 1)^2 = 0 \Rightarrow \lambda = 1$.

49. $\det(A - \lambda I_3) = \det\left(\begin{bmatrix} 1 & 0 & 0 \\ 5 & 3 & 0 \\ -4 & 7 & -2 \end{bmatrix} - \lambda\begin{bmatrix} 1 & 0 & 0 \\ 0 & 1 & 0 \\ 0 & 0 & 1 \end{bmatrix}\right) = \det\left(\begin{bmatrix} 1-\lambda & 0 & 0 \\ 5 & 3-\lambda & 0 \\ -4 & 7 & -2-\lambda \end{bmatrix}\right)$
$= (1-\lambda)(3-\lambda)(-2-\lambda) = 0 \Rightarrow \lambda = 1, \lambda = 3,$ or $\lambda = -2$.

51. $\det(A - \lambda I_3) = \det\left(\begin{bmatrix} 0 & 2 & 0 \\ 1 & 0 & 2 \\ 2 & -1 & 0 \end{bmatrix} - \lambda\begin{bmatrix} 1 & 0 & 0 \\ 0 & 1 & 0 \\ 0 & 0 & 1 \end{bmatrix}\right) = \det\left(\begin{bmatrix} -\lambda & 2 & 0 \\ 1 & -\lambda & 2 \\ 2 & -1 & -\lambda \end{bmatrix}\right) = 8 - \lambda^3 =$
$0 \Rightarrow \lambda = 2$.

53. (a) $\det\left(\begin{bmatrix} 3 & 5 \\ -2 & 4 \end{bmatrix}\right) = 22$; interchanging rows, $\det\left(\begin{bmatrix} -2 & 4 \\ 3 & 5 \end{bmatrix}\right) = -22$. Interchanging rows changes the sign of the determinant.

(b) $\det\left(\begin{bmatrix} 1 & 2 & -1 \\ 3 & 0 & 2 \\ 0 & 1 & -1 \end{bmatrix}\right) = 1$; interchanging rows 1 and 2, $\det\left(\begin{bmatrix} 3 & 0 & 2 \\ 1 & 2 & -1 \\ 0 & 1 & -1 \end{bmatrix}\right) = -1$. Interchanging rows changes the sign of the determinant.

55. (a) $\det\left(\begin{bmatrix} 3 & 1 & 0 \\ 1 & 2 & 3 \\ 0 & 2 & 1 \end{bmatrix}\right) = -13$; interchanging rows 1 and 2, $\det\left(\begin{bmatrix} 1 & 2 & 3 \\ 3 & 1 & 0 \\ 0 & 2 & 1 \end{bmatrix}\right) = 13$. Interchanging rows changes the sign of the determinant.

(b) $\det\left(\begin{bmatrix} 4 & -1 & 0 \\ 0 & 2 & 1 \\ 1 & 1 & 1 \end{bmatrix}\right) = 3$; interchanging rows 1 and 3, $\det\left(\begin{bmatrix} 1 & 1 & 1 \\ 0 & 2 & 1 \\ 4 & -1 & 0 \end{bmatrix}\right) = -3$. Interchanging rows changes the sign of the determinant.

57. (a) $\det\left(\begin{bmatrix} 3 & 5 \\ -2 & 4 \end{bmatrix}\right) = 22$; multiplying row 1 by 3, $\det\left(\begin{bmatrix} 9 & 15 \\ -2 & 4 \end{bmatrix}\right) = 66$. Multiplying a row by 3 causes the determinant to be multiplied by 3, and we conjecture that multiplying a row by c multiplies the determinant by c.

(b) $\det\left(\begin{bmatrix} 1 & 2 & -1 \\ 3 & 0 & 2 \\ 0 & 1 & -1 \end{bmatrix}\right) = 1$; multiplying row 1 by 3, $\det\left(\begin{bmatrix} 3 & 6 & -3 \\ 3 & 0 & 2 \\ 0 & 1 & -1 \end{bmatrix}\right) = 3$. Multiplying a row by 3 causes the determinant to be multiplied by 3, and we conjecture that multiplying a row by c multiplies the determinant by c.

59. (a) $\det\left(\begin{bmatrix} 3 & 1 & 0 \\ 1 & 2 & 3 \\ 0 & 2 & 1 \end{bmatrix}\right) = -13$; multiplying row 2 by 3, $\det\left(\begin{bmatrix} 3 & 1 & 0 \\ 3 & 6 & 9 \\ 0 & 2 & 1 \end{bmatrix}\right) = -39$. Multiplying a row by 3 causes the determinant to be multiplied by 3, and we conjecture that multiplying a row by c multiplies the determinant by c.

(b) $\det\left(\begin{bmatrix} 4 & -1 & 0 \\ 0 & 2 & 1 \\ 1 & 1 & 1 \end{bmatrix}\right) = 3$; multiplying row 2 by 3, $\det\left(\begin{bmatrix} 4 & -1 & 0 \\ 0 & 6 & 3 \\ 1 & 1 & 1 \end{bmatrix}\right) = 9$. Multiplying a row by 3 causes the determinant to be multiplied by 3, and we conjecture that multiplying a row by c multiplies the determinant by c.

61. For example, $A = \begin{bmatrix} 12 & 0 \\ 0 & 1 \end{bmatrix}$.

63. For example, $A = \begin{bmatrix} 1 & 4 \\ 1 & 1 \end{bmatrix}$.

65. For example, $A = \begin{bmatrix} 5 & -1 & \pi \\ e & 0 & 4 \\ 2 & 6 & -3 \end{bmatrix}$.

67. For example, $A = \begin{bmatrix} \pi & 0 & 5 \\ 8 & 1 & 0 \\ 0 & e & 1 \end{bmatrix}$.

69. False. Every *square* matrix has a determinant, but nonsquare matrices such as $A = \begin{bmatrix} 1 & 3 & 1 \\ 2 & 1 & 7 \end{bmatrix}$ do not.

71. False. Consider $A = \begin{bmatrix} 1 & 2 \\ 1 & 1 \end{bmatrix}$, which has $\det(A) = -1$.

73. False. Consider $A = \begin{bmatrix} 0 & 1 \\ 0 & 0 \end{bmatrix}$, with $\det(A) = 0$.

75. False. Consider $A = \begin{bmatrix} 1 & 1 \\ 1 & 1 \end{bmatrix}$. Then $C_{11} = 1$, $C_{12} = -1$, $C_{21} = -1$, and $C_{22} = 1$, but $\det(A) = 0$.

77. $\det\left(\begin{bmatrix} x & y & 1 \\ x_1 & y_1 & 1 \\ x_2 & y_2 & 1 \end{bmatrix}\right) = xy_1 - yx_1 - xy_2 + yx_2 + x_1y_2 - x_2y_1 = x(y_1 - y_2) + y(-x_1 + x_2) + (x_1y_2 - x_2y_1) = 0$ is linear in x and y, and hence gives a line. Each point (x_1, y_1) and (x_2, y_2) satisfies this equation, either by direct substitution, or by observing that $\begin{bmatrix} x_1 & y_1 & 1 \\ x_1 & y_1 & 1 \\ x_2 & y_2 & 1 \end{bmatrix}$ and $\begin{bmatrix} x_2 & y_2 & 1 \\ x_1 & y_1 & 1 \\ x_2 & y_2 & 1 \end{bmatrix}$ have identical rows, hence zero determinant. Thus, the line must pass through (x_1, y_1) and (x_2, y_2).

79. Let B be the matrix which has the same rows as A, except row j of B is the ith row of A. Then $\det(B) = 0$, since B has two identical rows. Using the cofactor expansion on row j of B, and noting that the cofactors of B along row j are the same as the cofactors of A along row j, we have $\det(B) = a_{i1}C_{j1} + a_{i2}C_{j2} + \cdots + a_{in}C_{jn}$. Hence, $a_{i1}C_{j1} + a_{i2}C_{j2} + \cdots + a_{in}C_{jn} = 0$.

81. Suppose the ith row is all zeroes, then the cofactor expansion along the ith row produces $\det(A) = a_{i1}C_{i1} + a_{i2}C_{i2} + \cdots + a_{in}C_{in} = (0)C_{i1} + (0)C_{i2} + \cdots + (0)C_{in} = 0$. If the jth column is all zeroes, then the cofactor expansion along the jth column produces $\det(A) = a_{1j}C_{1j} + a_{2j}C_{2j} + \cdots + a_{nj}C_{nj} = (0)C_{1j} + (0)C_{2j} + \cdots + (0)C_{nj} = 0$.

83. Using a computer algebra system, $\det\left(\begin{bmatrix} 3 & -4 & 0 & 5 \\ 2 & 1 & -7 & 1 \\ 0 & -3 & 2 & 2 \\ 5 & 8 & -2 & -1 \end{bmatrix}\right) = -26.$

85. Using a computer algebra system, $\det\left(\begin{bmatrix} 3 & 5 & 0 & 0 & 2 \\ 0 & 1 & -2 & -3 & -2 \\ 7 & -2 & -1 & 0 & 0 \\ 4 & 1 & 1 & 1 & 4 \\ -5 & -1 & 0 & 5 & 3 \end{bmatrix}\right) = 1215.$

5.2 Properties of the Determinant

1. Row-reduce to echelon form:

$$\begin{bmatrix} 2 & 8 \\ -1 & -3 \end{bmatrix} \xrightarrow[\sim]{(1/2)R_1+R_2\Rightarrow R_2} \begin{bmatrix} 2 & 8 \\ 0 & 1 \end{bmatrix}$$

Thus $\det(A) = (2)(1) = 2$.

3. Row-reduce to echelon form:

$$\begin{bmatrix} 1 & -1 & -3 \\ -2 & 2 & 6 \\ -3 & -3 & 10 \end{bmatrix} \xrightarrow[\sim]{\substack{2R_1+R_2\Rightarrow R_2 \\ 3R_1+R_3\Rightarrow R_3}} \begin{bmatrix} 1 & -1 & -3 \\ 0 & 0 & 0 \\ 0 & -6 & 1 \end{bmatrix}$$
$$\xrightarrow[\sim]{R_2\Leftrightarrow R_3} \begin{bmatrix} 1 & -1 & -3 \\ 0 & -6 & 1 \\ 0 & 0 & 0 \end{bmatrix}$$

Thus $\det(A) = (-1)^1 (1)(-6)(0) = 0$.

5. Row-reduce to echelon form:

$$\begin{bmatrix} 0 & 1 & 0 & 0 \\ 1 & 0 & 0 & 0 \\ 0 & 0 & 0 & 1 \\ 0 & 0 & 1 & 0 \end{bmatrix} \xrightarrow[\sim]{\substack{R_1\Leftrightarrow R_2 \\ R_3\Leftrightarrow R_4}} \begin{bmatrix} 1 & 0 & 0 & 0 \\ 0 & 1 & 0 & 0 \\ 0 & 0 & 1 & 0 \\ 0 & 0 & 0 & 1 \end{bmatrix}$$

Thus $\det(A) = (-1)^2 (1)(1)(1)(1) = 1$.

7. $\det(A) = (-1)^1 (1)(-4) = 4$. Since $\det(A) \neq 0$, A is invertible.

9. $\det(A) = (-1)^1 (1)(-3)(7) = 21$. Since $\det(A) \neq 0$, A is invertible.

11. $\det(A) = \frac{1}{-7}(-6)(0)(-4) = 0$. Since $\det(A) = 0$, A is not invertible.

13. $\det(A) = (-1)^2 (1)(1)(2)(4) = 8$. Since $\det(A) \neq 0$, A is invertible.

15. There is one row exchange, hence the value of the determinant is $(-1)^1 \det(A) = (-1)(3) = -3$.

17. The row operations are $-2R_2 \Rightarrow R_2$ and $R_1 + R_3 \Rightarrow R_3$, hence the value of the determinant is $-2\det(A) = -2(3) = -6$.

19. $\det(A) = \det\left(\begin{bmatrix} 2 & 3 \\ 1 & -4 \end{bmatrix}\right) = -11$, $\det(B) = \det\left(\begin{bmatrix} 0 & -1 \\ 3 & 7 \end{bmatrix}\right) = 3$,

$\det(AB) = \det\left(\begin{bmatrix} 2 & 3 \\ 1 & -4 \end{bmatrix}\begin{bmatrix} 0 & -1 \\ 3 & 7 \end{bmatrix}\right) = \det\left(\begin{bmatrix} 9 & 19 \\ -12 & -29 \end{bmatrix}\right) = -33$; so $\det(A)\det(B)$

$= (-11)(3) = -33 = \det(AB)$. $\det\left(\begin{bmatrix} 2 & 3 \\ 1 & -4 \end{bmatrix} + \begin{bmatrix} 0 & -1 \\ 3 & 7 \end{bmatrix}\right) = \det\left(\begin{bmatrix} 2 & 2 \\ 4 & 3 \end{bmatrix}\right) = -2$; so

$\det(A) + \det(B) = -11 + 3 = -8 \neq -2 = \det(A+B)$.

21. $\det(A) = \det\left(\begin{bmatrix} 2 & 0 & -1 \\ 1 & 1 & 0 \\ 0 & 1 & 1 \end{bmatrix}\right) = 1$, $\det(B) = \det\left(\begin{bmatrix} 2 & 2 & 1 \\ 2 & 0 & 4 \\ -1 & 3 & 1 \end{bmatrix}\right) = -30$, $\det(AB)$

$= \det\left(\begin{bmatrix} 2 & 0 & -1 \\ 1 & 1 & 0 \\ 0 & 1 & 1 \end{bmatrix}\begin{bmatrix} 2 & 2 & 1 \\ 2 & 0 & 4 \\ -1 & 3 & 1 \end{bmatrix}\right) = \det\left(\begin{bmatrix} 5 & 1 & 1 \\ 4 & 2 & 5 \\ 1 & 3 & 5 \end{bmatrix}\right) = -30$; so $\det(A)\det(B) = (1)(-30) =$

$-30 = \det(AB)$. $\det\left(\begin{bmatrix} 2 & 0 & -1 \\ 1 & 1 & 0 \\ 0 & 1 & 1 \end{bmatrix} + \begin{bmatrix} 2 & 2 & 1 \\ 2 & 0 & 4 \\ -1 & 3 & 1 \end{bmatrix}\right) = \det\left(\begin{bmatrix} 4 & 2 & 0 \\ 3 & 1 & 4 \\ -1 & 4 & 2 \end{bmatrix}\right) = -76$; so $\det(A) +$

$\det(B) = 1 + (-30) = -29 \neq -76 = \det(A + B)$.

23.   (a) $\det(A^2) = (\det(A))^2 = (3)^2 = 9$

    (b) $\det(A^4) = (\det(A))^4 = (3)^4 = 81$

    (c) $\det(A^2 A^T) = (\det(A^2))(\det(A^T)) = (\det(A))^2 \det(A) = (3)^2(3) = 27$

    (d) $\det(A^{-1}) = \frac{1}{\det(A)} = \frac{1}{3}$

25.   (a) $\det(A^2 B^3) = (\det(A))^2(\det(B))^3 = (3)^2(-2)^3 = -72$

    (b) $\det(AB^{-1}) = \det(A)\det(B^{-1}) = \det(A)\frac{1}{\det(B)} = 3\frac{1}{-2} = -\frac{3}{2}$

    (c) $\det(B^3 A^T) = (\det(B))^3 \det(A^T) = (\det(B))^3 \det(A) = (-2)^3(3) = -24$

    (d) $\det(A^2 B^3 B^T) = (\det(A))^2(\det(B))^3(\det(B^T)) = (\det(A))^2(\det(B))^3(\det(B)) = (3)^2(-2)^3(-2) =$
      144

27. $\det(A) = \det\left(\left[\begin{array}{cc|cc} 1 & -3 & 0 & 0 \\ 2 & 5 & 0 & 0 \\ \hline 0 & 0 & 5 & -1 \\ 0 & 0 & 3 & 3 \end{array}\right]\right) = \det\left(\begin{bmatrix} 1 & -3 \\ 2 & 5 \end{bmatrix}\right)\det\left(\begin{bmatrix} 5 & -1 \\ 3 & 3 \end{bmatrix}\right)$

$= (11)(18) = 198$

29. $\det(A) = \det\left(\left[\begin{array}{cc|cc} 4 & 2 & 1 & 1 \\ 3 & 1 & 5 & 9 \\ \hline 0 & 0 & 3 & 2 \\ 0 & 0 & 1 & 0 \end{array}\right]\right) = \det\left(\begin{bmatrix} 4 & 2 \\ 3 & 1 \end{bmatrix}\right)\det\left(\begin{bmatrix} 3 & 2 \\ 1 & 0 \end{bmatrix}\right) = (-2)(-2)$

$= 4$

31. $\det(A) = \det\left(\left[\begin{array}{cc|cc} 9 & 3 & 0 & 0 \\ 3 & 1 & 0 & 0 \\ \hline 3 & 9 & 2 & 2 \\ 8 & 1 & 3 & 1 \end{array}\right]\right) = \det\left(\begin{bmatrix} 9 & 3 \\ 3 & 1 \end{bmatrix}\right)\det\left(\begin{bmatrix} 2 & 2 \\ 3 & 1 \end{bmatrix}\right) = (0)(-4)$

$= 0$

33. $\det\left(\left[\begin{array}{c|c} A & B \\ \hline C & D \end{array}\right]\right) = \det\left(\left[\begin{array}{cc|cc} 1 & 2 & -1 & 0 \\ 0 & 4 & 5 & 1 \\ \hline 0 & -2 & 2 & 1 \\ -1 & 3 & 4 & 0 \end{array}\right]\right) = -3$. $\det(A)\det(D) - \det(B)\det(C) =$

$\det\left(\begin{bmatrix} 1 & 2 \\ 0 & 4 \end{bmatrix}\right)\det\left(\begin{bmatrix} 2 & 1 \\ 4 & 0 \end{bmatrix}\right) - \det\left(\begin{bmatrix} -1 & 0 \\ 5 & 1 \end{bmatrix}\right)\det\left(\begin{bmatrix} 0 & -2 \\ -1 & 3 \end{bmatrix}\right)$

$= (4)(-4) - (-1)(-2) = -18$.

35. $\det(A) = \det\left(\begin{bmatrix} 6 & -5 \\ -2 & 7 \end{bmatrix}\right) = 32 \neq 0$, hence and the system of equations $A\mathbf{x} = \mathbf{b}$ has a unique solution by The Big Theorem.

37. $\det(A) = \det\left(\begin{bmatrix} 3 & 2 & 7 \\ 0 & 0 & -3 \\ 0 & -1 & -4 \end{bmatrix}\right) = -9 \neq 0$, hence and the system of equations $A\mathbf{x} = \mathbf{b}$ has a unique solution by The Big Theorem.

39. $\det(A) = \det\left(\begin{bmatrix} 1 & 1 & -2 \\ 3 & -2 & 2 \\ 6 & -7 & -1 \end{bmatrix}\right) = 49 \neq 0$, hence and the system of equations $A\mathbf{x} = \mathbf{b}$ has a unique solution by The Big Theorem.

41. For example, $3\det\left(\begin{bmatrix} 1 & 2 \\ 2 & 4 \end{bmatrix}\right) = \det\left(3\begin{bmatrix} 1 & 2 \\ 2 & 4 \end{bmatrix}\right) = \det\left(\begin{bmatrix} 3 & 6 \\ 6 & 12 \end{bmatrix}\right) = 0.$

43. For example, $\det\left(\begin{bmatrix} 1 & 2 \\ 2 & 4 \end{bmatrix} + \begin{bmatrix} -1 & -2 \\ -2 & -4 \end{bmatrix}\right) = \det\left(\begin{bmatrix} 0 & 0 \\ 0 & 0 \end{bmatrix}\right) = 0$, and

$\det\left(\begin{bmatrix} 1 & 2 \\ 2 & 4 \end{bmatrix}\right) + \det\left(\begin{bmatrix} -1 & -2 \\ -2 & -4 \end{bmatrix}\right) = 0 + 0 = 0.$

45. For example, $\det\left(\begin{bmatrix} 1 & 1 & 1 \\ 1 & 2 & 2 \\ 1 & 1 & 2 \end{bmatrix}\right) = 1.$

47. False. Interchanging two rows will change the sign of the determinant. For instance, $\det\left(\begin{bmatrix} 1 & 1 \\ 0 & 2 \end{bmatrix}\right) = 2$ while $\det\left(\begin{bmatrix} 0 & 2 \\ 1 & 1 \end{bmatrix}\right) = -2.$

49. False. For example, $E = \begin{bmatrix} 0 & 1 \\ 1 & 0 \end{bmatrix}$ is elementary with $\det(E) = -1.$

51. False. For example, $\text{rank}\left(\begin{bmatrix} 1 & 0 & 0 \\ 0 & 1 & 0 \\ 0 & 0 & 0 \end{bmatrix}\right) = 2$ but $\det\left(\begin{bmatrix} 1 & 0 & 0 \\ 0 & 1 & 0 \\ 0 & 0 & 0 \end{bmatrix}\right) = 0.$

53. True, since $\det(B) = \det(S^{-1}AS) = \det(S^{-1})\det(A)\det(S) = \frac{1}{\det(S)}\det(A)\det(S) = \det(A).$

55. True. B^T is obtained by multiplying the third row of A^T by 2, so $\det(B) = \det(B^T) = 2\det(A^T) = 2\det(A).$

57. Let rows i and j of matrix A be identical. Let B be the matrix obtained from A by the row operation $-R_i + R_j \Rightarrow R_j$. Then row j of B is a row of zeroes, and therefore by considering the cofactor expansion along row j of B, we have $\det(A) = \det(B) = 0.$

59. $\det(A^TA) = \det(A^T)\det(A) = \det(A)\det(A) = (\det(A))^2 \geq 0.$

61. By Theorem 5.13(b) applied to the n rows of A, we obtain n factors of -1; so $\det(-A) = \det((-1)A) = (-1^n)\det(A).$

63. $\det(A^2) = \det(A)\det(A) = \det(A)^2$. Since $\det(A^2) = \det(A)$ we have $\det(A)^2 = \det(A) \Rightarrow \det(A)(\det(A) - 1) = 0$. Thus either $\det(A) = 0$ or $\det(A) = 1.$

65. $\det(-I_n) = (-1)^n\det(I_n) = (-1)\det(I_n) = -1$, since n is odd. But $\det(A^2) = \det(A)^2 \geq 0$, so it is impossible for $A^2 = -I_n$ if n is odd.

67. A has an echelon form B which is obtained from A using row operations of the form $R_i \Leftrightarrow R_j$ or $cR_i + R_j \Rightarrow R_j$. Since the first of these operations only changes the sign of the determinant, and the second has no effect on the determinant, we conclude that $\det(A) = \pm\det(B).$

69. (a) Suppose E is the matrix corresponding to the row operation $cR_i \Rightarrow R_i$. $\det(E) = c$, since E is a diagonal matrix with all ones, expect for c in the ith diagonal entry. Consider the cofactor expansion of row i of the matrix EB:

$$\begin{aligned}
\det(EB) &= (cb_{i1})\,C_{i1} + (cb_{i2})\,C_{i2} + \cdots + (cb_{in})\,C_{in} \\
&= c\,(b_{i1}C_{i1} + b_{i2}C_{i2} + \cdots + b_{in}C_{in}) \\
&= c\det(B) \\
&= \det(E)\det(B)
\end{aligned}$$

where we have used that the cofactors C_{ij} are the same for B and EB.

(b) Suppose E is the matrix corresponding to the row operation $cR_i + R_j \Rightarrow R_j$. $\det(E) = 1$, since E is a triangular matrix with all ones on the diagonal. Consider the cofactor expansion of row j of the matrix EB:

$$\begin{aligned}
\det(EB) &= (cb_{i1} + b_{j1})\,C_{j1} + (cb_{i2} + b_{j2})\,C_{j2} + \cdots + (cb_{jn})\,C_{jn} \\
&= c\,(b_{i1}C_{j1} + b_{i2}C_{j2} + \cdots + b_{in}C_{jn}) + (b_{j1}C_{j1} + b_{j2}C_{j2} + \cdots + b_{jn}C_{jn})
\end{aligned}$$

The first term, $c\,(b_{i1}C_{j1} + b_{i2}C_{j2} + \cdots + b_{in}C_{jn}) = 0$, since $b_{i1}C_{j1} + b_{i2}C_{j2} + \cdots + b_{in}C_{jn}$ is the cofactor expansion along row j of the matrix obtained from B by replacing row j with row i. Since this matrix has identical rows, its determinant is zero. The second term, $b_{j1}C_{j1} + b_{j2}C_{j2} + \cdots + b_{jn}C_{jn}$, is the determinant of B. Thus, $\det(EB) = 0 + \det(B) = (1)\det(B) = \det(E)\det(B)$.

71. Let $M = \left[\begin{array}{c|c} A & B \\ \hline 0 & D \end{array}\right]$, A a $j \times j$ square matrix, and D any square matrix. We prove $\det(M) = \det(A)\det(D)$ by induction on j. If $j = 1$, then by expansion along the first column we have

$$\begin{aligned}
\det(M) &= \det\left(\left[\begin{array}{c|c} a & B \\ \hline 0 & D \end{array}\right]\right) \\
&= (a)\det(D) \\
&= \det(A)\det(D)
\end{aligned}$$

Now suppose that $\det(M) = \det(A)\det(D)$ for all matrices A of size $(j-1) \times (j-1)$. Then expansion along the first column produces

$$\det(M) = a_{11}\det\left(\left[\begin{array}{c|c} C_{11} & * \\ \hline 0 & D \end{array}\right]\right) + a_{21}\det\left(\left[\begin{array}{c|c} C_{21} & * \\ \hline 0 & D \end{array}\right]\right) + \cdots + a_{j1}\det\left(\left[\begin{array}{c|c} C_{j1} & * \\ \hline 0 & D \end{array}\right]\right)$$

where each C_{i1} is a $(j-1) \times (j-1)$ submatrix of A, each $*$ represents B with one row removed, and the lower left matrix 0 has one fewer column. Apply the induction hypothesis to obtain

$$\begin{aligned}
\det(M) &= a_{11}\det(C_{11})\det(D) + a_{21}\det(C_{21})\det(D) + \cdots + a_{j1}\det(C_{j1})\det(D) \\
&= (a_{11}\det(C_{11}) + a_{21}\det(C_{21}) + \cdots + a_{j1}\det(C_{j1}))\det(D) \\
&= \det(A)\det(D)
\end{aligned}$$

If instead $M = \left[\begin{array}{c|c} A & 0 \\ \hline C & D \end{array}\right]$, we apply the above result to obtain $\det(M) = \det(M^T)$

$$= \det\left(\left[\begin{array}{c|c} A^T & C^T \\ \hline 0 & D^T \end{array}\right]\right) = \det(A^T)\det(D^T) = \det(A)\det(D).$$

73. $\det(I_4 + AB) = \det\left(\begin{bmatrix} 1 & 0 & 0 & 0 \\ 0 & 1 & 0 & 0 \\ 0 & 0 & 1 & 0 \\ 0 & 0 & 0 & 1 \end{bmatrix} + \begin{bmatrix} 3 & -2 & 6 \\ 4 & 0 & 5 \\ 2 & -9 & 1 \\ 5 & -1 & -4 \end{bmatrix} \begin{bmatrix} 0 & -3 & 2 & 6 \\ 1 & 4 & 4 & 2 \\ -8 & 3 & 0 & 5 \end{bmatrix}\right)$

$= \det\left(\begin{bmatrix} -49 & 1 & -2 & 44 \\ -40 & 4 & 8 & 49 \\ -17 & -39 & -31 & -1 \\ 31 & -31 & 6 & 9 \end{bmatrix}\right) = -45,780.$

$$\det\left(I_3 + BA\right) = \det\left(\begin{bmatrix} 1 & 0 & 0 \\ 0 & 1 & 0 \\ 0 & 0 & 1 \end{bmatrix} + \begin{bmatrix} 0 & -3 & 2 & 6 \\ 1 & 4 & 4 & 2 \\ -8 & 3 & 0 & 5 \end{bmatrix}\begin{bmatrix} 3 & -2 & 6 \\ 4 & 0 & 5 \\ 2 & -9 & 1 \\ 5 & -1 & -4 \end{bmatrix}\right)$$

$$= \det\left(\begin{bmatrix} 23 & -24 & -37 \\ 37 & -39 & 22 \\ 13 & 11 & -52 \end{bmatrix}\right) = -45,780.$$

5.3 Applications of the Determinant

1. Let $A = \begin{bmatrix} 6 & -5 \\ -2 & 7 \end{bmatrix}$ and $\mathbf{b} = \begin{bmatrix} 12 \\ 0 \end{bmatrix}$. Then $A_1 = \begin{bmatrix} 12 & -5 \\ 0 & 7 \end{bmatrix}$, and $A_2 = \begin{bmatrix} 6 & 12 \\ -2 & 0 \end{bmatrix}$. By Cramer's Rule, $x_1 = \frac{\det(A_1)}{\det(A)} = \frac{84}{32} = \frac{21}{8}$ and $x_2 = \frac{\det(A_2)}{\det(A)} = \frac{24}{32} = \frac{3}{4}$.

3. Let $A = \begin{bmatrix} 3 & 2 & 7 \\ 0 & 0 & -3 \\ 0 & -1 & -4 \end{bmatrix}$, and $\mathbf{b} = \begin{bmatrix} 0 \\ -3 \\ 13 \end{bmatrix}$. Then $A_1 = \begin{bmatrix} 0 & 2 & 7 \\ -3 & 0 & -3 \\ 13 & -1 & -4 \end{bmatrix}$, $A_2 = \begin{bmatrix} 3 & 0 & 7 \\ 0 & -3 & -3 \\ 0 & 13 & -4 \end{bmatrix}$, and $A_3 = \begin{bmatrix} 3 & 2 & 0 \\ 0 & 0 & -3 \\ 0 & -1 & 13 \end{bmatrix}$. By Cramer's Rule, $x_1 = \frac{\det(A_1)}{\det(A)} = \frac{-81}{-9} = 9$, $x_2 = \frac{\det(A_2)}{\det(A)} = \frac{153}{-9} = -17$, and $x_3 = \frac{\det(A_3)}{\det(A)} = \frac{-9}{-9} = 1$.

5. Let $A = \begin{bmatrix} 1 & 1 & -2 \\ 3 & -2 & 2 \\ 6 & -7 & -1 \end{bmatrix}$, and $\mathbf{b} = \begin{bmatrix} -3 \\ 9 \\ 4 \end{bmatrix}$. Then $A_1 = \begin{bmatrix} -3 & 1 & -2 \\ 9 & -2 & 2 \\ 4 & -7 & -1 \end{bmatrix}$, $A_2 = \begin{bmatrix} 1 & -3 & -2 \\ 3 & 9 & 2 \\ 6 & 4 & -1 \end{bmatrix}$, and $A_3 = \begin{bmatrix} 1 & 1 & -3 \\ 3 & -2 & 9 \\ 6 & -7 & 4 \end{bmatrix}$. By Cramer's Rule, $x_1 = \frac{\det(A_1)}{\det(A)} = \frac{79}{49}$, $x_2 = \frac{\det(A_2)}{\det(A)} = \frac{22}{49}$, and $x_3 = \frac{\det(A_3)}{\det(A)} = \frac{124}{49}$.

7. $x_2 = \frac{\det\left(A_2\right)}{\det\left(A\right)} = \dfrac{\det\left(\begin{bmatrix} -2 & 3 \\ -3 & -1 \end{bmatrix}\right)}{\det\left(\begin{bmatrix} -2 & 3 \\ -3 & -7 \end{bmatrix}\right)} = \frac{11}{23}.$

9. $x_2 = \frac{\det\left(A_2\right)}{\det\left(A\right)} = \dfrac{\det\left(\begin{bmatrix} 3 & 1 & 2 \\ 0 & 3 & 2 \\ 2 & -4 & 1 \end{bmatrix}\right)}{\det\left(\begin{bmatrix} 3 & 0 & 2 \\ 0 & 3 & 2 \\ 2 & 3 & 1 \end{bmatrix}\right)} = \frac{25}{-21} = -\frac{25}{21}.$

11. $x_2 = \frac{\det\left(A_2\right)}{\det\left(A\right)} = \dfrac{\det\left(\begin{bmatrix} 0 & 1 & 0 & -3 \\ 2 & -2 & 3 & -3 \\ -2 & 0 & 2 & 2 \\ 2 & -1 & 2 & 1 \end{bmatrix}\right)}{\det\left(\begin{bmatrix} 0 & 3 & 0 & -3 \\ 2 & -1 & 3 & -3 \\ -2 & 3 & 2 & 2 \\ 2 & 0 & 2 & 1 \end{bmatrix}\right)} = \frac{-56}{-156} = \frac{14}{39}.$

13. $\mathrm{adj}\left(A\right) = \mathrm{adj}\left(\begin{bmatrix} 2 & 5 \\ 3 & 7 \end{bmatrix}\right) = \begin{bmatrix} 7 & -5 \\ -3 & 2 \end{bmatrix}$,

$A^{-1} = \frac{1}{\det(A)}\mathrm{adj}\left(A\right) = \frac{1}{-1}\begin{bmatrix} 7 & -5 \\ -3 & 2 \end{bmatrix} = \begin{bmatrix} -7 & 5 \\ 3 & -2 \end{bmatrix}.$

15. $\text{adj}\,(A) = \text{adj}\left(\begin{bmatrix} 0 & 1 & 0 \\ 0 & 0 & 1 \\ 1 & 0 & 0 \end{bmatrix}\right) = \begin{bmatrix} 0 & 0 & 1 \\ 1 & 0 & 0 \\ 0 & 1 & 0 \end{bmatrix}$,

$A^{-1} = \frac{1}{\det(A)}\text{adj}\,(A) = \frac{1}{1}\begin{bmatrix} 0 & 0 & 1 \\ 1 & 0 & 0 \\ 0 & 1 & 0 \end{bmatrix} = \begin{bmatrix} 0 & 0 & 1 \\ 1 & 0 & 0 \\ 0 & 1 & 0 \end{bmatrix}$.

17. $\text{adj}\,(A) = \text{adj}\left(\begin{bmatrix} 1 & 2 & 1 \\ 0 & 1 & 2 \\ 0 & 0 & 1 \end{bmatrix}\right) = \begin{bmatrix} 1 & -2 & 3 \\ 0 & 1 & -2 \\ 0 & 0 & 1 \end{bmatrix}$,

$A^{-1} = \frac{1}{\det(A)}\text{adj}\,(A) = \frac{1}{1}\begin{bmatrix} 1 & -2 & 3 \\ 0 & 1 & -2 \\ 0 & 0 & 1 \end{bmatrix} = \begin{bmatrix} 1 & -2 & 3 \\ 0 & 1 & -2 \\ 0 & 0 & 1 \end{bmatrix}$.

19. 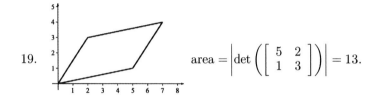 $\text{area} = \left|\det\left(\begin{bmatrix} 5 & 2 \\ 1 & 3 \end{bmatrix}\right)\right| = 13.$

21. $\text{area} = \left|\det\left(\begin{bmatrix} 3 & -3 \\ 4 & 1 \end{bmatrix}\right)\right| = 15.$

23. $\text{area}\,(T\,(\mathcal{D})) = |\det\,(A)|\,\text{area}\,(\mathcal{D}) = \left|\det\left(\begin{bmatrix} 3 & -1 \\ 5 & 2 \end{bmatrix}\right)\right|(7-2)(5-2) = 165.$

25. Since $\mathcal{D}$ is the image of the unit square under the mapping $T\,(\mathbf{x}) = B\mathbf{x} = \begin{bmatrix} 5 & 2 \\ 1 & 4 \end{bmatrix}\mathbf{x}$, $\text{area}\,(\mathcal{D}) = |\det\,(B)| = \left|\det\left(\begin{bmatrix} 5 & 2 \\ 1 & 4 \end{bmatrix}\right)\right| = 18.$ Hence $\text{area}\,(T\,(\mathcal{D})) = |\det\,(A)|\,\text{area}\,(\mathcal{D}) = \left|\det\left(\begin{bmatrix} 1 & 2 \\ 4 & 5 \end{bmatrix}\right)\right|(18) = 54.$

27. Since $\mathcal{D}$ is the image of the unit square under the mapping $T\,(\mathbf{x}) = B\mathbf{x}+\begin{bmatrix} 1 \\ 2 \end{bmatrix} = \begin{bmatrix} 5 & 1 \\ 2 & 4 \end{bmatrix}\mathbf{x}+\begin{bmatrix} 1 \\ 2 \end{bmatrix}$,

$\text{area}\,(\mathcal{D}) = |\det\,(B)| = \left|\det\left(\begin{bmatrix} 5 & 1 \\ 2 & 4 \end{bmatrix}\right)\right| = 18.$ Hence $\text{area}\,(T\,(\mathcal{D})) = |\det\,(A)|\,\text{area}\,(\mathcal{D})$

$= \left|\det\left(\begin{bmatrix} 1 & 4 \\ 2 & 5 \end{bmatrix}\right)\right|(18) = 54.$

29. $T\,(\mathbf{x}) = A\mathbf{x} = \begin{bmatrix} 5 & 0 \\ 0 & 3 \end{bmatrix}\mathbf{x}$

31. $T\,(\mathbf{x}) = RA\mathbf{x} = \begin{bmatrix} \cos\,(45°) & \sin\,(45°) \\ -\sin\,(45°) & \cos\,(45°) \end{bmatrix}\begin{bmatrix} 3 & 0 \\ 0 & 6 \end{bmatrix}\mathbf{x} = \begin{bmatrix} \frac{3}{2}\sqrt{2} & 3\sqrt{2} \\ -\frac{3}{2}\sqrt{2} & 3\sqrt{2} \end{bmatrix}\mathbf{x}.$

33. $\text{volume} = \left| \det \left(\begin{bmatrix} 4 & 0 & 0 \\ 0 & 3 & 0 \\ 0 & 0 & 5 \end{bmatrix} \right) \right| \left(\frac{4}{3}\pi \right) = 60 \left(\frac{4}{3}\pi \right) = 80\pi$

35. $\text{volume} = \left| \det \left(\begin{bmatrix} 3 & 6 & 2 \\ 5 & 1 & 0 \\ 2 & 3 & 4 \end{bmatrix} \right) \right| = 82$

37. For example,

$$x_1 + x_2 = 1$$
$$2x_1 + 2x_2 = 2$$

39. For example, let the parallelogram have vertices $(0,0)$, $(5,0)$, $(5,1)$, and $(0,1)$.

41. Let $A = \begin{bmatrix} 1 & 3 \\ 2 & 5 \end{bmatrix}$, then $\text{adj}(A) = \text{adj} \left(\begin{bmatrix} 1 & 3 \\ 2 & 5 \end{bmatrix} \right) = \begin{bmatrix} 5 & -3 \\ -2 & 1 \end{bmatrix}$.

43. False. Cramer's rule requires that A is invertible, so for example would not apply to the system

$$x_1 + x_2 = 1$$
$$2x_1 + 2x_2 = 2$$

45. False. If A is an invertible 3×3 matrix, then $\text{adj}(2A) = \det(2A)(2A)^{-1} = (2^3)\det(A)(2^{-1})A^{-1} = 2^2\left(\det(A)A^{-1}\right) = 2^2\text{adj}(A)$.

47. False. If $\det(A) = 0$, then $T(\mathcal{S})$ will be a line segment (or a point). For instance, if $A = \begin{bmatrix} 1 & 0 \\ 0 & 0 \end{bmatrix}$, then $T(\mathcal{S})$ is the line segment from $(0,0)$ to $(1,0)$.

49. True, by Theorem 5.18, and the fact that $\text{adj}(A)$ will have integer entries.

51. If $T(\mathbf{x}) = T(\mathbf{y})$ where $\mathbf{x} = \begin{bmatrix} x_1 \\ x_2 \end{bmatrix}$ and $\mathbf{y} = \begin{bmatrix} y_1 \\ y_2 \end{bmatrix}$ are two points in the interior of $\mathcal{S}$, then $B\mathbf{x} = B\mathbf{y}$, so $\begin{bmatrix} b_1 x_1 \\ b_2 x_2 \end{bmatrix} = \begin{bmatrix} b_1 y_1 \\ b_2 y_2 \end{bmatrix}$. As b_1 and b_2 are both nonzero, we conclude that $x_1 = y_1$ and $x_2 = y_2$, and hence $\mathbf{x} = \mathbf{y}$. Thus T is one-to-one. If $\mathbf{z} = \begin{bmatrix} z_1 \\ z_2 \end{bmatrix}$ is in the interior of $\mathcal{R}$, then $0 < z_1 < b_1$ and $0 < z_2 < b_2$. Let $\mathbf{x} = \begin{bmatrix} z_1/b_1 \\ z_2/b_2 \end{bmatrix}$. Then since $0 < z_1/b_1 < 1$ and $0 < z_2/b_2 < 1$, we have $\mathbf{x}$ is in the interior of $\mathcal{S}$. Also $T(\mathbf{x}) = B\mathbf{x} = \begin{bmatrix} z_1 \\ z_2 \end{bmatrix} = \mathbf{z}$, which shows that T maps the interior of $\mathcal{S}$ onto the interior of $\mathcal{R}$. Thus T gives a one-to-one correspondence between $\mathcal{S}$ and $\mathcal{R}$.

53. Since $\det(A) = -2 \neq 0$, A is invertible and $A^{-1} = \frac{1}{\det(A)}\text{adj}(A)$. Multiply both sides by $\det(A)A$ to obtain $\det(A)I_3 = A\text{adj}(A)$. Thus $A\text{adj}(A)$ is a 3×3 diagonal matrix, with each diagonal term $\det(A) = -2$.

55. From Exercise 50, $(\text{adj}(A))^T = \text{adj}(A^T)$. Since A is symmetric, $A^T = A$, and hence $(\text{adj}(A))^T = \text{adj}(A)$. We conclude that $\text{adj}(A)$ is symmetric.

57. Let the cofactors of A be given by C_{ij}, and consider a cofactor G_{ij} of the matrix cA. The corresponding N_{ij} is an $(n-1) \times (n-1)$ matrix whose entries are c times the entries of M_{ij} of A. Thus $G_{ij} = (-1)^{i+j}\det(N_{ij}) = (-1)^{i+j}c^{n-1}\det(M_{ij}) = c^{n-1}C_{ij}$. Consequently, C, the cofactor matrix of A, satisfies $G = c^{n-1}C$, and hence $\text{adj}(cA) = G^T = (c^{n-1}C)^T = c^{n-1}C^T = c^{n-1}\text{adj}(A)$.

59. Since $A \operatorname{adj}(A) = \det(A) I_n$, $\left(\frac{1}{\det(A)} A\right) \operatorname{adj}(A) = I_n \quad \Rightarrow \quad (\operatorname{adj}(A))^{-1} = \frac{1}{\det(A)} A$. Also, $A^{-1} \operatorname{adj}\left(A^{-1}\right) = \det\left(A^{-1}\right) I_n$, so $\operatorname{adj}\left(A^{-1}\right) = \det\left(A^{-1}\right) A = \frac{1}{\det(A)} A$. Hence, $(\operatorname{adj}(A))^{-1} = \operatorname{adj}\left(A^{-1}\right)$.

61. Suppose $i \neq j$. Since A is diagonal, it follows that column i (and row j) of M_{ij} consists completely of zeroes, so the cofactor $C_{ij} = 0$. Hence $\operatorname{adj}(A)$ is also diagonal.

63. $x_1 = \frac{\det(A_1)}{\det(A)} = \frac{1221}{752}$, $x_2 = \frac{\det(A_2)}{\det(A)} = \frac{811}{752}$, and $x_3 = \frac{\det(A_3)}{\det(A)} = \frac{1064}{752} = \frac{133}{94}$.

65. $x_1 = \frac{\det(A_1)}{\det(A)} = \frac{704}{245}$, $x_2 = \frac{\det(A_2)}{\det(A)} = \frac{-686}{245} = -\frac{14}{5}$, $x_3 = \frac{\det(A_3)}{\det(A)} = \frac{247}{245}$, and $x_4 = \frac{\det(A_4)}{\det(A)} = \frac{-85}{245} = -\frac{17}{49}$.

67. $\operatorname{adj}\left(\begin{bmatrix} 4 & -2 & 5 \\ 8 & 3 & 0 \\ -1 & 7 & 9 \end{bmatrix}\right) = \begin{bmatrix} 27 & 53 & -15 \\ -72 & 41 & 40 \\ 59 & -26 & 28 \end{bmatrix};$

$A^{-1} = \frac{1}{\det(A)} \operatorname{adj}(A) = \frac{1}{547} \begin{bmatrix} 27 & 53 & -15 \\ -72 & 41 & 40 \\ 59 & -26 & 28 \end{bmatrix} = \begin{bmatrix} \frac{27}{547} & \frac{53}{547} & -\frac{15}{547} \\ -\frac{72}{547} & \frac{41}{547} & \frac{40}{547} \\ \frac{59}{547} & -\frac{26}{547} & \frac{28}{547} \end{bmatrix}.$

69. $\operatorname{adj}\left(\begin{bmatrix} 4 & 2 & 5 & -1 \\ -2 & 3 & 0 & 6 \\ 5 & 7 & 2 & 11 \\ 3 & 0 & 1 & -5 \end{bmatrix}\right) = \begin{bmatrix} -21 & -126 & 60 & -15 \\ -36 & 207 & -18 & 216 \\ 118 & 3 & -35 & -97 \\ 11 & -75 & 29 & -113 \end{bmatrix};$

$A^{-1} = \frac{1}{\det(A)} \operatorname{adj}(A) = \frac{1}{423} \begin{bmatrix} -21 & -126 & 60 & -15 \\ -36 & 207 & -18 & 216 \\ 118 & 3 & -35 & -97 \\ 11 & -75 & 29 & -113 \end{bmatrix} = \begin{bmatrix} -\frac{7}{141} & -\frac{14}{47} & \frac{20}{141} & -\frac{5}{141} \\ -\frac{4}{47} & \frac{23}{47} & -\frac{2}{47} & \frac{24}{47} \\ \frac{118}{423} & \frac{1}{141} & -\frac{35}{423} & -\frac{97}{423} \\ \frac{11}{423} & -\frac{25}{141} & \frac{29}{423} & -\frac{113}{423} \end{bmatrix}.$

Chapter 6

Eigenvalues and Eigenvectors

6.1 Eigenvalues and Eigenvectors

1. $A\mathbf{x}_1 = \begin{bmatrix} 1 & 3 \\ 2 & 2 \end{bmatrix} \begin{bmatrix} -3 \\ 2 \end{bmatrix} = \begin{bmatrix} 3 \\ -2 \end{bmatrix} = (-1)\mathbf{x}_1$, so $\mathbf{x}_1$ is an eigenvector with associated eigenvalue $\lambda = -1$.

    $A\mathbf{x}_2 = \begin{bmatrix} 1 & 3 \\ 2 & 2 \end{bmatrix} \begin{bmatrix} 1 \\ -1 \end{bmatrix} = \begin{bmatrix} -2 \\ 0 \end{bmatrix} \neq \lambda\mathbf{x}_2$ for any λ, so $\mathbf{x}_2$ is not an eigenvector.

    $A\mathbf{x}_3 = \begin{bmatrix} 1 & 3 \\ 2 & 2 \end{bmatrix} \begin{bmatrix} -2 \\ -2 \end{bmatrix} = \begin{bmatrix} -8 \\ -8 \end{bmatrix} = 4\begin{bmatrix} -2 \\ -2 \end{bmatrix} = 4\mathbf{x}_3$, so $\mathbf{x}_3$ is an eigenvector with associated eigenvalue $\lambda = 4$.

3. $A\mathbf{x}_1 = \begin{bmatrix} 2 & 7 & 2 \\ 0 & -1 & 0 \\ 0 & -2 & 1 \end{bmatrix} \begin{bmatrix} -3 \\ 1 \\ 1 \end{bmatrix} = \begin{bmatrix} 3 \\ -1 \\ -1 \end{bmatrix} = (-1) \begin{bmatrix} -3 \\ 1 \\ 1 \end{bmatrix} = (-1)\mathbf{x}_1$,

    so $\mathbf{x}_1$ is an eigenvector with associated eigenvalue $\lambda = -1$.

    $A\mathbf{x}_2 = \begin{bmatrix} 2 & 7 & 2 \\ 0 & -1 & 0 \\ 0 & -2 & 1 \end{bmatrix} \begin{bmatrix} -2 \\ 0 \\ 1 \end{bmatrix} = \begin{bmatrix} -2 \\ 0 \\ 1 \end{bmatrix} = 1\mathbf{x}_2$,

    so $\mathbf{x}_2$ is an eigenvector with associated eigenvalue $\lambda = 1$.

    $A\mathbf{x}_3 = \begin{bmatrix} 2 & 7 & 2 \\ 0 & -1 & 0 \\ 0 & -2 & 1 \end{bmatrix} \begin{bmatrix} 1 \\ 0 \\ 0 \end{bmatrix} = \begin{bmatrix} 2 \\ 0 \\ 0 \end{bmatrix} = 2 \begin{bmatrix} 1 \\ 0 \\ 0 \end{bmatrix} = 2\mathbf{x}_3$,

    so $\mathbf{x}_3$ is an eigenvector with associated eigenvalue $\lambda = 2$.

5. $A\mathbf{x}_1 = \begin{bmatrix} 6 & -3 & 1 & 0 \\ 0 & 3 & 1 & 0 \\ -6 & 6 & 0 & 0 \\ -3 & 3 & -2 & 3 \end{bmatrix} \begin{bmatrix} 1 \\ 1 \\ 0 \\ 0 \end{bmatrix} = \begin{bmatrix} 3 \\ 3 \\ 0 \\ 0 \end{bmatrix} = 3 \begin{bmatrix} 1 \\ 1 \\ 0 \\ 0 \end{bmatrix} = 3\mathbf{x}_1$,

    so $\mathbf{x}_1$ is an eigenvector with associated eigenvalue $\lambda = 3$.

    $A\mathbf{x}_2 = \begin{bmatrix} 6 & -3 & 1 & 0 \\ 0 & 3 & 1 & 0 \\ -6 & 6 & 0 & 0 \\ -3 & 3 & -2 & 3 \end{bmatrix} \begin{bmatrix} 1 \\ 2 \\ -1 \\ 0 \end{bmatrix} = \begin{bmatrix} -1 \\ 5 \\ 6 \\ 5 \end{bmatrix} \neq \lambda\mathbf{x}_2$ for any λ,

    so $\mathbf{x}_2$ is not an eigenvector.

    $A\mathbf{x}_3 = \begin{bmatrix} 6 & -3 & 1 & 0 \\ 0 & 3 & 1 & 0 \\ -6 & 6 & 0 & 0 \\ -3 & 3 & -2 & 3 \end{bmatrix} \begin{bmatrix} 1 \\ 1 \\ -3 \\ -2 \end{bmatrix} = \begin{bmatrix} 0 \\ 0 \\ 0 \\ 0 \end{bmatrix} = 0 \begin{bmatrix} 1 \\ 1 \\ -3 \\ -2 \end{bmatrix} = 0\mathbf{x}_3$,

    so $\mathbf{x}_3$ is an eigenvector with associated eigenvalue $\lambda = 0$.

7. $\det\left(A - 3I_2\right) = \det\left(\begin{bmatrix} 2 & 7 \\ -1 & 6 \end{bmatrix} - 3\begin{bmatrix} 1 & 0 \\ 0 & 1 \end{bmatrix}\right) = \det\left(\begin{bmatrix} -1 & 7 \\ -1 & 3 \end{bmatrix}\right) = 4 \neq 0$,
   so $\lambda = 3$ is not an eigenvalue of A.

9. $\det\left(A - (-2)\,I_3\right) = \det\left(\begin{bmatrix} 0 & 2 & 0 \\ 2 & 0 & 0 \\ 2 & 2 & -2 \end{bmatrix} - (-2)\begin{bmatrix} 1 & 0 & 0 \\ 0 & 1 & 0 \\ 0 & 0 & 1 \end{bmatrix}\right) = \det\left(\begin{bmatrix} 2 & 2 & 0 \\ 2 & 2 & 0 \\ 2 & 2 & 0 \end{bmatrix}\right) = 0$, so $\lambda = -2$
   is an eigenvalue of A.

11. We row-reduce to obtain the null space of $A - 4I_2 = \begin{bmatrix} -3 & -3 \\ 1 & 1 \end{bmatrix} \sim \begin{bmatrix} -3 & -3 \\ 0 & 0 \end{bmatrix}$. Solving, we obtain
    $\mathbf{x} = s\begin{bmatrix} -1 \\ 1 \end{bmatrix}$. A basis for the $\lambda = 4$ eigenspace is $\left\{\begin{bmatrix} -1 \\ 1 \end{bmatrix}\right\}$.

13. We row-reduce to obtain the null space of $A - 2I_2 = \begin{bmatrix} 4 & -10 \\ 2 & -5 \end{bmatrix} \sim \begin{bmatrix} 4 & -10 \\ 0 & 0 \end{bmatrix}$. Solving, we obtain
    $\mathbf{x} = s\begin{bmatrix} \frac{5}{2} \\ 1 \end{bmatrix}$. A basis for the $\lambda = 2$ eigenspace is $\left\{\begin{bmatrix} \frac{5}{2} \\ 1 \end{bmatrix}\right\}$.

15. We row-reduce to obtain the null space of $A - 4I_3 = \begin{bmatrix} 2 & -3 & 7 \\ 4 & -3 & 5 \\ 4 & -3 & 5 \end{bmatrix} \sim \begin{bmatrix} 2 & -3 & 7 \\ 0 & 3 & -9 \\ 0 & 0 & 0 \end{bmatrix}$. Solving, we
    obtain $\mathbf{x} = s\begin{bmatrix} 1 \\ 3 \\ 1 \end{bmatrix}$. A basis for the $\lambda = 4$ eigenspace is $\left\{\begin{bmatrix} 1 \\ 3 \\ 1 \end{bmatrix}\right\}$.

17. We row-reduce to obtain the null space of $A - 6I_3 = \begin{bmatrix} -1 & -1 & 2 \\ 2 & -4 & 2 \\ 2 & -1 & -1 \end{bmatrix} \sim \begin{bmatrix} -1 & -1 & 2 \\ 0 & -6 & 6 \\ 0 & 0 & 0 \end{bmatrix}$. Solving,
    we obtain $\mathbf{x} = s\begin{bmatrix} 1 \\ 1 \\ 1 \end{bmatrix}$. A basis for the $\lambda = 6$ eigenspace is $\left\{\begin{bmatrix} 1 \\ 1 \\ 1 \end{bmatrix}\right\}$.

19. We row-reduce to obtain the null space of $A - (-4)\,I_4 =$
    $\begin{bmatrix} 15 & -3 & -3 & 8 \\ 13 & -1 & -5 & 8 \\ 2 & -2 & 2 & 0 \\ -3 & 3 & 3 & 4 \end{bmatrix} \sim \begin{bmatrix} 2 & -2 & 2 & 0 \\ 0 & 12 & -18 & 8 \\ 0 & 0 & 6 & 4 \\ 0 & 0 & 0 & 0 \end{bmatrix}$. Solving, we obtain $\mathbf{x} = s\begin{bmatrix} -1 \\ -\frac{5}{3} \\ -\frac{2}{3} \\ 1 \end{bmatrix}$.

    A basis for the eigenspace of A associated with $\lambda = -4$ is $\left\{\begin{bmatrix} -1 \\ -\frac{5}{3} \\ -\frac{2}{3} \\ 1 \end{bmatrix}\right\}$.

21. Characteristic polynomial: $\det\left(A - \lambda I_2\right) = \det\left(\begin{bmatrix} 2 & 0 \\ 4 & -3 \end{bmatrix} - \lambda\begin{bmatrix} 1 & 0 \\ 0 & 1 \end{bmatrix}\right)$
    $= \det\left(\begin{bmatrix} 2-\lambda & 0 \\ 4 & -\lambda-3 \end{bmatrix}\right) = \lambda^2 + \lambda - 6$.
    Eigenvalues: $\lambda^2 + \lambda - 6 = (\lambda + 3)(\lambda - 2) = 0 \quad \Rightarrow \quad \lambda = -3$ and $\lambda = 2$.
    Eigenspace of $\lambda = -3$:
    $$A - (-3)\,I_2 = \begin{bmatrix} 5 & 0 \\ 4 & 0 \end{bmatrix} \sim \begin{bmatrix} 5 & 0 \\ 0 & 0 \end{bmatrix},$$

so a basis for this eigenspace is $\left\{ \begin{bmatrix} 0 \\ 1 \end{bmatrix} \right\}$.

Eigenspace of $\lambda = 2$:

$$A - 2I_2 = \begin{bmatrix} 0 & 0 \\ 4 & -5 \end{bmatrix} \sim \begin{bmatrix} 4 & -5 \\ 0 & 0 \end{bmatrix},$$

so a basis for this eigenspace is $\left\{ \begin{bmatrix} \frac{5}{4} \\ 1 \end{bmatrix} \right\}$.

23. Characteristic polynomial: $\det(A - \lambda I_2) = \det\left(\begin{bmatrix} 1 & -2 \\ 2 & -3 \end{bmatrix} - \lambda \begin{bmatrix} 1 & 0 \\ 0 & 1 \end{bmatrix} \right)$

$= \det\left(\begin{bmatrix} 1 - \lambda & -2 \\ 2 & -\lambda - 3 \end{bmatrix} \right) = \lambda^2 + 2\lambda + 1.$

Eigenvalues: $\lambda^2 + 2\lambda + 1 = (\lambda + 1)^2 = 0 \quad \Rightarrow \quad \lambda = -1.$

Eigenspace of $\lambda = -1$:

$$A - (-1)I_2 = \begin{bmatrix} 2 & -2 \\ 2 & -2 \end{bmatrix} \sim \begin{bmatrix} 2 & -2 \\ 0 & 0 \end{bmatrix},$$

so a basis for this eigenspace is $\left\{ \begin{bmatrix} 1 \\ 1 \end{bmatrix} \right\}$.

25. Characteristic polynomial: $\det(A - \lambda I_3) = \det\left(\begin{bmatrix} 3 & 0 & 0 \\ 1 & 2 & 0 \\ -4 & 5 & -1 \end{bmatrix} - \lambda \begin{bmatrix} 1 & 0 & 0 \\ 0 & 1 & 0 \\ 0 & 0 & 1 \end{bmatrix} \right)$

$= \det\left(\begin{bmatrix} 3 - \lambda & 0 & 0 \\ 1 & 2 - \lambda & 0 \\ -4 & 5 & -1 - \lambda \end{bmatrix} \right) = -(\lambda - 2)(\lambda - 3)(\lambda + 1).$

Eigenvalues: $-(\lambda - 2)(\lambda - 3)(\lambda + 1) = 0 \quad \Rightarrow \quad \lambda = 2, \lambda = 3,$ and $\lambda = -1.$

Eigenspace of $\lambda = 2$:

$$A - 2I_3 = \begin{bmatrix} 1 & 0 & 0 \\ 1 & 0 & 0 \\ -4 & 5 & -3 \end{bmatrix} \sim \begin{bmatrix} 1 & 0 & 0 \\ 0 & 5 & -3 \\ 0 & 0 & 0 \end{bmatrix},$$

so a basis for this eigenspace is $\left\{ \begin{bmatrix} 0 \\ \frac{3}{5} \\ 1 \end{bmatrix} \right\}$.

Eigenspace of $\lambda = 3$:

$$A - 3I_3 = \begin{bmatrix} 0 & 0 & 0 \\ 1 & -1 & 0 \\ -4 & 5 & -4 \end{bmatrix} \sim \begin{bmatrix} 1 & -1 & 0 \\ 0 & 1 & -4 \\ 0 & 0 & 0 \end{bmatrix},$$

so a basis for this eigenspace is $\left\{ \begin{bmatrix} 4 \\ 4 \\ 1 \end{bmatrix} \right\}$.

Eigenspace of $\lambda = -1$:

$$A - (-1)I_3 = \begin{bmatrix} 4 & 0 & 0 \\ 1 & 3 & 0 \\ -4 & 5 & 0 \end{bmatrix} \sim \begin{bmatrix} 4 & 0 & 0 \\ 0 & 3 & 0 \\ 0 & 0 & 0 \end{bmatrix},$$

so a basis for this eigenspace is $\left\{ \begin{bmatrix} 0 \\ 0 \\ 1 \end{bmatrix} \right\}$.

27. Characteristic polynomial: $\det(A - \lambda I_3) = \det\left(\begin{bmatrix} 2 & 5 & 1 \\ 0 & -3 & -1 \\ 2 & 14 & 4 \end{bmatrix} - \lambda \begin{bmatrix} 1 & 0 & 0 \\ 0 & 1 & 0 \\ 0 & 0 & 1 \end{bmatrix} \right)$

$$= \det\left(\begin{bmatrix} 2-\lambda & 5 & 1 \\ 0 & -\lambda-3 & -1 \\ 2 & 14 & 4-\lambda \end{bmatrix}\right) = -\lambda^3 + 3\lambda^2 - 2\lambda.$$

Eigenvalues: $-\lambda^3 + 3\lambda^2 - 2\lambda = -\lambda(\lambda-1)(\lambda-2) = 0 \Rightarrow \lambda = 0, \lambda = 1,$ and $\lambda = 2.$

Eigenspace of $\lambda = 0$:

$$A - 0I_3 = \begin{bmatrix} 2 & 5 & 1 \\ 0 & -3 & -1 \\ 2 & 14 & 4 \end{bmatrix} \sim \begin{bmatrix} 2 & 5 & 1 \\ 0 & -3 & -1 \\ 0 & 0 & 0 \end{bmatrix},$$

so a basis for this eigenspace is $\left\{ \begin{bmatrix} \frac{1}{3} \\ -\frac{1}{3} \\ 1 \end{bmatrix} \right\}.$

Eigenspace of $\lambda = 1$:

$$A - (1)I_3 = \begin{bmatrix} 1 & 5 & 1 \\ 0 & -4 & -1 \\ 2 & 14 & 3 \end{bmatrix} \sim \begin{bmatrix} 1 & 5 & 1 \\ 0 & -4 & -1 \\ 0 & 0 & 0 \end{bmatrix},$$

so a basis for this eigenspace is $\left\{ \begin{bmatrix} \frac{1}{4} \\ -\frac{1}{4} \\ 1 \end{bmatrix} \right\}.$

Eigenspace of $\lambda = 2$:

$$A - 2I_3 = \begin{bmatrix} 0 & 5 & 1 \\ 0 & -5 & -1 \\ 2 & 14 & 2 \end{bmatrix} \sim \begin{bmatrix} 2 & 14 & 2 \\ 0 & -5 & -1 \\ 0 & 0 & 0 \end{bmatrix},$$

so a basis for this eigenspace is $\left\{ \begin{bmatrix} \frac{2}{5} \\ -\frac{1}{5} \\ 1 \end{bmatrix} \right\}.$

29. Characteristic polynomial: $\det(A - \lambda I_4) = \det\left(\begin{bmatrix} -1 & 0 & 0 & 0 \\ 5 & -2 & 0 & 0 \\ 0 & 3 & 1 & 0 \\ 2 & 0 & 1 & 1 \end{bmatrix} - \lambda \begin{bmatrix} 1 & 0 & 0 & 0 \\ 0 & 1 & 0 & 0 \\ 0 & 0 & 1 & 0 \\ 0 & 0 & 0 & 1 \end{bmatrix}\right)$

$= \det\left(\begin{bmatrix} -1-\lambda & 0 & 0 & 0 \\ 5 & -2-\lambda & 0 & 0 \\ 0 & 3 & 1-\lambda & 0 \\ 2 & 0 & 1 & 1-\lambda \end{bmatrix}\right) = (\lambda+2)(\lambda+1)(\lambda-1)^2.$

Eigenvalues: $(\lambda+2)(\lambda+1)(\lambda-1)^2 = 0 \Rightarrow \lambda = -2, \lambda = -1,$ and $\lambda = 1.$

Eigenspace of $\lambda = -2$:

$$A - (-2)I_4 = \begin{bmatrix} 1 & 0 & 0 & 0 \\ 5 & 0 & 0 & 0 \\ 0 & 3 & 3 & 0 \\ 2 & 0 & 1 & 3 \end{bmatrix} \sim \begin{bmatrix} 1 & 0 & 0 & 0 \\ 0 & 3 & 3 & 0 \\ 0 & 0 & 1 & 3 \\ 0 & 0 & 0 & 0 \end{bmatrix}$$

so a basis for this eigenspace is $\left\{ \begin{bmatrix} 0 \\ 3 \\ -3 \\ 1 \end{bmatrix} \right\}.$

Eigenspace of $\lambda = -1$:

$$A - (-1)I_4 = \begin{bmatrix} 0 & 0 & 0 & 0 \\ 5 & -1 & 0 & 0 \\ 0 & 3 & 2 & 0 \\ 2 & 0 & 1 & 2 \end{bmatrix} \sim \begin{bmatrix} 5 & -1 & 0 & 0 \\ 0 & 3 & 2 & 0 \\ 0 & 0 & \frac{11}{15} & 2 \\ 0 & 0 & 0 & 0 \end{bmatrix}$$

so a basis for this eigenspace is $\left\{ \begin{bmatrix} \frac{4}{11} \\ \frac{20}{11} \\ -\frac{30}{11} \\ 1 \end{bmatrix} \right\}$.

Eigenspace of $\lambda = 1$:

$$A - (1)\,I_4 = \begin{bmatrix} -2 & 0 & 0 & 0 \\ 5 & -3 & 0 & 0 \\ 0 & 3 & 0 & 0 \\ 2 & 0 & 1 & 0 \end{bmatrix} \sim \begin{bmatrix} -2 & 0 & 0 & 0 \\ 0 & -3 & 0 & 0 \\ 0 & 0 & 1 & 0 \\ 0 & 0 & 0 & 0 \end{bmatrix},$$

so a basis for this eigenspace is $\left\{ \begin{bmatrix} 0 \\ 0 \\ 0 \\ 1 \end{bmatrix} \right\}$.

31. For example, $A = \begin{bmatrix} 1 & 0 \\ 0 & 2 \end{bmatrix}$.

33. For example, $A = \begin{bmatrix} 1 & 0 & 0 \\ 0 & -2 & 0 \\ 0 & 0 & 3 \end{bmatrix}$.

35. For example, $A = \begin{bmatrix} 0 & 1 \\ -1 & 0 \end{bmatrix}$, has characteristic polynomial $\lambda^2 + 1$, so it has no real eigenvalues.

37. False. An eigenvalue may be 0, as with the matrix $A = \begin{bmatrix} 1 & 0 \\ 0 & 0 \end{bmatrix}$, which eigenvalues $\lambda = 0$ and $\lambda = 1$. Moreover, by Definition 6.1, an eigenvector must be a nonzero vector.

39. False, if $\lambda < 0$, then $A\mathbf{u}$ and $\mathbf{u}$ point in opposite directions. For instance, if $A = \begin{bmatrix} -1 & 0 \\ 0 & 0 \end{bmatrix}$ and $\mathbf{u} = \begin{bmatrix} 1 \\ 0 \end{bmatrix}$, then $A\mathbf{u} = -\mathbf{u}$.

41. True, since $A - \lambda I$ will be a diagonal matrix with $a_{ii} - \lambda$ along the diagonal, so $\det(A - \lambda I)$ have have $a_{ii} - \lambda$ as a factor.

43. True. Since 0 is an eigenvalue, by the Big Theorem there exists a nonzero vector $\mathbf{x}$ such that $A\mathbf{x} = \mathbf{0}$, and thus nullity $(A) > 0$.

45. False, $\begin{bmatrix} 0 & 0 \\ 1 & 0 \end{bmatrix}$ has only the eigenvalue 0.

47. (a) A is 6×6.
    (b) $\lambda = 3$, $\lambda = 2$, and $\lambda = -1$.
    (c) A is invertible, since 0 is not an eigenvalue.
    (d) The largest possible dimension of an eigenspace is 3, corresponding to $\lambda = 3$.

49. If 0 is not an eigenvalue, then T is onto, by The Big Theorem - Version 8, (m)→(d).

51. Since $\lambda = 1$ is an eigenvalue of A, $\det(A - \lambda I) = \det(A - I) = 0$. Thus $A - I$ is singular.

53. If $A\mathbf{u} = \lambda_1\mathbf{u} = \lambda_2\mathbf{u}$, then $(\lambda_1 - \lambda_2)\mathbf{u} = \mathbf{0}$. Since $\mathbf{u}$ is an eigenvector, $\mathbf{u} \neq \mathbf{0}$, and hence $\lambda_1 - \lambda_2 = 0$. So $\lambda_1 = \lambda_2$, and $\mathbf{u}$ cannot be associated with distinct eigenvalues.

55. Since $A\mathbf{u} = A\mathbf{0} = \mathbf{0} = \lambda\mathbf{0}$ for every λ, we would have that every value λ would be an associated eigenvalue of $\mathbf{u}$.

57. Since A is invertible, $\lambda \neq 0$. Let $\mathbf{u}$ be an eigenvector of A associated with λ, so $A\mathbf{u} = \lambda\mathbf{u}$. Multiply by $\lambda^{-1}A^{-1}$, $\left(\lambda^{-1}A^{-1}\right)(A\mathbf{u}) = \left(\lambda^{-1}A^{-1}\right)(\lambda\mathbf{u}) \quad \Rightarrow \quad \lambda^{-1}\mathbf{u} = A^{-1}\mathbf{u}$, from which we conclude that λ^{-1} is an eigenvalue of A^{-1} with associated eigenvector $\mathbf{u}$.

59. There exist λ_1 and λ_2 such that $A\mathbf{u} = \lambda_1\mathbf{u}$ and $B\mathbf{u} = \lambda_2\mathbf{u}$. Thus $(AB)\mathbf{u} = A(B\mathbf{u}) = A(\lambda_2\mathbf{u}) = \lambda_2(A\mathbf{u}) = \lambda_2(\lambda_1\mathbf{u}) = (\lambda_1\lambda_2)\mathbf{u}$, and so $\mathbf{u}$ is an eigenvector of AB with associated eigenvalue $\lambda_1\lambda_2$.

61. Since each row of A adds to zero, we have $A\begin{bmatrix} 1 \\ 1 \\ \vdots \\ 1 \end{bmatrix} = \mathbf{0} = 0\begin{bmatrix} 1 \\ 1 \\ \vdots \\ 1 \end{bmatrix}$. This implies that A is not invertible since $\mathrm{null}(A)$ is not trivial. Also, we see directly that $\lambda = 0$ is an eigenvalue of A.

63. Since λ is an eigenvalue of A, $\det(A - \lambda I_n) = 0$. Thus $0 = \det\left(\left(A - \lambda I_n\right)^T\right) = \det\left(A^T - \lambda I_n\right)$, and hence λ is an eigenvalue of A^T.

65. Since each row of A adds to c, we obtain $A\begin{bmatrix} 1 \\ 1 \\ \vdots \\ 1 \end{bmatrix} = \begin{bmatrix} c \\ c \\ \vdots \\ c \end{bmatrix} = c\begin{bmatrix} 1 \\ 1 \\ \vdots \\ 1 \end{bmatrix}$. Thus $\lambda = c$ is an eigenvalue of A.

67. $\det\left(\begin{bmatrix} 0 & 0 & -2 & -1 \\ 1 & 1 & 6 & 5 \\ 2 & 0 & 4 & 1 \\ -2 & 0 & -2 & 1 \end{bmatrix} - \lambda\begin{bmatrix} 1 & 0 & 0 & 0 \\ 0 & 1 & 0 & 0 \\ 0 & 0 & 1 & 0 \\ 0 & 0 & 0 & 1 \end{bmatrix}\right) = \lambda^4 - 6\lambda^3 + 13\lambda^2 - 12\lambda + 4$
$= (\lambda - 1)^2(\lambda - 2)^2 = 0 \quad \Rightarrow \quad \lambda = 1$ and $\lambda = 2$ are the eigenvalues, with bases for their eigenspaces:

$$\lambda = 1 \Rightarrow \left\{\begin{bmatrix} 0 \\ 1 \\ 0 \\ 0 \end{bmatrix}, \begin{bmatrix} 1 \\ 0 \\ -1 \\ 1 \end{bmatrix}\right\}, \lambda = 2 \Rightarrow \left\{\begin{bmatrix} -1 \\ 5 \\ 1 \\ 0 \end{bmatrix}, \begin{bmatrix} -\frac{1}{2} \\ \frac{9}{2} \\ 0 \\ 1 \end{bmatrix}\right\}.$$

69. $\det\left(\begin{bmatrix} 10 & 0 & 1 & -3 & 3 \\ 23 & -1 & 6 & -3 & 2 \\ -24 & 0 & -1 & 9 & -9 \\ 14 & 0 & 1 & -5 & 5 \\ -10 & 0 & -1 & 3 & -3 \end{bmatrix} - \lambda\begin{bmatrix} 1 & 0 & 0 & 0 & 0 \\ 0 & 1 & 0 & 0 & 0 \\ 0 & 0 & 1 & 0 & 0 \\ 0 & 0 & 0 & 1 & 0 \\ 0 & 0 & 0 & 0 & 1 \end{bmatrix}\right) = -\lambda^5 + 5\lambda^3 - 4\lambda$
$= -\lambda(\lambda - 1)(\lambda - 2)(\lambda + 2)(\lambda + 1) = 0 \quad \Rightarrow \quad \lambda = 0, \lambda = 1, \lambda = 2, \lambda = -2$, and $\lambda = -1$ are the eigenvalues, with bases for their eigenspaces:

$$\lambda = 0 \Rightarrow \left\{\begin{bmatrix} 0 \\ -1 \\ 0 \\ 1 \\ 1 \end{bmatrix}\right\}, \lambda = 1 \Rightarrow \left\{\begin{bmatrix} -1 \\ 0 \\ 3 \\ -1 \\ 1 \end{bmatrix}\right\}, \lambda = 2 \Rightarrow \left\{\begin{bmatrix} -1 \\ -2 \\ 2 \\ -1 \\ 1 \end{bmatrix}\right\},$$

$$\lambda = -2 \Rightarrow \left\{\begin{bmatrix} -1 \\ -3 \\ 3 \\ -2 \\ 1 \end{bmatrix}\right\}, \lambda = -1 \Rightarrow \left\{\begin{bmatrix} 0 \\ 1 \\ 0 \\ 0 \\ 0 \end{bmatrix}\right\}.$$

6.2 Approximation Methods

Note: Many of the solutions in this section are decimal approximations related by an "=" even though the equality is only approximate.

1. $\mathbf{x}_1 = A\mathbf{x}_0 = \begin{bmatrix} 1 & -3 \\ 1 & 5 \end{bmatrix} \begin{bmatrix} 1 \\ 0 \end{bmatrix} = \begin{bmatrix} 1 \\ 1 \end{bmatrix}$, $\mathbf{x}_2 = A\mathbf{x}_1 = \begin{bmatrix} 1 & -3 \\ 1 & 5 \end{bmatrix} \begin{bmatrix} 1 \\ 1 \end{bmatrix} = \begin{bmatrix} -2 \\ 6 \end{bmatrix}$, $\mathbf{x}_3 = A\mathbf{x}_2 = \begin{bmatrix} 1 & -3 \\ 1 & 5 \end{bmatrix} \begin{bmatrix} -2 \\ 6 \end{bmatrix} = \begin{bmatrix} -20 \\ 28 \end{bmatrix}$.

3. $\mathbf{x}_1 = A\mathbf{x}_0 = \begin{bmatrix} 6 & -3 & 7 \\ 4 & 1 & 5 \\ 4 & -3 & 9 \end{bmatrix} \begin{bmatrix} 1 \\ 0 \\ 0 \end{bmatrix} = \begin{bmatrix} 6 \\ 4 \\ 4 \end{bmatrix}$, $\mathbf{x}_2 = A\mathbf{x}_1 = \begin{bmatrix} 6 & -3 & 7 \\ 4 & 1 & 5 \\ 4 & -3 & 9 \end{bmatrix} \begin{bmatrix} 6 \\ 4 \\ 4 \end{bmatrix} = \begin{bmatrix} 52 \\ 48 \\ 48 \end{bmatrix}$, $\mathbf{x}_3 = A\mathbf{x}_2 = \begin{bmatrix} 6 & -3 & 7 \\ 4 & 1 & 5 \\ 4 & -3 & 9 \end{bmatrix} \begin{bmatrix} 52 \\ 48 \\ 48 \end{bmatrix} = \begin{bmatrix} 504 \\ 496 \\ 496 \end{bmatrix}$.

5. $\mathbf{x}_1 = A\mathbf{x}_0 = \begin{bmatrix} 5 & -1 & 2 \\ 2 & 2 & 2 \\ 2 & -1 & 5 \end{bmatrix} \begin{bmatrix} 1 \\ 0 \\ -1 \end{bmatrix} = \begin{bmatrix} 3 \\ 0 \\ -3 \end{bmatrix}$, $\mathbf{x}_2 = A\mathbf{x}_1 = \begin{bmatrix} 5 & -1 & 2 \\ 2 & 2 & 2 \\ 2 & -1 & 5 \end{bmatrix} \begin{bmatrix} 3 \\ 0 \\ -3 \end{bmatrix} = \begin{bmatrix} 9 \\ 0 \\ -9 \end{bmatrix}$,

$\mathbf{x}_3 = A\mathbf{x}_2 = \begin{bmatrix} 5 & -1 & 2 \\ 2 & 2 & 2 \\ 2 & -1 & 5 \end{bmatrix} \begin{bmatrix} 9 \\ 0 \\ -9 \end{bmatrix} = \begin{bmatrix} 27 \\ 0 \\ -27 \end{bmatrix}$.

7. $A\mathbf{x}_0 = \begin{bmatrix} 2 & -1 \\ 0 & 1 \end{bmatrix} \begin{bmatrix} 0 \\ 1 \end{bmatrix} = \begin{bmatrix} -1 \\ 1 \end{bmatrix}$, $s_1 = 1$, $\mathbf{x}_1 = \frac{1}{s_1} A\mathbf{x}_0 = \frac{1}{1} \begin{bmatrix} -1 \\ 1 \end{bmatrix} = \begin{bmatrix} -1 \\ 1 \end{bmatrix}$.

$A\mathbf{x}_1 = \begin{bmatrix} 2 & -1 \\ 0 & 1 \end{bmatrix} \begin{bmatrix} -1 \\ 1 \end{bmatrix} = \begin{bmatrix} -3 \\ 1 \end{bmatrix}$, $s_2 = -3$, $\mathbf{x}_2 = \frac{1}{s_2} A\mathbf{x}_1 = \frac{1}{-3} \begin{bmatrix} -3 \\ 1 \end{bmatrix} = \begin{bmatrix} 1 \\ -\frac{1}{3} \end{bmatrix} = \begin{bmatrix} 1.00 \\ -0.33 \end{bmatrix}$.

9. $A\mathbf{x}_0 = \begin{bmatrix} -1 & 0 & 2 \\ 1 & 1 & 0 \\ 0 & -2 & 1 \end{bmatrix} \begin{bmatrix} 0 \\ 1 \\ 0 \end{bmatrix} = \begin{bmatrix} 0 \\ 1 \\ -2 \end{bmatrix}$, $s_1 = -2$, $\mathbf{x}_1 = \frac{1}{s_1} A\mathbf{x}_0 = \frac{1}{-2} \begin{bmatrix} 0 \\ 1 \\ -2 \end{bmatrix} = \begin{bmatrix} 0 \\ -\frac{1}{2} \\ 1 \end{bmatrix} =$

$\begin{bmatrix} 0.00 \\ -0.50 \\ 1.00 \end{bmatrix}$. $A\mathbf{x}_1 = \begin{bmatrix} -1 & 0 & 2 \\ 1 & 1 & 0 \\ 0 & -2 & 1 \end{bmatrix} \begin{bmatrix} 0.00 \\ -0.50 \\ 1.00 \end{bmatrix} = \begin{bmatrix} 2.00 \\ -0.50 \\ 2.00 \end{bmatrix}$, $s_2 = 2$, $\mathbf{x}_2 = \frac{1}{s_2} A\mathbf{x}_1 = \frac{1}{2} \begin{bmatrix} 2.00 \\ -0.50 \\ 2.00 \end{bmatrix} =$

$\begin{bmatrix} 1.00 \\ -0.25 \\ 1.00 \end{bmatrix}$.

11. $A\mathbf{x}_0 = \begin{bmatrix} 1 & 0 & 0 \\ -1 & 3 & 0 \\ 2 & -1 & 1 \end{bmatrix} \begin{bmatrix} 0 \\ 1 \\ -1 \end{bmatrix} = \begin{bmatrix} 0 \\ 3 \\ -2 \end{bmatrix}$, $s_1 = 3$, $\mathbf{x}_1 = \frac{1}{s_1} A\mathbf{x}_0 = \frac{1}{3} \begin{bmatrix} 0 \\ 3 \\ -2 \end{bmatrix} = \begin{bmatrix} 0 \\ 1 \\ -\frac{2}{3} \end{bmatrix} = \begin{bmatrix} 0.00 \\ 1.00 \\ -0.67 \end{bmatrix}$.

$A\mathbf{x}_1 = \begin{bmatrix} 1 & 0 & 0 \\ -1 & 3 & 0 \\ 2 & -1 & 1 \end{bmatrix} \begin{bmatrix} 0.00 \\ 1.00 \\ -0.67 \end{bmatrix} = \begin{bmatrix} 0.00 \\ 3.00 \\ -1.67 \end{bmatrix}$, $s_2 = 3.00$, $\mathbf{x}_2 = \frac{1}{s_2} A\mathbf{x}_1 = \frac{1}{3.00} \begin{bmatrix} 0.00 \\ 3.00 \\ -1.67 \end{bmatrix} =$

$\begin{bmatrix} 0.00 \\ 1.00 \\ -0.56 \end{bmatrix}$.

13. The Power Method will converge, with eigenvalue $\lambda = 7$.

15. The Power Method will converge, with eigenvalue $\lambda = -6$.

17. The Power Method will converge, with eigenvalue $\lambda = 6$.

19. $B = A - 4I_2 = \begin{bmatrix} 1 & 2 \\ 3 & 2 \end{bmatrix} - 4 \begin{bmatrix} 1 & 0 \\ 0 & 1 \end{bmatrix} = \begin{bmatrix} -3 & 2 \\ 3 & -2 \end{bmatrix}$.

21. $B = A - 9I_3 = \begin{bmatrix} -1 & 2 & -7 \\ -10 & 2 & 2 \\ -10 & 2 & 2 \end{bmatrix} - 9 \begin{bmatrix} 1 & 0 & 0 \\ 0 & 1 & 0 \\ 0 & 0 & 1 \end{bmatrix} = \begin{bmatrix} -10 & 2 & -7 \\ -10 & -7 & 2 \\ -10 & 2 & -7 \end{bmatrix}.$

23. $B = A - 4I_2 = \begin{bmatrix} -3 & 1 \\ 5 & 2 \end{bmatrix} - 4 \begin{bmatrix} 1 & 0 \\ 0 & 1 \end{bmatrix} = \begin{bmatrix} -7 & 1 \\ 5 & -2 \end{bmatrix}.$

25. $B = A - (-1)I_3 = \begin{bmatrix} 3 & 1 & 4 \\ 1 & 5 & 9 \\ 2 & 6 & 1 \end{bmatrix} - (-1) \begin{bmatrix} 1 & 0 & 0 \\ 0 & 1 & 0 \\ 0 & 0 & 1 \end{bmatrix} = \begin{bmatrix} 4 & 1 & 4 \\ 1 & 6 & 9 \\ 2 & 6 & 2 \end{bmatrix}.$

27. We have $\lim_{k \to \infty} s_k = \frac{1}{4}$. Since this is the Inverse Power Method, the eigenvalue is $\frac{1}{1/4} = 4$. The eigenvector is $\lim_{k \to \infty} \mathbf{x}_k = \begin{bmatrix} 1 \\ 1/2 \\ 0 \end{bmatrix}.$

29. For example, $A = \begin{bmatrix} 1 & 0 \\ 0 & 0 \end{bmatrix}$ and $\mathbf{x}_0 = \begin{bmatrix} 1 \\ 0 \end{bmatrix}.$

31. For example, $A = \begin{bmatrix} 0 & 1 \\ 1 & 0 \end{bmatrix}$ and $\mathbf{x}_0 = \begin{bmatrix} 1 \\ 0 \end{bmatrix}$. Then $\mathbf{x}_1 = \begin{bmatrix} 0 \\ 1 \end{bmatrix}$, $\mathbf{x}_2 = \begin{bmatrix} 1 \\ 0 \end{bmatrix}$, and hence we obtain $\mathbf{x}_0 = \mathbf{x}_2 = \cdots$, $\mathbf{x}_1 = \mathbf{x}_3 = \cdots$, and $\mathbf{x}_0 \neq \mathbf{x}_1$.

33. For example, $A = \begin{bmatrix} 0 & -1 \\ 1 & -1 \end{bmatrix}$ and $\mathbf{x}_0 = \begin{bmatrix} 1 \\ 0 \end{bmatrix}$. Then the Power Method without scaling produces $\mathbf{x}_0 = \begin{bmatrix} 1 \\ 0 \end{bmatrix}$, $\mathbf{x}_1 = \begin{bmatrix} 0 \\ 1 \end{bmatrix}$, $\mathbf{x}_2 = \begin{bmatrix} -1 \\ -1 \end{bmatrix}$, $\mathbf{x}_3 = \begin{bmatrix} 1 \\ 0 \end{bmatrix}$, $\cdots$.

35. False. For instance, if $A = \begin{bmatrix} 2 & 0 \\ 0 & -1 \end{bmatrix}$, then $\lambda = 2$ is the dominant eigenvalue. But if we start with $\mathbf{x}_0 = \begin{bmatrix} 0 \\ 1 \end{bmatrix}$, then $\mathbf{x}_1 = \begin{bmatrix} 0 \\ -1 \end{bmatrix}$, $\mathbf{x}_2 = \begin{bmatrix} 0 \\ 1 \end{bmatrix}$, ..., which does not converge.

37. True, as the Inverse Power Method requires that A^{-1} exists.

39. True.

41. $A\mathbf{x}_0 = \begin{bmatrix} 1 & 1 \\ 0 & -1 \end{bmatrix} \begin{bmatrix} 0 \\ 1 \end{bmatrix} = \begin{bmatrix} 1 \\ -1 \end{bmatrix}$, $s_1 = 1$, $\mathbf{x}_1 = \frac{1}{s_1} A\mathbf{x}_0 = \frac{1}{1} \begin{bmatrix} 1 \\ -1 \end{bmatrix} = \begin{bmatrix} 1 \\ -1 \end{bmatrix}.$

$A\mathbf{x}_1 = \begin{bmatrix} 1 & 1 \\ 0 & -1 \end{bmatrix} \begin{bmatrix} 1 \\ -1 \end{bmatrix} = \begin{bmatrix} 0 \\ 1 \end{bmatrix}$, $s_2 = 1$, $\mathbf{x}_2 = \frac{1}{s_2} A\mathbf{x}_1 = \frac{1}{1} \begin{bmatrix} 0 \\ 1 \end{bmatrix} = \begin{bmatrix} 0 \\ 1 \end{bmatrix}.$

$A\mathbf{x}_2 = \begin{bmatrix} 1 & 1 \\ 0 & -1 \end{bmatrix} \begin{bmatrix} 0 \\ 1 \end{bmatrix} = \begin{bmatrix} 1 \\ -1 \end{bmatrix}$, $s_3 = 1$, $\mathbf{x}_3 = \frac{1}{s_3} A\mathbf{x}_2 = \frac{1}{1} \begin{bmatrix} 1 \\ -1 \end{bmatrix} = \begin{bmatrix} 1 \\ -1 \end{bmatrix}.$

$A\mathbf{x}_3 = \begin{bmatrix} 1 & 1 \\ 0 & -1 \end{bmatrix} \begin{bmatrix} 1 \\ -1 \end{bmatrix} = \begin{bmatrix} 0 \\ 1 \end{bmatrix}$, $s_4 = 1$, $\mathbf{x}_4 = \frac{1}{s_4} A\mathbf{x}_3 = \frac{1}{1} \begin{bmatrix} 0 \\ 1 \end{bmatrix} = \begin{bmatrix} 0 \\ 1 \end{bmatrix}.$

The sequence $\mathbf{x}_k$ does not converge, it alternates. The eigenvalues of A are $\lambda = 1$ and $\lambda = -1$, and so there is no dominant eigenvalue, and convergence is not assured.

43. $A\mathbf{x}_0 = \begin{bmatrix} 1 & 0 \\ 2 & 2 \end{bmatrix} \begin{bmatrix} -1 \\ 2 \end{bmatrix} = \begin{bmatrix} -1 \\ 2 \end{bmatrix}$, $s_1 = 2$, $\mathbf{x}_1 = \frac{1}{s_1} A\mathbf{x}_0 = \frac{1}{2} \begin{bmatrix} -1 \\ 2 \end{bmatrix} = \begin{bmatrix} -\frac{1}{2} \\ 1 \end{bmatrix}.$

$A\mathbf{x}_1 = \begin{bmatrix} 1 & 0 \\ 2 & 2 \end{bmatrix} \begin{bmatrix} -\frac{1}{2} \\ 1 \end{bmatrix} = \begin{bmatrix} -\frac{1}{2} \\ 1 \end{bmatrix}$, $s_2 = 1$, $\mathbf{x}_2 = \frac{1}{s_2} A\mathbf{x}_1 = \frac{1}{1} \begin{bmatrix} -\frac{1}{2} \\ 1 \end{bmatrix} = \begin{bmatrix} -\frac{1}{2} \\ 1 \end{bmatrix}.$

Since $\mathbf{x}_2 = \mathbf{x}_1$, the sequence converges, and we will obtain $s_k = 1$ for all $k > 1$. Thus, an eigenvalue is given by $\lim_{k \to \infty} s_k = 1$. We did not obtain the dominant eigenvalue because $\mathbf{x}_0$ is an eigenvector of A associated with eigenvalue $\lambda = 1$.

45. $\mathbf{x}_1 = \begin{bmatrix} 1 & -3 \\ 1 & 5 \end{bmatrix} \begin{bmatrix} 1 \\ 0 \end{bmatrix} = \begin{bmatrix} 1 \\ 1 \end{bmatrix}$, $\mathbf{x}_2 = \begin{bmatrix} 1 & -3 \\ 1 & 5 \end{bmatrix} \begin{bmatrix} 1 \\ 1 \end{bmatrix} = \begin{bmatrix} -2 \\ 6 \end{bmatrix}$,

$\mathbf{x}_3 = \begin{bmatrix} 1 & -3 \\ 1 & 5 \end{bmatrix} \begin{bmatrix} -2 \\ 6 \end{bmatrix} = \begin{bmatrix} -20 \\ 28 \end{bmatrix}$, $\mathbf{x}_4 = \begin{bmatrix} 1 & -3 \\ 1 & 5 \end{bmatrix} \begin{bmatrix} -20 \\ 28 \end{bmatrix} = \begin{bmatrix} -104 \\ 120 \end{bmatrix}$,

$\mathbf{x}_5 = \begin{bmatrix} 1 & -3 \\ 1 & 5 \end{bmatrix} \begin{bmatrix} -104 \\ 120 \end{bmatrix} = \begin{bmatrix} -464 \\ 496 \end{bmatrix}$, $\mathbf{x}_6 = \begin{bmatrix} 1 & -3 \\ 1 & 5 \end{bmatrix} \begin{bmatrix} -464 \\ 496 \end{bmatrix} = \begin{bmatrix} -1952 \\ 2016 \end{bmatrix}$.

47. $\mathbf{x}_1 = \begin{bmatrix} 6 & -3 & 7 \\ 4 & 1 & 5 \\ 4 & -3 & 9 \end{bmatrix} \begin{bmatrix} 1 \\ 0 \\ 0 \end{bmatrix} = \begin{bmatrix} 6 \\ 4 \\ 4 \end{bmatrix}$, $\mathbf{x}_2 = \begin{bmatrix} 6 & -3 & 7 \\ 4 & 1 & 5 \\ 4 & -3 & 9 \end{bmatrix} \begin{bmatrix} 6 \\ 4 \\ 4 \end{bmatrix} = \begin{bmatrix} 52 \\ 48 \\ 48 \end{bmatrix}$,

$\mathbf{x}_3 = \begin{bmatrix} 6 & -3 & 7 \\ 4 & 1 & 5 \\ 4 & -3 & 9 \end{bmatrix} \begin{bmatrix} 52 \\ 48 \\ 48 \end{bmatrix} = \begin{bmatrix} 504 \\ 496 \\ 496 \end{bmatrix}$, $\mathbf{x}_4 = \begin{bmatrix} 6 & -3 & 7 \\ 4 & 1 & 5 \\ 4 & -3 & 9 \end{bmatrix} \begin{bmatrix} 504 \\ 496 \\ 496 \end{bmatrix} = \begin{bmatrix} 5008 \\ 4992 \\ 4992 \end{bmatrix}$,

$\mathbf{x}_5 = \begin{bmatrix} 6 & -3 & 7 \\ 4 & 1 & 5 \\ 4 & -3 & 9 \end{bmatrix} \begin{bmatrix} 5008 \\ 4992 \\ 4992 \end{bmatrix} = \begin{bmatrix} 50016 \\ 49984 \\ 49984 \end{bmatrix}$, $\mathbf{x}_6 = \begin{bmatrix} 6 & -3 & 7 \\ 4 & 1 & 5 \\ 4 & -3 & 9 \end{bmatrix} \begin{bmatrix} 50016 \\ 49984 \\ 49984 \end{bmatrix} = \begin{bmatrix} 500032 \\ 499968 \\ 499968 \end{bmatrix}$.

49. $\mathbf{x}_1 = \begin{bmatrix} 5 & -1 & 2 \\ 2 & 2 & 2 \\ 2 & -1 & 5 \end{bmatrix} \begin{bmatrix} 1 \\ 0 \\ -1 \end{bmatrix} = \begin{bmatrix} 3 \\ 0 \\ -3 \end{bmatrix}$, $\mathbf{x}_2 = \begin{bmatrix} 5 & -1 & 2 \\ 2 & 2 & 2 \\ 2 & -1 & 5 \end{bmatrix} \begin{bmatrix} 3 \\ 0 \\ -3 \end{bmatrix} = \begin{bmatrix} 9 \\ 0 \\ -9 \end{bmatrix}$,

$\mathbf{x}_3 = \begin{bmatrix} 5 & -1 & 2 \\ 2 & 2 & 2 \\ 2 & -1 & 5 \end{bmatrix} \begin{bmatrix} 9 \\ 0 \\ -9 \end{bmatrix} = \begin{bmatrix} 27 \\ 0 \\ -27 \end{bmatrix}$, $\mathbf{x}_4 = \begin{bmatrix} 5 & -1 & 2 \\ 2 & 2 & 2 \\ 2 & -1 & 5 \end{bmatrix} \begin{bmatrix} 27 \\ 0 \\ -27 \end{bmatrix} = \begin{bmatrix} 81 \\ 0 \\ -81 \end{bmatrix}$,

$\mathbf{x}_5 = \begin{bmatrix} 5 & -1 & 2 \\ 2 & 2 & 2 \\ 2 & -1 & 5 \end{bmatrix} \begin{bmatrix} 81 \\ 0 \\ -81 \end{bmatrix} = \begin{bmatrix} 243 \\ 0 \\ -243 \end{bmatrix}$, $\mathbf{x}_6 = \begin{bmatrix} 5 & -1 & 2 \\ 2 & 2 & 2 \\ 2 & -1 & 5 \end{bmatrix} \begin{bmatrix} 243 \\ 0 \\ -243 \end{bmatrix} = \begin{bmatrix} 729 \\ 0 \\ -729 \end{bmatrix}$.

51. Applying the Power Method, we obtain

k	$\mathbf{x}_k$	s_k
5	$(-1.0000, 0.0323)$	2.0667
10	$(-1.0000, 0.0010)$	2.0020
15	$(-1.0000, 0.0000)$	2.0001

We conclude that there is an eigenvalue $\lambda = \lim\limits_{k \to \infty} s_k = 2$, with eigenvector

$$\lim_{k \to \infty} \mathbf{x}_k = \begin{bmatrix} -1 \\ 0 \end{bmatrix}.$$

53. Applying the Power Method, we obtain

k	$\mathbf{x}_k$	s_k
5	$(-0.0042, 1.0000, -0.6983)$	4.0170
10	$(-0.0649, 1.0000, -0.6446)$	4.2776
15	$(-0.0569, 1.0000, -0.6529)$	4.2412
20	$(-0.0580, 1.0000, -0.6517)$	4.2465
25	$(-0.0579, 1.0000, -0.6519)$	4.2457
30	$(-0.0579, 1.0000, -0.6518)$	4.2458

We conclude that there is an eigenvalue $\lambda = \lim\limits_{k \to \infty} s_k = 4.2458$, with eigenvector

$$\lim_{k \to \infty} \mathbf{x}_k = \begin{bmatrix} -0.0579 \\ 1.0000 \\ -0.6518 \end{bmatrix}.$$

55. Applying the Power Method, we obtain

k	$\mathbf{x}_k$	s_k
5	$(0, 1.0000, -0.5021)$	3
10	$(0, 1.0000, -0.5000)$	3
15	$(0, 1.0000, -0.5000)$	3

We conclude that there is an eigenvalue $\lambda = \lim\limits_{k\to\infty} s_k = 3$, with eigenvector

$$\lim_{k\to\infty} \mathbf{x}_k = \begin{bmatrix} 0 \\ 1 \\ -1/2 \end{bmatrix}.$$

6.3   Change of Basis

1. $\mathbf{x} = U\mathbf{x}_B = \begin{bmatrix} 3 & 2 \\ -2 & 5 \end{bmatrix} \begin{bmatrix} 1 \\ -1 \end{bmatrix}_B = \begin{bmatrix} 1 \\ -7 \end{bmatrix}$

3. $\mathbf{x} = U\mathbf{x}_B = \begin{bmatrix} 4 & 2 \\ 3 & 1 \end{bmatrix} \begin{bmatrix} 2 \\ -4 \end{bmatrix}_B = \begin{bmatrix} 0 \\ 2 \end{bmatrix}$

5. $\mathbf{x} = U\mathbf{x}_B = \begin{bmatrix} 1 & -1 & 2 \\ -2 & 2 & -1 \\ -1 & 0 & 3 \end{bmatrix} \begin{bmatrix} 1 \\ 2 \\ 1 \end{bmatrix}_B = \begin{bmatrix} 1 \\ 1 \\ 2 \end{bmatrix}$

7. $\mathbf{x}_B = U^{-1}\mathbf{x} = \begin{bmatrix} 1 & 1 \\ 2 & 3 \end{bmatrix}^{-1} \begin{bmatrix} 3 \\ -1 \end{bmatrix} = \begin{bmatrix} 3 & -1 \\ -2 & 1 \end{bmatrix} \begin{bmatrix} 3 \\ -1 \end{bmatrix} = \begin{bmatrix} 10 \\ -7 \end{bmatrix}$

9. $\mathbf{x}_B = U^{-1}\mathbf{x} = \begin{bmatrix} -2 & 5 \\ 1 & -3 \end{bmatrix}^{-1} \begin{bmatrix} 1 \\ -1 \end{bmatrix} = \begin{bmatrix} -3 & -5 \\ -1 & -2 \end{bmatrix} \begin{bmatrix} 1 \\ -1 \end{bmatrix} = \begin{bmatrix} 2 \\ 1 \end{bmatrix}$

11. $\mathbf{x}_B = U^{-1}\mathbf{x} = \begin{bmatrix} 1 & -1 & 1 \\ 0 & 2 & -1 \\ 0 & 1 & 0 \end{bmatrix}^{-1} \begin{bmatrix} 1 \\ 2 \\ -1 \end{bmatrix} = \begin{bmatrix} 1 & 1 & -1 \\ 0 & 0 & 1 \\ 0 & -1 & 2 \end{bmatrix} \begin{bmatrix} 1 \\ 2 \\ -1 \end{bmatrix} = \begin{bmatrix} 4 \\ -1 \\ -4 \end{bmatrix}$

13. $\begin{bmatrix} 5 & 3 \\ 7 & 4 \end{bmatrix}^{-1} \begin{bmatrix} 2 & 1 \\ 1 & 1 \end{bmatrix} = \begin{bmatrix} -5 & -1 \\ 9 & 2 \end{bmatrix}$

15. $\begin{bmatrix} 1 & 0 & 3 \\ 3 & 1 & 7 \\ -1 & -1 & 0 \end{bmatrix}^{-1} \begin{bmatrix} -1 & 2 & 1 \\ 0 & 3 & -1 \\ 4 & 3 & -2 \end{bmatrix} = \begin{bmatrix} -19 & -4 & 16 \\ 15 & 1 & -14 \\ 6 & 2 & -5 \end{bmatrix}$

17. We row-reduce $[\; \mathcal{B}_2 \mid \mathcal{B}_1\;]$ to row echelon form to solve for C, as $[\; \mathcal{B}_2 \mid \mathcal{B}_1\;] \sim \left[\begin{array}{c|c} I & C \\ \hline 0 & 0 \end{array}\right]$:

$$\left[\begin{array}{rr|rr} 2 & 1 & 3 & -1 \\ -1 & 3 & -5 & 4 \\ 2 & -1 & 5 & -3 \end{array}\right] \sim \left[\begin{array}{rr|rr} 1 & 0 & 2 & -1 \\ 0 & 1 & -1 & 1 \\ 0 & 0 & 0 & 0 \end{array}\right]$$

Thus the change of basis matrix from $\mathcal{B}_1$ to $\mathcal{B}_2$ is $C = \begin{bmatrix} 2 & -1 \\ -1 & 1 \end{bmatrix}$.

19. $\begin{bmatrix} 2 & 1 \\ 1 & 1 \end{bmatrix}^{-1} \begin{bmatrix} 5 & 3 \\ 7 & 4 \end{bmatrix} = \begin{bmatrix} -2 & -1 \\ 9 & 5 \end{bmatrix}$

21. $\begin{bmatrix} -1 & 1 & -2 \\ 1 & 0 & 5 \\ -1 & 2 & 0 \end{bmatrix}^{-1} \begin{bmatrix} -2 & 2 & 4 \\ 1 & 0 & 1 \\ 3 & 1 & -1 \end{bmatrix} = \begin{bmatrix} 31 & -15 & -49 \\ 17 & -7 & -25 \\ -6 & 3 & 10 \end{bmatrix}$

23. We row-reduce $[\ \mathcal{B}_1\ |\ \mathcal{B}_2\ ]$ to row echelon form to solve for C, as $[\ \mathcal{B}_1\ |\ \mathcal{B}_2\ ] \sim \left[\begin{array}{c|c} I & C \\ \hline 0 & 0 \end{array}\right]$:

$$\left[\begin{array}{cc|cc} -1 & 2 & 1 & -4 \\ 4 & 3 & 7 & 5 \\ 2 & 1 & 3 & 3 \end{array}\right] \sim \left[\begin{array}{cc|cc} 1 & 0 & 1 & 2 \\ 0 & 1 & 1 & -1 \\ 0 & 0 & 0 & 0 \end{array}\right]$$

Thus the change of basis matrix from $\mathcal{B}_2$ to $\mathcal{B}_1$ is $C = \begin{bmatrix} 1 & 2 \\ 1 & -1 \end{bmatrix}$.

25. Using the change of basis matrix from Exercise 13, we have $\mathbf{x}_{\mathcal{B}_2} = \begin{bmatrix} -5 & -1 \\ 9 & 2 \end{bmatrix} \mathbf{x}_{\mathcal{B}_1}$

$$= \begin{bmatrix} -5 & -1 \\ 9 & 2 \end{bmatrix} \begin{bmatrix} 2 \\ -1 \end{bmatrix}_{\mathcal{B}_1} = \begin{bmatrix} -9 \\ 16 \end{bmatrix}_{\mathcal{B}_2}.$$

27. From Exercise 17 we know that the change of basis matrix from $\mathcal{B}_1$ to $\mathcal{B}_2$ is $C = \begin{bmatrix} 2 & -1 \\ -1 & 1 \end{bmatrix}$. Therefore

$$\mathbf{x}_{\mathcal{B}_2} = C\mathbf{x}_{\mathcal{B}_1} = \begin{bmatrix} -1 \\ 3 \end{bmatrix}_{\mathcal{B}_2}.$$

29. Using the change of basis matrix from Exercise 21, we have $\mathbf{x}_{\mathcal{B}_1} = \begin{bmatrix} 31 & -15 & -49 \\ 17 & -7 & -25 \\ -6 & 3 & 10 \end{bmatrix} \mathbf{x}_{\mathcal{B}_2} =$

$$\begin{bmatrix} 31 & -15 & -49 \\ 17 & -7 & -25 \\ -6 & 3 & 10 \end{bmatrix} \begin{bmatrix} -1 \\ 1 \\ 3 \end{bmatrix}_{\mathcal{B}_2} = \begin{bmatrix} -193 \\ -99 \\ 39 \end{bmatrix}_{\mathcal{B}_1}.$$

31. The change of basis matrix from $\mathcal{B}_1$ to $\mathcal{B}_2$ is given by

$$[\ \mathbf{u}_2\ \ \mathbf{u}_1\ ]^{-1}[\ \mathbf{u}_1\ \ \mathbf{u}_2\ ] = \left([\ \mathbf{u}_1\ \ \mathbf{u}_2\ ]\begin{bmatrix} 0 & 1 \\ 1 & 0 \end{bmatrix}\right)^{-1}[\ \mathbf{u}_1\ \ \mathbf{u}_2\ ]$$

$$= \begin{bmatrix} 0 & 1 \\ 1 & 0 \end{bmatrix}^{-1}\left([\ \mathbf{u}_1\ \ \mathbf{u}_2\ ]^{-1}[\ \mathbf{u}_1\ \ \mathbf{u}_2\ ]\right)$$

$$= \begin{bmatrix} 0 & 1 \\ 1 & 0 \end{bmatrix}I_2 = \begin{bmatrix} 0 & 1 \\ 1 & 0 \end{bmatrix}$$

Thus $\mathbf{x}_{\mathcal{B}_2} = \begin{bmatrix} 0 & 1 \\ 1 & 0 \end{bmatrix}\mathbf{x}_{\mathcal{B}_1} = \begin{bmatrix} 0 & 1 \\ 1 & 0 \end{bmatrix}\begin{bmatrix} a \\ b \end{bmatrix}_{\mathcal{B}_1} = \begin{bmatrix} b \\ a \end{bmatrix}_{\mathcal{B}_2}.$

33. For example, $\mathcal{B} = \left\{ \begin{bmatrix} -2 \\ 0 \end{bmatrix}, \begin{bmatrix} 0 \\ 1/3 \end{bmatrix} \right\}.$

35. For example, $\mathcal{B}_1 = \left\{ \begin{bmatrix} 2 \\ 0 \end{bmatrix}, \begin{bmatrix} 0 \\ -1/2 \end{bmatrix} \right\}$ and $\mathcal{B}_2 = \left\{ \begin{bmatrix} 1 \\ 0 \end{bmatrix}, \begin{bmatrix} 0 \\ 1 \end{bmatrix} \right\}.$

37. For example, $\mathcal{B}_1 = \left\{ \begin{bmatrix} 1 \\ 2 \end{bmatrix}, \begin{bmatrix} 3 \\ 7 \end{bmatrix} \right\}$ and $\mathcal{B}_2 = \left\{ \begin{bmatrix} 1 \\ 0 \end{bmatrix}, \begin{bmatrix} 0 \\ 1 \end{bmatrix} \right\}.$

39. True, since U is a matrix corresponding to a transformation from a vector $\mathbf{x}_{\mathcal{B}_1}$ to $\mathbf{x}_{\mathcal{B}_2}$, both of which have n components.

41. True. Since $U = W^{-1}V$ where V has columns given by the basis vectors of $\mathcal{B}_2$, and W has columns given by the basis vectors of $\mathcal{B}_1$, it follows that U is invertible, with $U^{-1} = V^{-1}W$.

43. False. The correct change of basis matrix is $\begin{bmatrix} a & c \\ b & d \end{bmatrix}$. To see why, note that if $\mathbf{u} = e\mathbf{u}_1 + f\mathbf{u}_2$, then we also have $\mathbf{u} = e(a\mathbf{v}_1 + b\mathbf{v}_2) + f(c\mathbf{v}_1 + d\mathbf{v}_2) = (ae + cf)\mathbf{v}_1 + (be + df)\mathbf{v}_2$. Therefore

$$\mathbf{u}_{\mathcal{B}_2} = \begin{bmatrix} a & c \\ b & d \end{bmatrix} \begin{bmatrix} e \\ f \end{bmatrix}_{\mathcal{B}_1}$$

45. Let $\mathcal{B} = \{\mathbf{w}_1, \ldots, \mathbf{w}_n\}$. Suppose $\mathbf{u}_{\mathcal{B}} = \begin{bmatrix} a_1 \\ \vdots \\ a_n \end{bmatrix}_{\mathcal{B}}$ and $\mathbf{v}_{\mathcal{B}} = \begin{bmatrix} b_1 \\ \vdots \\ b_n \end{bmatrix}_{\mathcal{B}}$. Then $\mathbf{u} = a_1\mathbf{w}_1 + \cdots + a_n\mathbf{w}_n$ and $\mathbf{v} = b_1\mathbf{w}_1 + \cdots + b_n\mathbf{w}_n$. Hence $\mathbf{u} + \mathbf{v} = (a_1\mathbf{w}_1 + \cdots + a_n\mathbf{w}_n) + (b_1\mathbf{w}_1 + \cdots + b_n\mathbf{w}_n) = (a_1 + b_1)\mathbf{w}_1 + \cdots + (a_n + b_n)\mathbf{w}_n$. This implies $[\mathbf{u} + \mathbf{v}]_{\mathcal{B}}$

$$= \begin{bmatrix} a_1 + b_1 \\ \vdots \\ a_n + b_n \end{bmatrix}_{\mathcal{B}} = \begin{bmatrix} a_1 \\ \vdots \\ a_n \end{bmatrix}_{\mathcal{B}} + \begin{bmatrix} b_1 \\ \vdots \\ b_n \end{bmatrix}_{\mathcal{B}} = \mathbf{u}_{\mathcal{B}} + \mathbf{v}_{\mathcal{B}}.$$

47. By Exercise 45, $T(\mathbf{x} + \mathbf{y}) = [\mathbf{x} + \mathbf{y}]_{\mathcal{B}} = \mathbf{x}_{\mathcal{B}} + \mathbf{y}_{\mathcal{B}} = T(\mathbf{x}) + T(\mathbf{y})$, and by Exercise 46, $T(c\mathbf{x}) = [c\mathbf{x}]_{\mathcal{B}} = c\mathbf{x}_{\mathcal{B}} = cT(\mathbf{x})$. Hence T is a linear transformation.

49. Let $\mathbf{u}_i = c_1\mathbf{v}_1 + \cdots + c_n\mathbf{v}_n$, then $[\mathbf{u}_i]_{\mathcal{B}_2} = \begin{bmatrix} c_1 \\ \vdots \\ c_n \end{bmatrix}_{\mathcal{B}_2}$. Thus $V[\mathbf{u}_i]_{\mathcal{B}_2} = \begin{bmatrix} \mathbf{v}_1 & \cdots & \mathbf{v}_n \end{bmatrix} \begin{bmatrix} c_1 \\ \vdots \\ c_n \end{bmatrix}_{\mathcal{B}_2}$

$= c_1\mathbf{v}_1 + \cdots + c_n\mathbf{v}_n = \mathbf{u}_i$. Hence $V \begin{bmatrix} [\mathbf{u}_1]_{\mathcal{B}_2} & \cdots & [\mathbf{u}_n]_{\mathcal{B}_2} \end{bmatrix} = \begin{bmatrix} \mathbf{u}_1 & \cdots & \mathbf{u}_n \end{bmatrix}$

$\Rightarrow \quad VC = U \quad \Rightarrow \quad C = V^{-1}U.$

51. Row-reduce $[\ V \mid U\ ]$:

$$\begin{bmatrix} 3 & 2 & -4 & | & 3 & 4 & 2 \\ 2 & 5 & -6 & | & 5 & -2 & 11 \\ -7 & 8 & 1 & | & 9 & 7 & -6 \end{bmatrix} \sim \begin{bmatrix} & & & | & \frac{61}{35} & \frac{48}{5} & -\frac{316}{35} \\ I_3 & & & | & \frac{17}{7} & 8 & -\frac{51}{7} \\ & & & | & \frac{62}{35} & \frac{51}{5} & -\frac{382}{35} \end{bmatrix}$$

Hence the change of basis matrix is $\begin{bmatrix} \frac{61}{35} & \frac{48}{5} & -\frac{316}{35} \\ \frac{17}{7} & 8 & -\frac{51}{7} \\ \frac{62}{35} & \frac{51}{5} & -\frac{382}{35} \end{bmatrix}$.

53. Row-reduce $[\ V \mid U\ ]$:

$$\begin{bmatrix} 3 & 7 & 5 & 4 & | & 1 & 2 & 5 & 3 \\ 4 & -2 & -3 & 0 & | & 0 & -2 & 0 & 7 \\ -2 & -1 & 2 & 3 & | & 1 & -5 & 0 & -9 \\ 1 & 0 & 3 & -1 & | & -1 & 3 & -2 & -2 \end{bmatrix} \sim \begin{bmatrix} & & & & | & -\frac{49}{538} & \frac{55}{269} & \frac{3}{538} & \frac{364}{269} \\ I_4 & & & & | & \frac{37}{538} & \frac{211}{269} & \frac{393}{538} & \frac{340}{269} \\ & & & & | & -\frac{45}{269} & \frac{112}{269} & -\frac{129}{269} & -\frac{369}{269} \\ & & & & | & \frac{219}{538} & -\frac{416}{269} & \frac{305}{538} & -\frac{205}{269} \end{bmatrix}$$

Hence the change of basis matrix is $\begin{bmatrix} -\frac{49}{538} & \frac{55}{269} & \frac{3}{538} & \frac{364}{269} \\ \frac{37}{538} & \frac{211}{269} & \frac{393}{538} & \frac{340}{269} \\ -\frac{45}{269} & \frac{112}{269} & -\frac{129}{269} & -\frac{369}{269} \\ \frac{219}{538} & -\frac{416}{269} & \frac{305}{538} & -\frac{205}{269} \end{bmatrix}$.

55. Row-reduce $[\ V \mid U\ ]$:

$$\begin{bmatrix} 2 & -3 & 6 & | & -10 & -5 & -8 \\ 1 & 0 & -3 & | & 4 & 5 & 5 \\ 0 & 2 & 2 & | & 2 & 0 & 2 \\ -1 & -2 & 0 & | & -5 & -4 & -6 \end{bmatrix} \sim \begin{bmatrix} 1 & 0 & 0 & | & 1 & 2 & 2 \\ 0 & 1 & 0 & | & 2 & 1 & 2 \\ 0 & 0 & 1 & | & -1 & -1 & -1 \\ 0 & 0 & 0 & | & 0 & 0 & 0 \end{bmatrix}$$

Hence the change of basis matrix is $\begin{bmatrix} 1 & 2 & 2 \\ 2 & 1 & 2 \\ -1 & -1 & -1 \end{bmatrix}$.

6.4 Diagonalization

1. $A^5 = PD^5P^{-1} = \begin{bmatrix} 4 & 3 \\ 1 & 1 \end{bmatrix} \begin{bmatrix} 2 & 0 \\ 0 & -1 \end{bmatrix}^5 \begin{bmatrix} 4 & 3 \\ 1 & 1 \end{bmatrix}^{-1}$

$= \begin{bmatrix} 4 & 3 \\ 1 & 1 \end{bmatrix} \begin{bmatrix} 2^5 & 0 \\ 0 & (-1)^5 \end{bmatrix} \begin{bmatrix} 1 & -3 \\ -1 & 4 \end{bmatrix} = \begin{bmatrix} 131 & -396 \\ 33 & -100 \end{bmatrix}$

3. $A^5 = PD^5P^{-1} = \begin{bmatrix} 1 & 3 & 1 \\ 0 & -1 & 2 \\ 0 & 0 & -1 \end{bmatrix} \begin{bmatrix} 1 & 0 & 0 \\ 0 & 2 & 0 \\ 0 & 0 & -1 \end{bmatrix}^5 \begin{bmatrix} 1 & 3 & 1 \\ 0 & -1 & 2 \\ 0 & 0 & -1 \end{bmatrix}^{-1}$

$= \begin{bmatrix} 1 & 3 & 1 \\ 0 & -1 & 2 \\ 0 & 0 & -1 \end{bmatrix} \begin{bmatrix} 1^5 & 0 & 0 \\ 0 & 2^5 & 0 \\ 0 & 0 & (-1)^5 \end{bmatrix} \begin{bmatrix} 1 & 3 & 7 \\ 0 & -1 & -2 \\ 0 & 0 & -1 \end{bmatrix} = \begin{bmatrix} 1 & -93 & -184 \\ 0 & 32 & 66 \\ 0 & 0 & -1 \end{bmatrix}$

5. $A = PDP^{-1} = \begin{bmatrix} 2 & 3 \\ 3 & 5 \end{bmatrix} \begin{bmatrix} 1 & 0 \\ 0 & -1 \end{bmatrix} \begin{bmatrix} 2 & 3 \\ 3 & 5 \end{bmatrix}^{-1}$

$= \begin{bmatrix} 2 & 3 \\ 3 & 5 \end{bmatrix} \begin{bmatrix} 1 & 0 \\ 0 & -1 \end{bmatrix} \begin{bmatrix} 5 & -3 \\ -3 & 2 \end{bmatrix} = \begin{bmatrix} 19 & -12 \\ 30 & -19 \end{bmatrix}$

7. $A = PDP^{-1} = \begin{bmatrix} 1 & 1 & -1 \\ 1 & 2 & 1 \\ 0 & 1 & 1 \end{bmatrix} \begin{bmatrix} -1 & 0 & 0 \\ 0 & 0 & 0 \\ 0 & 0 & 1 \end{bmatrix} \begin{bmatrix} 1 & 1 & -1 \\ 1 & 2 & 1 \\ 0 & 1 & 1 \end{bmatrix}^{-1}$

$= \begin{bmatrix} 1 & 1 & -1 \\ 1 & 2 & 1 \\ 0 & 1 & 1 \end{bmatrix} \begin{bmatrix} -1 & 0 & 0 \\ 0 & 0 & 0 \\ 0 & 0 & 1 \end{bmatrix} \begin{bmatrix} -1 & 2 & -3 \\ 1 & -1 & 2 \\ -1 & 1 & -1 \end{bmatrix} = \begin{bmatrix} 2 & -3 & 4 \\ 0 & -1 & 2 \\ -1 & 1 & -1 \end{bmatrix}$

9. $\det\left(\begin{bmatrix} 1 & -2 \\ 0 & 1 \end{bmatrix} - \lambda \begin{bmatrix} 1 & 0 \\ 0 & 1 \end{bmatrix}\right) = \lambda^2 - 2\lambda + 1 = (\lambda - 1)^2 = 0 \quad \Rightarrow \quad \lambda = 1$.

Since $\begin{bmatrix} 1 & -2 \\ 0 & 1 \end{bmatrix} - (1) \begin{bmatrix} 1 & 0 \\ 0 & 1 \end{bmatrix} = \begin{bmatrix} 0 & -2 \\ 0 & 0 \end{bmatrix}$, a basis for the eigenspace of $\lambda = 1$ is $\left\{ \begin{bmatrix} 1 \\ 0 \end{bmatrix} \right\}$. The multiplicity of the eigenvalue $\lambda = 1$ is two, which exceeds the dimension of its eigenspace. Thus $\begin{bmatrix} 1 & -2 \\ 0 & 1 \end{bmatrix}$ is not diagonalizable.

11. $\det\left(\begin{bmatrix} 7 & -8 \\ 4 & -5 \end{bmatrix} - \lambda \begin{bmatrix} 1 & 0 \\ 0 & 1 \end{bmatrix}\right) = \lambda^2 - 2\lambda - 3 = : (\lambda + 1)(\lambda - 3) = 0 \quad \Rightarrow \quad \lambda_1 = -1$ and $\lambda_2 = 3$.

Since $\begin{bmatrix} 7 & -8 \\ 4 & -5 \end{bmatrix} - (-1) \begin{bmatrix} 1 & 0 \\ 0 & 1 \end{bmatrix} = \begin{bmatrix} 8 & -8 \\ 4 & -4 \end{bmatrix} \sim \begin{bmatrix} 8 & -8 \\ 0 & 0 \end{bmatrix}$, a basis for the eigenspace of $\lambda_1 = -1$ is $\left\{ \begin{bmatrix} 1 \\ 1 \end{bmatrix} \right\}$. Since $\begin{bmatrix} 7 & -8 \\ 4 & -5 \end{bmatrix} - 3 \begin{bmatrix} 1 & 0 \\ 0 & 1 \end{bmatrix} = \begin{bmatrix} 4 & -8 \\ 4 & -8 \end{bmatrix} \sim \begin{bmatrix} 4 & -8 \\ 0 & 0 \end{bmatrix}$, a basis for the eigenspace of $\lambda_2 = 3$ is $\left\{ \begin{bmatrix} 2 \\ 1 \end{bmatrix} \right\}$.

We thus have $\begin{bmatrix} 7 & -8 \\ 4 & -5 \end{bmatrix} = \begin{bmatrix} 1 & 2 \\ 1 & 1 \end{bmatrix} \begin{bmatrix} -1 & 0 \\ 0 & 3 \end{bmatrix} \begin{bmatrix} 1 & 2 \\ 1 & 1 \end{bmatrix}^{-1}$.

13. $\det\left(\begin{bmatrix} 1 & 2 & 1 \\ 0 & -3 & -2 \\ 2 & 4 & 2 \end{bmatrix} - \lambda \begin{bmatrix} 1 & 0 & 0 \\ 0 & 1 & 0 \\ 0 & 0 & 1 \end{bmatrix}\right) = \lambda - \lambda^3 = -\lambda(\lambda - 1)(\lambda + 1) = 0 \quad \Rightarrow \quad \lambda_1 = 0, \lambda_2 = 1,$ and $\lambda_3 = -1$.

Since $\begin{bmatrix} 1 & 2 & 1 \\ 0 & -3 & -2 \\ 2 & 4 & 2 \end{bmatrix} - (0) \begin{bmatrix} 1 & 0 & 0 \\ 0 & 1 & 0 \\ 0 & 0 & 1 \end{bmatrix} = \begin{bmatrix} 1 & 2 & 1 \\ 0 & -3 & -2 \\ 2 & 4 & 2 \end{bmatrix} \sim \begin{bmatrix} 1 & 0 & -\frac{1}{3} \\ 0 & 1 & \frac{2}{3} \\ 0 & 0 & 0 \end{bmatrix}$,

a basis for the eigenspace of $\lambda_1 = 0$ is $\left\{ \begin{bmatrix} \frac{1}{3} \\ -\frac{2}{3} \\ 1 \end{bmatrix} \right\}$.

Since $\begin{bmatrix} 1 & 2 & 1 \\ 0 & -3 & -2 \\ 2 & 4 & 2 \end{bmatrix} - 1 \begin{bmatrix} 1 & 0 & 0 \\ 0 & 1 & 0 \\ 0 & 0 & 1 \end{bmatrix} = \begin{bmatrix} 0 & 2 & 1 \\ 0 & -4 & -2 \\ 2 & 4 & 1 \end{bmatrix} \sim \begin{bmatrix} 1 & 0 & -\frac{1}{2} \\ 0 & 1 & \frac{1}{2} \\ 0 & 0 & 0 \end{bmatrix}$,

a basis for the eigenspace of $\lambda_2 = 1$ is $\left\{ \begin{bmatrix} \frac{1}{2} \\ -\frac{1}{2} \\ 1 \end{bmatrix} \right\}$.

Since $\begin{bmatrix} 1 & 2 & 1 \\ 0 & -3 & -2 \\ 2 & 4 & 2 \end{bmatrix} - (-1) \begin{bmatrix} 1 & 0 & 0 \\ 0 & 1 & 0 \\ 0 & 0 & 1 \end{bmatrix} = \begin{bmatrix} 2 & 2 & 1 \\ 0 & -2 & -2 \\ 2 & 4 & 3 \end{bmatrix} \sim \begin{bmatrix} 1 & 0 & -\frac{1}{2} \\ 0 & 1 & 1 \\ 0 & 0 & 0 \end{bmatrix}$,

a basis for the eigenspace of $\lambda_3 = -1$ is $\left\{ \begin{bmatrix} \frac{1}{2} \\ -1 \\ 1 \end{bmatrix} \right\}$.

We thus have $\begin{bmatrix} 1 & 2 & 1 \\ 0 & -3 & -2 \\ 2 & 4 & 2 \end{bmatrix} = \begin{bmatrix} \frac{1}{3} & \frac{1}{2} & \frac{1}{2} \\ -\frac{2}{3} & -\frac{1}{2} & -1 \\ 1 & 1 & 1 \end{bmatrix} \begin{bmatrix} 0 & 0 & 0 \\ 0 & 1 & 0 \\ 0 & 0 & -1 \end{bmatrix} \begin{bmatrix} \frac{1}{3} & \frac{1}{2} & \frac{1}{2} \\ -\frac{2}{3} & -\frac{1}{2} & -1 \\ 1 & 1 & 1 \end{bmatrix}^{-1}$.

15. $\det \left(\begin{bmatrix} 0 & 1 & -1 \\ 1 & 0 & 1 \\ 1 & -1 & 2 \end{bmatrix} - \lambda \begin{bmatrix} 1 & 0 & 0 \\ 0 & 1 & 0 \\ 0 & 0 & 1 \end{bmatrix} \right) = -\lambda^3 + 2\lambda^2 - \lambda = -\lambda(\lambda - 1)^2 = 0 \quad \Rightarrow \quad \lambda_1 = 0$, and $\lambda_2 = 1$.

Since $\begin{bmatrix} 0 & 1 & -1 \\ 1 & 0 & 1 \\ 1 & -1 & 2 \end{bmatrix} - 0 \begin{bmatrix} 1 & 0 & 0 \\ 0 & 1 & 0 \\ 0 & 0 & 1 \end{bmatrix} = \begin{bmatrix} 0 & 1 & -1 \\ 1 & 0 & 1 \\ 1 & -1 & 2 \end{bmatrix} \sim \begin{bmatrix} 1 & 0 & 1 \\ 0 & 1 & -1 \\ 0 & 0 & 0 \end{bmatrix}$,

a basis for the eigenspace of $\lambda_1 = 0$ is $\left\{ \begin{bmatrix} -1 \\ 1 \\ 1 \end{bmatrix} \right\}$.

Since $\begin{bmatrix} 0 & 1 & -1 \\ 1 & 0 & 1 \\ 1 & -1 & 2 \end{bmatrix} - 1 \begin{bmatrix} 1 & 0 & 0 \\ 0 & 1 & 0 \\ 0 & 0 & 1 \end{bmatrix} = \begin{bmatrix} -1 & 1 & -1 \\ 1 & -1 & 1 \\ 1 & -1 & 1 \end{bmatrix} \sim \begin{bmatrix} 1 & -1 & 1 \\ 0 & 0 & 0 \\ 0 & 0 & 0 \end{bmatrix}$,

a basis for the eigenspace of $\lambda_2 = 1$ is $\left\{ \begin{bmatrix} 1 \\ 1 \\ 0 \end{bmatrix}, \begin{bmatrix} -1 \\ 0 \\ 1 \end{bmatrix} \right\}$.

We thus have $\begin{bmatrix} 0 & 1 & -1 \\ 1 & 0 & 1 \\ 1 & -1 & 2 \end{bmatrix} = \begin{bmatrix} -1 & 1 & -1 \\ 1 & 1 & 0 \\ 1 & 0 & 1 \end{bmatrix} \begin{bmatrix} 0 & 0 & 0 \\ 0 & 1 & 0 \\ 0 & 0 & 1 \end{bmatrix} \begin{bmatrix} -1 & 1 & -1 \\ 1 & 1 & 0 \\ 1 & 0 & 1 \end{bmatrix}^{-1}$.

17. $\det \left(\begin{bmatrix} 1 & 0 & 0 & 1 \\ 0 & 2 & 0 & 0 \\ 0 & 0 & 3 & 0 \\ 0 & 0 & 0 & 4 \end{bmatrix} - \lambda \begin{bmatrix} 1 & 0 & 0 & 0 \\ 0 & 1 & 0 & 0 \\ 0 & 0 & 1 & 0 \\ 0 & 0 & 0 & 1 \end{bmatrix} \right) = \lambda^4 - 10\lambda^3 + 35\lambda^2 - 50\lambda + 24$

$= (\lambda - 1)(\lambda - 2)(\lambda - 3)(\lambda - 4) = 0 \quad \Rightarrow \quad \lambda_1 = 1, \lambda_2 = 2, \lambda_3 = 3$, and $\lambda_4 = 4$.

Since $\begin{bmatrix} 1 & 0 & 0 & 1 \\ 0 & 2 & 0 & 0 \\ 0 & 0 & 3 & 0 \\ 0 & 0 & 0 & 4 \end{bmatrix} - 1 \begin{bmatrix} 1 & 0 & 0 & 0 \\ 0 & 1 & 0 & 0 \\ 0 & 0 & 1 & 0 \\ 0 & 0 & 0 & 1 \end{bmatrix} = \begin{bmatrix} 0 & 0 & 0 & 1 \\ 0 & 1 & 0 & 0 \\ 0 & 0 & 2 & 0 \\ 0 & 0 & 0 & 3 \end{bmatrix} \sim \begin{bmatrix} 0 & 1 & 0 & 0 \\ 0 & 0 & 1 & 0 \\ 0 & 0 & 0 & 1 \\ 0 & 0 & 0 & 0 \end{bmatrix}$, a basis for the eigenspace of $\lambda_1 = 1$ is $\left\{ \begin{bmatrix} 1 \\ 0 \\ 0 \\ 0 \end{bmatrix} \right\}$.

Since $\begin{bmatrix} 1 & 0 & 0 & 1 \\ 0 & 2 & 0 & 0 \\ 0 & 0 & 3 & 0 \\ 0 & 0 & 0 & 4 \end{bmatrix} - 2 \begin{bmatrix} 1 & 0 & 0 & 0 \\ 0 & 1 & 0 & 0 \\ 0 & 0 & 1 & 0 \\ 0 & 0 & 0 & 1 \end{bmatrix} = \begin{bmatrix} -1 & 0 & 0 & 1 \\ 0 & 0 & 0 & 0 \\ 0 & 0 & 1 & 0 \\ 0 & 0 & 0 & 2 \end{bmatrix} \sim \begin{bmatrix} 1 & 0 & 0 & 0 \\ 0 & 0 & 1 & 0 \\ 0 & 0 & 0 & 1 \\ 0 & 0 & 0 & 0 \end{bmatrix}$,

a basis for the eigenspace of $\lambda_2 = 2$ is $\left\{ \begin{bmatrix} 0 \\ 1 \\ 0 \\ 0 \end{bmatrix} \right\}$.

Since $\begin{bmatrix} 1 & 0 & 0 & 1 \\ 0 & 2 & 0 & 0 \\ 0 & 0 & 3 & 0 \\ 0 & 0 & 0 & 4 \end{bmatrix} - 3 \begin{bmatrix} 1 & 0 & 0 & 0 \\ 0 & 1 & 0 & 0 \\ 0 & 0 & 1 & 0 \\ 0 & 0 & 0 & 1 \end{bmatrix} = \begin{bmatrix} -2 & 0 & 0 & 1 \\ 0 & -1 & 0 & 0 \\ 0 & 0 & 0 & 0 \\ 0 & 0 & 0 & 1 \end{bmatrix} \sim \begin{bmatrix} 1 & 0 & 0 & 0 \\ 0 & 1 & 0 & 0 \\ 0 & 0 & 0 & 1 \\ 0 & 0 & 0 & 0 \end{bmatrix}$,

a basis for the eigenspace of $\lambda_3 = 3$ is $\left\{ \begin{bmatrix} 0 \\ 0 \\ 1 \\ 0 \end{bmatrix} \right\}$.

Since $\begin{bmatrix} 1 & 0 & 0 & 1 \\ 0 & 2 & 0 & 0 \\ 0 & 0 & 3 & 0 \\ 0 & 0 & 0 & 4 \end{bmatrix} - 4 \begin{bmatrix} 1 & 0 & 0 & 0 \\ 0 & 1 & 0 & 0 \\ 0 & 0 & 1 & 0 \\ 0 & 0 & 0 & 1 \end{bmatrix} = \begin{bmatrix} -3 & 0 & 0 & 1 \\ 0 & -2 & 0 & 0 \\ 0 & 0 & -1 & 0 \\ 0 & 0 & 0 & 0 \end{bmatrix} \sim \begin{bmatrix} 1 & 0 & 0 & -\frac{1}{3} \\ 0 & 1 & 0 & 0 \\ 0 & 0 & 1 & 0 \\ 0 & 0 & 0 & 0 \end{bmatrix}$,

a basis for the eigenspace of $\lambda_4 = 4$ is $\left\{ \begin{bmatrix} \frac{1}{3} \\ 0 \\ 0 \\ 1 \end{bmatrix} \right\}$.

We thus have $\begin{bmatrix} 1 & 0 & 0 & 1 \\ 0 & 2 & 0 & 0 \\ 0 & 0 & 3 & 0 \\ 0 & 0 & 0 & 4 \end{bmatrix} = \begin{bmatrix} 1 & 0 & 0 & \frac{1}{3} \\ 0 & 1 & 0 & 0 \\ 0 & 0 & 1 & 0 \\ 0 & 0 & 0 & 1 \end{bmatrix} \begin{bmatrix} 1 & 0 & 0 & 0 \\ 0 & 2 & 0 & 0 \\ 0 & 0 & 3 & 0 \\ 0 & 0 & 0 & 4 \end{bmatrix} \begin{bmatrix} 1 & 0 & 0 & \frac{1}{3} \\ 0 & 1 & 0 & 0 \\ 0 & 0 & 1 & 0 \\ 0 & 0 & 0 & 1 \end{bmatrix}^{-1}$.

19. $\det \left(\begin{bmatrix} -3 & 4 \\ -2 & 3 \end{bmatrix} - \lambda \begin{bmatrix} 1 & 0 \\ 0 & 1 \end{bmatrix} \right) = (\lambda^2 - 1) = (\lambda - 1)(\lambda + 1) = 0 \quad \Rightarrow \quad \lambda_1 = -1 \text{ and } \lambda_2 = 1.$

Since $\begin{bmatrix} -3 & 4 \\ -2 & 3 \end{bmatrix} - (-1) \begin{bmatrix} 1 & 0 \\ 0 & 1 \end{bmatrix} = \begin{bmatrix} -2 & 4 \\ -2 & 4 \end{bmatrix} \sim \begin{bmatrix} -1 & 2 \\ 0 & 0 \end{bmatrix}$, a basis for the eigenspace of $\lambda_1 = -1$ is $\left\{ \begin{bmatrix} 2 \\ 1 \end{bmatrix} \right\}$.

Since $\begin{bmatrix} -3 & 4 \\ -2 & 3 \end{bmatrix} - 1 \begin{bmatrix} 1 & 0 \\ 0 & 1 \end{bmatrix} = \begin{bmatrix} -4 & 4 \\ -2 & 2 \end{bmatrix} \sim \begin{bmatrix} -1 & 1 \\ 0 & 0 \end{bmatrix}$, a basis for the eigenspace of $\lambda_2 = 1$ is $\left\{ \begin{bmatrix} 1 \\ 1 \end{bmatrix} \right\}$.

The matrix A is diagonalizable, with $A = \begin{bmatrix} 2 & 1 \\ 1 & 1 \end{bmatrix} \begin{bmatrix} -1 & 0 \\ 0 & 1 \end{bmatrix} \begin{bmatrix} 2 & 1 \\ 1 & 1 \end{bmatrix}^{-1}$. Thus

$A^{1000} = \begin{bmatrix} 2 & 1 \\ 1 & 1 \end{bmatrix} \begin{bmatrix} -1 & 0 \\ 0 & 1 \end{bmatrix}^{1000} \begin{bmatrix} 2 & 1 \\ 1 & 1 \end{bmatrix}^{-1}$

$= \begin{bmatrix} 2 & 1 \\ 1 & 1 \end{bmatrix} \begin{bmatrix} (-1)^{1000} & 0 \\ 0 & 1^{1000} \end{bmatrix} \begin{bmatrix} 1 & -1 \\ -1 & 2 \end{bmatrix} = \begin{bmatrix} 1 & 0 \\ 0 & 1 \end{bmatrix}$.

21. $\det \left(\begin{bmatrix} 7 & -8 \\ 4 & -5 \end{bmatrix} - \lambda \begin{bmatrix} 1 & 0 \\ 0 & 1 \end{bmatrix} \right) = \lambda^2 - 2\lambda - 3 = (\lambda + 1)(\lambda - 3) = 0 \quad \Rightarrow \quad \lambda_1 = -1 \text{ and } \lambda_2 = 3.$

Since $\begin{bmatrix} 7 & -8 \\ 4 & -5 \end{bmatrix} - (-1)\begin{bmatrix} 1 & 0 \\ 0 & 1 \end{bmatrix} = \begin{bmatrix} 8 & -8 \\ 4 & -4 \end{bmatrix} \sim \begin{bmatrix} 1 & -1 \\ 0 & 0 \end{bmatrix}$, a basis for the eigenspace of $\lambda_1 = -1$ is $\left\{ \begin{bmatrix} 1 \\ 1 \end{bmatrix} \right\}$.

Since $\begin{bmatrix} 7 & -8 \\ 4 & -5 \end{bmatrix} - 3\begin{bmatrix} 1 & 0 \\ 0 & 1 \end{bmatrix} = \begin{bmatrix} 4 & -8 \\ 4 & -8 \end{bmatrix} \sim \begin{bmatrix} 1 & -2 \\ 0 & 0 \end{bmatrix}$, a basis for the eigenspace of $\lambda_2 = 3$ is $\left\{ \begin{bmatrix} 2 \\ 1 \end{bmatrix} \right\}$.

The matrix A is diagonalizable, with $A = \begin{bmatrix} 1 & 2 \\ 1 & 1 \end{bmatrix}\begin{bmatrix} -1 & 0 \\ 0 & 3 \end{bmatrix}\begin{bmatrix} 1 & 2 \\ 1 & 1 \end{bmatrix}^{-1}$. Thus

$$A^{1000} = \begin{bmatrix} 1 & 2 \\ 1 & 1 \end{bmatrix}\begin{bmatrix} -1 & 0 \\ 0 & 3 \end{bmatrix}^{1000}\begin{bmatrix} 1 & 2 \\ 1 & 1 \end{bmatrix}^{-1}$$
$$= \begin{bmatrix} 1 & 2 \\ 1 & 1 \end{bmatrix}\begin{bmatrix} 1 & 0 \\ 0 & 3^{1000} \end{bmatrix}\begin{bmatrix} -1 & 2 \\ 1 & -1 \end{bmatrix} = \begin{bmatrix} 2\left(3^{1000}\right) - 1 & 2 - 2\left(3^{1000}\right) \\ 3^{1000} - 1 & 2 - 3^{1000} \end{bmatrix}.$$

23. Since the matrix is diagonalizable, there is a basis of eigenvectors of $\mathbf{R}^4$. The one eigenspace has dimension 2, hence the other eigenspace must have dimension $4 - 2 = 2$.

25. For example, $A = \begin{bmatrix} 0 & 0 \\ 0 & 1 \end{bmatrix}$.

27. For example, $A = \begin{bmatrix} 1 & 1 \\ 0 & 1 \end{bmatrix}$.

29. For example, $A = \begin{bmatrix} 0 & 1 & 0 \\ 0 & 1 & 0 \\ 0 & 0 & 2 \end{bmatrix}$ has eigenvalues $0, 1$, and 2.

31. True, by Theorem 6.15.

33. False. For example $\begin{bmatrix} 1 & 0 \\ 0 & 0 \end{bmatrix}$ is not invertible, but is diagonalizable.

35. False. For example $\begin{bmatrix} 1 & 0 \\ 0 & 0 \end{bmatrix}$ is diagonalizable, but $\text{rank}\left(\begin{bmatrix} 1 & 0 \\ 0 & 0 \end{bmatrix}\right) = 1 < 2$.

37. False. Let $A = \begin{bmatrix} 1 & 1 \\ 0 & -1 \end{bmatrix}$ and $B = \begin{bmatrix} -1 & 1 \\ 0 & 1 \end{bmatrix}$. Then both A and B are diagonalizable, but $A + B = \begin{bmatrix} 1 & 1 \\ 0 & -1 \end{bmatrix} + \begin{bmatrix} -1 & 1 \\ 0 & 1 \end{bmatrix} = \begin{bmatrix} 0 & 2 \\ 0 & 0 \end{bmatrix}$ is not diagonalizable.

39. By Theorem 6.14, $\mathbf{u}_1$ and $\mathbf{u}_2$ are linearly independent, as eigenvectors associated with distinct eigenvalues. Hence the matrix $U = [\ \mathbf{u}_1 \quad \mathbf{u}_2\ ]$ is invertible, and $\det(U) \neq 0$.

41. By the proof of Theorem 6.13, the columns of P must be eigenvectors of A. By scaling these eigenvectors by a non-zero scalar c we can form infinitely many matrices P. There are finitely many distinct matrices D, determined by rearrangements on the diagonal.

43. If A is diagonalizable, then $A = PDP^{-1}$ for a diagonal matrix D and invertible matrix P. Thus $A^T = \left(PDP^{-1}\right)^T = \left(P^{-1}\right)^T D^T P^T = \left(P^T\right)^{-1} D\left(P^T\right) = QDQ^{-1}$, where $Q = \left(P^T\right)^{-1}$ is an invertible matrix. This shows A^T is diagonalizable.

45. We are given that $A = PD_1P^{-1}$ and $B = PD_2P^{-1}$. Thus $AB = \left(PD_1P^{-1}\right)\left(PD_2P^{-1}\right)$
$= PD_1\left(P^{-1}P\right)D_2P^{-1} = P\left(D_1D_2\right)P^{-1}$. Also, $BA = \left(PD_2P^{-1}\right)\left(PD_1P^{-1}\right)$
$= PD_2\left(P^{-1}P\right)D_1P^{-1} = P\left(D_2D_1\right)P^{-1}$. Since both D_1 and D_2 are diagonal matrices, $D_1D_2 = D_2D_1$, hence $AB = P\left(D_1D_2\right)P^{-1} = P\left(D_2D_1\right)P^{-1} = BA$.

47. $\begin{bmatrix} 3 & 0 & -2 & -1 \\ -1 & 2 & 5 & 4 \\ 6 & 0 & -5 & -3 \\ -6 & 0 & 4 & 2 \end{bmatrix} = \begin{bmatrix} -1 & -\frac{2}{3} & -\frac{1}{3} & 0 \\ 2 & \frac{2}{3} & \frac{1}{3} & 1 \\ -3 & -2 & -\frac{2}{3} & 0 \\ 2 & 2 & \frac{2}{3} & 0 \end{bmatrix} \begin{bmatrix} -1 & 0 & 0 & 0 \\ 0 & 0 & 0 & 0 \\ 0 & 0 & 1 & 0 \\ 0 & 0 & 0 & 2 \end{bmatrix} \begin{bmatrix} -1 & -\frac{2}{3} & -\frac{1}{3} & 0 \\ 2 & \frac{2}{3} & \frac{1}{3} & 1 \\ -3 & -2 & -\frac{2}{3} & 0 \\ 2 & 2 & \frac{2}{3} & 0 \end{bmatrix}^{-1}$

49. $\begin{bmatrix} 2 & -1 & 0 & 1 & 0 \\ -4 & 3 & 0 & -5 & -4 \\ 5 & -2 & -1 & 1 & -1 \\ -6 & 2 & 0 & -4 & -2 \\ 2 & -1 & 0 & 1 & 0 \end{bmatrix}$

$= \begin{bmatrix} 0 & 0 & \frac{2}{3} & 2 & 4 \\ 4 & 0 & -\frac{2}{3} & -2 & -8 \\ 4 & 1 & \frac{4}{3} & 4 & 8 \\ 4 & 0 & -2 & -4 & -8 \\ 0 & 0 & \frac{4}{3} & 2 & 4 \end{bmatrix} \begin{bmatrix} -2 & 0 & 0 & 0 & 0 \\ 0 & -1 & 0 & 0 & 0 \\ 0 & 0 & 0 & 0 & 0 \\ 0 & 0 & 0 & 1 & 0 \\ 0 & 0 & 0 & 0 & 2 \end{bmatrix} \begin{bmatrix} 0 & 0 & \frac{2}{3} & 2 & 4 \\ 4 & 0 & -\frac{2}{3} & -2 & -8 \\ 4 & 1 & \frac{4}{3} & 4 & 8 \\ 4 & 0 & -2 & -4 & -8 \\ 0 & 0 & \frac{4}{3} & 2 & 4 \end{bmatrix}^{-1}$

6.5   Complex Eigenvalues

1. (a) $z_1 + z_2 = (5 - 2i) + (3 + 4i) = (5 + 3) + (-2 + 4)i = 8 + 2i$

   (b) $2z_1 + 3z_2 = 2(5 - 2i) + 3(3 + 4i) = (10 - 4i) + (9 + 12i) = (10 + 9) + (-4 + 12)i = 19 + 8i$

   (c) $z_1 - z_2 = (5 - 2i) - (3 + 4i) = (5 - 3) + (-2 - 4)i = 2 - 6i$

   (d) $z_1 z_2 = (5 - 2i)(3 + 4i) = (5(3) - (-2)(4)) + (5(4) + (-2)(3))i = 23 + 14i$

3. Characteristic polynomial: $\det\left(\begin{bmatrix} 3 & 1 \\ -5 & 1 \end{bmatrix} - \lambda \begin{bmatrix} 1 & 0 \\ 0 & 1 \end{bmatrix}\right) = \det\left(\begin{bmatrix} 3 - \lambda & 1 \\ -5 & 1 - \lambda \end{bmatrix}\right) = \lambda^2 - 4\lambda + 8.$
   Eigenvalues: $\lambda^2 - 4\lambda + 8 = 0 \Rightarrow \lambda_1 = 2 + 2i$ and $\lambda_2 = 2 - 2i$.
   Eigenspace of $\lambda_1 = 2 + 2i$:

$$A - (2 + 2i)I_2 = \begin{bmatrix} 1 - 2i & 1 \\ -5 & -1 - 2i \end{bmatrix} \sim \begin{bmatrix} -5 & -1 - 2i \\ 0 & 0 \end{bmatrix},$$

   so a basis for this eigenspace is $\left\{\begin{bmatrix} 1 + 2i \\ -5 \end{bmatrix}\right\}$.
   A basis for the eigenspace of $\lambda_2 = 2 - 2i = \overline{\lambda_1}$ is $\left\{\begin{bmatrix} 1 - 2i \\ -5 \end{bmatrix}\right\}$.

5. Characteristic polynomial: $\det\left(\begin{bmatrix} 1 & -2 \\ 1 & 3 \end{bmatrix} - \lambda \begin{bmatrix} 1 & 0 \\ 0 & 1 \end{bmatrix}\right) = \det\left(\begin{bmatrix} 1 - \lambda & -2 \\ 1 & 3 - \lambda \end{bmatrix}\right) = \lambda^2 - 4\lambda + 5.$
   Eigenvalues: $\lambda^2 - 4\lambda + 5 = 0 \Rightarrow \lambda_1 = 2 + i$ and $\lambda_2 = 2 - i$.
   Eigenspace of $\lambda_1 = 2 + i$:

$$A - (2 + i)I_2 = \begin{bmatrix} -1 - i & -2 \\ 1 & 1 - i \end{bmatrix} \sim \begin{bmatrix} 1 & 1 - i \\ 0 & 0 \end{bmatrix},$$

   so a basis for this eigenspace is $\left\{\begin{bmatrix} 1 - i \\ -1 \end{bmatrix}\right\}$.
   A basis for the eigenspace of $\lambda_2 = 2 - i = \overline{\lambda_1}$ is $\left\{\begin{bmatrix} 1 + i \\ -1 \end{bmatrix}\right\}$.

7. Characteristic polynomial: $\det\left(\begin{bmatrix} 4 & 2 \\ -1 & 2 \end{bmatrix} - \lambda \begin{bmatrix} 1 & 0 \\ 0 & 1 \end{bmatrix}\right) = \det\left(\begin{bmatrix} 4-\lambda & 2 \\ -1 & 2-\lambda \end{bmatrix}\right) = \lambda^2 - 6\lambda + 10.$

   Eigenvalues: $\lambda^2 - 6\lambda + 10 = 0 \quad \Rightarrow \quad \lambda_1 = 3+i$ and $\lambda_2 = 3-i$.

   Eigenspace of $\lambda_1 = 3+i$:

$$A - (3+i)I_2 = \begin{bmatrix} 1-i & 2 \\ -1 & -1-i \end{bmatrix} \sim \begin{bmatrix} 1 & 1+i \\ 0 & 0 \end{bmatrix},$$

   so a basis for this eigenspace is $\left\{\begin{bmatrix} 1+i \\ -1 \end{bmatrix}\right\}$.

   A basis for the eigenspace of $\lambda_2 = 3-i = \overline{\lambda_1}$ is $\left\{\begin{bmatrix} 1-i \\ -1 \end{bmatrix}\right\}$.

9. An eigenvalue is $\lambda = 2+i$. So the rotation is by $\tan^{-1}(1/2) \approx 0.4636$ radians, the dilation is by $\sqrt{2^2 + 1^2} = \sqrt{5}$.

11. An eigenvalue is $\lambda = 1+i$. So the rotation is by $\tan^{-1}(1/1) = \frac{\pi}{4}$ radians, the dilation is by $\sqrt{1^2 + 1^2} = \sqrt{2}$.

13. An eigenvalue is $\lambda = 4-3i$. So the rotation is by $\tan^{-1}(-3/4) \approx -0.6435$ radians, the dilation is by $\sqrt{4^2 + 3^2} = 5$.

15. Solve $\det\left(\begin{bmatrix} 3 & 1 \\ -5 & 1 \end{bmatrix} - \lambda \begin{bmatrix} 1 & 0 \\ 0 & 1 \end{bmatrix}\right) = \det\left(\begin{bmatrix} 3-\lambda & 1 \\ -5 & 1-\lambda \end{bmatrix}\right) = \lambda^2 - 4\lambda + 8 = 0 \quad \Rightarrow \quad \lambda = 2 \pm 2i.$

    The rotation-dilation matrix is $B = \begin{bmatrix} 2 & -2 \\ 2 & 2 \end{bmatrix}$.

17. Solve $\det\left(\begin{bmatrix} 1 & -2 \\ 1 & 3 \end{bmatrix} - \lambda \begin{bmatrix} 1 & 0 \\ 0 & 1 \end{bmatrix}\right) = \det\left(\begin{bmatrix} 1-\lambda & -2 \\ 1 & 3-\lambda \end{bmatrix}\right) = \lambda^2 - 4\lambda + 5 = 0 \quad \Rightarrow \quad \lambda = 2 \pm i.$

    The rotation-dilation matrix is $B = \begin{bmatrix} 2 & -1 \\ 1 & 2 \end{bmatrix}$.

19. Solve $\det\left(\begin{bmatrix} 4 & 2 \\ -1 & 2 \end{bmatrix} - \lambda \begin{bmatrix} 1 & 0 \\ 0 & 1 \end{bmatrix}\right) = \det\left(\begin{bmatrix} 4-\lambda & 2 \\ -1 & 2-\lambda \end{bmatrix}\right) = \lambda^2 - 6\lambda + 10 = 0 \quad \Rightarrow \quad \lambda = 3 \pm i.$

    The rotation-dilation matrix is $B = \begin{bmatrix} 3 & -1 \\ 1 & 3 \end{bmatrix}$.

21. Since roots occur in complex conjugate pairs, the other roots are $\overline{(1+2i)} = 1-2i$ and $\overline{(3-i)} = 3+i$. Since there are 5 roots and the polynomial has degree 5, the multiplicity of each root must be one.

23. Set $z = a + ib$ and $a = 2b$. Then $|z| = |2b + ib| = \sqrt{5b^2} = \sqrt{5}\,|b|$. Set this equal to 3 to obtain $b = \pm\frac{3}{\sqrt{5}} = \pm\frac{3\sqrt{5}}{5}$. So, for example, $z = \frac{6\sqrt{5}}{5} + \frac{3\sqrt{5}}{5}i$.

25. We set $B = \begin{bmatrix} a & -b \\ b & a \end{bmatrix}$ and require that $a = 0$ and $b > 0$ for a rotation of $90°$. For a dilation by 2 we need $\sqrt{a^2 + b^2} = \sqrt{0^2 + b^2} = 2$, hence $b = 2$. Thus $B = \begin{bmatrix} 0 & -2 \\ 2 & 0 \end{bmatrix}$.

27. We may obtain A as $\begin{bmatrix} 1 & 1 \\ 0 & 1 \end{bmatrix}\begin{bmatrix} 1 & -2 \\ 2 & 1 \end{bmatrix}\begin{bmatrix} 1 & 1 \\ 0 & 1 \end{bmatrix}^{-1} = \begin{bmatrix} 3 & -4 \\ 2 & -1 \end{bmatrix}$.

29. For example, $A = \begin{bmatrix} i & i \\ -i & -i \end{bmatrix}$ has characteristic polynomial $\det(A - \lambda I_2) = \lambda^2$, hence real eigenvalue 0 of multiplicity 2.

31. True, by considering z and w in polar form.

33. False. One can have real eigenvalues and complex eigenvectors by multiplying any real eigenvector by a complex scalar z.

35. False. For example, $|(1 + i) + (1 - i)| = |2| = 2$, and $|1 + i| + |1 - i| = \sqrt{2} + \sqrt{2} = 2\sqrt{2}$.

37. True, since $\overline{\overline{z}} = \overline{\overline{x + iy}} = \overline{x - iy} = x + iy = z$.

39. True, since a rotation-dilation matrix has eigenvalues which occur as complex conjugates, $\lambda_2 = \overline{\lambda_1}$, hence $|\lambda_1| = |\lambda_2|$.

41. (a) Let $z = x + iy$ and $w = u + iv$. Then $\overline{z + w} = \overline{(x + iy) + (u + iv)} = \overline{(x + u) + i(y + v)} = (x + u) - i(y + v) = (x - iy) + (u - iv) = \overline{(x + iy)} + \overline{(u + iv)} = \overline{z} + \overline{w}$.

    (b) Let $z = x + iy$ and $w = u + iv$. Then $\overline{zw} = \overline{(x + iy)(u + iv)} = \overline{(xu - yv) + i(xv + yu)} = (xu - yv) - i(xv + yu) = (x - iy)(u - iv) = \overline{(x + iy)(u + iv)} = \overline{z}\,\overline{w}$.

43. Let $\mathbf{v} = \begin{bmatrix} v_1 \\ \vdots \\ v_n \end{bmatrix}$, where each v_j is complex. Then $\overline{c} \cdot \overline{\mathbf{v}} = \overline{c} \cdot \begin{bmatrix} \overline{v_1} \\ \vdots \\ \overline{v_n} \end{bmatrix} = \begin{bmatrix} \overline{c}\,\overline{v_1} \\ \vdots \\ \overline{c}\,\overline{v_n} \end{bmatrix} = \begin{bmatrix} \overline{cv_1} \\ \vdots \\ \overline{cv_n} \end{bmatrix} =$

$\overline{\left(\begin{bmatrix} cv_1 \\ \vdots \\ cv_n \end{bmatrix} \right)} = \overline{\left(c \begin{bmatrix} v_1 \\ \vdots \\ v_n \end{bmatrix} \right)} = \overline{c\mathbf{v}}$.

45. Let $\lambda = a + ib$. Then $\overline{e^{\lambda t}} = \overline{e^{(a+ib)t}} = e^{(a-ib)t} = e^{at - i(bt)} = e^{at}(\cos(-bt) + \sin(-bt)\,i)$
$= e^{at}(\cos(bt) - i\sin(bt)) = e^{at}\cos(bt) - ie^{at}\sin(bt) = \overline{e^{at}\cos(bt) + ie^{at}\sin(bt)} = \overline{e^{at}(\cos(bt) + i\sin(bt))} =$
$\overline{e^{at + i(bt)}} = \overline{e^{(a+ib)t}} = \overline{e^{\lambda t}}$.

47. $\det(A - (a + ib)I_2) = \det\left(\begin{bmatrix} a & -b \\ b & a \end{bmatrix} - (a + ib) \begin{bmatrix} 1 & 0 \\ 0 & 1 \end{bmatrix} \right) = \det\left(\begin{bmatrix} -ib & -b \\ b & -ib \end{bmatrix} \right) = 0$, so $\lambda = a + ib$ is an eigenvalue of A.

49. (a) Let $\mathbf{u} = \mathbf{x} + i\mathbf{y}$. Then $A(\operatorname{Re}(\mathbf{u})) = A(\operatorname{Re}(\mathbf{x} + i\mathbf{y})) = A\mathbf{x} = \operatorname{Re}(A\mathbf{x} + iA\mathbf{y}) = \operatorname{Re}(A(\mathbf{x} + i\mathbf{y})) = \operatorname{Re}(A\mathbf{u})$. And $A(\operatorname{Im}(\mathbf{u})) = A(\operatorname{Im}(\mathbf{x} + i\mathbf{y})) = A\mathbf{y} = \operatorname{Im}(A\mathbf{x} + iA\mathbf{y})$
$= \operatorname{Im}(A(\mathbf{x} + i\mathbf{y})) = \operatorname{Im}(A\mathbf{u})$.

    (b) $A\mathbf{u} = \lambda\mathbf{u} \Rightarrow A(\operatorname{Re}(\mathbf{u}) + i\operatorname{Im}(\mathbf{u})) = (a - ib)(\operatorname{Re}(\mathbf{u}) + i\operatorname{Im}(\mathbf{u})) \Rightarrow A(\operatorname{Re}(\mathbf{u})) + iA(\operatorname{Im}(\mathbf{u})) = (a\operatorname{Re}(\mathbf{u}) + b\operatorname{Im}(\mathbf{u})) + (-b\operatorname{Re}(\mathbf{u}) + a\operatorname{Im}(\mathbf{u}))$. Now equate the real and imaginary parts to obtain $A(\operatorname{Re}(\mathbf{u})) = a\operatorname{Re}(\mathbf{u}) + b\operatorname{Im}(\mathbf{u})$ and $A(\operatorname{Im}(\mathbf{u})) = -b\operatorname{Re}(\mathbf{u}) + a\operatorname{Im}(\mathbf{u})$.

    (c) $AP = A[\ \operatorname{Re}(\mathbf{u})\ \ \operatorname{Im}(\mathbf{u})\ ] = [\ A(\operatorname{Re}(\mathbf{u}))\ \ A(\operatorname{Im}(\mathbf{u}))\ ]$
$= [\ a\operatorname{Re}(\mathbf{u}) + b\operatorname{Im}(\mathbf{u})\ \ -b\operatorname{Re}(\mathbf{u}) + a\operatorname{Im}(\mathbf{u})\ ] = [\ \operatorname{Re}(\mathbf{u})\ \ \operatorname{Im}(\mathbf{u})\ ] \begin{bmatrix} a & -b \\ b & a \end{bmatrix} = PC$.

51. Characteristic polynomial: $\det(A - \lambda I_3) = -\lambda^3 + \lambda^2 + 21\lambda + 55$. Eigenvalues: $-\lambda^3 + \lambda^2 + 21\lambda + 55 = 0 \Rightarrow \lambda_1 = -2.507 + 1.692i$, $\lambda_2 = -2.507 - 1.692i$, and $\lambda_3 = 6.013$, with bases for their eigenspaces:

$$\lambda_{1,2} = -2.507 \pm 1.692i \Rightarrow \left\{ \begin{bmatrix} 0.2373 \pm 0.3607i \\ 0.0862 \mp 0.2878i \\ -0.8505 \end{bmatrix} \right\}, \ \lambda_3 = 6.013 \Rightarrow \left\{ \begin{bmatrix} 0.5837 \\ 0.6889 \\ 0.4298 \end{bmatrix} \right\}.$$

53. Characteristic polynomial: $\det(A - \lambda I_4) = \lambda^4 - 13\lambda^3 - 25\lambda^2 + 322\lambda - 500$. Eigenvalues: $\lambda^4 - 13\lambda^3 - 25\lambda^2 + 322\lambda - 500 = 0 \Rightarrow \lambda_1 = 2.5948 + 0.4119i$, $\lambda_2 = 2.5948 - 0.4119i$, $\lambda_3 = 13.2693$, and

$\lambda_4 = -5.4589$, with bases for their eigenspaces:

$$\lambda_{1,2} = 2.5948 \pm 0.4119i \Rightarrow \left\{ \begin{bmatrix} 0.6638 \\ 0.1906 \pm 0.2156i \\ 0.0472 \mp 0.2804i \\ -0.6280 \mp 0.0368i \end{bmatrix} \right\},$$

$$\lambda_3 = 13.2693 \Rightarrow \left\{ \begin{bmatrix} 0.0639 \\ 0.1441 \\ 0.3504 \\ 0.9232 \end{bmatrix} \right\}, \; \lambda_4 = -5.4589 \Rightarrow \left\{ \begin{bmatrix} 0.6812 \\ -0.3472 \\ -0.5476 \\ 0.3399 \end{bmatrix} \right\}.$$

6.6   Systems of Differential Equations

1. $\mathbf{y} = c_1 e^{-t} \begin{bmatrix} 1 \\ 1 \end{bmatrix} + c_2 e^{2t} \begin{bmatrix} 1 \\ -1 \end{bmatrix}$ $\Rightarrow$ $y_1 = c_1 e^{-t} + c_2 e^{2t}$ and $y_2 = c_1 e^{-t} - c_2 e^{2t}$.

3. $\mathbf{y} = c_1 e^{2t} \begin{bmatrix} 4 \\ 3 \\ 1 \end{bmatrix} + c_2 e^{-2t} \begin{bmatrix} 1 \\ 2 \\ 0 \end{bmatrix} + c_3 e^{-2t} \begin{bmatrix} 2 \\ 3 \\ 1 \end{bmatrix}$ $\Rightarrow$ $y_1 = 4c_1 e^{2t} + c_2 e^{-2t} + 2c_3 e^{-2t}$, $y_2 = 3c_1 e^{2t} + 2c_2 e^{-2t} + 3c_3 e^{-2t}$, and $y_3 = c_1 e^{2t} + c_3 e^{-2t}$.

5. $\mathbf{y} = c_1 \left(\cos(2t) \begin{bmatrix} 1 \\ 2 \end{bmatrix} - \sin(2t) \begin{bmatrix} 1 \\ -1 \end{bmatrix} \right) + c_2 \left(\sin(2t) \begin{bmatrix} 1 \\ 2 \end{bmatrix} + \cos(2t) \begin{bmatrix} 1 \\ -1 \end{bmatrix} \right)$
$\Rightarrow$ $y_1 = c_1 (\cos 2t - \sin 2t) + c_2 (\cos 2t + \sin 2t)$ and $y_2 = c_1 (2 \cos 2t + \sin 2t) - c_2 (\cos 2t - 2 \sin 2t)$.

7. $\mathbf{y} = c_1 e^{4t} \begin{bmatrix} 3 \\ 1 \\ 5 \end{bmatrix} + c_2 e^t \left(\cos(t) \begin{bmatrix} 4 \\ 0 \\ 3 \end{bmatrix} - \sin(t) \begin{bmatrix} 1 \\ -2 \\ 1 \end{bmatrix} \right) + c_3 e^t \left(\sin(t) \begin{bmatrix} 4 \\ 0 \\ 3 \end{bmatrix} + \cos(t) \begin{bmatrix} 1 \\ -2 \\ 1 \end{bmatrix} \right)$ $\Rightarrow$
$y_1 = 3c_1 e^{4t} - c_2 (\sin t - 4 \cos t) e^t + c_3 (\cos t + 4 \sin t) e^t$, $y_2 = c_1 e^{4t} - 2c_3 (\cos t) e^t + 2c_2 (\sin t) e^t$, and
$y_3 = 5c_1 e^{4t} - c_2 (\sin t - 3 \cos t) e^t + c_3 (\cos t + 3 \sin t) e^t$.

9. $\mathbf{y} = c_1 e^t \begin{bmatrix} 6 \\ 2 \\ 5 \\ 0 \end{bmatrix} + c_2 e^{4t} \begin{bmatrix} 1 \\ 2 \\ 3 \\ 2 \end{bmatrix}$

$+ c_3 e^t \left(\cos(t) \begin{bmatrix} 3 \\ 6 \\ 2 \\ 0 \end{bmatrix} - \sin(t) \begin{bmatrix} 2 \\ 0 \\ -3 \\ -5 \end{bmatrix} \right) + c_4 e^t \left(\sin(t) \begin{bmatrix} 3 \\ 6 \\ 2 \\ 0 \end{bmatrix} + \cos(t) \begin{bmatrix} 2 \\ 0 \\ -3 \\ -5 \end{bmatrix} \right)$

$\Rightarrow$ $y_1 = 6c_1 e^t + c_2 e^{4t} + c_3 (3 \cos t - 2 \sin t) e^t + c_4 (2 \cos t + 3 \sin t) e^t$, $y_2 = 2c_1 e^t + 2c_2 e^{4t} + 6c_3 (\cos t) e^t + 6c_4 (\sin t) e^t$, $y_3 = 5c_1 e^t + 3c_2 e^{4t} + c_3 (2 \cos t + 3 \sin t) e^t - c_4 (3 \cos t - 2 \sin t) e^t$, and $y_4 = 2c_2 e^{4t} + 5c_3 (\sin t) e^t - 5c_4 (\cos t) e^t$.

11. To solve $\mathbf{y}' = A\mathbf{y} = \begin{bmatrix} 1 & 4 \\ 1 & 1 \end{bmatrix} \mathbf{y}$, we determine the eigenvalues and eigenvectors of A, from the characteristic polynomial, $\det(A - \lambda I_2) = (\lambda + 1)(\lambda - 3)$.

$$\lambda = -1 \Rightarrow \mathbf{u} = \begin{bmatrix} 2 \\ 1 \end{bmatrix}, \lambda = 3 \Rightarrow \mathbf{u} = \begin{bmatrix} 2 \\ -1 \end{bmatrix}$$

Hence $y_1 = 2c_1 e^{-t} + 2c_2 e^{3t}$, and $y_2 = -c_1 e^{-t} + c_2 e^{3t}$.

13. To solve $\mathbf{y}' = A\mathbf{y} = \begin{bmatrix} 7 & -8 \\ 4 & -5 \end{bmatrix} \mathbf{y}$, we determine the eigenvalues and eigenvectors of A, from the characteristic polynomial, $\det(A - \lambda I_2) = (\lambda + 1)(\lambda - 3)$.

$$\lambda = -1 \Rightarrow \mathbf{u} = \begin{bmatrix} 1 \\ 1 \end{bmatrix}, \lambda = 3 \Rightarrow \mathbf{u} = \begin{bmatrix} 2 \\ 1 \end{bmatrix}$$

Hence $y_1 = c_1 e^{-t} + 2c_2 e^{3t}$, and $y_2 = c_1 e^{-t} + c_2 e^{3t}$.

15. To solve $\mathbf{y}' = A\mathbf{y} = \begin{bmatrix} 3 & 1 \\ -5 & 1 \end{bmatrix} \mathbf{y}$, we determine the eigenvalues and eigenvectors of A, from the characteristic polynomial, $\det(A - \lambda I_2) = \lambda^2 - 4\lambda + 8$

$$\lambda = 2 + 2i \Rightarrow \mathbf{u} = \begin{bmatrix} -1 - 2i \\ 5 \end{bmatrix}, \lambda = 2 - 2i \Rightarrow \mathbf{u} = \begin{bmatrix} -1 + 2i \\ 5 \end{bmatrix}$$

Hence

$$\mathbf{y} = c_1 e^{2t} \left(\cos(2t) \begin{bmatrix} -1 \\ 5 \end{bmatrix} - \sin(2t) \begin{bmatrix} -2 \\ 0 \end{bmatrix} \right) + c_2 e^{2t} \left(\sin(2t) \begin{bmatrix} -1 \\ 5 \end{bmatrix} + \cos(2t) \begin{bmatrix} -2 \\ 0 \end{bmatrix} \right).$$

Thus $y_1 = -c_1 (\cos 2t - 2\sin 2t) e^{2t} - c_2 (2\cos 2t + \sin 2t) e^{2t}$, and $y_2 = 5c_1 (\cos 2t) e^{2t} + 5c_2 (\sin 2t) e^{2t}$.

17. To solve $\mathbf{y}' = A\mathbf{y} = \begin{bmatrix} 7 & 2 & -8 \\ -3 & 0 & 3 \\ 6 & 2 & -7 \end{bmatrix} \mathbf{y}$, we determine the eigenvalues and eigenvectors of A, from the characteristic polynomial, $\det(A - \lambda I_3) = \lambda(\lambda + 1)(\lambda - 1)$

$$\lambda = 0 \Rightarrow \mathbf{u} = \begin{bmatrix} 2 \\ 1 \\ 2 \end{bmatrix}, \lambda = -1 \Rightarrow \mathbf{u} = \begin{bmatrix} 1 \\ 0 \\ 1 \end{bmatrix}, \lambda = 1 \Rightarrow \mathbf{u} = \begin{bmatrix} -1 \\ 3 \\ 0 \end{bmatrix}$$

Thus $y_1 = 2c_1 + c_2 e^{-t} - c_3 e^t$, $y_2 = c_1 + 3c_3 e^t$, $y_3 = 2c_1 + c_2 e^{-t}$.

19. To solve $\mathbf{y}' = A\mathbf{y} = \begin{bmatrix} 8 & -10 \\ 5 & -7 \end{bmatrix} \mathbf{y}$, we determine the eigenvalues and eigenvectors of A, from the characteristic polynomial, $\det(A - \lambda I_2) = (\lambda + 2)(\lambda - 3)$.

$$\lambda = -2 \Rightarrow \mathbf{u} = \begin{bmatrix} 1 \\ 1 \end{bmatrix}, \lambda = 3 \Rightarrow \mathbf{u} = \begin{bmatrix} 2 \\ 1 \end{bmatrix}$$

Hence $y_1 = c_1 e^{-2t} + 2c_2 e^{3t}$, and $y_2 = c_1 e^{-2t} + c_2 e^{3t}$.
Setting $y_1(0) = 4$ and $y_2(0) = 1$, we solve the system

$$c_1 + 2c_2 = 4$$
$$c_1 + c_2 = 1$$

to obtain $c_1 = -2$ and $c_2 = 3$. Thus $y_1 = -2e^{-2t} + 6e^{3t}$, and $y_2 = -2e^{-2t} + 3e^{3t}$.

21. To solve $\mathbf{y}' = A\mathbf{y} = \begin{bmatrix} 1 & 3 \\ -3 & 1 \end{bmatrix} \mathbf{y}$, we determine the eigenvalues and eigenvectors of A, from the characteristic polynomial, $\det(A - \lambda I_2) = \lambda^2 - 2\lambda + 10$.

$$\lambda = 1 + 3i \Rightarrow \mathbf{u} = \begin{bmatrix} -i \\ 1 \end{bmatrix}, \lambda = 1 - 3i \Rightarrow \mathbf{u} = \begin{bmatrix} i \\ 1 \end{bmatrix}$$

Hence

$$\mathbf{y} = c_1 e^t \left(\cos(3t) \begin{bmatrix} 0 \\ 1 \end{bmatrix} - \sin(3t) \begin{bmatrix} -1 \\ 0 \end{bmatrix} \right) + c_2 e^t \left(\sin(3t) \begin{bmatrix} 0 \\ 1 \end{bmatrix} + \cos(3t) \begin{bmatrix} -1 \\ 0 \end{bmatrix} \right).$$

Thus $y_1 = c_1 (\sin 3t) e^t - c_2 (\cos 3t) e^t$ and $y_2 = c_1 (\cos 3t) e^t + c_2 (\sin 3t) e^t$.
Setting $y_1(0) = 2$ and $y_2(0) = -1$, we obtain

$$-c_2 = 2$$
$$c_1 = -1$$

Thus $c_1 = -1$ and $c_2 = -2$. Therefore $y_1 = -(\sin 3t) e^t + 2(\cos 3t) e^t$, and $y_2 = -(\cos 3t) e^t - 2(\sin 3t) e^t$.

23. To solve $\mathbf{y}' = A\mathbf{y} = \begin{bmatrix} 2 & -1 & -1 \\ 6 & 3 & -2 \\ 6 & -2 & -3 \end{bmatrix} \mathbf{y}$, we determine the eigenvalues and eigenvectors of A, from the

characteristic polynomial, $\det(A - \lambda I_2) = (\lambda - 1)(\lambda - 2)(\lambda + 1)$.

$$\lambda = 1 \Rightarrow \mathbf{u} = \begin{bmatrix} 1 \\ -1 \\ 2 \end{bmatrix}, \lambda = 2 \Rightarrow \mathbf{u} = \begin{bmatrix} 1 \\ -2 \\ 2 \end{bmatrix}, \lambda = -1 \Rightarrow \mathbf{u} = \begin{bmatrix} 1 \\ 0 \\ 3 \end{bmatrix}$$

Hence $y_1 = c_1 e^t + c_2 e^{2t} + c_3 e^{-t}$, $y_2 = -c_1 e^t - 2c_2 e^{2t}$, and $y_3 = 2c_1 e^t + 2c_2 e^{2t} + 3c_3 e^{-t}$.
Setting $y_1(0) = -1$, $y_2(0) = 0$, and $y_3 = -4$ we obtain

$$c_1 + c_2 + c_3 = -1$$
$$-c_1 - 2c_2 = 0$$
$$2c_1 + 2c_2 + 3c_3 = -4$$

Thus $c_1 = 2$, $c_2 = -1$, and $c_3 = -2$. Therefore $y_1 = 2e^t - e^{2t} - 2e^{-t}$, $y_2 = -2e^t + 2e^{2t}$, and $y_3 = 4e^t - 2e^{2t} - 6e^{-t}$.

25. To solve $\mathbf{y}' = A\mathbf{y} = \begin{bmatrix} -0.1 & 0.2 \\ -0.3 & -0.6 \end{bmatrix} \mathbf{y}$, we determine the eigenvalues and eigenvectors of A, from the

characteristic polynomial, $\det(A - \lambda I_2) = \lambda^2 + 0.7\lambda + 0.12 = (\lambda + 0.3)(\lambda + 0.4)$.

$$\lambda = -0.3 \Rightarrow \mathbf{u} = \begin{bmatrix} 1 \\ -1 \end{bmatrix}, \lambda = -0.4 \Rightarrow \mathbf{u} = \begin{bmatrix} 2 \\ -3 \end{bmatrix}$$

Hence $y_1 = c_1 e^{(-0.3)t} + 2c_2 e^{(-0.4)t}$, and $y_2 = -c_1 e^{(-0.3)t} - 3c_2 e^{(-0.4)t}$.

27. Using the result from Exercise 25, $y_1 = c_1 e^{(-0.3)t} + 2c_2 e^{(-0.4)t}$, and $y_2 = -c_1 e^{(-0.3)t} - 3c_2 e^{(-0.4)t}$, and the initial condition $y_1(0) = 10$ and $y_2(0) = 20$, we have the equations

$$c_1 + 2c_2 = 10$$
$$-c_1 - 3c_2 = 20$$

Hence $c_1 = 70$ and $c_2 = -30$. Thus $y_1 = 70e^{(-0.3)t} + 2(-30)e^{(-0.4)t} = 70e^{-0.3t} - 60e^{-0.4t}$, and $y_2 = -70e^{(-0.3)t} - 3(-30)e^{(-0.4)t} = 90e^{-0.4t} - 70e^{-0.3t}$.

29. To solve $\mathbf{y}' = A\mathbf{y} = \begin{bmatrix} -3 & 5 \\ 4 & -4 \end{bmatrix} \mathbf{y}$, we determine the eigenvalues and eigenvectors of A, from the

characteristic polynomial, $\det(A - \lambda I_2) = \lambda^2 + 7\lambda - 8 = (\lambda + 8)(\lambda - 1)$.

$$\lambda = -8 \Rightarrow \mathbf{u} = \begin{bmatrix} -1 \\ 1 \end{bmatrix}, \lambda = 1 \Rightarrow \mathbf{u} = \begin{bmatrix} 5 \\ 4 \end{bmatrix}$$

Hence $y_1 = -c_1 e^{-8t} + 5c_2 e^t$, and $y_2 = c_1 e^{-8t} + 4c_2 e^t$. As t gets large, $y_1 \approx 5c_2 e^t$ and $y_2 \approx 4c_2 e^t$, and hence the ratio $y_1/y_2 \approx 5/4$.

31. Using the result from Exercise 29, $y_1 = -c_1 e^{-8t} + 5c_2 e^t$, and $y_2 = c_1 e^{-8t} + 4c_2 e^t$, and the initial condition $y_1(0) = 1$ and $y_2(0) = 2$, we have the equations

$$-c_1 + 5c_2 = 1$$
$$c_1 + 4c_2 = 2$$

Hence $c_1 = \frac{2}{3}$ and $c_2 = \frac{1}{3}$. Thus $y_1 = -\frac{2}{3}e^{-8t} + 5\left(\frac{1}{3}\right)e^t = \frac{5}{3}e^t - \frac{2}{3}e^{-8t}$, and $y_2 = \frac{2}{3}e^{-8t} + 4\left(\frac{1}{3}\right)e^t = \frac{4}{3}e^t + \frac{2}{3}e^{-8t}$.

33. For example, $y_1' = -3y_1$ and $y_2' = 2y_2$.

35. For example,

$$y_1' = 10y_1 + 6y_2$$
$$y_2' = -18y_1 - 11y_2$$

37. For example,

$$y_1' = -6y_1 + 4y_2 + 7y_3$$
$$y_2' = -7y_1 + 5y_2 + 7y_3$$
$$y_3' = -4y_1 + 4y_2 + 5y_3$$

39. True, $y = ce^{kt}$ is established as a solution, and it can be shown using calculus that every solution has this form for some c.

41. False. If each eigenvalue is pure imaginary, then solutions involve only combinations of trigonometric functions.

43. $\det(A - \lambda I_3) = -\lambda^3 + \lambda^2 + 27\lambda + 112 = 0 \quad \Rightarrow \quad \lambda_1 \approx 7.0652, \lambda_2 \approx -3.0326 + 2.5798i, \lambda_3 \approx -3.0326 - 2.5798i$.

$$\lambda_1 \approx 7.0652 \Rightarrow \mathbf{u} \approx \begin{bmatrix} -0.7811 \\ 0.4471 \\ -0.4359 \end{bmatrix}, \lambda_{2,3} \approx -3.0326 \pm 2.5798i \Rightarrow \mathbf{u} \approx \begin{bmatrix} -0.7041 \\ -0.1528 \pm 0.4597i \\ 0.5141 \mp 0.0729i \end{bmatrix}$$

Hence

$$\mathbf{y} \approx c_1 e^{(7.0652)t} \begin{bmatrix} -0.7811 \\ 0.4471 \\ -0.4359 \end{bmatrix}$$

$$+ c_2 e^{(-3.0326)t} \left(\cos(2.5798t) \begin{bmatrix} -0.7041 \\ -0.1528 \\ 0.5141 \end{bmatrix} - \sin(2.5798t) \begin{bmatrix} 0 \\ 0.4597 \\ -0.0729 \end{bmatrix} \right)$$

$$+ c_3 e^{(-3.0326)t} \left(\sin(2.5798t) \begin{bmatrix} -0.7041 \\ -0.1528 \\ 0.5141 \end{bmatrix} + \cos(2.5798t) \begin{bmatrix} 0 \\ 0.4597 \\ -0.0729 \end{bmatrix} \right).$$

Thus

$$y_1 \approx -0.7811 c_1 e^{7.065t} - 0.7041 c_2 (\cos 2.580t) e^{(-3.033t)} - 0.7041 c_3 (\sin 2.580t) e^{(-3.033t)}$$
$$y_2 \approx 0.4471 c_1 e^{7.065t} - c_2 (0.1528 \cos(2.580t) + 0.4597 \sin(2.580t)) e^{(-3.033t)}$$
$$- c_3 (0.1528 \sin(2.580t) - 0.4597 \cos(2.580t)) e^{(-3.033t)}$$
$$y_3 \approx -0.4359 c_1 e^{7.065t} + c_2 (0.5141 \cos(2.580t) + 0.0729 \sin(2.580t)) e^{(-3.033t)}$$
$$+ c_3 (0.5141 \sin(2.580t) - 0.0729 \cos(2.580t)) e^{(-3.033t)}$$

45. $\det(A - \lambda I_4) = \lambda^4 - 6\lambda^3 + 3\lambda^2 + 8\lambda - 722 = 0 \quad \Rightarrow \quad \lambda_1 \approx -4.114, \lambda_2 \approx 7.297, \lambda_3 \approx 1.408 + 4.698i, \lambda_4 \approx 1.408 - 4.698i$

$$\lambda_1 \approx -4.114 \Rightarrow \mathbf{u} \approx \begin{bmatrix} -0.8167 \\ -0.8101 \\ 0.5896 \\ 1 \end{bmatrix}, \lambda_2 \approx 7.297 \Rightarrow \mathbf{u} \approx \begin{bmatrix} 1.139 \\ 0.1064 \\ 0.39 \\ 1 \end{bmatrix}$$

$$\lambda_{3,4} \approx 1.408 \pm 4.698i \Rightarrow \mathbf{u} \approx \begin{bmatrix} -0.5848 \mp 0.2576i \\ 2.858 \pm 2.336i \\ -2.385 \pm 1.314i \\ 1 \end{bmatrix}.$$

Hence

$$\mathbf{y} \approx c_1 e^{(-4.114)t} \begin{bmatrix} -0.8167 \\ -0.8101 \\ 0.5896 \\ 1 \end{bmatrix} + c_2 e^{(7.297)t} \begin{bmatrix} 1.139 \\ 0.1064 \\ 0.39 \\ 1 \end{bmatrix}$$

$$+ c_3 e^{(1.408)t} \left(\cos(4.698t) \begin{bmatrix} -0.5848 \\ 2.858 \\ -2.385 \\ 1 \end{bmatrix} - \sin(4.698t) \begin{bmatrix} -0.2576 \\ 2.336 \\ 1.314 \\ 0 \end{bmatrix} \right)$$

$$+ c_4 e^{(1.408)t} \left(\sin(4.698t) \begin{bmatrix} -0.5848 \\ 2.858 \\ -2.385 \\ 1 \end{bmatrix} + \cos(4.698t) \begin{bmatrix} -0.2576 \\ 2.336 \\ 1.314 \\ 0 \end{bmatrix} \right)$$

Thus

$$y_1 \approx -0.8167 c_1 e^{-4.114t} + 1.139 c_2 e^{7.297t} + c_3 \left(0.2576 \sin(4.698t) - 0.5848 (\cos 4.698t)\right) e^{1.408t}$$
$$- c_4 \left(0.2576 \cos(4.698t) + 0.5848 \sin(4.698t)\right) e^{1.408t}$$

$$y_2 \approx -0.8101 c_1 e^{-4.114t} + 0.1064 c_2 e^{7.297t} - c_3 \left(2.336 \sin(4.698t) - 2.858 \cos(4.698t)\right) e^{1.408t}$$
$$+ c_4 \left(2.336 \cos(4.698t) + 2.858 \sin(4.698t)\right) e^{1.408t}$$

$$y_3 \approx 0.5896 c_1 e^{-4.114t} + 0.39 c_2 e^{7.297t} - c_3 e^{1.408t} \left(2.385 \cos(4.698t) + 1.314 \sin(4.698t)\right)$$
$$- c_4 e^{1.408t} \left(2.385 \sin(4.698t) - 1.314 \cos(4.698t)\right)$$

$$y_4 \approx c_1 e^{-4.114t} + c_2 e^{7.297t} + c_3 \cos(4.698t) e^{1.408t} + c_4 \sin(4.698t) e^{1.408t}$$

47. $\det(A - \lambda I_3) = \lambda^3 - 2\lambda^2 - 56\lambda + 63 = 0 \quad \Rightarrow \quad \lambda_1 \approx 8.0096, \lambda_2 \approx 1.1055, \lambda_3 \approx -7.1151.$

$$\lambda_1 \approx 8.0096 \Rightarrow \mathbf{u} \approx \begin{bmatrix} -0.6278 \\ 0.0341 \\ -0.7777 \end{bmatrix}, \lambda_2 \approx 1.1055 \Rightarrow \mathbf{u} \approx \begin{bmatrix} 0.8310 \\ -0.2998 \\ -0.4685 \end{bmatrix},$$

$$\lambda_3 \approx -7.1151 \Rightarrow \mathbf{u} \approx \begin{bmatrix} 0.2591 \\ 0.8596 \\ -0.4403 \end{bmatrix}.$$

Thus $\mathbf{y} \approx c_1 e^{(8.0096)t} \begin{bmatrix} -0.6278 \\ 0.0341 \\ -0.7777 \end{bmatrix} + c_2 e^{(1.1055)t} \begin{bmatrix} 0.8310 \\ -0.2998 \\ -0.4685 \end{bmatrix} + c_3 e^{(-7.1151)t} \begin{bmatrix} 0.2591 \\ 0.8596 \\ -0.4403 \end{bmatrix}$. Hence

$$y_1 \approx -0.6278 c_1 e^{8.01t} + 0.831 c_2 e^{1.106t} + 0.2591 c_3 e^{-7.115t}$$
$$y_2 \approx 0.0341 c_1 e^{8.01t} - 0.2998 c_2 e^{1.106t} + 0.8596 c_3 e^{-7.115t}$$
$$y_3 \approx -0.7777 c_1 e^{8.01t} - 0.4685 c_2 e^{1.106t} - 0.4403 c_3 e^{-7.115t}$$

Using the initial conditions we have

$$-0.6278 c_1 + 0.831 c_2 + 0.2591 c_3 = -1$$
$$0.0341 c_1 - 0.2998 c_2 + 0.8596 c_3 = -4$$
$$-0.7777 c_1 - 0.4685 c_2 - 0.4403 c_3 = 3$$

and obtain $c_1 \approx -0.9152, c_2 \approx -0.4106, c_3 \approx -4.76$. Substituting, we obtain

$$y_1 \approx 0.5746 e^{8.01t} - 0.3412 e^{1.106t} - 1.233 e^{-7.115t}$$
$$y_2 \approx -0.03121 e^{8.01t} + 0.1231 e^{1.106t} - 4.092 e^{-7.115t}$$
$$y_3 \approx 0.7118 e^{8.01t} + 0.1924 e^{1.106t} + 2.096 e^{-7.115t}$$

49. $\det(A - \lambda I_4) = \lambda^4 - 6\lambda^3 - 66\lambda^2 - 196\lambda - 32 = 0 \quad \Rightarrow \quad \lambda_1 \approx -0.1732, \lambda_2 \approx 12.5312, \lambda_3 \approx -3.1790 + 2.1534i, \lambda_4 = -3.1790 - 2.1534i.$

$$\lambda_1 \approx -0.1732 \Rightarrow \mathbf{u} \approx \begin{bmatrix} -0.9559 \\ -1.3511 \\ 0.6732 \\ 1 \end{bmatrix}, \quad \lambda_2 \approx 12.5312 \Rightarrow \mathbf{u} \approx \begin{bmatrix} 0.3788 \\ 0.3130 \\ 0.1085 \\ 1 \end{bmatrix}$$

$$\lambda_{3,4} \approx -3.1790 \pm 2.1534i \Rightarrow \mathbf{u} \approx \begin{bmatrix} -1.255 \mp 0.1926i \\ -0.9193 \mp 0.1079i \\ 0.0871 \pm 0.5377i \\ 1 \end{bmatrix}.$$

Hence

$$\mathbf{y} \approx c_1 e^{(-0.1732)t} \begin{bmatrix} -0.9559 \\ -1.3511 \\ 0.6732 \\ 1 \end{bmatrix} + c_2 e^{(12.5312)t} \begin{bmatrix} 0.3788 \\ 0.3130 \\ 0.1085 \\ 1 \end{bmatrix}$$

$$+ c_3 e^{(-3.1790)t} \left(\cos(2.153t) \begin{bmatrix} -1.255 \\ -0.9193 \\ 0.0871 \\ 1 \end{bmatrix} - \sin(2.153t) \begin{bmatrix} -0.1926 \\ -0.1079 \\ 0.5377 \\ 0 \end{bmatrix} \right)$$

$$+ c_4 e^{(-3.1790)t} \left(\sin(2.153t) \begin{bmatrix} -1.255 \\ -0.9193 \\ 0.0871 \\ 1 \end{bmatrix} + \cos(2.153t) \begin{bmatrix} -0.1926 \\ -0.1079 \\ 0.5377 \\ 0 \end{bmatrix} \right).$$

Thus

$$y_1 \approx -0.9559 c_1 e^{-0.1732t} + 0.3788 c_2 e^{(12.53t)} + c_3 \left(-1.255 \cos(2.153t) + 0.1926 \sin(2.153t) \right) e^{-3.179t}$$
$$- c_4 \left(1.255 \sin(2.153t) + 0.1926 \cos(2.153t) \right) e^{-3.179t}$$

$$y_2 \approx -1.3511 c_1 e^{-0.1732t} + 0.313 c_2 e^{(12.53t)} + c_3 \left(-0.9193 \cos(2.153t) + 0.1079 \sin(2.153t) \right) e^{-3.179t}$$
$$- c_4 \left(0.1079 \cos(2.153t) + 0.9193 \sin(2.153t) \right) e^{-3.179t}$$

$$y_3 \approx 0.6732 c_1 e^{-0.1732t} + 0.1085 c_2 e^{(12.53t)} + c_3 \left(0.0871 \cos(2.153t) - 0.5377 \sin(2.153t) \right) e^{-3.179t}$$
$$+ c_4 \left(0.5377 \cos(2.153t) + 0.0871 \sin(2.153t) \right) e^{-3.179t}$$

$$y_4 \approx c_1 e^{-0.1732t} + c_2 e^{(12.53t)} + c_3 \left(\cos(2.153t) \right) e^{-3.179t} + c_4 \left(\sin(2.153t) \right) e^{-3.179t}$$

Using the initial conditions we have

$$-0.9559 c_1 + 0.3788 c_2 - 1.255 c_3 - 0.1926 c_4 = 7$$
$$-1.3511(4.3494) + 0.313 c_2 - 0.9193 c_3 - 0.1079 c_4 = 2$$
$$0.6732(4.3494) + 0.1085 c_2 + 0.0871 c_3 + 0.5377 c_4 = -2$$
$$(4.3494) + c_2 + c_3 = -5$$

and obtain $c_1 \approx 4.3494, c_2 \approx -1.249, c_3 \approx -8.101, c_4 \approx -7.601$. Substituting, we obtain

$$y_1 \approx 11.63 \left(\cos 2.153t \right) e^{-3.179t} - 0.4729 e^{12.53t} - 4.158 e^{-0.1732t} + 7.978 \left(\sin 2.153t \right) e^{-3.179t}$$

$$y_2 \approx 8.266 \left(\cos 2.153t \right) e^{-3.179t} - 0.3908 e^{12.53t} - 5.876 e^{-0.1732t} + 6.113 \left(\sin 2.153t \right) e^{-3.179t}$$

$$y_3 \approx 2.928 e^{-0.1732t} - 0.1355 e^{12.53t} - 4.792 \left(\cos 2.153t \right) e^{-3.179t} + 3.693 \left(\sin 2.153t \right) e^{-3.179t}$$

$$y_4 \approx 4.349 e^{-0.1732t} - 1.249 e^{12.53t} - 8.101 \left(\cos 2.153t \right) e^{-3.179t} - 7.601 \left(\sin 2.153t \right) e^{-3.179t}$$

Chapter 7

Vector Spaces

7.1 Vector Spaces and Subspaces

1. Let $A = \begin{bmatrix} a_{11} & a_{12} \\ a_{21} & a_{22} \end{bmatrix}$, $B = \begin{bmatrix} b_{11} & b_{12} \\ b_{21} & b_{22} \end{bmatrix}$, and $U = \begin{bmatrix} u_{11} & u_{12} \\ u_{21} & u_{22} \end{bmatrix}$ be 2×2 real matrices, and c_1 and c_2 real scalars.

(a) $A + B = \begin{bmatrix} a_{11} & a_{12} \\ a_{21} & a_{22} \end{bmatrix} + \begin{bmatrix} b_{11} & b_{12} \\ b_{21} & b_{22} \end{bmatrix} = \begin{bmatrix} a_{11} + b_{11} & a_{12} + b_{12} \\ a_{21} + b_{21} & a_{22} + b_{22} \end{bmatrix} = \begin{bmatrix} b_{11} + a_{11} & b_{12} + a_{12} \\ b_{21} + a_{21} & b_{22} + a_{22} \end{bmatrix}$

$= \begin{bmatrix} b_{11} & b_{12} \\ b_{21} & b_{22} \end{bmatrix} + \begin{bmatrix} a_{11} & a_{12} \\ a_{21} & a_{22} \end{bmatrix} = B + A.$

(b) $(A + B) + U = \left(\begin{bmatrix} a_{11} & a_{12} \\ a_{21} & a_{22} \end{bmatrix} + \begin{bmatrix} b_{11} & b_{12} \\ b_{21} & b_{22} \end{bmatrix} \right) + \begin{bmatrix} u_{11} & u_{12} \\ u_{21} & u_{22} \end{bmatrix} = \begin{bmatrix} a_{11} + b_{11} & a_{12} + b_{12} \\ a_{21} + b_{21} & a_{22} + b_{22} \end{bmatrix} +$

$\begin{bmatrix} u_{11} & u_{12} \\ u_{21} & u_{22} \end{bmatrix} = \begin{bmatrix} (a_{11} + b_{11}) + u_{11} & (a_{12} + b_{12}) + u_{12} \\ (a_{21} + b_{21}) + u_{21} & (a_{22} + b_{22}) + u_{22} \end{bmatrix} =$

$\begin{bmatrix} a_{11} + (b_{11} + u_{11}) & a_{12} + (b_{12} + u_{12}) \\ a_{21} + (b_{21} + u_{21}) & a_{22} + (b_{22} + u_{22}) \end{bmatrix} = \begin{bmatrix} a_{11} & a_{12} \\ a_{21} & a_{22} \end{bmatrix} + \begin{bmatrix} b_{11} + u_{11} & b_{12} + u_{12} \\ b_{21} + u_{21} & b_{22} + u_{22} \end{bmatrix} =$

$\begin{bmatrix} a_{11} & a_{12} \\ a_{21} & a_{22} \end{bmatrix} + \left(\begin{bmatrix} b_{11} & b_{12} \\ b_{21} & b_{22} \end{bmatrix} + \begin{bmatrix} u_{11} & u_{12} \\ u_{21} & u_{22} \end{bmatrix} \right) = A + (B + U).$

(c) $c_1 (A + B) = c_1 \left(\begin{bmatrix} a_{11} & a_{12} \\ a_{21} & a_{22} \end{bmatrix} + \begin{bmatrix} b_{11} & b_{12} \\ b_{21} & b_{22} \end{bmatrix} \right) = c_1 \left(\begin{bmatrix} a_{11} + b_{11} & a_{12} + b_{12} \\ a_{21} + b_{21} & a_{22} + b_{22} \end{bmatrix} \right)$

$= \begin{bmatrix} c_1 (a_{11} + b_{11}) & c_1 (a_{12} + b_{12}) \\ c_1 (a_{21} + b_{21}) & c_1 (a_{22} + b_{22}) \end{bmatrix} = \begin{bmatrix} c_1 a_{11} + c_1 b_{11} & c_1 a_{12} + c_1 b_{12} \\ c_1 a_{21} + c_1 b_{21} & c_1 a_{22} + c_1 b_{22} \end{bmatrix} = \begin{bmatrix} c_1 a_{11} & c_1 a_{12} \\ c_1 a_{21} & c_1 a_{22} \end{bmatrix} +$

$\begin{bmatrix} c_1 b_{11} & c_1 b_{12} \\ c_1 b_{21} & c_1 b_{22} \end{bmatrix} = c_1 \begin{bmatrix} a_{11} & a_{12} \\ a_{21} & a_{22} \end{bmatrix} + c_1 \begin{bmatrix} b_{11} & b_{12} \\ b_{21} & b_{22} \end{bmatrix} = c_1 A + c_2 B.$

(d) $(c_1 + c_2) A = (c_1 + c_2) \begin{bmatrix} a_{11} & a_{12} \\ a_{21} & a_{22} \end{bmatrix} = \begin{bmatrix} (c_1 + c_2) a_{11} & (c_1 + c_2) a_{12} \\ (c_1 + c_2) a_{21} & (c_1 + c_2) a_{22} \end{bmatrix} =$

$\begin{bmatrix} c_1 a_{11} + c_2 a_{11} & c_1 a_{12} + c_2 a_{12} \\ c_1 a_{21} + c_2 a_{21} & c_1 a_{22} + c_2 a_{22} \end{bmatrix} = \begin{bmatrix} c_1 a_{11} & c_1 a_{12} \\ c_1 a_{21} & c_1 a_{22} \end{bmatrix} + \begin{bmatrix} c_2 a_{11} & c_2 a_{12} \\ c_2 a_{21} & c_2 a_{22} \end{bmatrix} =$

$c_1 \begin{bmatrix} a_{11} & a_{12} \\ a_{21} & a_{22} \end{bmatrix} + c_2 \begin{bmatrix} a_{11} & a_{12} \\ a_{21} & a_{22} \end{bmatrix} = c_1 A + c_2 A.$

(e) $(c_1 c_2) A = (c_1 c_2) \begin{bmatrix} a_{11} & a_{12} \\ a_{21} & a_{22} \end{bmatrix} = \begin{bmatrix} (c_1 c_2) a_{11} & (c_1 c_2) a_{12} \\ (c_1 c_2) a_{21} & (c_1 c_2) a_{22} \end{bmatrix} = \begin{bmatrix} c_1 (c_2 a_{11}) & c_1 (c_2 a_{12}) \\ c_1 (c_2 a_{21}) & c_1 (c_2 a_{22}) \end{bmatrix}$

$= c_1 \begin{bmatrix} c_2 a_{11} & c_2 a_{12} \\ c_2 a_{21} & c_2 a_{22} \end{bmatrix} = c_1 \left(c_2 \begin{bmatrix} a_{11} & a_{12} \\ a_{21} & a_{22} \end{bmatrix} \right) = c_1 (c_2 A).$

(f) $1 \cdot A = 1 \cdot \begin{bmatrix} a_{11} & a_{12} \\ a_{21} & a_{22} \end{bmatrix} = \begin{bmatrix} 1(a_{11}) & 1(a_{12}) \\ 1(a_{21}) & 1(a_{22}) \end{bmatrix} = \begin{bmatrix} a_{11} & a_{12} \\ a_{21} & a_{22} \end{bmatrix} = A$

3. We need to establish properties (4) and (5). To establish (4), let $f \in C[a,b]$, and define $g(x) = -f(x)$. Then $g \in C[a,b]$, and $(f+g)(x) = f(x) + g(x) = f(x) - f(x) = 0 = z(x)$ for all x. Hence every $f \in C[a,b]$ has an additive inverse. Now let f, g, and h belong to $C[a,b]$, and c_1 and c_2 two scalars. Then for $x \in [a,b]$,

   (a) $(f+g)(x) = f(x) + g(x) = g(x) + f(x) = (g+f)(x)$, so $f + g = g + f$.

   (b) $((f+g)+h)(x) = (f+g)(x) + h(x) = (f(x) + g(x)) + h(x) = f(x) + (g(x) + h(x)) = (f+(g+h))(x)$, so $f + (g+h) = (f+g) + h$.

   (c) $(c_1(f+g))(x) = c_1((f+g)(x)) = c_1(f(x) + g(x)) = c_1f(x) + c_1g(x) = (c_1f)(x) + (c_1g)(x) = (c_1f + c_1g)(x)$, so $c_1(f+g) = c_1f + c_1g$.

   (d) $((c_1+c_2)f)(x) = (c_1+c_2)(f(x)) = c_1(f(x)) + c_2(f(x)) = (c_1f)(x) + (c_2f)(x) = (c_1f + c_2f)(x)$, so $(c_1+c_2)f = c_1f + c_2f$.

   (e) $((c_1c_2)f)(x) = (c_1c_2)(f(x)) = c_1(c_2(f(x))) = c_1((c_2f)(x)) = (c_1(c_2f))(x)$, so $(c_1c_2)f = c_1(c_2f)$.

   (f) $(1 \cdot f)(x) = 1 \cdot f(x) = f(x)$, so $1 \cdot f = f$.

5. Property (1): If q_1 and q_2 are polynomials with real coefficients and degree no greater than n, then $q_1 + q_2$ is a polynomial with real coefficients and degree no greater than n, hence V is closed under addition.
   Property (2): If c is a real scalar and q is a polynomial with real coefficients and degree no greater than n, then cq is a polynomial with real coefficients and degree no greater than n, hence V is closed under scalar multiplication.
   Property (3): Let $z(x) = 0$, then z is in V, and $(z+q)(x) = z(x) + q(x) = 0 + q(x) = q(x)$, so $z + q = q$ for every q in V, hence z is the zero vector in V.
   Property (4): Given q_1 in V, let $q_2(x) = -q_1(x)$. Then q_2 is in V, and $(q_1 + q_2)(x) = q_1(x) + q_2(x) = q_1(x) - q_1(x) = 0 = z(x)$. Thus $q_1 + q_2 = z$, hence q_2 is an additive inverse of q_1.
   Property (5): Let q_1, q_2, and q_3 belong to V, and c_1 and c_2 scalars. Then,

   (a) $(q_1 + q_2)(x) = q_1(x) + q_2(x) = (q_2 + q_1)(x)$, hence $q_1 + q_2 = q_2 + q_1$.

   (b) $((q_1 + q_2) + q_3)(x) = (q_1 + q_2)(x) + q_3(x) = (q_1(x) + q_2(x)) + q_3(x) = q_1(x) + (q_2(x) + q_3(x)) = q_1(x) + (q_2 + q_3)(x) = (q_1 + (q_2 + q_3))(x)$, hence $(q_1 + q_2) + q_3 = q_1 + (q_2 + q_3)$.

   (c) $(c_1(q_1 + q_2))(x) = c_1((q_1 + q_2)(x)) = c_1(q_1(x) + q_2(x)) = c_1(q_1(x)) + c_1(q_2(x)) = (c_1q_1)(x) + (c_1q_2)(x) = (c_1q_1 + c_1q_2)(x)$, hence $c_1(q_1 + q_2) = c_1q_1 + c_1q_2$.

   (d) $((c_1 + c_2)q_1)(x) = (c_1 + c_2)(q_1(x)) = c_1(q_1(x)) + c_2(q_1(x)) = (c_1q_1)(x) + (c_2q_1)(x) = (c_1q_1 + c_2q_1)(x)$, hence $(c_1 + c_2)q_1 = c_1q_1 + c_2q_1$.

   (e) $((c_1c_2)q_1)(x) = (c_1c_2)(q_1(x)) = c_1(c_2(q_1(x))) = c_1((c_2q_1)(x)) = (c_1(c_2q_1))(x)$, hence $(c_1c_2)q_1 = c_1(c_2q_1)$.

   (f) $(1 \cdot q_1)(x) = (1)(q_1(x)) = q_1(x)$, hence $1 \cdot q_1 = q_1$.

7. Property (1): If f and g are real-valued continuous functions defined on $\mathbf{R}$, then $f + g$ is a real-valued continuous function defined on $\mathbf{R}$, hence V is closed under addition.
   Property (2): If c is a real scalar and q is a real-valued continuous function defined on $\mathbf{R}$, then cq is a real-valued continuous function defined on $\mathbf{R}$, hence V is closed under scalar multiplication.
   Property (3): Let $z(x) = 0$, then z is in V, and $(z+f)(x) = z(x) + f(x) = 0 + f(x) = f(x)$, so $z + f = f$ for every f in V, hence z is the zero vector in V.
   Property (4): Given f in V, let $g(x) = -f(x)$. Then g is in V, and $(f+g)(x) = f(x) + g(x) = f(x) - f(x) = 0 = z(x)$. Thus $f + g = z$, hence g is an additive inverse of f.
   Property (5): Let f, g, and h belong to V, and c_1 and c_2 scalars. Then for $x \in [a,b]$,

   (a) $(f+g)(x) = f(x) + g(x) = g(x) + f(x) = (g+f)(x)$, so $f + g = g + f$.

(b) $((f + g) + h)(x) = (f + g)(x) + h(x) = (f(x) + g(x)) + h(x) = f(x) + (g(x) + h(x)) = (f + (g + h))(x)$, so $f + (g + h) = (f + g) + h$.

(c) $(c_1(f + g))(x) = c_1((f + g)(x)) = c_1(f(x) + g(x)) = c_1(f(x)) + c_1(g(x)) = (c_1 f)(x) + (c_1 g)(x) = (c_1 f + c_1 g)(x)$, so $c_1(f + g) = c_1 f + c_1 g$.

(d) $((c_1 + c_2)f)(x) = (c_1 + c_2)(f(x)) = c_1(f(x)) + c_2(f(x)) = (c_1 f)(x) + (c_2 f)(x) = (c_1 f + c_2 f)(x)$, so $(c_1 + c_2)f = c_1 f + c_2 f$.

(e) $((c_1 c_2)f)(x) = (c_1 c_2)(f(x)) = c_1(c_2(f(x))) = c_1((c_2 f)(x)) = (c_1(c_2 f))(x)$, so $(c_1 c_2)f = c_1(c_2 f)$.

(f) $(1 \cdot f)(x) = 1 \cdot f(x) = f(x)$, so $1 \cdot f = f$.

9. V is not a vector space, as there is no zero vector. Suppose there exists a zero vector, $z(x) = z_2 x^2 + z_1 x + z_0$. Consider $q(x) = 0$, then $(q + z)(x) = (0x^2 + 0x + 0) + (z_2 x^2 + z_1 x + z_0) = (0 + z_0)x^2 + (0 + z_1)x + (0 + z_2) = z_0 x^2 + z_1 x + z_2 = q(x) \Rightarrow z_0 = 0, z_1 = 0$, and $z_2 = 0$. Hence $z(x) = 0$. But if $p(x) = 1$, then $(p + z)(x) = (0x^2 + 0x + 1) + (0x^2 + 0x + 0) = (1 + 0)x^2 + (0 + 0)x + (0 + 0) = x^2 \neq p(x)$, hence $z(x)$ can not be the zero vector.

11. V is not a vector space, as Property 5(d) does not hold. For example,
$$(1 + 0)\begin{bmatrix} 1 \\ 0 \end{bmatrix} = (1)\begin{bmatrix} 1 \\ 0 \end{bmatrix} = \begin{bmatrix} (1)1 \\ 0 \end{bmatrix} = \begin{bmatrix} 1 \\ 0 \end{bmatrix}, \text{ but } (1)\begin{bmatrix} 1 \\ 0 \end{bmatrix} + (0)\begin{bmatrix} 1 \\ 0 \end{bmatrix} = \begin{bmatrix} (1)1 \\ 0 \end{bmatrix} + \begin{bmatrix} (0)1 \\ 0 \end{bmatrix} =$$
$$\begin{bmatrix} 1 \\ 0 \end{bmatrix} + \begin{bmatrix} 0 \\ 0 \end{bmatrix} = \begin{bmatrix} 0 \\ 0 \end{bmatrix}.$$

13. S contains the zero vector, $\begin{bmatrix} 0 & 0 & 0 \\ 0 & 0 & 0 \\ 0 & 0 & 0 \end{bmatrix}$. Suppose U and V belong to S, then $U + V = \begin{bmatrix} u_{11} & u_{12} & u_{13} \\ 0 & u_{22} & u_{23} \\ 0 & 0 & u_{33} \end{bmatrix} +$
$\begin{bmatrix} v_{11} & v_{12} & v_{13} \\ 0 & v_{22} & v_{23} \\ 0 & 0 & v_{33} \end{bmatrix} = \begin{bmatrix} u_{11} + v_{11} & u_{12} + v_{12} & u_{13} + v_{13} \\ 0 & u_{22} + v_{22} & u_{23} + v_{23} \\ 0 & 0 & u_{33} + v_{33} \end{bmatrix}$ is also upper triangular, and therefore in S.
Also, $cU = \begin{bmatrix} cu_{11} & cu_{12} & cu_{13} \\ 0 & cu_{22} & cu_{23} \\ 0 & 0 & cu_{33} \end{bmatrix}$ is upper triangular, and hence in S. Consequently, S is a subspace.

15. S contains the zero vector, since $z(x) = 0$ satisfies $z(4) = 0$, so z is in S. Suppose f and g are in S, then $f(4) = 0$ and $g(4) = 0$. Hence $(f + g)(4) = f(4) + g(4) = 0 + 0 = 0$, and $f + g$ is in S. Also, $(cf)(4) = cf(4) = c(0) = 0$, hence cf is in S. Thus, S is a subspace.

17. S contains the zero vector, since if $a_1 = a_2 = 0$, then $T\left(\begin{bmatrix} x_1 \\ x_2 \end{bmatrix}\right) = \begin{bmatrix} (0)x_1 \\ (0)x_2 \end{bmatrix} = \begin{bmatrix} 0 \\ 0 \end{bmatrix}$, hence T is the zero vector in $T(2, 2)$. Suppose $T_1\left(\begin{bmatrix} x_1 \\ x_2 \end{bmatrix}\right) = \begin{bmatrix} a_1 x_1 \\ a_2 x_2 \end{bmatrix}$ and $T_2\left(\begin{bmatrix} x_1 \\ x_2 \end{bmatrix}\right) = \begin{bmatrix} b_1 x_1 \\ b_2 x_2 \end{bmatrix}$. Then
$(T_1 + T_2)\left(\begin{bmatrix} x_1 \\ x_2 \end{bmatrix}\right) = T_1\left(\begin{bmatrix} x_1 \\ x_2 \end{bmatrix}\right) + T_2\left(\begin{bmatrix} x_1 \\ x_2 \end{bmatrix}\right) = \begin{bmatrix} a_1 x_1 \\ a_2 x_2 \end{bmatrix} + \begin{bmatrix} b_1 x_1 \\ b_2 x_2 \end{bmatrix} = \begin{bmatrix} (a_1 + b_1)x_1 \\ (a_2 + b_2)x_2 \end{bmatrix}$, hence $T_1 + T_2$ is in S. Also, $(cT_1)\left(\begin{bmatrix} x_1 \\ x_2 \end{bmatrix}\right) = cT_1\left(\begin{bmatrix} x_1 \\ x_2 \end{bmatrix}\right) = c\begin{bmatrix} a_1 x_1 \\ a_2 x_2 \end{bmatrix} = \begin{bmatrix} c(a_1 x_1) \\ c(a_2 x_2) \end{bmatrix} = \begin{bmatrix} (ca_1)x_1 \\ (ca_2)x_2 \end{bmatrix}$, so cT_1 belongs to S. Thus, S is a subspace.

19. S is a subspace. The zero function, $z(x) = 0$, is a polynomial, hence in $\mathbf{P} = S$. If p and q are in S, then both p and q are polynomials, and hence the sum $p + q$ is a polynomial, and in $\mathbf{P} = S$. Also, for scalars c, cp is a polynomial, hence in $\mathbf{P} = S$. Therefore, S is a subspace.

21. S is a subspace. The zero vector, $(0, 0, \ldots)$ is in S, since the second component is 0. If $\mathbf{u}$ and $\mathbf{v}$ are in S, then $\mathbf{u} = (u_1, 0, u_3, \ldots)$ and $\mathbf{v} = (v_1, 0, v_3, \ldots)$, so $\mathbf{u} + \mathbf{v} = (u_1 + v_1, 0 + 0, u_3 + v_3, \ldots) = (u_1 + v_1, 0, u_3 + v_3, \ldots)$ belongs to S. Also, $c\mathbf{u} = c(u_1, 0, u_3, \ldots) = (cu_1, c(0), cu_3, \ldots) = (cu_1, 0, cu_3, \ldots)$ belongs to S. Thus S is a subspace.

23. S is a subspace. S contains the zero vector, $(0, 0, 0, \ldots)$, since it has zero nonzero components. Suppose $\mathbf{u} = (u_1, u_2, \ldots)$ and $\mathbf{v} = (v_1, v_2, \ldots)$ belong to S. Then there exists integers m and n such that $u_p = 0$ for $p > m$ and $v_q = 0$ for $q > n$. Let $r = \max(m, n)$, and $s > r$. Then $u_s + v_s = 0 + 0 = 0$, since $s > r \geq m \;\; \Rightarrow \;\; u_s = 0$ and $s > r \geq n \;\; \Rightarrow \;\; v_s = 0$. Hence $\mathbf{u} + \mathbf{v} = (u_1 + v_1, u_2 + v_2, \ldots)$ has finitely many nonzero components, and therefore $\mathbf{u} + \mathbf{v}$ belongs to S. Also, $cu_p = 0$ for $p > m$, so $c\mathbf{u} = (cu_1, cu_2, \ldots)$ has finitely many nonzero components. Therefore S is a subspace.

25. S is a subspace. The zero function, $z(x) = 0$ belongs to S, since $z(2) + z(3) = 0 + 0 = 0$. Suppose p and q are in S, then $p + q$ is in $\mathbf{P}^3$, and $(p + q)(2) + (p + q)(3) = (p(2) + q(2)) + (p(3) + q(3)) = (p(2) + p(3)) + (q(2) + q(3)) = 0 + 0 = 0$. Hence $p + q$ is in S. Also, $(cp)(2) + (cp)(3) = c(p(2)) + c(p(3)) = c(p(2) + p(3)) = c(0) = 0$, so cp belongs to S. Therefore S is a subspace.

27. S is not a subspace. The zero transformation, $T(\mathbf{x}) = \mathbf{0}$, is not invertible, and thus does not belong to S.

29. S is a subspace. The zero function, $z(x) = 0$ has n continuous derivatives, hence is in S. If f and g are in S, then $f + g$ has n continuous derivatives, hence is in S. Also, if c is a scalar, then cf has n continuous derivatives, hence is in S. Thus S is a subspace.

31. S is not a subspace. The zero function, $z(x) = 0$, does not belong to S since $\int_a^b z(x)\,dx = \int_a^b 0\,dx = 0 \neq b - a$, since $a < b$.

33. For example, the set of vectors $\left\{ \begin{bmatrix} x \\ y \end{bmatrix} \right\}$ such that $x \geq 0$, $y \geq 0$, with the usual definitions of addition and scalar multiplication. This is not closed under scalar multiplication, so is not a vector space.

35. For example, let $V = \mathbf{R}^2$, and $S = \left\{ \begin{bmatrix} x \\ y \end{bmatrix} \right\}$ such that $x \geq 0$, $y \geq 0$.

37. Aside from $V_1 = \mathbf{R}^n$ with the usual definition of addition and scalar multiplication, we can also have $V_2 = \mathbf{R}^n$, but let $\mathbf{w}$ be a fixed nonzero vector and then define addition by $\mathbf{u} \oplus \mathbf{v} = \mathbf{u} + \mathbf{v} - \mathbf{w}$ and scalar multiplication by $c \odot \mathbf{u} = c(\mathbf{u} - \mathbf{w}) + \mathbf{w}$. In this case, $\mathbf{w}$ is the zero vector for V_2. (*"Two plus two equals five" has been used to illustrate how people can be coerced to believe the absurd - for example see George Orwell's* Nineteen Eighty-Four.)

39. True, provided these definitions satisfy the properties of Definition 7.1.

41. True, since $\mathbf{v}_1 - \mathbf{v}_2 = \mathbf{v}_1 + (-1)\mathbf{v}_2$, and V is closed under scalar multiplication, so $(-1)\mathbf{v}_2 \in V$, and vector addition, so $\mathbf{v}_1 + (-1)\mathbf{v}_2 \in V$.

43. True. Since the zero vector belongs to both S_1 and S_2, it belongs to $S_1 \cap S_2$. If $\mathbf{u}$ and $\mathbf{v}$ belong to $S_1 \cap S_2$, then $\mathbf{u}$ and $\mathbf{v}$ belong to S_1, and since S_1 is a subspace $\mathbf{u} + \mathbf{v}$ is in S_1. Likewise, $\mathbf{u} + \mathbf{v}$ is in S_2, and hence $\mathbf{u} + \mathbf{v}$ belongs to $S_1 \cap S_2$. If c is a scalar, then $c\mathbf{u}$ is in S_1 since S_1 is a subspace; and likewise $c\mathbf{u}$ belongs to S_2. Hence $c\mathbf{u}$ is in $S_1 \cap S_2$. Thus $S_1 \cap S_2$ is a subspace.

45. True. Since S_1 and S_2 are subspaces, the zero vector $\mathbf{0}$ belongs to both S_1 and S_2, hence $\mathbf{0} + \mathbf{0} = \mathbf{0}$ belongs to $S_1 + S_2$. If $\mathbf{u}$ and $\mathbf{v}$ belong to $S_1 + S_2$, then $\mathbf{u} = \mathbf{s}_1 + \mathbf{s}_2$ and $\mathbf{v} = \mathbf{t}_1 + \mathbf{t}_2$, where $\mathbf{s}_1$ and $\mathbf{t}_1$ are in S_1, and $\mathbf{s}_2$ and $\mathbf{t}_2$ are in S_2. Thus, $\mathbf{u} + \mathbf{v} = (\mathbf{s}_1 + \mathbf{s}_2) + (\mathbf{t}_1 + \mathbf{t}_2) = (\mathbf{s}_1 + \mathbf{t}_1) + (\mathbf{s}_2 + \mathbf{t}_2)$ is in $S_1 + S_2$, since $\mathbf{s}_1 + \mathbf{t}_1$ belongs to S_1 and $\mathbf{s}_2 + \mathbf{t}_2$ belongs to S_2. If c is a scalar, then $c\mathbf{u} = c(\mathbf{s}_1 + \mathbf{s}_2) = c\mathbf{s}_1 + c\mathbf{s}_2$ is in $S_1 + S_2$, since $c\mathbf{s}_1$ is in S_1 and $c\mathbf{s}_2$ is in S_2. Hence $c\mathbf{u}$ is in $S_1 + S_2$. Thus $S_1 + S_2$ is a subspace.

47. If S is nonempty, then there exists a vector $\mathbf{u}$ in S. Thus, from Definition 7.3 (b) and (c), $\mathbf{u} + (-1)\mathbf{u} = \mathbf{u} - \mathbf{u} = \mathbf{0}$ belongs to S. Conversely, if S contains $\mathbf{0}$, then S is nonempty.

49. (a) Since $-\mathbf{v}$ is an additive inverse of $\mathbf{v}$, $\mathbf{v} + (-\mathbf{v}) = \mathbf{0}$. But by Property 5(a) of Definition 7.1, $-\mathbf{v} + \mathbf{v} = \mathbf{v} + (-\mathbf{v}) = \mathbf{0}$.

   (b) Suppose $\mathbf{0} + \mathbf{v} = \mathbf{v}$ and $\mathbf{z} + \mathbf{v} = \mathbf{v}$ for all vectors $\mathbf{v}$. Then $\mathbf{z} = \mathbf{0} + \mathbf{z} = \mathbf{z} + \mathbf{0} = \mathbf{0}$, and therefore the zero vector is unique.

(c) $0 \cdot \mathbf{v} = (0 + 0) \cdot \mathbf{v} = 0 \cdot \mathbf{v} + 0 \cdot \mathbf{v}$, from Property 5(d) of Definition 7.1. Thus, using Properties 4 and 5(d), we have $\mathbf{0} = 0 \cdot \mathbf{v} + (-0 \cdot \mathbf{v}) = (0 \cdot \mathbf{v} + 0 \cdot \mathbf{v}) + (-0 \cdot \mathbf{v}) = 0 \cdot \mathbf{v} + (0 \cdot \mathbf{v} + (-0 \cdot \mathbf{v})) = 0 \cdot \mathbf{v} + \mathbf{0} = 0 \cdot \mathbf{v}$.

(d) From part (c), and Properties 4, 5(d), and 5(f) of Definition 5.1, we have $\mathbf{0} = 0 \cdot \mathbf{v} = (1 + (-1)) \cdot \mathbf{v} = 1 \cdot \mathbf{v} + (-1) \cdot \mathbf{v} = \mathbf{v} + (-1) \cdot \mathbf{v}$, hence $-\mathbf{v} = (-1) \cdot \mathbf{v}$.

7.2 Span and Linear Independence

1. $\mathbf{v} = 3x^2 + 11x - 8$ is in span $\left\{ 3x^2 + x - 1, x^2 - 3x + 2 \right\}$ if there exist c_1 and c_2 such that $3x^2 + 11x - 8 = c_1 \left(3x^2 + x - 1 \right) + c_2 \left(x^2 - 3x + 2 \right) = (3c_1 + c_2) x^2 + (c_1 - 3c_2) x + (-c_1 + 2c_2)$. We consider the system

$$3c_1 + c_2 = 3$$
$$c_1 - 3c_2 = 11$$
$$-c_1 + 2c_2 = -8$$

and obtain $c_1 = 2$, and $c_2 = -3$. Hence $\mathbf{v} \in \text{span} \left\{ 3x^2 + x - 1, x^2 - 3x + 2 \right\}$.

3. $\mathbf{v} = 10x - 7$ is in span $\left\{ 3x^2 + x - 1, x^2 - 3x + 2 \right\}$ if there exist c_1 and c_2 such that $10x - 7 = c_1 \left(3x^2 + x - 1 \right) + c_2 \left(x^2 - 3x + 2 \right) = (3c_1 + c_2) x^2 + (c_1 - 3c_2) x + (-c_1 + 2c_2)$. We consider the system

$$3c_1 + c_2 = 0$$
$$c_1 - 3c_2 = 10$$
$$-c_1 + 2c_2 = -7$$

and obtain $c_1 = 1$, and $c_2 = -3$. Hence $\mathbf{v} \in \text{span} \left\{ 3x^2 + x - 1, x^2 - 3x + 2 \right\}$.

5. $\mathbf{v} = x^3 + 2x^2 - 3x$ is in span $\left\{ x^3 + x - 2, x^2 + 2x + 1, x^3 - x^2 + x \right\}$ if there exist c_1, c_2, and c_3 such that $x^3 + 2x^2 - 3x = c_1 \left(x^3 + x - 2 \right) + c_2 \left(x^2 + 2x + 1 \right) + c_3 \left(x^3 - x^2 + x \right) = (c_1 + c_3) x^3 + (c_2 - c_3) x^2 + (c_1 + 2c_2 + c_3) x + (-2c_1 + c_2)$. We consider the system

$$c_1 + c_3 = 1$$
$$c_2 - c_3 = 2$$
$$c_1 + 2c_2 + c_3 = -3$$
$$-2c_1 + c_2 = 0$$

and obtain no solutions for c_1, c_2, and c_3. Hence $\mathbf{v} \notin \text{span} \left\{ x^3 + x - 2, x^2 + 2x + 1, x^3 - x^2 + x \right\}$.

7. $\mathbf{v} = x^2 + 4x + 4$ is in span $\left\{ x^3 + x - 2, x^2 + 2x + 1, x^3 - x^2 + x \right\}$ if there exist c_1, c_2, and c_3 such that $x^2 + 4x + 4 = c_1 \left(x^3 + x - 2 \right) + c_2 \left(x^2 + 2x + 1 \right) + c_3 \left(x^3 - x^2 + x \right) = (c_1 + c_3) x^3 + (c_2 - c_3) x^2 + (c_1 + 2c_2 + c_3) x + (-2c_1 + c_2)$. We consider the system

$$c_1 + c_3 = 0$$
$$c_2 - c_3 = 1$$
$$c_1 + 2c_2 + c_3 = 4$$
$$-2c_1 + c_2 = 4$$

and obtain $c_1 = -1$, $c_2 = 2$, and $c_3 = 1$. Hence $\mathbf{v} \in \text{span} \left\{ x^3 + x - 2, x^2 + 2x + 1, x^3 - x^2 + x \right\}$.

9. $\mathbf{v} = \begin{bmatrix} -1 & 4 & 1 \\ -2 & 1 & -3 \end{bmatrix}$ is in span $\left\{ \begin{bmatrix} 1 & 2 & 1 \\ 0 & 1 & 3 \end{bmatrix}, \begin{bmatrix} 0 & 3 & 1 \\ -1 & 1 & 0 \end{bmatrix} \right\}$ if there exist c_1 and c_2 such that $\begin{bmatrix} -1 & 4 & 1 \\ -2 & 1 & -3 \end{bmatrix} = c_1 \begin{bmatrix} 1 & 2 & 1 \\ 0 & 1 & 3 \end{bmatrix} + c_2 \begin{bmatrix} 0 & 3 & 1 \\ -1 & 1 & 0 \end{bmatrix} = \begin{bmatrix} c_1 & 2c_1 + 3c_2 & c_1 + c_2 \\ -c_2 & c_1 + c_2 & 3c_1 \end{bmatrix}$. We consider the system

$$c_1 = -1, \quad 2c_1 + 3c_2 = 4, \quad c_1 + c_2 = 1$$
$$-c_2 = -2, \quad c_1 + c_2 = 1, \quad 3c_1 = -3$$

and obtain $c_1 = -1$, and $c_2 = 2$. Hence $\mathbf{v} \in \text{span} \left\{ \begin{bmatrix} 1 & 2 & 1 \\ 0 & 1 & 3 \end{bmatrix}, \begin{bmatrix} 0 & 3 & 1 \\ -1 & 1 & 0 \end{bmatrix} \right\}$.

11. $\mathbf{v} = \begin{bmatrix} 2 & -5 & -1 \\ 3 & -1 & 6 \end{bmatrix}$ is in $\text{span} \left\{ \begin{bmatrix} 1 & 2 & 1 \\ 0 & 1 & 3 \end{bmatrix}, \begin{bmatrix} 0 & 3 & 1 \\ -1 & 1 & 0 \end{bmatrix} \right\}$ if there exist c_1 and c_2 such that

$\begin{bmatrix} 2 & -5 & -1 \\ 3 & -1 & 6 \end{bmatrix} = c_1 \begin{bmatrix} 1 & 2 & 1 \\ 0 & 1 & 3 \end{bmatrix} + c_2 \begin{bmatrix} 0 & 3 & 1 \\ -1 & 1 & 0 \end{bmatrix} = \begin{bmatrix} c_1 & 2c_1 + 3c_2 & c_1 + c_2 \\ -c_2 & c_1 + c_2 & 3c_1 \end{bmatrix}$. We consider the system

$$c_1 = 2, \quad 2c_1 + 3c_2 = -5, \quad c_1 + c_2 = -1$$
$$-c_2 = 3, \quad c_1 + c_2 = -1, \quad 3c_1 = 6$$

and obtain $c_1 = 2$, and $c_2 = -3$. Hence $\mathbf{v} \in \text{span} \left\{ \begin{bmatrix} 1 & 2 & 1 \\ 0 & 1 & 3 \end{bmatrix}, \begin{bmatrix} 0 & 3 & 1 \\ -1 & 1 & 0 \end{bmatrix} \right\}$.

13. $\mathbf{v} = \begin{bmatrix} -4 & 3 \\ 5 & 5 \end{bmatrix}$ is in $\text{span} \left\{ \begin{bmatrix} -1 & 3 \\ 4 & 1 \end{bmatrix}, \begin{bmatrix} 0 & 2 \\ 5 & -3 \end{bmatrix}, \begin{bmatrix} 1 & 4 \\ 2 & 1 \end{bmatrix} \right\}$ if there exist c_1, c_2, and c_3 such that

$\begin{bmatrix} -4 & 3 \\ 5 & 5 \end{bmatrix} = c_1 \begin{bmatrix} -1 & 3 \\ 4 & 1 \end{bmatrix} + c_2 \begin{bmatrix} 0 & 2 \\ 5 & -3 \end{bmatrix} + c_3 \begin{bmatrix} 1 & 4 \\ 2 & 1 \end{bmatrix} =$
$\begin{bmatrix} -c_1 + c_3 & 3c_1 + 2c_2 + 4c_3 \\ 4c_1 + 5c_2 + 2c_3 & c_1 - 3c_2 + c_3 \end{bmatrix}$. We consider the system

$$-c_1 + c_3 = -4, \quad 3c_1 + 2c_2 + 4c_3 = 3$$
$$4c_1 + 5c_2 + 2c_3 = 5, \quad c_1 - 3c_2 + c_3 = 5$$

and obtain $c_1 = 3$, $c_2 = -1$, and $c_3 = -1$. Hence $\mathbf{v} \in \text{span} \left\{ \begin{bmatrix} -1 & 3 \\ 4 & 1 \end{bmatrix}, \begin{bmatrix} 0 & 2 \\ 5 & -3 \end{bmatrix}, \begin{bmatrix} 1 & 4 \\ 2 & 1 \end{bmatrix} \right\}$.

15. $\mathbf{v} = \begin{bmatrix} -2 & -1 \\ 2 & 0 \end{bmatrix}$ is in $\text{span} \left\{ \begin{bmatrix} -1 & 3 \\ 4 & 1 \end{bmatrix}, \begin{bmatrix} 0 & 2 \\ 5 & -3 \end{bmatrix}, \begin{bmatrix} 1 & 4 \\ 2 & 1 \end{bmatrix} \right\}$ if there exist c_1, c_2, and c_3 such that

$\begin{bmatrix} -2 & -1 \\ 2 & 0 \end{bmatrix} = c_1 \begin{bmatrix} -1 & 3 \\ 4 & 1 \end{bmatrix} + c_2 \begin{bmatrix} 0 & 2 \\ 5 & -3 \end{bmatrix} + c_3 \begin{bmatrix} 1 & 4 \\ 2 & 1 \end{bmatrix} =$
$\begin{bmatrix} -c_1 + c_3 & 3c_1 + 2c_2 + 4c_3 \\ 4c_1 + 5c_2 + 2c_3 & c_1 - 3c_2 + c_3 \end{bmatrix}$. We consider the system

$$-c_1 + c_3 = -2, \quad 3c_1 + 2c_2 + 4c_3 = -1$$
$$4c_1 + 5c_2 + 2c_3 = 2, \quad c_1 - 3c_2 + c_3 = 0$$

and obtain $c_1 = 1$, $c_2 = 0$, and $c_3 = -1$. Hence $\mathbf{v} \in \text{span} \left\{ \begin{bmatrix} -1 & 3 \\ 4 & 1 \end{bmatrix}, \begin{bmatrix} 0 & 2 \\ 5 & -3 \end{bmatrix}, \begin{bmatrix} 1 & 4 \\ 2 & 1 \end{bmatrix} \right\}$.

17. $\left\{ x^2 - 3, 3x^2 + 1 \right\}$ is linearly independent in $\mathbf{P}^2$ if the equation $c_1 \left(x^2 - 3 \right) + c_2 \left(3x^2 + 1 \right) = 0$ has only the trivial solution. Expanding, we obtain $(c_1 + 3c_2) x^2 + (-3c_1 + c_2) = 0$. Equate coefficients and solve the system

$$c_1 + 3c_2 = 0$$
$$-3c_1 + c_2 = 0$$

to obtain $c_1 = 0$, and $c_2 = 0$. Hence $\left\{ x^2 - 3, 3x^2 + 1 \right\}$ is linearly independent in $\mathbf{P}^2$.

19. $\left\{ x^3 + 2x + 4, x^2 - x - 1, x^3 + 2x^2 + 2 \right\}$ is linearly independent in $\mathbf{P}^3$ if the equation $c_1 \left(x^3 + 2x + 4 \right) + c_2 \left(x^2 - x - 1 \right) + c_3 \left(x^3 + 2x^2 + 2 \right) = 0$ has only the trivial solution. Expanding, we obtain $(c_1 + c_3) x^3 + (c_2 + 2c_3) x^2 + (2c_1 - c_2) x + (4c_1 - c_2 + 2c_3) = 0$. Equate coefficients and solve

the system

$$c_1 + c_3 = 0$$
$$c_2 + 2c_3 = 0$$
$$2c_1 - c_2 = 0$$
$$4c_1 - c_2 + 2c_3 = 0$$

and obtain solutions $\begin{bmatrix} c_1 \\ c_2 \\ c_3 \end{bmatrix} \in \text{span} \left\{ \begin{bmatrix} -1 \\ -2 \\ 1 \end{bmatrix} \right\}$. Hence
$\{ x^3 + 2x + 4, \ x^2 - x - 1, \ x^3 + 2x^2 + 2 \}$ is not linearly independent in $\mathbf{P}^3$.

21. $\left\{ \begin{bmatrix} 2 & -1 \\ 1 & 3 \end{bmatrix}, \begin{bmatrix} -4 & 2 \\ -2 & -6 \end{bmatrix} \right\}$ is linearly independent in $\mathbf{R}^{2 \times 2}$ if the equation $c_1 \begin{bmatrix} 2 & -1 \\ 1 & 3 \end{bmatrix} + c_2 \begin{bmatrix} -4 & 2 \\ -2 & -6 \end{bmatrix} =$
$\begin{bmatrix} 0 & 0 \\ 0 & 0 \end{bmatrix}$ has only the trivial solution. Expanding, we obtain
$\begin{bmatrix} 2c_1 - 4c_2 & -c_1 + 2c_2 \\ c_1 - 2c_2 & 3c_1 - 6c_2 \end{bmatrix} = \begin{bmatrix} 0 & 0 \\ 0 & 0 \end{bmatrix}$. Equate entries and solve the system

$$2c_1 - 4c_2 = 0, \quad -c_1 + 2c_2 = 0$$
$$c_1 - 2c_2 = 0, \quad 3c_1 - 6c_2 = 0$$

and obtain solutions $\begin{bmatrix} c_1 \\ c_2 \end{bmatrix} \in \text{span} \left\{ \begin{bmatrix} 2 \\ 1 \end{bmatrix} \right\}$. Hence $\left\{ \begin{bmatrix} 2 & -1 \\ 1 & 3 \end{bmatrix}, \begin{bmatrix} -4 & 2 \\ -2 & -6 \end{bmatrix} \right\}$ is not linearly independent in $\mathbf{R}^{2 \times 2}$.

23. $\left\{ \begin{bmatrix} 1 & 0 & 1 \\ 2 & 1 & 4 \end{bmatrix}, \begin{bmatrix} 3 & 1 & 2 \\ 0 & 3 & 3 \end{bmatrix} \right\}$ is linearly independent in $\mathbf{R}^{2 \times 3}$ if the equation
$c_1 \begin{bmatrix} 1 & 0 & 1 \\ 2 & 1 & 4 \end{bmatrix} + c_2 \begin{bmatrix} 3 & 1 & 2 \\ 0 & 3 & 3 \end{bmatrix} = \begin{bmatrix} 0 & 0 & 0 \\ 0 & 0 & 0 \end{bmatrix}$ has only the trivial solution. Expanding, we obtain
$\begin{bmatrix} c_1 + 3c_2 & c_2 & c_1 + 2c_2 \\ 2c_1 & c_1 + 3c_2 & 4c_1 + 3c_2 \end{bmatrix} = \begin{bmatrix} 0 & 0 & 0 \\ 0 & 0 & 0 \end{bmatrix}$. Equate entries and solve the system

$$c_1 + 3c_2 = 0, \quad c_2 = 0, \quad c_1 + 2c_2 = 0$$
$$2c_1 = 0, \quad c_1 + 3c_2 = 0, \quad 4c_1 + 3c_2 = 0$$

to obtain $c_1 = 0$, and $c_2 = 0$. Hence $\left\{ \begin{bmatrix} 1 & 0 & 1 \\ 2 & 1 & 4 \end{bmatrix}, \begin{bmatrix} 3 & 1 & 2 \\ 0 & 3 & 3 \end{bmatrix} \right\}$ is linearly independent in $\mathbf{R}^{2 \times 3}$.

25. $\{ \sin^2(x), \cos^2(x), 1 \}$ is linearly independent in $C[0, \pi]$ if the equation
$c_1 \sin^2(x) + c_2 \cos^2(x) + c_3(1) = 0$ has only the trivial solution. From the identity $\sin^2(x) + \cos^2(x) = 1$ we obtain $(1)\sin^2(x) + (1)\cos^2(x) + (-1)1 = 0$. Hence $\{ \sin^2(x), \cos^2(x), 1 \}$ is not linearly independent in $C[0, \pi]$.

27. For example, $\left\{ \begin{bmatrix} 1 & 0 \\ 0 & 0 \end{bmatrix}, \begin{bmatrix} 0 & 1 \\ 0 & 0 \end{bmatrix}, \begin{bmatrix} 0 & 0 \\ 1 & 0 \end{bmatrix}, \begin{bmatrix} 0 & 0 \\ 0 & 1 \end{bmatrix}, \begin{bmatrix} 1 & 1 \\ 1 & 1 \end{bmatrix} \right\}$ spans $\mathbf{R}^{2 \times 2}$ but is not linearly independent.

29. For example, let $V = \mathbf{P}$, then $\{ 1, x, x^2, x^3, \ldots \}$ is an infinite linearly independent subset.

31. Let $\mathcal{V}_1 = \{ (1, 0, 0, \ldots), (0, 0, 1, 0, 0 \ldots), (0, 0, 0, 0, 1, 0, 0, \ldots) \}$ and
$\mathcal{V}_2 = \{ (0, 1, 0, 0, \ldots), (0, 0, 0, 1, 0, 0 \ldots), (0, 0, 0, 0, 0, 1, 0, 0, \ldots) \}$. Then $\mathcal{V}_1$ and $\mathcal{V}_2$ are infinite linearly independent subsets of $\mathbf{R}^\infty$ and $\text{span}(\mathcal{V}_1) \cap \text{span}(\mathcal{V}_2) = \{ \mathbf{0} \}$.

33. False. Considering the examples presented so far, we realize that vectors can be functions, matrices, linear transformations, and so on.

35. False. For example, $\left\{ \begin{bmatrix} 1 \\ 0 \end{bmatrix}, \begin{bmatrix} 0 \\ 1 \end{bmatrix} \right\}$ is linearly independent and spans $\mathbf{R}^2$.

37. False. For example, if $\mathcal{V}_1 = \left\{ \begin{bmatrix} 1 \\ 0 \end{bmatrix} \right\}$ and $\mathcal{V}_2 = \left\{ \begin{bmatrix} 1 \\ 0 \end{bmatrix}, \begin{bmatrix} 2 \\ 0 \end{bmatrix} \right\}$, then $\mathcal{V}_1$ is linearly independent, $\mathcal{V}_1 \subset \mathcal{V}_2$, but $\mathcal{V}_2$ is linearly dependent.

39. False. For example, if $\mathcal{V}_1 = \left\{ \begin{bmatrix} 1 \\ 0 \end{bmatrix} \right\}$ and $\mathcal{V}_2 = \left\{ \begin{bmatrix} 1 \\ 0 \end{bmatrix}, \begin{bmatrix} 0 \\ 1 \end{bmatrix} \right\}$, then $\mathcal{V}_2$ spans $\mathbf{R}^2$, $\mathcal{V}_1 \subset \mathcal{V}_2$, but $\mathcal{V}_1$ does not span $\mathbf{R}^2$.

41. True. Consider $c_1\mathbf{v}_1 + c_2(\mathbf{v}_2 - \mathbf{v}_1) + c_3(\mathbf{v}_3 - \mathbf{v}_2 + \mathbf{v}_1) = \mathbf{0}$, then $(c_1 - c_2 + c_3)\mathbf{v}_1 + (c_2 - c_3)\mathbf{v}_2 + c_3\mathbf{v}_3 = \mathbf{0}$. Since $\{\mathbf{v}_1, \mathbf{v}_2, \mathbf{v}_3\}$ is linearly independent, we must have

$$c_1 - c_2 + c_3 = 0$$
$$c_2 - c_3 = 0$$
$$c_3 = 0$$

Solving, we obtain $c_1 = c_2 = c_3 = 0$. Thus $\{\mathbf{v}_1, \mathbf{v}_2 - \mathbf{v}_1, \mathbf{v}_3 - \mathbf{v}_2 + \mathbf{v}_1\}$ is linearly independent.

43. Let $p \in \mathbf{P}$. Since p is a polynomial, there exist a_k such that $p(x) = a_0 + a_1 x + \cdots + a_n x^n$ for some n. Thus $p \in \text{span}\left\{1, x, x^2, \ldots, x^n\right\}$. Since we have shown that every polynomial p in $\mathbf{P}$ is in the span of finitely many vectors in $\left\{1, x, x^2, \ldots\right\}$, it follows that $\left\{1, x, x^2, \ldots\right\}$ spans $\mathbf{P}$.

45. Since $(1)\mathbf{0} + (0)\mathbf{v}_2 + \cdots + (0)\mathbf{v}_m = \mathbf{0}$, $\{\mathbf{0}, \mathbf{v}_2, \ldots, \mathbf{v}_m\}$ is linearly dependent.

47. If $\mathbf{v}_1 = \mathbf{0}$, then $\{\mathbf{v}_1\}$ is linearly dependent, since $(1)\mathbf{v}_1 = (1)\mathbf{0} = \mathbf{0}$. If $\{\mathbf{v}_1\}$ is linearly dependent, then there exists $c \neq 0$ such that $c\mathbf{v}_1 = \mathbf{0}$. Thus $\mathbf{v}_1 = \left(c^{-1}\right)\mathbf{0} = \mathbf{0}$.

49. Let $\{\mathbf{v}_1, \mathbf{v}_2, \ldots, \mathbf{v}_m\}$ be a finite subset of $\mathbf{R}^\infty$. For each $i = 1, \ldots, m$, let $\mathbf{v}_i^* \in \mathbf{R}^{m+1}$ be formed from the first $m + 1$ components of $\mathbf{v}_i$. Then $S = \text{span}\{\mathbf{v}_1^*, \ldots, \mathbf{v}_m^*\} \neq \mathbf{R}^{m+1}$, because $\dim(S) \leq m$. Therefore $\text{span}\{\mathbf{v}_1, \ldots, \mathbf{v}_m\} \neq \mathbf{R}^\infty$.

51. Let $\{\mathbf{v}_1, \mathbf{v}_2, \ldots, \mathbf{v}_m\}$ be a finite subset of $\mathcal{V}_1$, and consider $c_1\mathbf{v}_1 + c_2\mathbf{v}_2 + \cdots + c_m\mathbf{v}_m = \mathbf{0}$. Since each $\mathbf{v}_i \in \mathcal{V}_1$ and $\mathcal{V}_1 \subset \mathcal{V}_2$, each $\mathbf{v}_i \in \mathcal{V}_2$, and $\{\mathbf{v}_1, \mathbf{v}_2, \ldots, \mathbf{v}_m\}$ is a finite subset of $\mathcal{V}_2$. Because $\mathcal{V}_2$ is linearly independent, we must have each $c_i = 0$. Therefore, $\mathcal{V}_1$ is linearly independent.

53. Since $\mathbf{v}$ is in V, and $\{\mathbf{v}, \mathbf{v}_1, \mathbf{v}_2, \ldots, \mathbf{v}_m\}$ spans V, there exist c_i such that $c_1\mathbf{v}_1 + c_2\mathbf{v}_2 + \cdots + c_m\mathbf{v}_m = \mathbf{v}$. Thus $(-1)\mathbf{v} + c_1\mathbf{v}_1 + c_2\mathbf{v}_2 + \cdots + c_m\mathbf{v}_m = \mathbf{0}$, which shows that $\{\mathbf{v}, \mathbf{v}_1, \mathbf{v}_2, \ldots, \mathbf{v}_m\}$ is linearly dependent.

55. We consider the equation $c_1 x + c_2 \sin(\pi x/2) + c_3 e^x = 0$. First let $x = 0$ to obtain $c_3 = 0$. Now let $x = 2$ to obtain $c_1(2) + c_2 \sin(\pi(2)/2) + (0)e^2 = 0 \Rightarrow c_1 = 0$. Finally, let $x = 1$ to obtain $(0)x + c_2 \sin(\pi(1)/2) + (0)e^1 = 0 \Rightarrow c_2 = 0$. We conclude that $\{x, \sin(\pi x/2), e^x\}$ is linearly independent.

57. Since $\cos^2(x) = \frac{1}{2} + \frac{1}{2}\cos(2x)$, we have $(0)e^x + (1)\cos^2(x) - \frac{1}{2}\cos(x) - \frac{1}{2}(1) = 0$, and hence $\left\{e^x, \cos^2(x), \cos(2x), 1\right\}$ is linearly dependent.

7.3   Basis and Dimension

1. $\mathcal{V}$ could not be a basis for $\mathbf{P}^2$, since $\dim\left(\mathbf{P}^2\right) = 3$, and $\mathcal{V}$ has only 2 vectors.

3. $\mathcal{V}$ could be a basis for $\mathbf{R}^{2\times 2}$, since $\dim\left(\mathbf{R}^{2\times 2}\right) = 4$, and $\mathcal{V}$ has 4 vectors.

5. $\mathcal{V}$ could not be a basis for $\mathbf{P}^4$, since $\dim\left(\mathbf{P}^4\right) = 5$, and $\mathcal{V}$ has only 4 vectors.

7. Since $\mathcal{V}$ has 3 vectors and $\dim\left(\mathbf{P}^2\right) = 3$, we need only show $\mathcal{V}$ is linearly independent. Consider $c_1\left(2x^2 + x - 3\right) + c_2\left(x + 1\right) + c_3\left(-5\right) = 0 \quad \Rightarrow \quad 2c_1x^2 + \left(c_1 + c_2\right)x + \left(-3c_1 + c_2 - 5c_3\right) = 0.$ We solve

$$2c_1 = 0$$
$$c_1 + c_2 = 0$$
$$-3c_1 + c_2 - 5c_3 = 0$$

and obtain $c_1 = c_2 = c_3 = 0$. Thus $\mathcal{V}$ is linearly independent, and hence $\mathcal{V}$ is a basis.

9. Since $\mathcal{V}$ has 4 vectors and $\dim\left(\mathbf{R}^{2\times 2}\right) = 4$, we need only determine if $\mathcal{V}$ is linearly independent. Consider $c_1\begin{bmatrix} 1 & 2 \\ 2 & 1 \end{bmatrix} + c_2\begin{bmatrix} 3 & 1 \\ 0 & 3 \end{bmatrix} + c_3\begin{bmatrix} 2 & 2 \\ 1 & 1 \end{bmatrix} + c_4\begin{bmatrix} 3 & 3 \\ 3 & 4 \end{bmatrix} = \begin{bmatrix} 0 & 0 \\ 0 & 0 \end{bmatrix}$

$$\Rightarrow \quad \begin{bmatrix} c_1 + 3c_2 + 2c_3 + 3c_4 & 2c_1 + c_2 + 2c_3 + 3c_4 \\ 2c_1 + c_3 + 3c_4 & c_1 + 3c_2 + c_3 + 4c_4 \end{bmatrix} = \begin{bmatrix} 0 & 0 \\ 0 & 0 \end{bmatrix}.$$ We solve

$$c_1 + 3c_2 + 2c_3 + 3c_4 = 0, \quad 2c_1 + c_2 + 2c_3 + 3c_4 = 0$$
$$2c_1 + c_3 + 3c_4 = 0, \quad c_1 + 3c_2 + c_3 + 4c_4 = 0$$

and obtain $\begin{bmatrix} c_1 \\ c_2 \\ c_3 \\ c_4 \end{bmatrix} = \mathrm{span}\left\{\begin{bmatrix} -2 \\ -1 \\ 1 \\ 1 \end{bmatrix}\right\}$. Thus $\mathcal{V}$ is not linearly independent, and hence $\mathcal{V}$ is not a basis.

11. $\mathcal{V}$ does not span $\mathbf{R}^\infty$, since $(1, 1, 1, \ldots)$ is not in the span of any finite subset of vectors in $\mathcal{V}$. Each vector in $\mathcal{V}$ has only finite nonzero entries, and hence a finite collection of these vectors would not be able to span a vector with infinitely many nonzero entries. Thus $\mathcal{V}$ is not a basis for $\mathbf{R}^\infty$.

13. $\dim\left(S\right) = 8$, and a basis for S is

$$\left\{\begin{bmatrix} -1 & 0 & 0 \\ 0 & 1 & 0 \\ 0 & 0 & 0 \end{bmatrix}, \begin{bmatrix} -1 & 0 & 0 \\ 0 & 0 & 0 \\ 0 & 0 & 1 \end{bmatrix}, \begin{bmatrix} 0 & 1 & 0 \\ 0 & 0 & 0 \\ 0 & 0 & 0 \end{bmatrix}, \begin{bmatrix} 0 & 0 & 1 \\ 0 & 0 & 0 \\ 0 & 0 & 0 \end{bmatrix}, \right.$$
$$\left.\begin{bmatrix} 0 & 0 & 0 \\ 1 & 0 & 0 \\ 0 & 0 & 0 \end{bmatrix}, \begin{bmatrix} 0 & 0 & 0 \\ 0 & 0 & 1 \\ 0 & 0 & 0 \end{bmatrix}, \begin{bmatrix} 0 & 0 & 0 \\ 0 & 0 & 0 \\ 1 & 0 & 0 \end{bmatrix}, \begin{bmatrix} 0 & 0 & 0 \\ 0 & 0 & 0 \\ 0 & 1 & 0 \end{bmatrix}\right\}.$$

15. $\dim\left(S\right) = 3$, and a basis for S is $\left\{\begin{bmatrix} -1 & 0 \\ 0 & 1 \end{bmatrix}, \begin{bmatrix} 0 & -1 \\ 0 & 1 \end{bmatrix}, \begin{bmatrix} 0 & 0 \\ -1 & 1 \end{bmatrix}\right\}.$

17. If $\mathbf{v} = \begin{bmatrix} v_1 \\ v_2 \end{bmatrix} \neq \mathbf{0}$, $T_1\left(\begin{bmatrix} x_1 \\ x_2 \end{bmatrix}\right) = \begin{bmatrix} -v_2x_1 + v_1x_2 \\ 0 \end{bmatrix}$, and $T_2\left(\begin{bmatrix} x_1 \\ x_2 \end{bmatrix}\right) = \begin{bmatrix} 0 \\ -v_2x_1 + v_1x_2 \end{bmatrix}$, then $\{T_1, T_2\}$ is a basis for S and $\dim(S) = 2$. If $\mathbf{v} = \mathbf{0}$, then let $T_1\left(\begin{bmatrix} x_1 \\ x_2 \end{bmatrix}\right) = \begin{bmatrix} x_1 \\ 0 \end{bmatrix}$, $T_2\left(\begin{bmatrix} x_1 \\ x_2 \end{bmatrix}\right) = \begin{bmatrix} x_2 \\ 0 \end{bmatrix}$, $T_3\left(\begin{bmatrix} x_1 \\ x_2 \end{bmatrix}\right) = \begin{bmatrix} 0 \\ x_1 \end{bmatrix}$, and $T_4\left(\begin{bmatrix} x_1 \\ x_2 \end{bmatrix}\right) = \begin{bmatrix} 0 \\ x_2 \end{bmatrix}$. Then $\dim\left(S\right) = 4$, and $\{T_1, T_2, T_3, T_4\}$ is a basis for S.

19. $\dim\left(S\right) = \infty$. For example, $\left\{x\left(x - 1\right)\left(x - 2\right), x^2\left(x - 1\right)\left(x - 2\right), x^3\left(x - 1\right)\left(x - 2\right), \ldots\right\}$ is an infinite set of linearly independent vectors in $C\left(\mathbf{R}\right)$, each of which vanishes at $k = 0, 1, 2$.

21. We extend $\mathcal{V}$ to $\left\{2x^2 + 1, 4x - 3, 1\right\}$ to obtain a basis for $\mathbf{P}^2$.

23. We extend $\mathcal{V}$ to $\left\{\begin{bmatrix} 1 & 0 \\ 0 & 1 \end{bmatrix}, \begin{bmatrix} 0 & 1 \\ 1 & 0 \end{bmatrix}, \begin{bmatrix} 1 & 1 \\ 1 & 0 \end{bmatrix}, \begin{bmatrix} 0 & 1 \\ 0 & 0 \end{bmatrix}\right\}$ to obtain a basis for $\mathbf{R}^{2\times 2}$.

25. We reduce the set $\mathcal{V}$ to $\{x + 1, x + 2\}$ to obtain a basis for $\mathbf{P}^1$.

27. Let $\mathcal{V} = \{\cos(t), \sin(t)\}$. Since every vector in S is a function of the form $y(t) = c_1 \cos(t) + c_2 \sin(t)$, we conclude that $\mathcal{V}$ spans S. Now let $c_1 \cos(t) + c_2 \sin(t) = 0$. Then if $t = 0$ we conclude that $c_1 = 0$, and if $t = \frac{\pi}{2}$, we conclude that $c_2 = 0$. Thus $\mathcal{V}$ is linearly independent, and hence a basis for S. Since $\mathcal{V}$ contains 2 vectors, $\dim(S) = 2$.

29. If $p \in \mathbf{P}^6$, then $p(x) = a_0 + a_1 x + a_2 x^2 + a_3 x^3 + a_4 x^4 + a_5 x^5 + a_6 x^6$, and so $p'(x) = a_1 + 2a_2 x + 3a_3 x^2 + 4a_4 x^3 + 5a_5 x^4 + 6a_6 x^5$, and $p''(x) = 2a_2 + 6a_3 x + 12a_4 x^2 + 20a_5 x^3 + 30a_6 x^4$. Since $p''(x) = 0$, we must have $a_2 = a_3 = a_4 = a_5 = a_6 = 0$, and thus $p(x) = a_0 + a_1 x$. Since S is spanned by the set $\{1, x\}$, which is linearly independent, a basis for S is the set $\{1, x\}$.

31. For example, let $V = \mathbf{P}$, and $S = \operatorname{span}\{1\}$.

33. For example, let $V = \mathbf{P}^1$, and $S = \operatorname{span}\{1\}$.

35. For example, let $V = \mathbf{P}$, and $S = \operatorname{span}\{1, x^2, x^4, x^6, \dots\}$.

37. False. Since the number of vectors in any basis for a vector space is always the same, the size, or dimension, of the vector space does not vary from one basis to another.

39. False. $\dim(\mathbf{R}^2) = 2 = \dim(\mathbf{P}^1)$.

41. True. If there existed an infinite linearly independent subset $\mathcal{V}$ of V, then $\dim(\operatorname{span}(\mathcal{V})) = \infty$, which contradicts Theorem 7.14, since $\dim(V)$ is finite.

43. False. If $\mathcal{V}$ is linearly dependent, then adding vectors to $\mathcal{V}$ will still produce a linearly dependent set, and hence not a basis.

45. False, if $c = 0$. True, if $c \neq 0$.

47. Consider $c_1 \mathbf{v}_1 + c_2(2\mathbf{v}_2) + \dots + c_k(k\mathbf{v}_k) = c_1 \mathbf{v}_1 + (2c_2)\mathbf{v}_2 + \dots + (kc_k)\mathbf{v}_k = \mathbf{0}$. Since $\{\mathbf{v}_1, \mathbf{v}_2, \dots, \mathbf{v}_k\}$ is linearly independent, each $ic_i = 0$, and hence each $c_i = 0$. Therefore, $\{\mathbf{v}_1, 2\mathbf{v}_2, \dots, k\mathbf{v}_k\}$ is linearly independent. Since there are k vectors in $\{\mathbf{v}_1, 2\mathbf{v}_2, \dots, k\mathbf{v}_k\}$, and $\dim(V) = k$, we conclude that $\{\mathbf{v}_1, 2\mathbf{v}_2, \dots, k\mathbf{v}_k\}$ is a basis for V.

49. Consider $c_1 \begin{bmatrix} 1 & 0 \\ 0 & 1 \end{bmatrix} + c_2 \begin{bmatrix} 0 & 1 \\ 0 & 0 \end{bmatrix} + c_3 \begin{bmatrix} 0 & 0 \\ 1 & 0 \end{bmatrix} + c_4 \begin{bmatrix} 0 & 0 \\ 0 & 1 \end{bmatrix} = \begin{bmatrix} 0 & 0 \\ 0 & 0 \end{bmatrix}$. Then $\begin{bmatrix} c_1 & c_2 \\ c_3 & c_4 \end{bmatrix} = \begin{bmatrix} 0 & 0 \\ 0 & 0 \end{bmatrix}$, and hence $c_1 = c_2 = c_3 = c_4 = 0$, and thus $\left\{ \begin{bmatrix} 1 & 0 \\ 0 & 1 \end{bmatrix}, \begin{bmatrix} 0 & 1 \\ 0 & 0 \end{bmatrix}, \begin{bmatrix} 0 & 0 \\ 1 & 0 \end{bmatrix}, \begin{bmatrix} 0 & 0 \\ 0 & 1 \end{bmatrix} \right\}$ is linearly independent. Now suppose $A = \begin{bmatrix} a & b \\ c & d \end{bmatrix}$ is in $\mathbf{R}^{2 \times 2}$. Then $A = \begin{bmatrix} a & b \\ c & d \end{bmatrix} = a \begin{bmatrix} 1 & 0 \\ 0 & 1 \end{bmatrix} + b \begin{bmatrix} 0 & 1 \\ 0 & 0 \end{bmatrix} + c \begin{bmatrix} 0 & 0 \\ 1 & 0 \end{bmatrix} + d \begin{bmatrix} 0 & 0 \\ 0 & 1 \end{bmatrix}$, so $\left\{ \begin{bmatrix} 1 & 0 \\ 0 & 1 \end{bmatrix}, \begin{bmatrix} 0 & 1 \\ 0 & 0 \end{bmatrix}, \begin{bmatrix} 0 & 0 \\ 1 & 0 \end{bmatrix}, \begin{bmatrix} 0 & 0 \\ 0 & 1 \end{bmatrix} \right\}$ spans $\mathbf{R}^{2 \times 2}$. Therefore $\left\{ \begin{bmatrix} 1 & 0 \\ 0 & 1 \end{bmatrix}, \begin{bmatrix} 0 & 1 \\ 0 & 0 \end{bmatrix}, \begin{bmatrix} 0 & 0 \\ 1 & 0 \end{bmatrix}, \begin{bmatrix} 0 & 0 \\ 0 & 1 \end{bmatrix} \right\}$ is a basis for $\mathbf{R}^{2 \times 2}$.

51. Suppose $m \geq 1$. Let $\{\mathbf{v}_1, \mathbf{v}_2, \dots, \mathbf{v}_m\}$ be a basis for V, and define $S_k = \operatorname{span}\{\mathbf{v}_1, \mathbf{v}_2, \dots, \mathbf{v}_k\}$. Then since $\{\mathbf{v}_1, \mathbf{v}_2, \dots, \mathbf{v}_k\}$ must be linearly independent, and since $\{\mathbf{v}_1, \mathbf{v}_2, \dots, \mathbf{v}_k\}$ also spans S_k, it follows that $\dim(S_k) = k$. If $m = 0$, then $V = \{\mathbf{0}\}$ and $S_0 = \{\mathbf{0}\}$.

53. If $c_1 \mathbf{v}_1 + c_2 \mathbf{v}_2 + \dots + c_m \mathbf{v}_m = \mathbf{v}$ has a unique solution for every $\mathbf{v}$, then in particular $c_1 \mathbf{v}_1 + c_2 \mathbf{v}_2 + \dots + c_m \mathbf{v}_m = \mathbf{0}$ has a unique solution, namely $c_1 = c_2 = \dots = c_m = 0$. Hence $\{\mathbf{v}_1, \dots, \mathbf{v}_m\}$ is linearly independent. And since $c_1 \mathbf{v}_1 + c_2 \mathbf{v}_2 + \dots + c_m \mathbf{v}_m = \mathbf{v}$ has a solution for every $\mathbf{v}$, $\{\mathbf{v}_1, \dots, \mathbf{v}_m\}$ spans V. Thus $\{\mathbf{v}_1, \dots, \mathbf{v}_m\}$ is a basis for V. Now suppose $\{\mathbf{v}_1, \dots, \mathbf{v}_m\}$ is a basis for V. Then $\{\mathbf{v}_1, \dots, \mathbf{v}_m\}$ spans V, and therefore $c_1 \mathbf{v}_1 + c_2 \mathbf{v}_2 + \dots + c_m \mathbf{v}_m = \mathbf{v}$ has a solution for every $\mathbf{v}$. To see

that this solution is unique, consider $c_1\mathbf{v}_1 + c_2\mathbf{v}_2 + \cdots + c_m\mathbf{v}_m = \mathbf{v} = d_1\mathbf{v}_1 + d_2\mathbf{v}_2 + \cdots + d_m\mathbf{v}_m$. Then $(c_1 - d_1)\mathbf{v}_1 + \cdots + (c_m - d_m)\mathbf{v}_m = \mathbf{0}$, and since $\{\mathbf{v}_1, \ldots, \mathbf{v}_m\}$ is linearly independent, each $c_i - d_i = 0$, and hence each $c_i = d_i$. This shows that $c_1\mathbf{v}_1 + c_2\mathbf{v}_2 + \cdots + c_m\mathbf{v}_m = \mathbf{v}$ has a unique solution for every $\mathbf{v}$ in V.

55. (a) If $V_1 = V_2$ then clearly $\dim(V_1) = \dim(V_2)$. Now suppose that $V_1 \neq V_2$. Let $\{\mathbf{v}_1, \ldots, \mathbf{v}_m\}$ be a basis for V_1, and suppose that $\mathbf{v}$ is in V_2 but not in V_1. Then $\{\mathbf{v}, \mathbf{v}_1, \ldots, \mathbf{v}_m\}$ must be a linearly independent set, and since each of these vectors is in V_2, it must be that $\dim(V_2) > m = \dim(V_1)$, so $\dim(V_1) \neq \dim(V_2)$.

   (b) Let $V_2 = \mathbf{P}$ and $V_1 = \operatorname{span}\{x, x^2, x^3, \ldots\}$. Then $\dim(V_2) = \dim(V_1) = \infty$, but $V_2 \neq V_1$.

57. (a) If $\mathcal{V}$ spans V, then by Exercise 56(a) vectors can be removed from $\mathcal{V}$ to form a basis for V. But this set would contain at most m vectors, which would imply that $\dim(V) \leq m < n = \dim(V)$, a contradiction. Hence $\mathcal{V}$ does not span V.

   (b) If $\mathcal{V}$ is linearly independent, then by Exercise 56(b) vectors can be added to $\mathcal{V}$ to form a basis for V. But this set would contain at least m vectors, which would imply that $\dim(V) \geq m > n = \dim(V)$, a contradiction. Hence $\mathcal{V}$ is linearly dependent.

Chapter 8

Orthogonality

8.1 Dot Products and Orthogonal Sets

1. (a) $\mathbf{u}_1 \cdot \mathbf{u}_5 = \begin{bmatrix} -3 \\ 1 \\ 2 \end{bmatrix} \cdot \begin{bmatrix} 2 \\ 1 \\ 1 \end{bmatrix} = (-3)(2) + (1)(1) + (2)(1) = -3$

   (b) $\mathbf{u}_3 \cdot (-3\mathbf{u}_2) = -3 \begin{bmatrix} 2 \\ 0 \\ -1 \end{bmatrix} \cdot \begin{bmatrix} 1 \\ 1 \\ 1 \end{bmatrix} = -3((2)(1) + (0)(1) + (-1)(1)) = -3$

   (c) $\mathbf{u}_4 \cdot \mathbf{u}_7 = \begin{bmatrix} 1 \\ -3 \\ 2 \end{bmatrix} \cdot \begin{bmatrix} 3 \\ -4 \\ -2 \end{bmatrix} = 1(3) + (-3)(-4) + 2(-2) = 11$

   (d) $2\mathbf{u}_4 \cdot \mathbf{u}_7 = 2 \begin{bmatrix} 1 \\ -3 \\ 2 \end{bmatrix} \cdot \begin{bmatrix} 3 \\ -4 \\ -2 \end{bmatrix} = \begin{bmatrix} 2 \\ -6 \\ 4 \end{bmatrix} \cdot \begin{bmatrix} 3 \\ -4 \\ -2 \end{bmatrix} = 2(3) + (-6)(-4) + 4(-2) = 22$

3. (a) $\|\mathbf{u}_7\| = \left\| \begin{bmatrix} 3 \\ -4 \\ -2 \end{bmatrix} \right\| = \sqrt{3^2 + (-4)^2 + (-2)^2} = \sqrt{29}.$

   (b) $\|-\mathbf{u}_7\| = \|\mathbf{u}_7\| = \sqrt{29}$ (from (a)).

   (c) $\|2\mathbf{u}_5\| = \left\| 2 \begin{bmatrix} 2 \\ 1 \\ 1 \end{bmatrix} \right\| = \left\| \begin{bmatrix} 4 \\ 2 \\ 2 \end{bmatrix} \right\| = \sqrt{4^2 + 2^2 + 2^2} = \sqrt{24} = 2\sqrt{6}$

   (d) $\|-3\mathbf{u}_5\| = \left\| -3 \begin{bmatrix} 2 \\ 1 \\ 1 \end{bmatrix} \right\| = \left\| \begin{bmatrix} -6 \\ -3 \\ -3 \end{bmatrix} \right\| = \sqrt{(-6)^2 + (-3)^2 + (-3)^2} = \sqrt{72} = 6\sqrt{2}$

5. (a) $\|\mathbf{u}_1 - \mathbf{u}_2\| = \left\| \begin{bmatrix} -3 \\ 1 \\ 2 \end{bmatrix} - \begin{bmatrix} 1 \\ 1 \\ 1 \end{bmatrix} \right\| = \left\| \begin{bmatrix} -4 \\ 0 \\ 1 \end{bmatrix} \right\| = \sqrt{(-4)^2 + 0^2 + 1^2} = \sqrt{17}.$

   (b) $\|\mathbf{u}_3 - \mathbf{u}_8\| = \left\| \begin{bmatrix} 2 \\ 0 \\ -1 \end{bmatrix} - \begin{bmatrix} -1 \\ -1 \\ 3 \end{bmatrix} \right\| = \left\| \begin{bmatrix} 3 \\ 1 \\ -4 \end{bmatrix} \right\| = \sqrt{3^2 + 1^2 + (-4)^2} = \sqrt{26}.$

   (c) $\|2\mathbf{u}_6 - (-\mathbf{u}_3)\| = \left\| 2 \begin{bmatrix} 0 \\ 3 \\ -1 \end{bmatrix} - \left(- \begin{bmatrix} 2 \\ 0 \\ -1 \end{bmatrix} \right) \right\| = \left\| \begin{bmatrix} 2 \\ 6 \\ -3 \end{bmatrix} \right\| = \sqrt{2^2 + 6^2 + (-3)^2} = \sqrt{49} = 7$

(d) $\|-3\mathbf{u}_2 - (2\mathbf{u}_5)\| = \left\| (-3) \begin{bmatrix} 1 \\ 1 \\ 1 \end{bmatrix} - \left(2 \begin{bmatrix} 2 \\ 1 \\ 1 \end{bmatrix} \right) \right\| = \left\| \begin{bmatrix} -7 \\ -5 \\ -5 \end{bmatrix} \right\| =$

$\sqrt{(-7)^2 + (-5)^2 + (-5)^2} = \sqrt{99} = 3\sqrt{11}$

7. (a) $\mathbf{u}_1 \cdot \mathbf{u}_3 = \begin{bmatrix} -3 \\ 1 \\ 2 \end{bmatrix} \cdot \begin{bmatrix} 2 \\ 0 \\ -1 \end{bmatrix} = (-3)(2) + (1)(0) + (2)(-1) = -8 \neq 0$, so $\mathbf{u}_1$ and $\mathbf{u}_3$ are not orthogonal.

(b) $\mathbf{u}_3 \cdot \mathbf{u}_4 = \begin{bmatrix} 2 \\ 0 \\ -1 \end{bmatrix} \cdot \begin{bmatrix} 1 \\ -3 \\ 2 \end{bmatrix} = (2)(1) + (0)(-3) + (-1)(2) = 0$, so $\mathbf{u}_3$ and $\mathbf{u}_4$ are orthogonal.

(c) $\mathbf{u}_2 \cdot \mathbf{u}_5 = \begin{bmatrix} 1 \\ 1 \\ 1 \end{bmatrix} \cdot \begin{bmatrix} 2 \\ 1 \\ 1 \end{bmatrix} = 1(2) + 1(1) + 1(1) = 4 \neq 0$, so $\mathbf{u}_2$ and $\mathbf{u}_5$ are not orthogonal.

(d) $\mathbf{u}_1 \cdot \mathbf{u}_8 = \begin{bmatrix} -3 \\ 1 \\ 2 \end{bmatrix} \cdot \begin{bmatrix} -1 \\ -1 \\ 3 \end{bmatrix} = (-3)(-1) + 1(-1) + 2(3) = 8 \neq 0$, so $\mathbf{u}_1$ and $\mathbf{u}_8$ are not orthogonal.

9. $\mathbf{u} \cdot \mathbf{v} = \begin{bmatrix} a \\ 2 \\ -3 \end{bmatrix} \cdot \begin{bmatrix} 4 \\ a \\ 3 \end{bmatrix} = a(4) + 2(a) + (-3)(3) = 6a - 9 = 0 \quad \Rightarrow \quad a = \frac{3}{2}.$

11. $\mathbf{u} \cdot \mathbf{v} = \begin{bmatrix} 2 \\ a \\ -3 \\ -1 \end{bmatrix} \cdot \begin{bmatrix} -5 \\ 4 \\ 6 \\ a \end{bmatrix} = (2)(-5) + a(4) + (-3)(6) + (-1)(a) = 3a - 28 = 0 \Rightarrow a = \frac{28}{3}.$

13. $\mathbf{u}_1 \cdot \mathbf{u}_2 = \begin{bmatrix} 1 \\ -2 \end{bmatrix} \cdot \begin{bmatrix} 4 \\ 3 \end{bmatrix} = -2 \neq 0$, so $\{\mathbf{u}_1, \mathbf{u}_2\}$ is not an orthogonal set.

15. $\mathbf{u}_2 \cdot \mathbf{u}_3 = \begin{bmatrix} -5 \\ 13 \\ 16 \end{bmatrix} \cdot \begin{bmatrix} 5 \\ -4 \\ 2 \end{bmatrix} = -45 \neq 0$, so $\{\mathbf{u}_1, \mathbf{u}_2, \mathbf{u}_3\}$ is not an orthogonal set.

17. $\mathbf{u}_1 \cdot \mathbf{u}_2 = \begin{bmatrix} -1 \\ 0 \\ 2 \end{bmatrix} \cdot \begin{bmatrix} 4 \\ 3 \\ 2 \end{bmatrix} = 0$, $\mathbf{u}_1 \cdot \mathbf{u}_3 = \begin{bmatrix} -1 \\ 0 \\ 2 \end{bmatrix} \cdot \begin{bmatrix} 6 \\ a \\ 3 \end{bmatrix} = 0$, and $\mathbf{u}_2 \cdot \mathbf{u}_3 = \begin{bmatrix} 4 \\ 3 \\ 2 \end{bmatrix} \cdot \begin{bmatrix} 6 \\ a \\ 3 \end{bmatrix} = 3a + 30.$
Hence we need $3a + 30 = 0$. Thus if $a = -10$ then $\{\mathbf{u}_1, \mathbf{u}_2, \mathbf{u}_3\}$ forms an orthogonal set.

19. $\mathbf{u}_1 \cdot \mathbf{u}_2 = \begin{bmatrix} 2 \\ 1 \\ -1 \end{bmatrix} \cdot \begin{bmatrix} 3 \\ -4 \\ 2 \end{bmatrix} = 0$, $\mathbf{u}_1 \cdot \mathbf{u}_3 = \begin{bmatrix} 2 \\ 1 \\ -1 \end{bmatrix} \cdot \begin{bmatrix} 2 \\ a \\ b \end{bmatrix} = a - b + 4$, and $\mathbf{u}_2 \cdot \mathbf{u}_3 = \begin{bmatrix} 3 \\ -4 \\ 2 \end{bmatrix} \cdot \begin{bmatrix} 2 \\ a \\ b \end{bmatrix} =$
$-4a + 2b + 6$. Hence we obtain the system

$$a - b + 4 = 0$$
$$-4a + 2b + 6 = 0$$

Solving, we obtain $a = 7$ and $b = 11$.

21. $\|\mathbf{u}_1\| = \left\| \begin{bmatrix} 3 \\ -1 \end{bmatrix} \right\| = \sqrt{10}$, $\|\mathbf{u}_2\| = \left\| \begin{bmatrix} 1 \\ 3 \end{bmatrix} \right\| = \sqrt{10}$, and $\|\mathbf{u}_1 + \mathbf{u}_2\| = \left\| \begin{bmatrix} 3 \\ -1 \end{bmatrix} + \begin{bmatrix} 1 \\ 3 \end{bmatrix} \right\| = \left\| \begin{bmatrix} 4 \\ 2 \end{bmatrix} \right\| =$
$2\sqrt{5}$. Thus $\|\mathbf{u}_1\|^2 + \|\mathbf{u}_2\|^2 = \left(\sqrt{10}\right)^2 + \left(\sqrt{10}\right)^2 = 20 = \left(2\sqrt{5}\right)^2 = \|\mathbf{u}_1 + \mathbf{u}_2\|^2.$

23. $\|\mathbf{u}_1\| = \left\| \begin{bmatrix} 2 \\ -3 \\ 1 \end{bmatrix} \right\| = \sqrt{14}$, $\|\mathbf{u}_2\| = \left\| \begin{bmatrix} 4 \\ 3 \\ 1 \end{bmatrix} \right\| = \sqrt{26}$, and $\|\mathbf{u}_1 + \mathbf{u}_2\| = \left\| \begin{bmatrix} 2 \\ -3 \\ 1 \end{bmatrix} + \begin{bmatrix} 4 \\ 3 \\ 1 \end{bmatrix} \right\| =$

$\left\| \begin{bmatrix} 6 \\ 0 \\ 2 \end{bmatrix} \right\| = 2\sqrt{10}$. Thus $\|\mathbf{u}_1\|^2 + \|\mathbf{u}_2\|^2 = \left(\sqrt{14}\right)^2 + \left(\sqrt{26}\right)^2 = 40 = \left(2\sqrt{10}\right)^2 = \|\mathbf{u}_1 + \mathbf{u}_2\|^2$.

25. $\|3\mathbf{u}_1 + 4\mathbf{u}_2\| = \sqrt{\|3\mathbf{u}_1\|^2 + \|4\mathbf{u}_2\|^2} = \sqrt{\left(3\|\mathbf{u}_1\|\right)^2 + \left(4\|\mathbf{u}_2\|\right)^2} = \sqrt{\left(3\,(2)\right)^2 + \left(4\,(5)\right)^2} = \sqrt{436} = 2\sqrt{109}$.

27. $\begin{bmatrix} 2 \\ -3 \\ 1 \end{bmatrix} \cdot \begin{bmatrix} 1 \\ 2 \\ -1 \end{bmatrix} = -5 \neq 0$, so $\mathbf{u}$ is not orthogonal to S.

29. We determine the null space of $\begin{bmatrix} 1 & -3 \end{bmatrix}$, and obtain
    span $\left\{ \begin{bmatrix} 3 \\ 1 \end{bmatrix} \right\}$. Thus a basis for $S^\perp$ is $\left\{ \begin{bmatrix} 3 \\ 1 \end{bmatrix} \right\}$.

31. We determine the null space of $\begin{bmatrix} 1 & 1 & -2 \end{bmatrix}$, and obtain
    span $\left\{ \begin{bmatrix} -1 \\ 1 \\ 0 \end{bmatrix}, \begin{bmatrix} 2 \\ 0 \\ 1 \end{bmatrix} \right\}$. Thus a basis for $S^\perp$ is $\left\{ \begin{bmatrix} -1 \\ 1 \\ 0 \end{bmatrix}, \begin{bmatrix} 2 \\ 0 \\ 1 \end{bmatrix} \right\}$.

33. Let $\mathbf{s}_1 = \begin{bmatrix} 1 \\ 1 \\ 0 \end{bmatrix}$ and $\mathbf{s}_2 = \begin{bmatrix} 1 \\ -1 \\ 4 \end{bmatrix}$. Then $\mathbf{s}_1 \cdot \mathbf{s}_2 = \begin{bmatrix} 1 \\ 1 \\ 0 \end{bmatrix} \cdot \begin{bmatrix} 1 \\ -1 \\ 4 \end{bmatrix} = 0$, so $\{\mathbf{s}_1, \mathbf{s}_2\}$ is an orthogonal
    basis for S. Thus $\mathbf{s} = c_1\mathbf{s}_1 + c_2\mathbf{s}_2$ where

$$c_1 = \frac{\mathbf{s}_1 \cdot \mathbf{s}}{\|\mathbf{s}_1\|^2} = \frac{1\,(1) + 1\,(2) + 0\,(-2)}{1^2 + 1^2 + 0^2} = \frac{3}{2}$$

$$c_2 = \frac{\mathbf{s}_2 \cdot \mathbf{s}}{\|\mathbf{s}_2\|^2} = \frac{1\,(1) + (-1)\,(2) + (4)\,(-2)}{1^2 + (-1)^2 + 4^2} = \frac{-9}{18} = -\frac{1}{2}$$

Thus $\mathbf{s} = \frac{3}{2}\mathbf{s}_1 - \frac{1}{2}\mathbf{s}_2$.

35. For example, $\mathbf{u} = \begin{bmatrix} 12 \\ 0 \end{bmatrix}$ and $\mathbf{v} = \begin{bmatrix} 1 \\ 0 \end{bmatrix}$.

37. For example, $\mathbf{u} = \begin{bmatrix} 1/\sqrt{5} \\ 2/\sqrt{5} \end{bmatrix}$.

39. For example, $\begin{bmatrix} 0 \\ 1 \\ 0 \end{bmatrix}$ and $\begin{bmatrix} 1 \\ 0 \\ 2 \end{bmatrix}$.

41. For example, $S = \text{span} \left\{ \begin{bmatrix} 1 \\ 0 \\ 0 \end{bmatrix} \right\}$. Then $S^\perp = \text{span} \left\{ \begin{bmatrix} 0 \\ 1 \\ 0 \end{bmatrix}, \begin{bmatrix} 0 \\ 0 \\ 1 \end{bmatrix} \right\}$ has $\dim\left(S^\perp\right) = 2$.

43. For example, $\left\{ \begin{bmatrix} 0 \\ 0 \\ 0 \end{bmatrix}, \begin{bmatrix} 1 \\ 0 \\ 0 \end{bmatrix}, \begin{bmatrix} 0 \\ 1 \\ 0 \end{bmatrix} \right\}$.

45. False, since $\|2\mathbf{u} - 2\mathbf{v}\| = \|2\,(\mathbf{u} - \mathbf{v})\| = |2|\,\|\mathbf{u} - \mathbf{v}\| = 2\,(3) = 6$.

47. False. For example, let $\mathbf{u} = \begin{bmatrix} 1 \\ 0 \end{bmatrix}$ and $\mathbf{v} = \begin{bmatrix} -1 \\ 0 \end{bmatrix}$.

49. True. We have a basis $\mathbf{u} = \begin{bmatrix} u_1 \\ u_2 \end{bmatrix}$ for S, and hence $\mathbf{v} = \begin{bmatrix} -u_2 \\ u_1 \end{bmatrix}$ is a basis for $S^\perp$, and thus $\dim\left(S^\perp\right) = 1$.

51. False. For example, $\mathbf{u}_1 = \begin{bmatrix} 1 \\ 0 \end{bmatrix}$, $\mathbf{u}_2 = \begin{bmatrix} 0 \\ 1 \end{bmatrix}$, and $\mathbf{u}_3 = \begin{bmatrix} -1 \\ 0 \end{bmatrix}$.

53. True. Since $\mathbf{u}$ and $\mathbf{v}$ are orthogonal, $\mathbf{u} \cdot \mathbf{v} = 0$, and hence $\|\mathbf{u} - \mathbf{v}\|^2 = (\mathbf{u} - \mathbf{v}) \cdot (\mathbf{u} - \mathbf{v}) = \|\mathbf{u}\|^2 - 2(\mathbf{u} \cdot \mathbf{v}) + \|\mathbf{v}\|^2 = \|\mathbf{u}\|^2 - 2(0) + \|\mathbf{v}\|^2 = \|\mathbf{u}\|^2 + \|\mathbf{v}\|^2$. Thus the distance between $\mathbf{u}$ and $\mathbf{v}$ is $\|\mathbf{u} - \mathbf{v}\| = \sqrt{\|\mathbf{u}\|^2 + \|\mathbf{v}\|^2}$.

55. False. $S^\perp = \mathrm{null}\left(A^T\right)$.

57. The proof of Theorem 8.9 carries over nearly word-for-word. The only significant change is that $\mathcal{S}$ only spans S (instead of being a basis), so that while we are guaranteed there are scalars $c_1, \ldots, c_k$ such that

$$\mathbf{s} = c_1 \mathbf{s}_1 + \cdots + c_k \mathbf{s}_k$$

the scalars may not be unique.

59. The dot product of $\mathbf{e}_i$ and $\mathbf{e}_j$ for $i \neq j$ is

$$\mathbf{e}_i \cdot \mathbf{e}_j = 0\,(0) + \cdots + \overbrace{1\,(0)}^{i^{\text{th}}\ \text{term}} + \cdots + \overbrace{0\,(1)}^{j^{\text{th}}\ \text{term}} + \cdots + 0\,(0) = 0$$

and hence $\{\mathbf{e}_1, \ldots, \mathbf{e}_n\}$ is an orthogonal basis.

61. We apply Theorem 8.2:

$$\begin{aligned}
(c_1 \mathbf{u}_1) \cdot (c_2 \mathbf{u}_2) &= c_1\,(\mathbf{u}_1 \cdot (c_2 \mathbf{u}_2)),\ \text{by part (c)} \\
&= c_1\,((c_2 \mathbf{u}_2) \cdot \mathbf{u}_1),\ \text{by part (a)} \\
&= c_1\,(c_2\,(\mathbf{u}_2 \cdot \mathbf{u}_1)),\ \text{by part (c)} \\
&= (c_1 c_2)\,(\mathbf{u}_1 \cdot \mathbf{u}_2),\ \text{by part (a)}
\end{aligned}$$

63. $(c_1 \mathbf{u}_1 + \cdots + c_k \mathbf{u}_k) \cdot \mathbf{w} = (c_1 \mathbf{u}_1 \cdot \mathbf{w}) + \cdots + (c_k \mathbf{u}_k \cdot \mathbf{w})$, by part (b)
$= c_1\,(\mathbf{u}_1 \cdot \mathbf{w}) + \cdots + c_k\,(\mathbf{u}_k \cdot \mathbf{w})$, by part (c)

65. $\|\mathbf{v}\| = \left\| \dfrac{1}{\|\mathbf{u}\|} \mathbf{u} \right\| = \left| \dfrac{1}{\|\mathbf{u}\|} \right| \|\mathbf{u}\| = \dfrac{1}{\|\mathbf{u}\|} \|\mathbf{u}\| = 1$

67. Let $\mathbf{s} \in S \cap S^\perp$. Then, since $\mathbf{s} \in S$ and $\mathbf{s} \in S^\perp$, it follows that $\mathbf{s} \cdot \mathbf{s} = 0$. Thus, by Theorem 8.2(d) $\mathbf{s} = \mathbf{0}$. Consequently, $S \cap S^\perp = \{\mathbf{0}\}$.

69. Let $\mathbf{u} \in (\mathrm{col}\,(A))^\perp$. Then $\mathbf{v} \cdot \mathbf{u} = 0$ for every $\mathbf{v} \in \mathrm{col}\,(A)$. In particular, for every column vector $\mathbf{a}_i$ of A, $\mathbf{a}_i^T \mathbf{u} = \mathbf{a}_i \cdot \mathbf{u} = 0$. Thus $A^T \mathbf{u} = \begin{bmatrix} \mathbf{a}_1^T \\ \vdots \\ \mathbf{a}_n^T \end{bmatrix} \mathbf{u} = \begin{bmatrix} \mathbf{a}_1^T \mathbf{u} \\ \vdots \\ \mathbf{a}_n^T \mathbf{u} \end{bmatrix} = \begin{bmatrix} 0 \\ \vdots \\ 0 \end{bmatrix} = \mathbf{0}$. Thus $\mathbf{u} \in \mathrm{Null}\left(A^T\right)$, and $(\mathrm{col}\,(A))^\perp \subset \mathrm{Null}\left(A^T\right)$. Now suppose $\mathbf{u} \in \mathrm{Null}\left(A^T\right)$, so that $A^T \mathbf{u} = \mathbf{0}$, which implies $\mathbf{a}_i^T \mathbf{u} = 0$ for each i. Let $\mathbf{v} \in \mathrm{col}\,(A)$, so that $\mathbf{v} = c_1 \mathbf{a}_1 + \cdots + c_n \mathbf{a}_n$ for some c_j. Then $\mathbf{v} \cdot \mathbf{u} = (c_1 \mathbf{a}_1 + \cdots + c_n \mathbf{a}_n) \cdot \mathbf{u} = c_1\,(\mathbf{a}_1 \cdot \mathbf{u}) + \cdots + c_n\,(\mathbf{a}_n \cdot \mathbf{u}) = c_1\,(\mathbf{a}_1^T \mathbf{u}) + \cdots + c_n\,(\mathbf{a}_n^T \mathbf{u}) = c_1\,(0) + \cdots + c_n\,(0) = 0$. Thus $\mathbf{u} \in (\mathrm{col}\,(A))^\perp$, and so $\mathrm{Null}\left(A^T\right) \subset (\mathrm{col}\,(A))^\perp$. Therefore $\mathrm{Null}\left(A^T\right) = (\mathrm{col}\,(A))^\perp$.

71. (a) $\mathbf{u} \cdot \mathbf{v} = u_1 v_1 + \cdots + u_n v_n$ and $\mathbf{u}^T \mathbf{v} = \begin{bmatrix} u_1 \\ \vdots \\ u_n \end{bmatrix}^T \begin{bmatrix} v_1 \\ \vdots \\ v_n \end{bmatrix} = \begin{bmatrix} u_1 & \cdots & u_n \end{bmatrix} \begin{bmatrix} v_1 \\ \vdots \\ v_n \end{bmatrix} = [u_1 v_1 + \cdots +$

$u_n v_n]$ (a 1×1 matrix), so $\mathbf{u} \cdot \mathbf{v}$ and $\mathbf{u}^T \mathbf{v}$ are essentially the same.

(b) $(A\mathbf{u}) \cdot \mathbf{v} = (A\mathbf{u})^T \mathbf{v} = (\mathbf{u}^T A^T) \mathbf{v} = \mathbf{u}^T (A^T \mathbf{v}) = \mathbf{u} \cdot (A^T \mathbf{v})$

73. (a) $\mathbf{u}_2 \cdot \mathbf{u}_3 = \begin{bmatrix} 7 \\ 4 \\ 0 \\ 2 \\ 8 \end{bmatrix} \cdot \begin{bmatrix} 0 \\ 3 \\ -4 \\ 4 \\ -3 \end{bmatrix} = -4$

(b) $\|\mathbf{u}_1\| = \left\| \begin{bmatrix} 3 \\ -1 \\ 5 \\ 0 \\ 2 \end{bmatrix} \right\| = \sqrt{39}$

(c) $\|2\mathbf{u}_1 + 5\mathbf{u}_3\| = \left\| 2 \begin{bmatrix} 3 \\ -1 \\ 5 \\ 0 \\ 2 \end{bmatrix} + 5 \begin{bmatrix} 0 \\ 3 \\ -4 \\ 4 \\ -3 \end{bmatrix} \right\| = \sqrt{826}$

(d) $\|3\mathbf{u}_1 - 4\mathbf{u}_2 - \mathbf{u}_3\| = \left\| 3 \begin{bmatrix} 3 \\ -1 \\ 5 \\ 0 \\ 2 \end{bmatrix} - 4 \begin{bmatrix} 7 \\ 4 \\ 0 \\ 2 \\ 8 \end{bmatrix} - \begin{bmatrix} 0 \\ 3 \\ -4 \\ 4 \\ -3 \end{bmatrix} \right\| = \sqrt{1879}$

75. To determine a basis for $S^\perp$, we compute a basis for the null space of:

$$\begin{bmatrix} 2 & -1 & 3 & 5 \\ 0 & 1 & 7 & 4 \end{bmatrix}$$

and obtain $\left\{ \begin{bmatrix} -5 \\ -7 \\ 1 \\ 0 \end{bmatrix}, \begin{bmatrix} -\frac{9}{2} \\ -4 \\ 0 \\ 1 \end{bmatrix} \right\}$.

8.2   Projection and the Gram-Schmidt Process

1. (a) $\operatorname{proj}_{\mathbf{u}_3} \mathbf{u}_2 = \dfrac{\mathbf{u}_3 \cdot \mathbf{u}_2}{\|\mathbf{u}_3\|^2} \mathbf{u}_3 = \left(\begin{bmatrix} 2 \\ 0 \\ -1 \end{bmatrix} \cdot \begin{bmatrix} 1 \\ 1 \\ 1 \end{bmatrix} \bigg/ \left\| \begin{bmatrix} 2 \\ 0 \\ -1 \end{bmatrix} \right\|^2 \right) \begin{bmatrix} 2 \\ 0 \\ -1 \end{bmatrix} = \begin{bmatrix} \frac{2}{5} \\ 0 \\ -\frac{1}{5} \end{bmatrix}$

(b) $\operatorname{proj}_{\mathbf{u}_1} \mathbf{u}_2 = \dfrac{\mathbf{u}_1 \cdot \mathbf{u}_2}{\|\mathbf{u}_1\|^2} \mathbf{u}_1 = \left(\begin{bmatrix} -3 \\ 1 \\ 2 \end{bmatrix} \cdot \begin{bmatrix} 1 \\ 1 \\ 1 \end{bmatrix} \bigg/ \left\| \begin{bmatrix} -3 \\ 1 \\ 2 \end{bmatrix} \right\|^2 \right) \begin{bmatrix} -3 \\ 1 \\ 2 \end{bmatrix} = \begin{bmatrix} 0 \\ 0 \\ 0 \end{bmatrix}$

3. $\operatorname{proj}_S \mathbf{u}_2 = \dfrac{\mathbf{u}_3 \cdot \mathbf{u}_2}{\|\mathbf{u}_3\|^2} \mathbf{u}_3 + \dfrac{\mathbf{u}_4 \cdot \mathbf{u}_2}{\|\mathbf{u}_4\|^2} \mathbf{u}_4 = \left(\begin{bmatrix} 2 \\ 0 \\ -1 \end{bmatrix} \cdot \begin{bmatrix} 1 \\ 1 \\ 1 \end{bmatrix} \bigg/ \left\| \begin{bmatrix} 2 \\ 0 \\ -1 \end{bmatrix} \right\|^2 \right) \begin{bmatrix} 2 \\ 0 \\ -1 \end{bmatrix} +$

$\left(\begin{bmatrix} 1 \\ -3 \\ 2 \end{bmatrix} \cdot \begin{bmatrix} 1 \\ 1 \\ 1 \end{bmatrix} \bigg/ \left\| \begin{bmatrix} 1 \\ -3 \\ 2 \end{bmatrix} \right\|^2 \right) \begin{bmatrix} 1 \\ -3 \\ 2 \end{bmatrix} = \begin{bmatrix} \frac{2}{5} \\ 0 \\ -\frac{1}{5} \end{bmatrix} + \begin{bmatrix} 0 \\ 0 \\ 0 \end{bmatrix} = \begin{bmatrix} \frac{2}{5} \\ 0 \\ -\frac{1}{5} \end{bmatrix}$

5. (a) $\mathbf{v}_1 = \dfrac{1}{\|\mathbf{u}_1\|}\mathbf{u}_1 = \begin{bmatrix} -3 \\ 1 \\ 2 \end{bmatrix} \Big/ \left\| \begin{bmatrix} -3 \\ 1 \\ 2 \end{bmatrix} \right\| = \dfrac{1}{\sqrt{14}} \begin{bmatrix} -3 \\ 1 \\ 2 \end{bmatrix} = \begin{bmatrix} -\frac{3}{14}\sqrt{14} \\ \frac{1}{14}\sqrt{14} \\ \frac{1}{7}\sqrt{14} \end{bmatrix}$

(b) $\mathbf{v}_4 = \dfrac{1}{\|\mathbf{u}_4\|}\mathbf{u}_4 = \begin{bmatrix} 1 \\ -3 \\ 2 \end{bmatrix} \Big/ \left\| \begin{bmatrix} 1 \\ -3 \\ 2 \end{bmatrix} \right\| = \dfrac{1}{\sqrt{14}} \begin{bmatrix} 1 \\ -3 \\ 2 \end{bmatrix} = \begin{bmatrix} \frac{1}{14}\sqrt{14} \\ -\frac{3}{14}\sqrt{14} \\ \frac{1}{7}\sqrt{14} \end{bmatrix}$

7. $\mathbf{v}_1 = \mathbf{s}_1 = \begin{bmatrix} 1 \\ 3 \end{bmatrix}$. $\mathbf{v}_2 = \mathbf{s}_2 - \text{proj}_{\mathbf{v}_1}\mathbf{s}_2 = \begin{bmatrix} 4 \\ 2 \end{bmatrix} - \dfrac{\mathbf{v}_1 \cdot \mathbf{s}_2}{\|\mathbf{v}_1\|^2}\mathbf{v}_1 =$

$\begin{bmatrix} 4 \\ 2 \end{bmatrix} - \left(\begin{bmatrix} 1 \\ 3 \end{bmatrix} \cdot \begin{bmatrix} 4 \\ 2 \end{bmatrix} \Big/ \left\| \begin{bmatrix} 1 \\ 3 \end{bmatrix} \right\|^2 \right) \begin{bmatrix} 1 \\ 3 \end{bmatrix} = \begin{bmatrix} 4 \\ 2 \end{bmatrix} - \begin{bmatrix} 1 \\ 3 \end{bmatrix} = \begin{bmatrix} 3 \\ -1 \end{bmatrix}$. Thus an orthogonal basis for

S is $\left\{ \begin{bmatrix} 1 \\ 3 \end{bmatrix}, \begin{bmatrix} 3 \\ -1 \end{bmatrix} \right\}$.

9. $\mathbf{v}_1 = \mathbf{s}_1 = \begin{bmatrix} -2 \\ 2 \\ 1 \end{bmatrix}$. $\mathbf{v}_2 = \mathbf{s}_2 - \text{proj}_{\mathbf{v}_1}\mathbf{s}_2 = \begin{bmatrix} 3 \\ 4 \\ -2 \end{bmatrix} - \dfrac{\mathbf{v}_1 \cdot \mathbf{s}_2}{\|\mathbf{v}_1\|^2}\mathbf{v}_1 =$

$\begin{bmatrix} 3 \\ 4 \\ -2 \end{bmatrix} - \left(\begin{bmatrix} -2 \\ 2 \\ 1 \end{bmatrix} \cdot \begin{bmatrix} 3 \\ 4 \\ -2 \end{bmatrix} \Big/ \left\| \begin{bmatrix} -2 \\ 2 \\ 1 \end{bmatrix} \right\|^2 \right) \begin{bmatrix} -2 \\ 2 \\ 1 \end{bmatrix} = \begin{bmatrix} 3 \\ 4 \\ -2 \end{bmatrix} - \begin{bmatrix} 0 \\ 0 \\ 0 \end{bmatrix} = \begin{bmatrix} 3 \\ 4 \\ -2 \end{bmatrix}$.

Thus an orthogonal basis for S is $\left\{ \begin{bmatrix} -2 \\ 2 \\ 1 \end{bmatrix}, \begin{bmatrix} 3 \\ 4 \\ -2 \end{bmatrix} \right\}$.

11. $\mathbf{v}_1 = \mathbf{s}_1 = \begin{bmatrix} 1 \\ -1 \\ 0 \\ 1 \end{bmatrix}$. $\mathbf{v}_2 = \mathbf{s}_2 - \text{proj}_{\mathbf{v}_1}\mathbf{s}_2 = \begin{bmatrix} 4 \\ 1 \\ 2 \\ 0 \end{bmatrix} - \dfrac{\mathbf{v}_1 \cdot \mathbf{s}_2}{\|\mathbf{v}_1\|^2}\mathbf{v}_1 = \begin{bmatrix} 4 \\ 1 \\ 2 \\ 0 \end{bmatrix} -$

$\left(\begin{bmatrix} 1 \\ -1 \\ 0 \\ 1 \end{bmatrix} \cdot \begin{bmatrix} 4 \\ 1 \\ 2 \\ 0 \end{bmatrix} \Big/ \left\| \begin{bmatrix} 1 \\ -1 \\ 0 \\ 1 \end{bmatrix} \right\|^2 \right) \begin{bmatrix} 1 \\ -1 \\ 0 \\ 1 \end{bmatrix} = \begin{bmatrix} 4 \\ 1 \\ 2 \\ 0 \end{bmatrix} - \begin{bmatrix} 1 \\ -1 \\ 0 \\ 1 \end{bmatrix} = \begin{bmatrix} 3 \\ 2 \\ 2 \\ -1 \end{bmatrix}$.

Thus an orthogonal basis for S is $\left\{ \begin{bmatrix} 1 \\ -1 \\ 0 \\ 1 \end{bmatrix}, \begin{bmatrix} 3 \\ 2 \\ 2 \\ -1 \end{bmatrix} \right\}$.

13. $\mathbf{v}_1 = \mathbf{s}_1 = \begin{bmatrix} -1 \\ 0 \\ 1 \end{bmatrix}$. $\mathbf{v}_2 = \mathbf{s}_2 - \text{proj}_{S_1}\mathbf{s}_2 = \begin{bmatrix} 3 \\ 4 \\ 1 \end{bmatrix} - \dfrac{\mathbf{v}_1 \cdot \mathbf{s}_2}{\|\mathbf{v}_1\|^2}\mathbf{v}_1 = \begin{bmatrix} 3 \\ 4 \\ 1 \end{bmatrix} -$

$\left(\begin{bmatrix} -1 \\ 0 \\ 1 \end{bmatrix} \cdot \begin{bmatrix} 3 \\ 4 \\ 1 \end{bmatrix} \Big/ \left\| \begin{bmatrix} -1 \\ 0 \\ 1 \end{bmatrix} \right\|^2 \right) \begin{bmatrix} -1 \\ 0 \\ 1 \end{bmatrix} = \begin{bmatrix} 3 \\ 4 \\ 1 \end{bmatrix} - \begin{bmatrix} 1 \\ 0 \\ -1 \end{bmatrix} = \begin{bmatrix} 2 \\ 4 \\ 2 \end{bmatrix}$.

$\mathbf{v}_3 = \mathbf{s}_3 - \text{proj}_{S_2}\mathbf{s}_3 = \begin{bmatrix} 4 \\ 1 \\ 6 \end{bmatrix} - \dfrac{\mathbf{v}_1 \cdot \mathbf{s}_3}{\|\mathbf{v}_1\|^2}\mathbf{v}_1 - \dfrac{\mathbf{v}_2 \cdot \mathbf{s}_3}{\|\mathbf{v}_2\|^2}\mathbf{v}_2 =$

$\begin{bmatrix} 4 \\ 1 \\ 6 \end{bmatrix} - \left(\begin{bmatrix} -1 \\ 0 \\ 1 \end{bmatrix} \cdot \begin{bmatrix} 4 \\ 1 \\ 6 \end{bmatrix} \Big/ \left\| \begin{bmatrix} -1 \\ 0 \\ 1 \end{bmatrix} \right\|^2 \right) \begin{bmatrix} -1 \\ 0 \\ 1 \end{bmatrix} - \left(\begin{bmatrix} 2 \\ 4 \\ 2 \end{bmatrix} \cdot \begin{bmatrix} 4 \\ 1 \\ 6 \end{bmatrix} \Big/ \left\| \begin{bmatrix} 2 \\ 4 \\ 2 \end{bmatrix} \right\|^2 \right) \begin{bmatrix} 2 \\ 4 \\ 2 \end{bmatrix}$

$= \begin{bmatrix} 4 \\ 1 \\ 6 \end{bmatrix} - \begin{bmatrix} -1 \\ 0 \\ 1 \end{bmatrix} - \begin{bmatrix} 2 \\ 4 \\ 2 \end{bmatrix} = \begin{bmatrix} 3 \\ -3 \\ 3 \end{bmatrix}$. Hence $\mathbf{s}_3 = \text{proj}_{S_2}\mathbf{s}_3$, and so $\mathbf{s}_3 \in \text{span}\{\mathbf{s}_1, \mathbf{s}_2\}$.

Thus an orthogonal basis for S is $\left\{ \begin{bmatrix} -1 \\ 0 \\ 1 \end{bmatrix}, \begin{bmatrix} 2 \\ 4 \\ 2 \end{bmatrix}, \begin{bmatrix} 3 \\ -3 \\ 3 \end{bmatrix} \right\}.$

15. Using the orthogonal basis $\{\mathbf{v}_1, \mathbf{v}_2\}$ obtained in Exercise 7, $\text{proj}_S \mathbf{u} = \dfrac{\mathbf{v}_1 \cdot \mathbf{u}}{\|\mathbf{v}_1\|^2} \mathbf{v}_1 + \dfrac{\mathbf{v}_2 \cdot \mathbf{u}}{\|\mathbf{v}_2\|^2} \mathbf{v}_2$

$= \left(\begin{bmatrix} 1 \\ 3 \end{bmatrix} \cdot \begin{bmatrix} 1 \\ 3 \end{bmatrix} / \left\| \begin{bmatrix} 1 \\ 3 \end{bmatrix} \right\|^2 \right) \begin{bmatrix} 1 \\ 3 \end{bmatrix} + \left(\begin{bmatrix} 3 \\ -1 \end{bmatrix} \cdot \begin{bmatrix} 1 \\ 1 \end{bmatrix} / \left\| \begin{bmatrix} 3 \\ -1 \end{bmatrix} \right\|^2 \right) \begin{bmatrix} 3 \\ -1 \end{bmatrix} =$

$\begin{bmatrix} \frac{2}{5} \\ \frac{6}{5} \end{bmatrix} + \begin{bmatrix} \frac{3}{5} \\ -\frac{1}{5} \end{bmatrix} = \begin{bmatrix} 1 \\ 1 \end{bmatrix}.$ (Since S spans $\mathbf{R}^2$, $\text{proj}_S \mathbf{u} = \mathbf{u}$, as determined.)

17. Using the orthogonal basis $\{\mathbf{v}_1, \mathbf{v}_2\}$ obtained in Exercise 9, $\text{proj}_S \mathbf{u} = \dfrac{\mathbf{v}_1 \cdot \mathbf{u}}{\|\mathbf{v}_1\|^2} \mathbf{v}_1 + \dfrac{\mathbf{v}_2 \cdot \mathbf{u}}{\|\mathbf{v}_2\|^2} \mathbf{v}_2 =$

$\left(\begin{bmatrix} -2 \\ 2 \\ 1 \end{bmatrix} \cdot \begin{bmatrix} 1 \\ 0 \\ 2 \end{bmatrix} / \left\| \begin{bmatrix} -2 \\ 2 \\ 1 \end{bmatrix} \right\|^2 \right) \begin{bmatrix} -2 \\ 2 \\ 1 \end{bmatrix} + \left(\begin{bmatrix} 3 \\ 4 \\ -2 \end{bmatrix} \cdot \begin{bmatrix} 1 \\ 0 \\ 2 \end{bmatrix} / \left\| \begin{bmatrix} 3 \\ 4 \\ -2 \end{bmatrix} \right\|^2 \right) \begin{bmatrix} 3 \\ 4 \\ -2 \end{bmatrix} =$

$\begin{bmatrix} 0 \\ 0 \\ 0 \end{bmatrix} + \begin{bmatrix} -\frac{3}{29} \\ -\frac{4}{29} \\ \frac{2}{29} \end{bmatrix} = \begin{bmatrix} -\frac{3}{29} \\ -\frac{4}{29} \\ \frac{2}{29} \end{bmatrix}.$

19. Using the orthogonal basis $\{\mathbf{v}_1, \mathbf{v}_2\}$ obtained in Exercise 11, $\text{proj}_S \mathbf{u} = \dfrac{\mathbf{v}_1 \cdot \mathbf{u}}{\|\mathbf{v}_1\|^2} \mathbf{v}_1 + \dfrac{\mathbf{v}_2 \cdot \mathbf{u}}{\|\mathbf{v}_2\|^2} \mathbf{v}_2 =$

$\left(\begin{bmatrix} 1 \\ -1 \\ 0 \\ 1 \end{bmatrix} \cdot \begin{bmatrix} 1 \\ -1 \\ 0 \\ 1 \end{bmatrix} / \left\| \begin{bmatrix} 1 \\ -1 \\ 0 \\ 1 \end{bmatrix} \right\|^2 \right) \begin{bmatrix} 1 \\ -1 \\ 0 \\ 1 \end{bmatrix} + \left(\begin{bmatrix} 3 \\ 2 \\ 2 \\ -1 \end{bmatrix} \cdot \begin{bmatrix} 1 \\ -1 \\ 0 \\ 1 \end{bmatrix} / \left\| \begin{bmatrix} 3 \\ 2 \\ 2 \\ -1 \end{bmatrix} \right\|^2 \right) \begin{bmatrix} 3 \\ 2 \\ 2 \\ -1 \end{bmatrix} =$

$\begin{bmatrix} 1 \\ -1 \\ 0 \\ 1 \end{bmatrix} + \begin{bmatrix} 0 \\ 0 \\ 0 \\ 0 \end{bmatrix} = \begin{bmatrix} 1 \\ -1 \\ 0 \\ 1 \end{bmatrix}.$ (Since $\mathbf{u} = \mathbf{v}_1$, $\text{proj}_S \mathbf{u} = \mathbf{u}$, as determined.)

21. Using the orthogonal basis $\{\mathbf{v}_1, \mathbf{v}_2\}$ obtained in Exercise 13, $\text{proj}_S \mathbf{u} = \dfrac{\mathbf{v}_1 \cdot \mathbf{u}}{\|\mathbf{v}_1\|^2} \mathbf{v}_1 + \dfrac{\mathbf{v}_2 \cdot \mathbf{u}}{\|\mathbf{v}_2\|^2} \mathbf{v}_2 =$

$\left(\begin{bmatrix} -1 \\ 0 \\ 1 \end{bmatrix} \cdot \begin{bmatrix} 1 \\ 0 \\ 2 \end{bmatrix} / \left\| \begin{bmatrix} -1 \\ 0 \\ 1 \end{bmatrix} \right\|^2 \right) \begin{bmatrix} -1 \\ 0 \\ 1 \end{bmatrix} + \left(\begin{bmatrix} 2 \\ 4 \\ 2 \end{bmatrix} \cdot \begin{bmatrix} 1 \\ 0 \\ 2 \end{bmatrix} / \left\| \begin{bmatrix} 2 \\ 4 \\ 2 \end{bmatrix} \right\|^2 \right) \begin{bmatrix} 2 \\ 4 \\ 2 \end{bmatrix}$

$+ \left(\begin{bmatrix} 3 \\ -3 \\ 3 \end{bmatrix} \cdot \begin{bmatrix} 1 \\ 0 \\ 2 \end{bmatrix} / \left\| \begin{bmatrix} 3 \\ -3 \\ 3 \end{bmatrix} \right\|^2 \right) \begin{bmatrix} 3 \\ -3 \\ 3 \end{bmatrix} = \begin{bmatrix} -\frac{1}{2} \\ 0 \\ \frac{1}{2} \end{bmatrix} + \begin{bmatrix} \frac{1}{2} \\ 1 \\ \frac{1}{2} \end{bmatrix} + \begin{bmatrix} 1 \\ -1 \\ 1 \end{bmatrix} = \begin{bmatrix} 1 \\ 0 \\ 2 \end{bmatrix}.$

23. We normalize the orthogonal basis $\{\mathbf{v}_1, \mathbf{v}_2\}$ obtained in Exercise 7.

$\mathbf{w}_1 = \dfrac{\mathbf{v}_1}{\|\mathbf{v}_1\|} = \begin{bmatrix} 1 \\ 3 \end{bmatrix} / \left\| \begin{bmatrix} 1 \\ 3 \end{bmatrix} \right\| = \begin{bmatrix} \frac{1}{10}\sqrt{10} \\ \frac{3}{10}\sqrt{10} \end{bmatrix}, \mathbf{w}_2 = \dfrac{\mathbf{v}_2}{\|\mathbf{v}_2\|} = \begin{bmatrix} 3 \\ -1 \end{bmatrix} / \left\| \begin{bmatrix} 3 \\ -1 \end{bmatrix} \right\| = \begin{bmatrix} \frac{3}{10}\sqrt{10} \\ -\frac{1}{10}\sqrt{10} \end{bmatrix}.$

Thus an orthonormal basis for S is $\left\{ \begin{bmatrix} \frac{1}{10}\sqrt{10} \\ \frac{3}{10}\sqrt{10} \end{bmatrix}, \begin{bmatrix} \frac{3}{10}\sqrt{10} \\ -\frac{1}{10}\sqrt{10} \end{bmatrix} \right\}.$

25. We normalize the orthogonal basis $\{\mathbf{v}_1, \mathbf{v}_2\}$ obtained in Exercise 9.

$\mathbf{w}_1 = \dfrac{\mathbf{v}_1}{\|\mathbf{v}_1\|} = \left(\begin{bmatrix} -2 \\ 2 \\ 1 \end{bmatrix} / \left\| \begin{bmatrix} -2 \\ 2 \\ 1 \end{bmatrix} \right\| \right) = \begin{bmatrix} -\frac{2}{3} \\ \frac{2}{3} \\ \frac{1}{3} \end{bmatrix}, \mathbf{w}_2 = \dfrac{\mathbf{v}_2}{\|\mathbf{v}_2\|} = \left(\begin{bmatrix} 3 \\ 4 \\ -2 \end{bmatrix} / \left\| \begin{bmatrix} 3 \\ 4 \\ -2 \end{bmatrix} \right\| \right) = \begin{bmatrix} \frac{3}{29}\sqrt{29} \\ \frac{4}{29}\sqrt{29} \\ -\frac{2}{29}\sqrt{29} \end{bmatrix}.$

Thus an orthonormal basis for S is $\left\{ \begin{bmatrix} -\frac{2}{3} \\ \frac{2}{3} \\ \frac{1}{3} \end{bmatrix}, \begin{bmatrix} \frac{3}{29}\sqrt{29} \\ \frac{4}{29}\sqrt{29} \\ -\frac{2}{29}\sqrt{29} \end{bmatrix} \right\}.$

27. We normalize the orthogonal basis $\{\mathbf{v}_1, \mathbf{v}_2\}$ obtained in Exercise 11.

$$\mathbf{w}_1 = \frac{\mathbf{v}_1}{\|\mathbf{v}_1\|} = \begin{bmatrix} 1 \\ -1 \\ 0 \\ 1 \end{bmatrix} / \left\| \begin{bmatrix} 1 \\ -1 \\ 0 \\ 1 \end{bmatrix} \right\| = \begin{bmatrix} \frac{1}{3}\sqrt{3} \\ -\frac{1}{3}\sqrt{3} \\ 0 \\ \frac{1}{3}\sqrt{3} \end{bmatrix}, \mathbf{w}_2 = \frac{\mathbf{v}_2}{\|\mathbf{v}_2\|} = \begin{bmatrix} 3 \\ 2 \\ 2 \\ -1 \end{bmatrix} / \left\| \begin{bmatrix} 3 \\ 2 \\ 2 \\ -1 \end{bmatrix} \right\| = \begin{bmatrix} \frac{1}{2}\sqrt{2} \\ \frac{1}{3}\sqrt{2} \\ \frac{1}{3}\sqrt{2} \\ -\frac{1}{6}\sqrt{2} \end{bmatrix}.$$

Thus an orthonormal basis for S is $\left\{ \begin{bmatrix} \frac{1}{3}\sqrt{3} \\ -\frac{1}{3}\sqrt{3} \\ 0 \\ \frac{1}{3}\sqrt{3} \end{bmatrix}, \begin{bmatrix} \frac{1}{2}\sqrt{2} \\ \frac{1}{3}\sqrt{2} \\ \frac{1}{3}\sqrt{2} \\ -\frac{1}{6}\sqrt{2} \end{bmatrix} \right\}.$

29. We normalize the orthogonal basis $\{\mathbf{v}_1, \mathbf{v}_2\}$ obtained in Exercise 13.

$$\mathbf{w}_1 = \frac{\mathbf{v}_1}{\|\mathbf{v}_1\|} = \begin{bmatrix} -1 \\ 0 \\ 1 \end{bmatrix} / \left\| \begin{bmatrix} -1 \\ 0 \\ 1 \end{bmatrix} \right\| = \begin{bmatrix} -\frac{1}{2}\sqrt{2} \\ 0 \\ \frac{1}{2}\sqrt{2} \end{bmatrix}, \mathbf{w}_2 = \frac{\mathbf{v}_2}{\|\mathbf{v}_2\|} = \begin{bmatrix} 2 \\ 4 \\ 2 \end{bmatrix} / \left\| \begin{bmatrix} 2 \\ 4 \\ 2 \end{bmatrix} \right\| = \begin{bmatrix} \frac{1}{6}\sqrt{6} \\ \frac{1}{3}\sqrt{6} \\ \frac{1}{6}\sqrt{6} \end{bmatrix},$$

$$\mathbf{w}_3 = \frac{\mathbf{v}_3}{\|\mathbf{v}_3\|} = \begin{bmatrix} 3 \\ -3 \\ 3 \end{bmatrix} / \left\| \begin{bmatrix} 3 \\ -3 \\ 3 \end{bmatrix} \right\| = \begin{bmatrix} \frac{1}{3}\sqrt{3} \\ -\frac{1}{3}\sqrt{3} \\ \frac{1}{3}\sqrt{3} \end{bmatrix}.$$

Thus an orthonormal basis for S is $\left\{ \begin{bmatrix} -\frac{1}{2}\sqrt{2} \\ 0 \\ \frac{1}{2}\sqrt{2} \end{bmatrix}, \begin{bmatrix} \frac{1}{6}\sqrt{6} \\ \frac{1}{3}\sqrt{6} \\ \frac{1}{6}\sqrt{6} \end{bmatrix}, \begin{bmatrix} \frac{1}{3}\sqrt{3} \\ -\frac{1}{3}\sqrt{3} \\ \frac{1}{3}\sqrt{3} \end{bmatrix} \right\}.$

31. For example, let $\mathbf{u} = \begin{bmatrix} 1 \\ 0 \end{bmatrix}$ and $\mathbf{v} = \begin{bmatrix} 1 \\ 0 \end{bmatrix}$.

33. For example, let $\mathbf{u} = \begin{bmatrix} 1 \\ 0 \end{bmatrix}$ and $\mathbf{v} = \begin{bmatrix} 0 \\ 1 \end{bmatrix}$.

35. For example, let $\mathbf{u} = \begin{bmatrix} 3 \\ 1 \end{bmatrix}$ and $\mathbf{v} = \begin{bmatrix} 1 \\ 2 \end{bmatrix}$. Then

$$\mathrm{proj}_{\mathbf{v}}\mathbf{u} = \frac{\mathbf{v} \cdot \mathbf{u}}{\|\mathbf{v}\|^2}\mathbf{v} = \left(\begin{bmatrix} 1 \\ 2 \end{bmatrix} \cdot \begin{bmatrix} 3 \\ 1 \end{bmatrix} / \left\| \begin{bmatrix} 1 \\ 2 \end{bmatrix} \right\|^2 \right) \begin{bmatrix} 1 \\ 2 \end{bmatrix} = \begin{bmatrix} 1 \\ 2 \end{bmatrix} = \mathbf{v}.$$

37. True, by applying the Gram-Schmidt Process, and normalizing.

39. True. By Definition 8.15, $\mathrm{proj}_S\mathbf{u}$ is a linear combination of vectors in the subspace S, hence $\mathrm{proj}_S\mathbf{u}$ is in S.

41. True, since $\mathrm{proj}_{\mathbf{v}}\mathbf{u} = \frac{\mathbf{v} \cdot \mathbf{u}}{\|\mathbf{v}\|^2}\mathbf{v} = \frac{0}{\|\mathbf{v}\|^2}\mathbf{v} = \mathbf{0}$.

43. True. If $\{\mathbf{w}_1, \ldots, \mathbf{w}_k\}$ is a basis for $S^\perp$, then since $\mathbf{u}$ is in S, $\mathbf{w}_i \cdot \mathbf{u} = 0$ for each i. Thus $\mathrm{proj}_{S^\perp}\mathbf{u} = \frac{\mathbf{w}_1 \cdot \mathbf{u}}{\|\mathbf{w}_1\|^2}\mathbf{w}_1 + \cdots + \frac{\mathbf{w}_k \cdot \mathbf{u}}{\|\mathbf{w}_k\|^2}\mathbf{w}_k = \frac{0}{\|\mathbf{w}_1\|^2}\mathbf{w}_1 + \cdots + \frac{0}{\|\mathbf{w}_k\|^2}\mathbf{w}_k = \mathbf{0}$.

45. False. For example, if $\mathbf{u} = \begin{bmatrix} 1 \\ 0 \end{bmatrix}$ and $\mathbf{v} = \begin{bmatrix} 0 \\ 1 \end{bmatrix}$, then $\mathrm{proj}_{\mathbf{u}}(\mathrm{proj}_{\mathbf{v}}\mathbf{u}) = \mathrm{proj}_{\mathbf{u}}(\mathbf{0}) = \mathbf{0} \neq \mathbf{u}$.

47. (a) Since $S_i = \text{span}\{\mathbf{v}_1, \ldots, \mathbf{v}_i\}$ and $S_j = \text{span}\{\mathbf{v}_1, \ldots, \mathbf{v}_j\}$, if $\mathbf{w} = c_1\mathbf{v}_1 + \cdots + c_i\mathbf{v}_i \in S_i$, then $\mathbf{w} = c_1\mathbf{v}_1 + \cdots + c_i\mathbf{v}_i + 0\mathbf{v}_{i+1} + \cdots + 0\mathbf{v}_j \in S_j$, and $\mathbf{w} \in S_j$. Thus S_i is a subspace of S_j.

   (b) Let $\mathbf{w} \in S_j^\perp$. Then $\mathbf{w} \cdot \mathbf{v}_k = 0$ for $1 \leq k \leq j$, by Theorem 8.9. In particular, $\mathbf{w} \cdot \mathbf{v}_k = 0$ for $1 \leq k \leq i$, and therefore $\mathbf{w} \in S_i^\perp$, again by Theorem 8.9. Hence $S_j^\perp$ is a subspace of $S_i^\perp$.

49. If $\mathbf{u} + \mathbf{v} \in S$, then since S is a subspace, $(\mathbf{u} + \mathbf{v}) - \mathbf{u} = \mathbf{v} \in S$. Thus $\mathbf{v} \in S \cap S^\perp = \{\mathbf{0}\}$. But $\mathbf{v} \neq \mathbf{0}$, hence $\mathbf{u} + \mathbf{v} \notin S$. If $\mathbf{u} + \mathbf{v} \in S^\perp$, then since $S^\perp$ is a subspace, $(\mathbf{u} + \mathbf{v}) - \mathbf{v} = \mathbf{u} \in S^\perp$. Thus $\mathbf{u} \in S \cap S^\perp = \{\mathbf{0}\}$. But $\mathbf{u} \neq \mathbf{0}$, hence $\mathbf{u} + \mathbf{v} \notin S^\perp$.

51. $T_\mathbf{v}(\mathbf{u}_1 + \mathbf{u}_2) = \text{proj}_\mathbf{v}(\mathbf{u}_1 + \mathbf{u}_2) = \dfrac{\mathbf{v} \cdot (\mathbf{u}_1 + \mathbf{u}_2)}{\|\mathbf{v}\|^2}\mathbf{v} = \dfrac{\mathbf{v} \cdot \mathbf{u}_1 + \mathbf{v} \cdot \mathbf{u}_2}{\|\mathbf{v}\|^2}\mathbf{v} = \dfrac{\mathbf{v} \cdot \mathbf{u}_1}{\|\mathbf{v}\|^2}\mathbf{v} + \dfrac{\mathbf{v} \cdot \mathbf{u}_2}{\|\mathbf{v}\|^2}\mathbf{v} = \text{proj}_\mathbf{v}\mathbf{u}_1 +$

   $\text{proj}_\mathbf{v}\mathbf{u}_2 = T_\mathbf{v}(\mathbf{u}_1) + T_\mathbf{v}(\mathbf{u}_2)$. And $T_\mathbf{v}(c\mathbf{u}) = \text{proj}_\mathbf{v}(c\mathbf{u}) = \dfrac{\mathbf{v} \cdot (c\mathbf{u})}{\|\mathbf{v}\|^2}\mathbf{v} = \dfrac{c(\mathbf{v} \cdot \mathbf{u})}{\|\mathbf{v}\|^2}\mathbf{v} = c\left(\dfrac{\mathbf{v} \cdot \mathbf{u}}{\|\mathbf{v}\|^2}\mathbf{v}\right) =$

   $c(\text{proj}_\mathbf{v}\mathbf{u}) = cT_\mathbf{v}(\mathbf{u})$. Thus $T_\mathbf{v}$ is a linear transformation.

53. (a) Since $\mathbf{v}_1 = \mathbf{u}_1$, $\{\mathbf{v}_1\}$ is an orthogonal set. Suppose $j < k$, $\{\mathbf{v}_1, \ldots, \mathbf{v}_j\}$ is orthogonal, and consider $\{\mathbf{v}_1, \ldots, \mathbf{v}_j, \mathbf{v}_{j+1}\}$, where $\mathbf{v}_{j+1} = \mathbf{u}_{j+1} - \dfrac{\mathbf{v}_1 \cdot \mathbf{u}_{j+1}}{\|\mathbf{v}_1\|^2}\mathbf{v}_1 - \dfrac{\mathbf{v}_2 \cdot \mathbf{u}_{j+1}}{\|\mathbf{v}_2\|^2}\mathbf{v}_2 - \cdots - \dfrac{\mathbf{v}_j \cdot \mathbf{u}_{j+1}}{\|\mathbf{v}_j\|^2}\mathbf{v}_j$. Since $\{\mathbf{v}_1, \ldots, \mathbf{v}_j\}$ is orthogonal, $\mathbf{v}_i \cdot \mathbf{v}_l$ for distinct $i, l \leq j$. Now for $i \leq j$ we also check

$$\mathbf{v}_i \cdot \mathbf{v}_{j+1} = \mathbf{v}_i \cdot \left(\mathbf{u}_{j+1} - \dfrac{\mathbf{v}_1 \cdot \mathbf{u}_{j+1}}{\|\mathbf{v}_1\|^2}\mathbf{v}_1 - \dfrac{\mathbf{v}_2 \cdot \mathbf{u}_{j+1}}{\|\mathbf{v}_2\|^2}\mathbf{v}_2 - \cdots - \dfrac{\mathbf{v}_j \cdot \mathbf{u}_{j+1}}{\|\mathbf{v}_j\|^2}\mathbf{v}_j\right)$$

$$= (\mathbf{v}_i \cdot \mathbf{u}_{j+1}) - \dfrac{\mathbf{v}_1 \cdot \mathbf{u}_{j+1}}{\|\mathbf{v}_1\|^2}(\mathbf{v}_i \cdot \mathbf{v}_1) - \dfrac{\mathbf{v}_2 \cdot \mathbf{u}_{j+1}}{\|\mathbf{v}_2\|^2}(\mathbf{v}_i \cdot \mathbf{v}_2) - \cdots$$

$$- \dfrac{\mathbf{v}_j \cdot \mathbf{u}_{j+1}}{\|\mathbf{v}_j\|^2}(\mathbf{v}_i \cdot \mathbf{v}_j)$$

$$= (\mathbf{v}_i \cdot \mathbf{u}_{j+1}) - \dfrac{\mathbf{v}_i \cdot \mathbf{u}_{j+1}}{\|\mathbf{v}_i\|^2}(\mathbf{v}_i \cdot \mathbf{v}_i)$$

$$= (\mathbf{v}_i \cdot \mathbf{u}_{j+1}) - \dfrac{\mathbf{v}_i \cdot \mathbf{u}_{j+1}}{\|\mathbf{v}_i\|^2}\|\mathbf{v}_i\|^2 = \mathbf{v}_i \cdot \mathbf{u}_{j+1} - \mathbf{v}_i \cdot \mathbf{u}_{j+1} = 0$$

Thus $\{\mathbf{v}_1, \ldots, \mathbf{v}_j, \mathbf{v}_{j+1}\}$ is an orthogonal set for each $j < k$, so $\{\mathbf{v}_1, \ldots, \mathbf{v}_j\}$ is an orthogonal set for each $j \leq k$.

   (b) Since $\mathbf{v}_1 = \mathbf{u}_1$, $\text{span}\{\mathbf{v}_1\} = \text{span}\{\mathbf{u}_1\}$. Suppose $j < k$, and $\text{span}\{\mathbf{v}_1, \ldots, \mathbf{v}_j\} = \text{span}\{\mathbf{u}_1, \ldots, \mathbf{u}_j\}$. Let $\mathbf{w} \in \text{span}\{\mathbf{u}_1, \ldots, \mathbf{u}_{j+1}\}$, then $\mathbf{w} = (c_1\mathbf{u}_1 + \cdots + c_j\mathbf{u}_j) + c_{j+1}\mathbf{u}_{j+1}$. So $\mathbf{w} - c_{j+1}\mathbf{u}_{j+1} = c_1\mathbf{u}_1 + \cdots + c_j\mathbf{u}_j \in \text{span}\{\mathbf{u}_1, \ldots, \mathbf{u}_j\} = \text{span}\{\mathbf{v}_1, \ldots, \mathbf{v}_j\}$, hence $\mathbf{w} - c_{j+1}\mathbf{u}_{j+1} = d_1\mathbf{v}_1 + \cdots + d_j\mathbf{v}_j$. There-fore $\mathbf{w} = c_{j+1}\mathbf{u}_{j+1} + d_1\mathbf{v}_1 + \cdots + d_j\mathbf{v}_j = c_{j+1}\left(\mathbf{v}_{j+1} + \dfrac{\mathbf{v}_1 \cdot \mathbf{u}_{j+1}}{\|\mathbf{v}_1\|^2}\mathbf{v}_1 + \dfrac{\mathbf{v}_2 \cdot \mathbf{u}_{j+1}}{\|\mathbf{v}_2\|^2}\mathbf{v}_2 + \cdots + \dfrac{\mathbf{v}_j \cdot \mathbf{u}_{j+1}}{\|\mathbf{v}_j\|^2}\mathbf{v}_j\right) +$

$d_1\mathbf{v}_1 + \cdots + d_j\mathbf{v}_j$, and we see that $\mathbf{w} \in \text{span}\{\mathbf{v}_1, \ldots, \mathbf{v}_{j+1}\}$. Thus $\text{span}\{\mathbf{u}_1, \ldots, \mathbf{u}_{j+1}\} \subset \text{span}\{\mathbf{v}_1, \ldots, \mathbf{v}_{j+1}\}$. Now suppose $\mathbf{w} \in \text{span}\{\mathbf{v}_1, \ldots, \mathbf{v}_{j+1}\}$, so $\mathbf{w} = (d_1\mathbf{v}_1 + \cdots + d_j\mathbf{v}_j) +$

$d_{j+1}\mathbf{v}_{j+1} = (d_1\mathbf{v}_1 + \cdots + d_j\mathbf{v}_j) + d_{j+1}\left(\mathbf{u}_{j+1} - \dfrac{\mathbf{v}_1 \cdot \mathbf{u}_{j+1}}{\|\mathbf{v}_1\|^2}\mathbf{v}_1 - \dfrac{\mathbf{v}_2 \cdot \mathbf{u}_{j+1}}{\|\mathbf{v}_2\|^2}\mathbf{v}_2 - \cdots - \dfrac{\mathbf{v}_j \cdot \mathbf{u}_{j+1}}{\|\mathbf{v}_j\|^2}\mathbf{v}_j\right)$.

Thus

$$\mathbf{w} - d_{j+1}\mathbf{u}_{j+1} = (d_1\mathbf{v}_1 + \cdots + d_j\mathbf{v}_j) +$$

$$d_{j+1}\left(-\dfrac{\mathbf{v}_1 \cdot \mathbf{u}_{j+1}}{\|\mathbf{v}_1\|^2}\mathbf{v}_1 - \dfrac{\mathbf{v}_2 \cdot \mathbf{u}_{j+1}}{\|\mathbf{v}_2\|^2}\mathbf{v}_2 - \cdots - \dfrac{\mathbf{v}_j \cdot \mathbf{u}_{j+1}}{\|\mathbf{v}_j\|^2}\mathbf{v}_j\right)$$

$\in \text{span}\{\mathbf{v}_1, \ldots, \mathbf{v}_j\} = \text{span}\{\mathbf{u}_1, \ldots, \mathbf{u}_j\}$. Hence $\mathbf{w} - d_{j+1}\mathbf{u}_{j+1} = c_1\mathbf{u}_1 + \cdots + c_j\mathbf{u}_j$, and so $\mathbf{w} = c_1\mathbf{u}_1 + \cdots + c_j\mathbf{u}_j + d_{j+1}\mathbf{u}_{j+1} \in \text{span}\{\mathbf{u}_1, \ldots, \mathbf{u}_{j+1}\}$. Thus $\text{span}\{\mathbf{v}_1, \ldots, \mathbf{v}_{j+1}\} \subset \text{span}\{\mathbf{u}_1, \ldots, \mathbf{u}_{j+1}\}$. We conclude that $\text{span}\{\mathbf{v}_1, \ldots, \mathbf{v}_j\} = \text{span}\{\mathbf{u}_1, \ldots, \mathbf{u}_j\}$ for all $j \leq k$.

(c) Since $\{\mathbf{v}_1, \ldots, \mathbf{v}_k\}$ is orthogonal, $\{\mathbf{v}_1, \ldots, \mathbf{v}_k\}$ is also linearly independent. And since $\{\mathbf{v}_1, \ldots, \mathbf{v}_k\}$ spans the same subspace as $\{\mathbf{u}_1, \ldots, \mathbf{u}_k\}$, we conclude that $\{\mathbf{v}_1, \ldots, \mathbf{v}_k\}$ is an orthogonal basis for the subspace spanned by $\{\mathbf{u}_1, \ldots, \mathbf{u}_k\}$.

55. By Theorem 8.16(b), $\mathbf{u} - \text{proj}_S\mathbf{u}$ is orthogonal to S. Since $\text{proj}_S\mathbf{u}$ is in S, $\mathbf{u} - \text{proj}_S\mathbf{u}$ is orthogonal to $\text{proj}_S\mathbf{u}$. Thus, by Theorem 8.6, $\|\mathbf{u}\|^2 = \|(\mathbf{u} - \text{proj}_S\mathbf{u}) + \text{proj}_S\mathbf{u}\|^2 = \|\mathbf{u} - \text{proj}_S\mathbf{u}\|^2 + \|\text{proj}_S\mathbf{u}\|^2$.

57. (a) From (6) in Exercise 54, $\|\mathbf{u}\|^2 = \|\mathbf{u} - \text{proj}_\mathbf{v}\mathbf{u}\|^2 + \|\text{proj}_\mathbf{v}\mathbf{u}\|^2 \quad \Rightarrow$
$\|\text{proj}_\mathbf{v}\mathbf{u}\|^2 \leq \|\mathbf{u}\|^2 \quad \Rightarrow \quad \|\text{proj}_\mathbf{v}\mathbf{u}\| \leq \|\mathbf{u}\|$.

(b) Since $\text{proj}_\mathbf{v}\mathbf{u} = \dfrac{\mathbf{v} \cdot \mathbf{u}}{\|\mathbf{v}\|^2}\mathbf{v}$, $\|\text{proj}_\mathbf{v}\mathbf{u}\| = \left\|\dfrac{\mathbf{v} \cdot \mathbf{u}}{\|\mathbf{v}\|^2}\mathbf{v}\right\| = \dfrac{|\mathbf{u} \cdot \mathbf{v}|}{\|\mathbf{v}\|} \leq \|\mathbf{u}\|$, and hence
$|\mathbf{u} \cdot \mathbf{v}| \leq \|\mathbf{u}\|\,\|\mathbf{v}\|$.

(c) If $\mathbf{u} = c\mathbf{v}$, then $\mathbf{u} \cdot \mathbf{v} = c\mathbf{v} \cdot \mathbf{v} = c\|\mathbf{v}\|^2 = (c\|\mathbf{v}\|)\|\mathbf{v}\|$, and $|\mathbf{u} \cdot \mathbf{v}| = (|c|\|\mathbf{v}\|)\|\mathbf{v}\| = \|\mathbf{u}\|\,\|\mathbf{v}\|$.
If $|\mathbf{u} \cdot \mathbf{v}| = \|\mathbf{u}\|\,\|\mathbf{v}\|$, then we must have equality with $\|\text{proj}_\mathbf{v}\mathbf{u}\| \leq \|\mathbf{u}\|$. Thus $\|\text{proj}_\mathbf{v}\mathbf{u}\|^2 = \|\mathbf{u}\|^2 \quad \Rightarrow \quad \|\mathbf{u} - \text{proj}_\mathbf{v}\mathbf{u}\|^2 = 0$ in (6) of Exercise 54. But this implies that $\mathbf{u} - \text{proj}_\mathbf{v}\mathbf{u} = \mathbf{0}$, and thus $\text{proj}_\mathbf{v}\mathbf{u} = \mathbf{u}$, and hence $\mathbf{u} \in \text{span}\{\mathbf{v}\}$. Thus $\mathbf{u} = c\mathbf{v}$ for some c.

59. An orthonormal basis for $S = \text{span}\left\{\begin{bmatrix} 1 \\ 2 \\ -4 \\ -1 \end{bmatrix}, \begin{bmatrix} -3 \\ 0 \\ 5 \\ -2 \end{bmatrix}, \begin{bmatrix} 0 \\ 7 \\ 2 \\ -6 \end{bmatrix}\right\}$ is

$$\left\{\begin{bmatrix} \frac{1}{22}\sqrt{22} \\ \frac{1}{11}\sqrt{22} \\ -\frac{2}{11}\sqrt{22} \\ -\frac{1}{22}\sqrt{22} \end{bmatrix}, \begin{bmatrix} -\frac{9}{1738}\sqrt{22}\sqrt{395} \\ \frac{21}{4345}\sqrt{22}\sqrt{395} \\ \frac{13}{4345}\sqrt{22}\sqrt{395} \\ -\frac{13}{1738}\sqrt{22}\sqrt{395} \end{bmatrix}, \begin{bmatrix} \frac{86}{69757}\sqrt{395}\sqrt{883} \\ \frac{929}{1046355}\sqrt{395}\sqrt{883} \\ \frac{782}{1046355}\sqrt{395}\sqrt{883} \\ \frac{4}{209271}\sqrt{395}\sqrt{883} \end{bmatrix}\right\}.$$

61. An orthonormal basis for $S = \text{span}\left\{\begin{bmatrix} 3 \\ -1 \\ -2 \\ 7 \end{bmatrix}, \begin{bmatrix} 2 \\ 1 \\ 6 \\ 3 \end{bmatrix}\right\}$ is $\left\{\begin{bmatrix} \frac{1}{7}\sqrt{7} \\ -\frac{1}{21}\sqrt{7} \\ -\frac{2}{21}\sqrt{7} \\ \frac{1}{3}\sqrt{7} \end{bmatrix}, \begin{bmatrix} \frac{2}{211}\sqrt{422} \\ \frac{11}{1266}\sqrt{422} \\ \frac{29}{633}\sqrt{422} \\ \frac{13}{1266}\sqrt{422} \end{bmatrix}\right\}.$

Thus,

$$\text{proj}_S\mathbf{u} = \begin{bmatrix} \frac{1}{7}\sqrt{7} \\ -\frac{1}{21}\sqrt{7} \\ -\frac{2}{21}\sqrt{7} \\ \frac{1}{3}\sqrt{7} \end{bmatrix} \cdot \begin{bmatrix} 3 \\ -1 \\ -1 \\ 4 \end{bmatrix} \begin{bmatrix} \frac{1}{7}\sqrt{7} \\ -\frac{1}{21}\sqrt{7} \\ -\frac{2}{21}\sqrt{7} \\ \frac{1}{3}\sqrt{7} \end{bmatrix}$$

$$+ \begin{bmatrix} \frac{2}{211}\sqrt{422} \\ \frac{11}{1266}\sqrt{422} \\ \frac{29}{633}\sqrt{422} \\ \frac{13}{1266}\sqrt{422} \end{bmatrix} \cdot \begin{bmatrix} 3 \\ -1 \\ -1 \\ 4 \end{bmatrix} \begin{bmatrix} \frac{2}{211}\sqrt{422} \\ \frac{11}{1266}\sqrt{422} \\ \frac{29}{633}\sqrt{422} \\ \frac{13}{1266}\sqrt{422} \end{bmatrix}$$

$$= \begin{bmatrix} \frac{2902}{1477} \\ -\frac{1713}{2954} \\ -\frac{1447}{1477} \\ \frac{1903}{422} \end{bmatrix}$$

8.3 Diagonalizing Symmetric Matrices and QR Factorization

1. Not symmetric, $A^T = \begin{bmatrix} 1 & -2 \\ 2 & 1 \end{bmatrix}^T = \begin{bmatrix} 1 & 2 \\ -2 & 1 \end{bmatrix} \neq A$.

3. Symmetric, $A^T = \begin{bmatrix} 3 & 2 & 1 \\ 2 & 1 & 3 \\ 1 & 3 & 2 \end{bmatrix}^T = \begin{bmatrix} 3 & 2 & 1 \\ 2 & 1 & 3 \\ 1 & 3 & 2 \end{bmatrix} = A$.

5. Not symmetric, $A^T = \begin{bmatrix} 3 & -1 & 4 \\ -1 & 4 & 3 \end{bmatrix}^T = \begin{bmatrix} 3 & -1 \\ -1 & 4 \\ 4 & 3 \end{bmatrix} \neq A$. *(Or, A is cannot be symmetric since it is not a square matrix.)*

7. Not symmetric, $A^T = \begin{bmatrix} 1 & 7 & -3 \\ 7 & 2 & 4 \\ -3 & 0 & -6 \\ 4 & -6 & -1 \end{bmatrix}^T = \begin{bmatrix} 1 & 7 & -3 & 4 \\ 7 & 2 & 0 & -6 \\ -3 & 4 & -6 & -1 \end{bmatrix} \neq A$. *(Or, A is cannot be symmetric since it is not a square matrix.)*

9. The matrix is not orthogonal, since $\begin{bmatrix} 1 & -2 \\ 2 & 1 \end{bmatrix}^T \begin{bmatrix} 1 & -2 \\ 2 & 1 \end{bmatrix} = \begin{bmatrix} 5 & 0 \\ 0 & 5 \end{bmatrix} \neq I_2$.

11. The matrix is orthogonal, since $\begin{bmatrix} -\frac{5}{13} & \frac{12}{13} \\ \frac{12}{13} & \frac{5}{13} \end{bmatrix}^T \begin{bmatrix} -\frac{5}{13} & \frac{12}{13} \\ \frac{12}{13} & \frac{5}{13} \end{bmatrix} = \begin{bmatrix} 1 & 0 \\ 0 & 1 \end{bmatrix} = I_2$.

13. The matrix is not orthogonal, since
$$\begin{bmatrix} \frac{1}{2} & \frac{1}{3} & \frac{1}{4} \\ -\frac{1}{2} & \frac{1}{3} & \frac{1}{4} \\ 0 & \frac{1}{3} & -\frac{1}{2} \end{bmatrix}^T \begin{bmatrix} \frac{1}{2} & \frac{1}{3} & \frac{1}{4} \\ -\frac{1}{2} & \frac{1}{3} & \frac{1}{4} \\ 0 & \frac{1}{3} & -\frac{1}{2} \end{bmatrix} = \begin{bmatrix} \frac{1}{2} & 0 & 0 \\ 0 & \frac{1}{3} & 0 \\ 0 & 0 & \frac{3}{8} \end{bmatrix} \neq I_3.$$

15. Normalize to obtain $\mathbf{v}_1 = \dfrac{\mathbf{u}_1}{\|\mathbf{u}_1\|} = \begin{bmatrix} 1 \\ 2 \end{bmatrix} / \left\| \begin{bmatrix} 1 \\ 2 \end{bmatrix} \right\| = \begin{bmatrix} \frac{1}{5}\sqrt{5} \\ \frac{2}{5}\sqrt{5} \end{bmatrix}$ and $\mathbf{v}_2 = \dfrac{\mathbf{u}_2}{\|\mathbf{u}_2\|} = \begin{bmatrix} -2 \\ 1 \end{bmatrix} / \left\| \begin{bmatrix} -2 \\ 1 \end{bmatrix} \right\| =$
$\begin{bmatrix} -\frac{2}{5}\sqrt{5} \\ \frac{1}{5}\sqrt{5} \end{bmatrix}$. Thus $P = \begin{bmatrix} \frac{1}{5}\sqrt{5} & -\frac{2}{5}\sqrt{5} \\ \frac{2}{5}\sqrt{5} & \frac{1}{5}\sqrt{5} \end{bmatrix}$, and $D = \begin{bmatrix} \lambda_1 & 0 \\ 0 & \lambda_2 \end{bmatrix} = \begin{bmatrix} 2 & 0 \\ 0 & -3 \end{bmatrix}$.

17. Normalize to obtain $\mathbf{v}_1 = \dfrac{\mathbf{u}_1}{\|\mathbf{u}_1\|} = \begin{bmatrix} 1 \\ 1 \\ 1 \end{bmatrix} / \left\| \begin{bmatrix} 1 \\ 1 \\ 1 \end{bmatrix} \right\| = \begin{bmatrix} \frac{1}{3}\sqrt{3} \\ \frac{1}{3}\sqrt{3} \\ \frac{1}{3}\sqrt{3} \end{bmatrix}$, $\mathbf{v}_2 = \dfrac{\mathbf{u}_2}{\|\mathbf{u}_2\|} =$

$\begin{bmatrix} 1 \\ -1 \\ 0 \end{bmatrix} / \left\| \begin{bmatrix} 1 \\ -1 \\ 0 \end{bmatrix} \right\| = \begin{bmatrix} \frac{1}{2}\sqrt{2} \\ -\frac{1}{2}\sqrt{2} \\ 0 \end{bmatrix}$, and $\mathbf{v}_3 = \dfrac{\mathbf{u}_3}{\|\mathbf{u}_3\|} = \begin{bmatrix} -1 \\ -1 \\ 2 \end{bmatrix} / \left\| \begin{bmatrix} -1 \\ -1 \\ 2 \end{bmatrix} \right\| = \begin{bmatrix} -\frac{1}{6}\sqrt{6} \\ -\frac{1}{6}\sqrt{6} \\ \frac{1}{3}\sqrt{6} \end{bmatrix}$.

Thus $P = \begin{bmatrix} \frac{1}{3}\sqrt{3} & \frac{1}{2}\sqrt{2} & -\frac{1}{6}\sqrt{6} \\ \frac{1}{3}\sqrt{3} & -\frac{1}{2}\sqrt{2} & -\frac{1}{6}\sqrt{6} \\ \frac{1}{3}\sqrt{3} & 0 & \frac{1}{3}\sqrt{6} \end{bmatrix}$, and $D = \begin{bmatrix} \lambda_1 & 0 & 0 \\ 0 & \lambda_2 & 0 \\ 0 & 0 & \lambda_3 \end{bmatrix} = \begin{bmatrix} 0 & 0 & 0 \\ 0 & 2 & 0 \\ 0 & 0 & -1 \end{bmatrix}$.

19. We determine the corresponding orthonormal eigenvectors,
$$\lambda_1 = 0 \quad \Rightarrow \quad \mathbf{u}_1 = \begin{bmatrix} -\frac{1}{5}\sqrt{5} \\ \frac{2}{5}\sqrt{5} \end{bmatrix} \text{ and } \lambda_2 = 5 \quad \Rightarrow \quad \mathbf{u}_2 = \begin{bmatrix} \frac{2}{5}\sqrt{5} \\ \frac{1}{5}\sqrt{5} \end{bmatrix}.$$
Thus $P = \begin{bmatrix} -\frac{1}{5}\sqrt{5} & \frac{2}{5}\sqrt{5} \\ \frac{2}{5}\sqrt{5} & \frac{1}{5}\sqrt{5} \end{bmatrix}$ and $D = \begin{bmatrix} \lambda_1 & 0 \\ 0 & \lambda_2 \end{bmatrix} = \begin{bmatrix} 0 & 0 \\ 0 & 5 \end{bmatrix}$.

21. We determine the corresponding orthonormal eigenvectors, $\lambda_1 = -2 \quad \Rightarrow$

$$\mathbf{u}_1 = \begin{bmatrix} -\frac{1}{2}\sqrt{2} \\ 0 \\ \frac{1}{2}\sqrt{2} \end{bmatrix}, \lambda_2 = 0 \quad \Rightarrow \quad \mathbf{u}_2 = \begin{bmatrix} \frac{1}{6}\sqrt{6} \\ -\frac{1}{3}\sqrt{6} \\ \frac{1}{6}\sqrt{6} \end{bmatrix}, \text{ and } \lambda_3 = 3 \quad \Rightarrow \quad \mathbf{u}_3 = \begin{bmatrix} \frac{1}{3}\sqrt{3} \\ \frac{1}{3}\sqrt{3} \\ \frac{1}{3}\sqrt{3} \end{bmatrix}.$$

Thus $P = \begin{bmatrix} -\frac{1}{2}\sqrt{2} & \frac{1}{6}\sqrt{6} & \frac{1}{3}\sqrt{3} \\ 0 & -\frac{1}{3}\sqrt{6} & \frac{1}{3}\sqrt{3} \\ \frac{1}{2}\sqrt{2} & \frac{1}{6}\sqrt{6} & \frac{1}{3}\sqrt{3} \end{bmatrix}$ and $D = \begin{bmatrix} \lambda_1 & 0 & 0 \\ 0 & \lambda_2 & 0 \\ 0 & 0 & \lambda_3 \end{bmatrix} = \begin{bmatrix} -2 & 0 & 0 \\ 0 & 0 & 0 \\ 0 & 0 & 3 \end{bmatrix}.$

23. We determine the corresponding orthonormal eigenvectors, $\lambda_1 = -1 \quad \Rightarrow$

$$\mathbf{u}_1 = \begin{bmatrix} -\frac{1}{2}\sqrt{2} \\ 0 \\ \frac{1}{2}\sqrt{2} \end{bmatrix}, \text{ and } \lambda_2 = 1 \quad \Rightarrow \quad \mathbf{u}_2 = \begin{bmatrix} 0 \\ 1 \\ 0 \end{bmatrix} \text{ and } \mathbf{u}_3 = \begin{bmatrix} \frac{1}{2}\sqrt{2} \\ 0 \\ \frac{1}{2}\sqrt{2} \end{bmatrix}.$$

Thus $P = \begin{bmatrix} -\frac{1}{2}\sqrt{2} & 0 & \frac{1}{2}\sqrt{2} \\ 0 & 1 & 0 \\ \frac{1}{2}\sqrt{2} & 0 & \frac{1}{2}\sqrt{2} \end{bmatrix}$ and $D = \begin{bmatrix} \lambda_1 & 0 & 0 \\ 0 & \lambda_2 & 0 \\ 0 & 0 & \lambda_2 \end{bmatrix} = \begin{bmatrix} -1 & 0 & 0 \\ 0 & 1 & 0 \\ 0 & 0 & 1 \end{bmatrix}.$

25. $A^T A = \begin{bmatrix} 1 & 1 \\ 2 & 1 \\ 1 & 2 \end{bmatrix}^T \begin{bmatrix} 1 & 1 \\ 2 & 1 \\ 1 & 2 \end{bmatrix} = \begin{bmatrix} 6 & 5 \\ 5 & 6 \end{bmatrix}.$ $\det\left(\lambda I_2 - A^T A\right) = \det\left(\lambda \begin{bmatrix} 1 & 0 \\ 0 & 1 \end{bmatrix} - \begin{bmatrix} 6 & 5 \\ 5 & 6 \end{bmatrix}\right)$

$= \lambda^2 - 12\lambda + 11 = (\lambda - 1)(\lambda - 11) \quad \Rightarrow \quad \lambda_1 = 1 \geq 0 \text{ and } \lambda_2 = 11 \geq 0.$

27. $A^T A = \begin{bmatrix} 0 & 2 & 1 \\ 1 & 0 & 0 \end{bmatrix}^T \begin{bmatrix} 0 & 2 & 1 \\ 1 & 0 & 0 \end{bmatrix} = \begin{bmatrix} 1 & 0 & 0 \\ 0 & 4 & 2 \\ 0 & 2 & 1 \end{bmatrix}.$ $\det\left(\lambda I_3 - A^T A\right) =$

$\det\left(\lambda \begin{bmatrix} 1 & 0 & 0 \\ 0 & 1 & 0 \\ 0 & 0 & 1 \end{bmatrix} - \begin{bmatrix} 1 & 0 & 0 \\ 0 & 4 & 2 \\ 0 & 2 & 1 \end{bmatrix}\right) = \lambda^3 - 6\lambda^2 + 5\lambda = \lambda(\lambda - 1)(\lambda - 5) \quad \Rightarrow \quad \lambda_1 = 0 \geq 0, \lambda_2 = 1 \geq 0, \text{ and } \lambda_3 = 5 \geq 0.$

29. $\|\mathbf{q}_1\|^2 = \left\| \begin{bmatrix} 1 \\ -2 \end{bmatrix} \right\|^2 = 5$, so $\mathbf{q}_1^T / \|\mathbf{q}_1\|^2 = \frac{1}{5} \begin{bmatrix} 1 \\ -2 \end{bmatrix}^T = \begin{bmatrix} \frac{1}{5} & -\frac{2}{5} \end{bmatrix}$ and $\|\mathbf{q}_2\|^2 = \left\| \begin{bmatrix} 2 \\ 1 \end{bmatrix} \right\|^2 = 5$, so

$\mathbf{q}_2^T / \|\mathbf{q}_2\|^2 = \frac{1}{5} \begin{bmatrix} 2 \\ 1 \end{bmatrix}^T = \begin{bmatrix} \frac{2}{5} & \frac{1}{5} \end{bmatrix}.$ Thus $Q^{-1} = \begin{bmatrix} \frac{1}{5} & -\frac{2}{5} \\ \frac{2}{5} & \frac{1}{5} \end{bmatrix}.$

31. $\|\mathbf{q}_1\|^2 = \left\| \begin{bmatrix} 0 \\ 1 \\ -1 \end{bmatrix} \right\|^2 = 2$, so $\mathbf{q}_1^T / \|\mathbf{q}_1\|^2 = \frac{1}{2} \begin{bmatrix} 0 \\ 1 \\ -1 \end{bmatrix}^T = \begin{bmatrix} 0 & \frac{1}{2} & -\frac{1}{2} \end{bmatrix}$, $\|\mathbf{q}_2\|^2 = \left\| \begin{bmatrix} 2 \\ 0 \\ 0 \end{bmatrix} \right\|^2 = 4$, so

$\mathbf{q}_2^T / \|\mathbf{q}_2\|^2 = \frac{1}{4} \begin{bmatrix} 2 \\ 0 \\ 0 \end{bmatrix}^T = \begin{bmatrix} \frac{1}{2} & 0 & 0 \end{bmatrix}$, and $\|\mathbf{q}_3\|^2 = \left\| \begin{bmatrix} 0 \\ 1 \\ 1 \end{bmatrix} \right\|^2 = 2$, so $\mathbf{q}_3^T / \|\mathbf{q}_3\|^2 = \frac{1}{2} \begin{bmatrix} 0 \\ 1 \\ 1 \end{bmatrix}^T =$

$\begin{bmatrix} 0 & \frac{1}{2} & \frac{1}{2} \end{bmatrix}.$ Thus $Q^{-1} = \begin{bmatrix} 0 & \frac{1}{2} & -\frac{1}{2} \\ \frac{1}{2} & 0 & 0 \\ 0 & \frac{1}{2} & \frac{1}{2} \end{bmatrix}.$

33. We normalize the columns to obtain an orthonormal set

$\mathbf{q}_1 = \begin{bmatrix} \frac{3}{13}\sqrt{13} \\ \frac{2}{13}\sqrt{13} \end{bmatrix}$, and $\mathbf{q}_2 = \begin{bmatrix} -\frac{2}{13}\sqrt{13} \\ \frac{3}{13}\sqrt{13} \end{bmatrix}.$ Thus $Q = \begin{bmatrix} \frac{3}{13}\sqrt{13} & -\frac{2}{13}\sqrt{13} \\ \frac{2}{13}\sqrt{13} & \frac{3}{13}\sqrt{13} \end{bmatrix}$, and

$R = Q^T A = \begin{bmatrix} \frac{3}{13}\sqrt{13} & -\frac{2}{13}\sqrt{13} \\ \frac{2}{13}\sqrt{13} & \frac{3}{13}\sqrt{13} \end{bmatrix}^T \begin{bmatrix} 3 & -2 \\ 2 & 3 \end{bmatrix} = \begin{bmatrix} \sqrt{13} & 0 \\ 0 & \sqrt{13} \end{bmatrix}.$

35. We apply to Gram-Schmidt process to obtain an orthonormal set
$$\mathbf{q}_1 = \begin{bmatrix} \frac{1}{10}\sqrt{10} \\ \frac{3}{10}\sqrt{10} \end{bmatrix}, \text{ and } \mathbf{q}_2 = \begin{bmatrix} \frac{3}{10}\sqrt{10} \\ -\frac{1}{10}\sqrt{10} \end{bmatrix}. \text{ Thus } Q = \begin{bmatrix} \frac{1}{10}\sqrt{10} & \frac{3}{10}\sqrt{10} \\ \frac{3}{10}\sqrt{10} & -\frac{1}{10}\sqrt{10} \end{bmatrix}, \text{ and}$$
$$R = Q^T A = \begin{bmatrix} \frac{1}{10}\sqrt{10} & \frac{3}{10}\sqrt{10} \\ \frac{3}{10}\sqrt{10} & -\frac{1}{10}\sqrt{10} \end{bmatrix}^T \begin{bmatrix} 1 & 4 \\ 3 & 2 \end{bmatrix} = \begin{bmatrix} \sqrt{10} & \sqrt{10} \\ 0 & \sqrt{10} \end{bmatrix}.$$

37. We apply to Gram-Schmidt process to obtain an orthonormal set
$$\mathbf{q}_1 = \begin{bmatrix} \frac{1}{9}\sqrt{3} \\ \frac{1}{9}\sqrt{3} \\ \frac{5}{9}\sqrt{3} \end{bmatrix} \text{ and } \mathbf{q}_2 = \begin{bmatrix} \frac{25}{1629}\sqrt{1086} \\ -\frac{85}{3258}\sqrt{1086} \\ \frac{7}{3258}\sqrt{1086} \end{bmatrix}. \text{ Thus } Q = \begin{bmatrix} \frac{1}{9}\sqrt{3} & \frac{25}{1629}\sqrt{1086} \\ \frac{1}{9}\sqrt{3} & -\frac{85}{3258}\sqrt{1086} \\ \frac{5}{9}\sqrt{3} & \frac{7}{3258}\sqrt{1086} \end{bmatrix}, \text{ and}$$
$$R = Q^T A = \begin{bmatrix} \frac{1}{9}\sqrt{3} & \frac{25}{1629}\sqrt{1086} \\ \frac{1}{9}\sqrt{3} & -\frac{85}{3258}\sqrt{1086} \\ \frac{5}{9}\sqrt{3} & \frac{7}{3258}\sqrt{1086} \end{bmatrix}^T \begin{bmatrix} 1 & 2 \\ 1 & -3 \\ 5 & 1 \end{bmatrix} = \begin{bmatrix} 3\sqrt{3} & \frac{4}{9}\sqrt{3} \\ 0 & \frac{1}{9}\sqrt{1086} \end{bmatrix}.$$

39. We apply to Gram-Schmidt process to obtain an orthonormal set
$$\mathbf{q}_1 = \begin{bmatrix} -\frac{2}{3} \\ \frac{2}{3} \\ \frac{1}{3} \end{bmatrix} \text{ and } \mathbf{q}_2 = \begin{bmatrix} \frac{3}{29}\sqrt{29} \\ \frac{4}{29}\sqrt{29} \\ -\frac{2}{29}\sqrt{29} \end{bmatrix}. \text{ Thus } Q = \begin{bmatrix} -\frac{2}{3} & \frac{3}{29}\sqrt{29} \\ \frac{2}{3} & \frac{4}{29}\sqrt{29} \\ \frac{1}{3} & -\frac{2}{29}\sqrt{29} \end{bmatrix}, \text{ and}$$
$$R = Q^T A = \begin{bmatrix} -\frac{2}{3} & \frac{3}{29}\sqrt{29} \\ \frac{2}{3} & \frac{4}{29}\sqrt{29} \\ \frac{1}{3} & -\frac{2}{29}\sqrt{29} \end{bmatrix}^T \begin{bmatrix} -2 & 3 \\ 2 & 4 \\ 1 & -2 \end{bmatrix} = \begin{bmatrix} 3 & 0 \\ 0 & \sqrt{29} \end{bmatrix}.$$

41. For example, $A = \begin{bmatrix} 1 & 0 \\ 0 & 2 \end{bmatrix}$.

43. For example, $A = \begin{bmatrix} \frac{7}{5} & -\frac{6}{5} \\ -\frac{6}{5} & -\frac{2}{5} \end{bmatrix}$.

45. For example, $A = \begin{bmatrix} 0 & 0 \\ 0 & 0 \end{bmatrix}$.

47. For example, $A = \begin{bmatrix} -2 & 2 \\ -6 & 5 \end{bmatrix}$ (See Example 1, Section 6.4.)

49. True, by Theorem 8.23.

51. True, by Theorem 8.22.

53. True, since $\left(A^T A\right)^T = (A)^T \left(A^T\right)^T = A^T A$.

55. False. Indeed, if A is $n \times m$, then R is $m \times m$ and the columns of R are in $\mathbf{R}^m$, not $\mathbf{R}^n$.

57. True. Since R is an upper triangular matrix with positive diagonal entries, $\det(R)$ is the product of the diagonal entries and must be positive. Hence $\det(R) \neq 0$, and R is invertible.

59. Since A is orthogonal, $A^T A = I$. Hence
$$1 = \det(A^T A) = \det(A^T)\det(A) = \left(\det(A)\right)^2$$
so that $\det(A) = \pm 1$.

61. The (i, k) entry of $Q^T A$ is the dot product of the ith row of Q^T with the kth column of A. The ith row of Q^T is the ith column of Q, which is $\mathbf{q}_i$. Thus, the (i, k) entry of $Q^T A$ is $\mathbf{q}_i \cdot \mathbf{a}_k$.

63. If A is orthogonally diagonalizable, then there exists an orthogonal matrix P and a diagonal matrix D such that $A = PDP^{-1}$. Let $Q = P^T$, then Q is orthogonal by Exercise 62. Thus $A^T = \left(PDP^{-1}\right)^T = \left(P^{-1}\right)^T D^T P^T = \left(P^T\right)^{-1} D P^T = Q^{-1} D Q$. This shows that A^T is orthogonally diagonalizable.

65. If A is orthogonally diagonalizable, then there exists an orthogonal matrix P and a diagonal matrix D such that $A = PDP^{-1}$. Thus $A^2 = \left(PDP^{-1}\right)\left(PDP^{-1}\right) = PD\left(P^{-1}P\right)DP^{-1} = PDIDP^{-1} = PD^2P^{-1}$. Since D^2 is diagonal, it follows that A^2 is orthogonally diagonalizable.

67. We use a computer algebra system to approximate $A = PDP^{-1}$, with

$$D \approx \begin{bmatrix} 8.0463 & 0 & 0 \\ 0 & 2.2795 & 0 \\ 0 & 0 & -3.3258 \end{bmatrix}, P \approx \begin{bmatrix} 0.3603 & -0.8287 & -0.4282 \\ -0.3790 & -0.5495 & 0.7446 \\ 0.8524 & 0.1060 & 0.5120 \end{bmatrix}.$$

69. We use a computer algebra system to approximate $A = PDP^{-1}$, with

$$D \approx \begin{bmatrix} 7.624 & 0 & 0 & 0 \\ 0 & -1.211 & 0 & 0 \\ 0 & 0 & 5.639 & 0 \\ 0 & 0 & 0 & -6.051 \end{bmatrix}, P \approx \begin{bmatrix} -0.1376 & -0.6216 & 0.7118 & 0.2968 \\ -0.7426 & 0.5164 & 0.1391 & 0.4038 \\ -0.0078 & 0.3699 & 0.6127 & -0.6984 \\ 0.6558 & 0.4585 & 0.3140 & 0.5110 \end{bmatrix}.$$

71. We use a computer algebra system to obtain $A = QR$,

$$\begin{bmatrix} -1 & 3 & 3 \\ 0 & 2 & 4 \\ 1 & 1 & 5 \end{bmatrix} = \begin{bmatrix} -\frac{1}{2}\sqrt{2} & \frac{1}{3}\sqrt{3} & \frac{1}{6}\sqrt{6} \\ 0 & \frac{1}{3}\sqrt{3} & -\frac{1}{3}\sqrt{6} \\ \frac{1}{2}\sqrt{2} & \frac{1}{3}\sqrt{3} & \frac{1}{6}\sqrt{6} \end{bmatrix} \begin{bmatrix} \sqrt{2} & -\sqrt{2} & \sqrt{2} \\ 0 & 2\sqrt{3} & 4\sqrt{3} \\ 0 & 0 & 0 \end{bmatrix}.$$

73. We use a computer algebra system to obtain $A = QR$,

$$\begin{bmatrix} 1 & 1 & 4 \\ 1 & 3 & 2 \\ 1 & 0 & 2 \\ -1 & 1 & 4 \end{bmatrix} = \begin{bmatrix} \frac{1}{2} & \frac{1}{70}\sqrt{35} & \frac{47}{6790}\sqrt{6790} \\ \frac{1}{2} & \frac{9}{70}\sqrt{35} & -\frac{16}{3395}\sqrt{6790} \\ \frac{1}{2} & -\frac{3}{70}\sqrt{35} & \frac{17}{3395}\sqrt{6790} \\ -\frac{1}{2} & \frac{1}{10}\sqrt{35} & \frac{7}{970}\sqrt{6790} \end{bmatrix} \begin{bmatrix} 2 & \frac{3}{2} & 2 \\ 0 & \frac{1}{2}\sqrt{35} & \frac{22}{35}\sqrt{35} \\ 0 & 0 & \frac{2}{35}\sqrt{6790} \end{bmatrix}.$$

8.4   The Singular Value Decomposition

1. $\det\left(\lambda \begin{bmatrix} 1 & 0 \\ 0 & 1 \end{bmatrix} - \begin{bmatrix} 1 & 2 \\ -1 & 2 \end{bmatrix}^T \begin{bmatrix} 1 & 2 \\ -1 & 2 \end{bmatrix}\right) = \lambda^2 - 10\lambda + 16 = (\lambda - 2)(\lambda - 8) \quad \Rightarrow \quad \lambda_1 = 8$ and $\lambda_2 = 2 \quad \Rightarrow \quad \sigma_1 = \sqrt{8}$ and $\sigma_2 = \sqrt{2}$.

3. $\det\left(\lambda \begin{bmatrix} 1 & 0 \\ 0 & 1 \end{bmatrix} - \begin{bmatrix} 3 & -1 \\ -1 & 3 \end{bmatrix}^T \begin{bmatrix} 3 & -1 \\ -1 & 3 \end{bmatrix}\right) = \lambda^2 - 20\lambda + 64 = (\lambda - 4)(\lambda - 16) \quad \Rightarrow \quad \lambda_1 = 16$ and $\lambda_2 = 4 \quad \Rightarrow \quad \sigma_1 = \sqrt{16} = 4$ and $\sigma_2 = \sqrt{4} = 2$.

5. $\det\left(\lambda \begin{bmatrix} 1 & 0 \\ 0 & 1 \end{bmatrix} - \begin{bmatrix} 1 & 2 \\ 0 & 2 \\ 2 & -1 \end{bmatrix}^T \begin{bmatrix} 1 & 2 \\ 0 & 2 \\ 2 & -1 \end{bmatrix}\right) = \lambda^2 - 14\lambda + 45 = (\lambda - 5)(\lambda - 9) \quad \Rightarrow \quad \lambda_1 = 9$ and $\lambda_2 = 5 \quad \Rightarrow \quad \sigma_1 = \sqrt{9} = 3$ and $\sigma_2 = \sqrt{5}$.

7. Since $n = 2 < 3 = m$, we determine the singular values of $B = A^T = \begin{bmatrix} 1 & 2 & 1 \\ 0 & 1 & -1 \end{bmatrix}^T = \begin{bmatrix} 1 & 0 \\ 2 & 1 \\ 1 & -1 \end{bmatrix}$.

So we consider $\det \left(\lambda \begin{bmatrix} 1 & 0 \\ 0 & 1 \end{bmatrix} - \begin{bmatrix} 1 & 0 \\ 2 & 1 \\ 1 & -1 \end{bmatrix}^T \begin{bmatrix} 1 & 0 \\ 2 & 1 \\ 1 & -1 \end{bmatrix} \right) = \lambda^2 - 8\lambda + 11 = 0 \quad \Rightarrow \quad \lambda_1 = 4 + \sqrt{5}$

and $\lambda_2 = 4 - \sqrt{5} \quad \Rightarrow \quad \sigma_1 = \sqrt{4 + \sqrt{5}} \approx 2.49721$ and $\sigma_2 = \sqrt{4 - \sqrt{5}} \approx 1.32813$.

9. We determine the eigenvalues and normalized eigenvectors of $A^T A$, $\lambda_1 = 9 \Rightarrow \mathbf{v}_1 = \begin{bmatrix} \frac{1}{2}\sqrt{2} \\ \frac{1}{2}\sqrt{2} \end{bmatrix}$, $\lambda_2 = 1 \Rightarrow$

$\mathbf{v}_2 = \begin{bmatrix} -\frac{1}{2}\sqrt{2} \\ \frac{1}{2}\sqrt{2} \end{bmatrix}$. Thus $V = \begin{bmatrix} \frac{1}{2}\sqrt{2} & -\frac{1}{2}\sqrt{2} \\ \frac{1}{2}\sqrt{2} & \frac{1}{2}\sqrt{2} \end{bmatrix}$. The singular values are $\sigma_1 = \sqrt{\lambda_1} = \sqrt{9} = 3$ and $\sigma_2 = $

$\sqrt{\lambda_2} = \sqrt{1} = 1$. Thus $\Sigma = \begin{bmatrix} \sigma_1 & 0 \\ 0 & \sigma_2 \end{bmatrix} = \begin{bmatrix} 3 & 0 \\ 0 & 1 \end{bmatrix}$. Also, $\mathbf{u}_1 = \frac{1}{\sigma_1} A\mathbf{v}_1 = \frac{1}{3} \begin{bmatrix} 1 & 2 \\ 2 & 1 \end{bmatrix} \begin{bmatrix} \frac{1}{2}\sqrt{2} \\ \frac{1}{2}\sqrt{2} \end{bmatrix} =$

$\begin{bmatrix} \frac{1}{2}\sqrt{2} \\ \frac{1}{2}\sqrt{2} \end{bmatrix}$, and $\mathbf{u}_2 = \frac{1}{\sigma_2} A\mathbf{v}_2 = \frac{1}{1} \begin{bmatrix} 1 & 2 \\ 2 & 1 \end{bmatrix} \begin{bmatrix} -\frac{1}{2}\sqrt{2} \\ \frac{1}{2}\sqrt{2} \end{bmatrix} = \begin{bmatrix} \frac{1}{2}\sqrt{2} \\ -\frac{1}{2}\sqrt{2} \end{bmatrix}$. So $U = \begin{bmatrix} \frac{1}{2}\sqrt{2} & \frac{1}{2}\sqrt{2} \\ \frac{1}{2}\sqrt{2} & -\frac{1}{2}\sqrt{2} \end{bmatrix}$.

11. We determine the eigenvalues and normalized eigenvectors of $A^T A$, $\lambda_1 \approx 10.2361 \Rightarrow \mathbf{v}_1 \approx \begin{bmatrix} 0.2298 \\ -0.9732 \end{bmatrix}$,

$\lambda_2 \approx 5.7639 \Rightarrow \mathbf{v}_2 = \begin{bmatrix} 0.9732 \\ 0.2298 \end{bmatrix}$. Thus $V \approx \begin{bmatrix} 0.2298 & 0.9732 \\ -0.9732 & 0.2298 \end{bmatrix}$. The singular values are $\sigma_1 \approx$

$\sqrt{10.23610} \approx 3.199$ and $\sigma_2 \approx \sqrt{5.7639} = 2.401$. Thus $\Sigma = \begin{bmatrix} \sigma_1 & 0 \\ 0 & \sigma_2 \\ 0 & 0 \end{bmatrix} \approx \begin{bmatrix} 3.199 & 0 \\ 0 & 2.401 \\ 0 & 0 \end{bmatrix}$. Also,

$\mathbf{u}_1 = \frac{1}{\sigma_1} A\mathbf{v}_1 \approx \frac{1}{3.199} \begin{bmatrix} 2 & 1 \\ -1 & 3 \\ 1 & 0 \end{bmatrix} \begin{bmatrix} 0.2298 \\ -0.9732 \end{bmatrix} = \begin{bmatrix} -0.1606 \\ -0.9845 \\ 0.07183 \end{bmatrix}$, and

$\mathbf{u}_2 = \frac{1}{\sigma_2} A\mathbf{v}_2 \approx \frac{1}{2.401} \begin{bmatrix} 2 & 1 \\ -1 & 3 \\ 1 & 0 \end{bmatrix} \begin{bmatrix} 0.9732 \\ 0.2298 \end{bmatrix} = \begin{bmatrix} 0.9064 \\ -0.1182 \\ 0.4053 \end{bmatrix}$. To obtain $\mathbf{u}_3$, we determine null $\left(A^T \right) =$

null $\left(\begin{bmatrix} 2 & -1 & 1 \\ 1 & 3 & 0 \end{bmatrix} \right) \approx$ span $\left\{ \begin{bmatrix} -0.3906 \\ 0.1302 \\ 0.9113 \end{bmatrix} \right\}$, so $\mathbf{u}_3 \approx \begin{bmatrix} -0.3906 \\ 0.1302 \\ 0.9113 \end{bmatrix}$. Thus

$U \approx \begin{bmatrix} -0.1606 & 0.9064 & -0.3906 \\ -0.9845 & -0.1182 & 0.1302 \\ 0.07183 & 0.4053 & 0.9113 \end{bmatrix}$.

13. Let $B = A^T$, and determine the eigenvalues and normalized eigenvectors of $B^T B = \begin{bmatrix} 2 & 0 \\ 0 & 9 \end{bmatrix}$, $\lambda_1 = 9 \Rightarrow$

$\mathbf{v}_1 = \begin{bmatrix} 0 \\ 1 \end{bmatrix}$, $\lambda_2 = 2 \Rightarrow \mathbf{v}_2 = \begin{bmatrix} 1 \\ 0 \end{bmatrix}$. Thus $V = \begin{bmatrix} 0 & 1 \\ 1 & 0 \end{bmatrix}$. The singular values are $\sigma_1 = \sqrt{\lambda_1} = \sqrt{9} = 3$

and $\sigma_2 = \sqrt{\lambda_2} = \sqrt{2}$. Hence $\Sigma = \begin{bmatrix} 3 & 0 \\ 0 & \sqrt{2} \\ 0 & 0 \end{bmatrix}$. Also, $\mathbf{u}_1 = \frac{1}{\sigma_1} B\mathbf{v}_1 = \frac{1}{3} \begin{bmatrix} -1 & 2 \\ 1 & 2 \\ 0 & 1 \end{bmatrix} \begin{bmatrix} 0 \\ 1 \end{bmatrix} = \begin{bmatrix} \frac{2}{3} \\ \frac{2}{3} \\ \frac{1}{3} \end{bmatrix}$,

and $\mathbf{u}_2 = \frac{1}{\sigma_2} B\mathbf{v}_2 = \frac{1}{\sqrt{2}} \begin{bmatrix} -1 & 2 \\ 1 & 2 \\ 0 & 1 \end{bmatrix} \begin{bmatrix} 1 \\ 0 \end{bmatrix} = \begin{bmatrix} -\frac{1}{2}\sqrt{2} \\ \frac{1}{2}\sqrt{2} \\ 0 \end{bmatrix}$. To obtain $\mathbf{u}_3$, we determine null $\left(B^T \right) =$

null $\left(\begin{bmatrix} -1 & 1 & 0 \\ 2 & 2 & 1 \end{bmatrix} \right) \approx$

$\text{span}\left\{\begin{bmatrix} -\frac{1}{6}\sqrt{2} \\ -\frac{1}{6}\sqrt{2} \\ \frac{2}{3}\sqrt{2} \end{bmatrix}\right\}$, so $\mathbf{u}_3 = \begin{bmatrix} -\frac{1}{6}\sqrt{2} \\ -\frac{1}{6}\sqrt{2} \\ \frac{2}{3}\sqrt{2} \end{bmatrix}$. Thus $U = \begin{bmatrix} \frac{2}{3} & -\frac{1}{2}\sqrt{2} & -\frac{1}{6}\sqrt{2} \\ \frac{2}{3} & \frac{1}{2}\sqrt{2} & -\frac{1}{6}\sqrt{2} \\ \frac{1}{3} & 0 & \frac{2}{3}\sqrt{2} \end{bmatrix}$. The singular value decomposition for A is $V\Sigma^T U^T$.

15. Let $B = A^T$, and determine the eigenvalues and normalized eigenvectors of $B^T B = \begin{bmatrix} 9 & 0 \\ 0 & 3 \end{bmatrix}$, $\lambda_1 = 9 \Rightarrow$ $\mathbf{v}_1 = \begin{bmatrix} 1 \\ 0 \end{bmatrix}$, $\lambda_2 = 3 \Rightarrow \mathbf{v}_2 = \begin{bmatrix} 0 \\ 1 \end{bmatrix}$. Thus $V = \begin{bmatrix} 1 & 0 \\ 0 & 1 \end{bmatrix}$. The singular values are $\sigma_1 = \sqrt{\lambda_1} = \sqrt{9} = 3$ and $\sigma_2 = \sqrt{\lambda_2} = \sqrt{3}$. Hence $\Sigma = \begin{bmatrix} 3 & 0 \\ 0 & \sqrt{3} \\ 0 & 0 \\ 0 & 0 \end{bmatrix}$. Also, $\mathbf{u}_1 = \frac{1}{\sigma_1}B\mathbf{v}_1 = \frac{1}{3}\begin{bmatrix} 2 & 1 \\ 2 & -1 \\ 1 & 0 \\ 0 & 1 \end{bmatrix}\begin{bmatrix} 1 \\ 0 \end{bmatrix} = \begin{bmatrix} \frac{2}{3} \\ \frac{2}{3} \\ \frac{1}{3} \\ 0 \end{bmatrix}$,

and $\mathbf{u}_2 = \frac{1}{\sigma_2}B\mathbf{v}_2 = \frac{1}{\sqrt{3}}\begin{bmatrix} 2 & 1 \\ 2 & -1 \\ 1 & 0 \\ 0 & 1 \end{bmatrix}\begin{bmatrix} 0 \\ 1 \end{bmatrix} = \begin{bmatrix} \frac{1}{3}\sqrt{3} \\ -\frac{1}{3}\sqrt{3} \\ 0 \\ \frac{1}{3}\sqrt{3} \end{bmatrix}$. To obtain $\mathbf{u}_3$ and $\mathbf{u}_4$, we determine an

orthonormal basis for $\text{null}\left(B^T\right) = \text{null}\left(\begin{bmatrix} 2 & 2 & 1 & 0 \\ 1 & -1 & 0 & 1 \end{bmatrix}\right) = \text{span}\left\{\begin{bmatrix} -\frac{1}{6}\sqrt{2} \\ -\frac{1}{6}\sqrt{2} \\ \frac{2}{3}\sqrt{2} \\ 0 \end{bmatrix}, \begin{bmatrix} -\frac{1}{6}\sqrt{6} \\ \frac{1}{6}\sqrt{6} \\ 0 \\ \frac{1}{3}\sqrt{6} \end{bmatrix}\right\}$, so

$\mathbf{u}_3 = \begin{bmatrix} -\frac{1}{6}\sqrt{2} \\ -\frac{1}{6}\sqrt{2} \\ \frac{2}{3}\sqrt{2} \\ 0 \end{bmatrix}$ and $\mathbf{u}_4 = \begin{bmatrix} -\frac{1}{6}\sqrt{6} \\ \frac{1}{6}\sqrt{6} \\ 0 \\ \frac{1}{3}\sqrt{6} \end{bmatrix}$. Thus $U = \begin{bmatrix} \frac{2}{3} & \frac{1}{3}\sqrt{3} & \frac{1}{3}\sqrt{3} & -\frac{1}{6}\sqrt{6} \\ \frac{2}{3} & -\frac{1}{3}\sqrt{3} & -\frac{1}{3}\sqrt{3} & \frac{1}{6}\sqrt{6} \\ \frac{1}{3} & 0 & 0 & 0 \\ 0 & \frac{1}{3}\sqrt{3} & \frac{1}{3}\sqrt{3} & \frac{1}{3}\sqrt{6} \end{bmatrix}$. The singular

value decomposition for A is $V\Sigma^T U^T$.

17. $b = \sigma_1 \cdot \varepsilon \cdot n = 10 \cdot 10^{-9} \cdot 3 = 3.0 \times 10^{-8}$. Since $\sigma_2 = 6 \geq b$, but $\sigma_3 = 10^{-8} < b$, it follows that the numerical rank of A is 2.

19. $b = \sigma_1 \cdot \varepsilon \cdot n = 12 \cdot 10^{-7} \cdot 4 = 4.8 \times 10^{-6}$. Since $\sigma_2 = 4 \geq b$, but $\sigma_3 = 10^{-6} < b$, it follows that the numerical rank of A is 2.

21. False. Every matrix A has a singular value decomposition, by Theorem 8.26.

23. True. Since A is invertible, each eigenvalue is nonzero, thus each singular value is nonzero. If $A\mathbf{x} = \lambda\mathbf{x} = \sigma^2\mathbf{x}$, then $\frac{1}{\sigma^2}\mathbf{x} = \frac{1}{\lambda}\mathbf{x} = A^{-1}\mathbf{x}$, and thus A^{-1} has singular value σ^{-1}.

25. True, since $\left|\det\left(A\right)\right| = \left|\det\left(U\Sigma V^T\right)\right| = \left|\det\left(U\right)\det\left(\Sigma\right)\det\left(V^T\right)\right| =$ $\left|\det\left(U\right)\right|\left|\det\left(\Sigma\right)\right|\left|\det\left(V\right)\right| = (1)\left|\sigma_1 \cdot \cdots \cdot \sigma_n\right|(1) = \left|\sigma_1 \cdot \cdots \cdot \sigma_n\right|$.

27. $\sigma_1\mathbf{u}_1\mathbf{v}_1^T = 3.199\begin{bmatrix} -0.1606 \\ -0.9845 \\ 0.07183 \end{bmatrix}\begin{bmatrix} 0.2298 \\ -0.9732 \end{bmatrix}^T = \begin{bmatrix} -0.1181 & 0.5000 \\ -0.7237 & 3.065 \\ 0.0528 & -0.2236 \end{bmatrix}$, and

$\sigma_1\mathbf{u}_1\mathbf{v}_1^T + \sigma_2\mathbf{u}_2\mathbf{v}_2^T = \begin{bmatrix} -0.1181 & 0.5000 \\ -0.7237 & 3.065 \\ 0.0528 & -0.2236 \end{bmatrix} + 2.401\begin{bmatrix} 0.9064 \\ -0.1182 \\ 0.4053 \end{bmatrix}\begin{bmatrix} 0.9732 \\ 0.2298 \end{bmatrix}^T =$

$\begin{bmatrix} 2.000 & 1.0 \\ -0.9999 & 3.000 \\ 0.9998 & 2.419 \times 10^{-5} \end{bmatrix}$. We see that $\sigma_1\mathbf{u}_1\mathbf{v}_1^T$ is not approximately A, but $\sigma_1\mathbf{u}_1\mathbf{v}_1^T + \sigma_2\mathbf{u}_2\mathbf{v}_2^T \approx A$.

29. In Exercise 15, we determined the singular value decomposition of A by determining the singular value decomposition of $B = A^T$. Hence the vectors $\mathbf{u}_i$ are the columns of V determined in Exercise 15, likewise the vectors $\mathbf{v}_i$ are the columns of U. Thus, $\sigma_1 \mathbf{u}_1 \mathbf{v}_1^T = 3 \begin{bmatrix} 1 \\ 0 \end{bmatrix} \begin{bmatrix} \frac{2}{3} \\ \frac{2}{3} \\ \frac{1}{3} \\ 0 \end{bmatrix}^T = \begin{bmatrix} 2 & 2 & 1 & 0 \\ 0 & 0 & 0 & 0 \end{bmatrix}$,

and $\sigma_1 \mathbf{u}_1 \mathbf{v}_1^T + \sigma_2 \mathbf{u}_2 \mathbf{v}_2^T = \begin{bmatrix} 2 & 2 & 1 & 0 \\ 0 & 0 & 0 & 0 \end{bmatrix} + \sqrt{3} \begin{bmatrix} 0 \\ 1 \end{bmatrix} \begin{bmatrix} \frac{1}{3}\sqrt{3} \\ -\frac{1}{3}\sqrt{3} \\ 0 \\ \frac{1}{3}\sqrt{3} \end{bmatrix}^T = \begin{bmatrix} 2 & 2 & 1 & 0 \\ 1 & -1 & 0 & 1 \end{bmatrix}$. We see that

$\sigma_1 \mathbf{u}_1 \mathbf{v}_1^T$ is not approximately A, but $\sigma_1 \mathbf{u}_1 \mathbf{v}_1^T + \sigma_2 \mathbf{u}_2 \mathbf{v}_2^T = A$.

31. Let $A = U\Sigma V^T$ be the singular value decomposition of A. Then $A^T = \left(U\Sigma V^T\right)^T = V\Sigma^T U^T$ is the singular value decomposition of A^T. Since the singular values along the diagonal of Σ and Σ^T are the same, it follows that A and A^T have the same singular values.

33. If A is an $n \times n$ invertible matrix with singular value decomposition $A = U\Sigma V^T$, then U and V are both square $n \times n$ orthogonal matrices, with $U^{-1} = U^T$ and $V^{-1} = V^T$. Thus, $A^{-1} = \left(U\Sigma V^T\right)^{-1} = \left(V^T\right)^{-1} \left(\Sigma\right)^{-1} \left(U\right)^{-1} = V\Sigma^{-1}U^T$. Also, if $\sigma_1 \geq \cdots \geq \sigma_n > 0$ are the singular values of A ($\sigma_n \neq 0$ since A is invertible), then

$$\Sigma^{-1} = \begin{bmatrix} \sigma_1 & 0 & \cdots & 0 \\ 0 & \sigma_2 & \cdots & 0 \\ \vdots & \vdots & \ddots & \vdots \\ 0 & \cdots & \cdots & \sigma_n \end{bmatrix}^{-1} = \begin{bmatrix} \sigma_1^{-1} & 0 & \cdots & 0 \\ 0 & \sigma_2^{-1} & \cdots & 0 \\ \vdots & \vdots & \ddots & \vdots \\ 0 & \cdots & \cdots & \sigma_n^{-1} \end{bmatrix}$$

Note that this is not a singular value decomposition for A^{-1}, as the singular values σ_i^{-1} for A^{-1} in Σ^{-1} are ascending. To rectify this, we reverse the order of the vectors; if $V = [\ \mathbf{v}_1 \ \cdots \ \mathbf{v}_n\ ]$, and $U = [\ \mathbf{u}_1 \ \cdots \ \mathbf{u}_n\ ]$, then let $\widetilde{V} = [\ \mathbf{v}_n \ \cdots \ \mathbf{v}_1\ ]$, and $\widetilde{U} = [\ \mathbf{u}_n \ \cdots \ \mathbf{u}_1\ ]$. Finally, set

$$\widetilde{\Sigma} = \begin{bmatrix} \sigma_n^{-1} & 0 & \cdots & 0 \\ 0 & \sigma_{n-1}^{-1} & \cdots & 0 \\ \vdots & \vdots & \ddots & \vdots \\ 0 & \cdots & \cdots & \sigma_1^{-1} \end{bmatrix}$$

Then $A^{-1} = \widetilde{V}\widetilde{\Sigma}\widetilde{U}^T$ is the desired singular value decomposition for the inverse of A. (One can achieve this directly using the matrix $P = \begin{bmatrix} 0 & 0 & \cdots & 0 & 1 \\ 0 & 0 & \cdots & 1 & 0 \\ \vdots & \vdots & \diagup & \vdots & \vdots \\ 0 & 1 & \cdots & 0 & 0 \\ 1 & 0 & \cdots & 0 & 0 \end{bmatrix}$. Then $A^{-1} = V\Sigma^{-1}U^T$

$= V\left(PP^{-1}\right)\Sigma^{-1}\left(PP^{-1}\right)U^T = (VP)\left(P^{-1}\Sigma^{-1}P\right)\left(P^{-1}U^T\right) = \widetilde{V}\widetilde{\Sigma}\widetilde{U}^T$.)

35. P is orthogonal, so $P^T P = I_n$. The singular values of PA are the eigenvalues of $(PA)^T (PA) = A^T \left(P^T P\right) A = A^T I_n A = A^T A$. Since the singular values of A are also the square roots of the eigenvalues of $A^T A$, PA and A have the same singular values.

37. (a) If $A\mathbf{x} = \mathbf{0}$, then $\left(A^T A\right)\mathbf{x} = A^T (A\mathbf{x}) = A^T \mathbf{0} = \mathbf{0}$.

   (b) Suppose $\mathbf{x}$ satisfies $\left(A^T A\right)\mathbf{x} = \mathbf{0}$, then $A^T (A\mathbf{x}) = \mathbf{0}$. Thus $A\mathbf{x}$ is orthogonal to every row of A^T, and hence $A\mathbf{x}$ is orthogonal to every column of A. But this now implies that $A\mathbf{x}$ is in $(\mathrm{col}(A))^\perp$, since $\mathrm{col}(A)$ is spanned by the columns of A. Writing $A = [\ \mathbf{a}_1\ \cdots\ \mathbf{a}_n\ ]$, we have $A\mathbf{x} = [\ \mathbf{a}_1\ \cdots\ \mathbf{a}_n\ ] \begin{bmatrix} x_1 \\ \vdots \\ x_n \end{bmatrix} = x_1\mathbf{a}_1 + \cdots + x_n\mathbf{a}_n$, and therefore $A\mathbf{x}$ is in $\mathrm{col}(A)$. So $A\mathbf{x} \in (\mathrm{col}(A))^\perp \cap \mathrm{col}(A) = \{\mathbf{0}\}$, and thus $A\mathbf{x} = \mathbf{0}$.

(c) From part (a), $\mathrm{null}(A) \subset \mathrm{null}(A^T A)$, and from part (b) $\mathrm{null}(A^T A) \subset \mathrm{null}(A)$. Thus $\mathrm{nullity}(A) = \mathrm{nullity}(A^T A)$. Therefore, by the Rank-Nullity Theorem, and using that A and $A^T A$ both have n columns, we have $\mathrm{rank}(A^T A) = n - \mathrm{nullity}(A^T A) = n - \mathrm{nullity}(A) = \mathrm{rank}(A)$.

39. $A^T A = \begin{bmatrix} 3 & 5 \\ -1 & 2 \end{bmatrix}^T \begin{bmatrix} 3 & 5 \\ -1 & 2 \end{bmatrix} = \begin{bmatrix} 10 & 13 \\ 13 & 29 \end{bmatrix}$. We determine $\lambda_1 \approx 35.601$, $\mathbf{v}_1 \approx \begin{bmatrix} 0.4527 \\ 0.8916 \end{bmatrix}$, $\lambda_2 \approx 3.3988$, $\mathbf{v}_2 \approx \begin{bmatrix} 0.8916 \\ -0.4528 \end{bmatrix}$. Thus $V \approx \begin{bmatrix} 0.4527 & 0.8916 \\ 0.8916 & -0.4528 \end{bmatrix}$. The singular values are $\sigma_1 = \sqrt{\lambda_1} \approx \sqrt{35.601} \approx 5.9667$ and $\sigma_2 = \sqrt{\lambda_2} \approx \sqrt{3.3988} \approx 1.8436$. Thus $\Sigma \approx \begin{bmatrix} 5.9667 & 0 \\ 0 & 1.8436 \end{bmatrix}$. Also, $\mathbf{u}_1 = \frac{1}{\sigma_1} A\mathbf{v}_1 \approx \frac{1}{5.9667} \begin{bmatrix} 3 & 5 \\ -1 & 2 \end{bmatrix} \begin{bmatrix} 0.4527 \\ 0.8916 \end{bmatrix} \approx \begin{bmatrix} 0.9748 \\ 0.2230 \end{bmatrix}$, and $\mathbf{u}_2 = \frac{1}{\sigma_2} A\mathbf{v}_2 \approx \frac{1}{1.8436} \begin{bmatrix} 3 & 5 \\ -1 & 2 \end{bmatrix} \begin{bmatrix} 0.8916 \\ -0.4528 \end{bmatrix} \approx \begin{bmatrix} 0.2228 \\ -0.9748 \end{bmatrix}$.
Thus $U \approx \begin{bmatrix} 0.9748 & 0.2228 \\ 0.2230 & -0.9748 \end{bmatrix}$.

41. $A^T A = \begin{bmatrix} -5 & 0 & 2 \\ 1 & -1 & 3 \\ 0 & 4 & 2 \end{bmatrix}^T \begin{bmatrix} -5 & 0 & 2 \\ 1 & -1 & 3 \\ 0 & 4 & 2 \end{bmatrix} = \begin{bmatrix} 26 & -1 & -7 \\ -1 & 17 & 5 \\ -7 & 5 & 17 \end{bmatrix}$. We determine $\lambda_1 \approx 30.660$, $\mathbf{v}_1 \approx \begin{bmatrix} 0.8224 \\ -0.2477 \\ -0.5121 \end{bmatrix}$, $\lambda_2 \approx 18.766$, $\mathbf{v}_2 \approx \begin{bmatrix} 0.4739 \\ 0.7963 \\ 0.3760 \end{bmatrix}$, and $\lambda_3 \approx 10.574$, $\mathbf{v}_3 \approx \begin{bmatrix} 0.3147 \\ -0.5519 \\ 0.7722 \end{bmatrix}$. Thus $V \approx \begin{bmatrix} 0.8224 & 0.4739 & 0.3147 \\ -0.2477 & 0.7963 & -0.5519 \\ -0.5121 & 0.3760 & 0.7722 \end{bmatrix}$. The singular values are $\sigma_1 = \sqrt{\lambda_1} \approx \sqrt{30.660} \approx 5.5371$, $\sigma_2 = \sqrt{\lambda_2} \approx \sqrt{18.766} \approx 4.3320$, and $\sigma_3 = \sqrt{\lambda_3} \approx \sqrt{10.574} \approx 3.2518$. Thus $\Sigma \approx \begin{bmatrix} 5.5371 & 0 & 0 \\ 0 & 4.3320 & 0 \\ 0 & 0 & 3.2518 \end{bmatrix}$.
Also, $\mathbf{u}_1 = \frac{1}{\sigma_1} A\mathbf{v}_1 \approx \frac{1}{5.5371} \begin{bmatrix} -5 & 0 & 2 \\ 1 & -1 & 3 \\ 0 & 4 & 2 \end{bmatrix} \begin{bmatrix} 0.8224 \\ -0.2477 \\ -0.5121 \end{bmatrix} \approx \begin{bmatrix} -0.9276 \\ -0.08420 \\ -0.3639 \end{bmatrix}$,
$\mathbf{u}_2 = \frac{1}{\sigma_2} A\mathbf{v}_2 \approx \frac{1}{4.3320} \begin{bmatrix} -5 & 0 & 2 \\ 1 & -1 & 3 \\ 0 & 4 & 2 \end{bmatrix} \begin{bmatrix} 0.4739 \\ 0.7963 \\ 0.3760 \end{bmatrix} \approx \begin{bmatrix} -0.3734 \\ 0.1860 \\ 0.9089 \end{bmatrix}$, and
$\mathbf{u}_3 = \frac{1}{\sigma_3} A\mathbf{v}_3 \approx \frac{1}{3.2518} \begin{bmatrix} -5 & 0 & 2 \\ 1 & -1 & 3 \\ 0 & 4 & 2 \end{bmatrix} \begin{bmatrix} 0.3147 \\ -0.5519 \\ 0.7722 \end{bmatrix} \approx \begin{bmatrix} -0.008949 \\ 0.9789 \\ -0.2039 \end{bmatrix}$.
Thus $U \approx \begin{bmatrix} -0.9276 & -0.3734 & -0.008949 \\ -0.08420 & 0.1860 & 0.9789 \\ -0.3639 & 0.9089 & -0.2039 \end{bmatrix}$.

8.5   Least Squares Regression

1. An orthonormal basis for S is given by $\mathbf{v} = \begin{bmatrix} 1 \\ -1 \end{bmatrix} / \left\| \begin{bmatrix} 1 \\ -1 \end{bmatrix} \right\| = \begin{bmatrix} \frac{1}{2}\sqrt{2} \\ -\frac{1}{2}\sqrt{2} \end{bmatrix}$. The vector in S closest

   to $\mathbf{y}$ is $\text{proj}_S \mathbf{y} = (\mathbf{v} \cdot \mathbf{y})\,\mathbf{v} = \left(\begin{bmatrix} \frac{1}{2}\sqrt{2} \\ -\frac{1}{2}\sqrt{2} \end{bmatrix} \cdot \begin{bmatrix} 1 \\ 2 \end{bmatrix} \right) \begin{bmatrix} \frac{1}{2}\sqrt{2} \\ -\frac{1}{2}\sqrt{2} \end{bmatrix} = \begin{bmatrix} -\frac{1}{2} \\ \frac{1}{2} \end{bmatrix}$.

3. An orthonormal basis for S is given by $\mathbf{v}_1 = \begin{bmatrix} 1 \\ 3 \\ -2 \end{bmatrix} / \left\| \begin{bmatrix} 1 \\ 3 \\ -2 \end{bmatrix} \right\| = \begin{bmatrix} \frac{1}{14}\sqrt{14} \\ \frac{3}{14}\sqrt{14} \\ -\frac{1}{7}\sqrt{14} \end{bmatrix}$ and

   $\mathbf{v}_2 = \begin{bmatrix} 5 \\ -1 \\ 1 \end{bmatrix} / \left\| \begin{bmatrix} 5 \\ -1 \\ 1 \end{bmatrix} \right\| = \begin{bmatrix} \frac{5}{9}\sqrt{3} \\ -\frac{1}{9}\sqrt{3} \\ \frac{1}{9}\sqrt{3} \end{bmatrix}$.

   The vector in S closest to $\mathbf{y}$ is $\text{proj}_S \mathbf{y} = (\mathbf{v}_1 \cdot \mathbf{y})\,\mathbf{v}_1 + (\mathbf{v}_2 \cdot \mathbf{y})\,\mathbf{v}_2 =$

   $\left(\begin{bmatrix} \frac{1}{14}\sqrt{14} \\ \frac{3}{14}\sqrt{14} \\ -\frac{1}{7}\sqrt{14} \end{bmatrix} \cdot \begin{bmatrix} 1 \\ -1 \\ 2 \end{bmatrix} \right) \begin{bmatrix} \frac{1}{14}\sqrt{14} \\ \frac{3}{14}\sqrt{14} \\ -\frac{1}{7}\sqrt{14} \end{bmatrix} + \left(\begin{bmatrix} \frac{5}{9}\sqrt{3} \\ -\frac{1}{9}\sqrt{3} \\ \frac{1}{9}\sqrt{3} \end{bmatrix} \cdot \begin{bmatrix} 1 \\ -1 \\ 2 \end{bmatrix} \right) \begin{bmatrix} \frac{5}{9}\sqrt{3} \\ -\frac{1}{9}\sqrt{3} \\ \frac{1}{9}\sqrt{3} \end{bmatrix} = \begin{bmatrix} \frac{199}{189} \\ -\frac{299}{189} \\ \frac{218}{189} \end{bmatrix}$.

5. $A = \begin{bmatrix} 2 & -1 \\ 1 & 2 \\ 3 & -1 \end{bmatrix}$, $\mathbf{x} = \begin{bmatrix} x_1 \\ x_2 \end{bmatrix}$, $\mathbf{y} = \begin{bmatrix} 4 \\ 3 \\ 4 \end{bmatrix}$. We calculate

   $A^T A = \begin{bmatrix} 2 & -1 \\ 1 & 2 \\ 3 & -1 \end{bmatrix}^T \begin{bmatrix} 2 & -1 \\ 1 & 2 \\ 3 & -1 \end{bmatrix} = \begin{bmatrix} 14 & -3 \\ -3 & 6 \end{bmatrix}$ and

   $A^T \mathbf{y} = \begin{bmatrix} 2 & -1 \\ 1 & 2 \\ 3 & -1 \end{bmatrix}^T \begin{bmatrix} 4 \\ 3 \\ 4 \end{bmatrix} = \begin{bmatrix} 23 \\ -2 \end{bmatrix}$. The normal equations, $A^T A\mathbf{x} = A^T \mathbf{y}$, are thus given by

   $\begin{bmatrix} 14 & -3 \\ -3 & 6 \end{bmatrix} \begin{bmatrix} x_1 \\ x_2 \end{bmatrix} = \begin{bmatrix} 23 \\ -2 \end{bmatrix}$, or

   $$14x_1 - 3x_2 = 23$$
   $$-3x_1 + 6x_2 = -2$$

7. $A = \begin{bmatrix} 1 & 1 & -1 \\ 2 & -1 & 2 \\ 1 & 4 & -5 \end{bmatrix}$, $\mathbf{x} = \begin{bmatrix} x_1 \\ x_2 \\ x_3 \end{bmatrix}$, $\mathbf{y} = \begin{bmatrix} 2 \\ -1 \\ 6 \end{bmatrix}$. We calculate

   $A^T A = \begin{bmatrix} 1 & 1 & -1 \\ 2 & -1 & 2 \\ 1 & 4 & -5 \end{bmatrix}^T \begin{bmatrix} 1 & 1 & -1 \\ 2 & -1 & 2 \\ 1 & 4 & -5 \end{bmatrix} = \begin{bmatrix} 6 & 3 & -2 \\ 3 & 18 & -23 \\ -2 & -23 & 30 \end{bmatrix}$ and

   $A^T \mathbf{y} = \begin{bmatrix} 1 & 1 & -1 \\ 2 & -1 & 2 \\ 1 & 4 & -5 \end{bmatrix}^T \begin{bmatrix} 2 \\ -1 \\ 6 \end{bmatrix} = \begin{bmatrix} 6 \\ 27 \\ -34 \end{bmatrix}$. The normal equations, $A^T A\mathbf{x} = A^T \mathbf{y}$, are thus given

   by

   $\begin{bmatrix} 6 & 3 & -2 \\ 3 & 18 & -23 \\ -2 & -23 & 30 \end{bmatrix} \begin{bmatrix} x_1 \\ x_2 \\ x_3 \end{bmatrix} = \begin{bmatrix} 6 \\ 27 \\ -34 \end{bmatrix}$, or

   $$6x_1 + 3x_2 - 2x_3 = 6$$
   $$3x_1 + 18x_2 - 23x_3 = 27$$
   $$-2x_1 - 23x_2 + 30x_3 = -34$$

9. We solve the normal equations, $A^T A\mathbf{x} = A^T \mathbf{y}$. $A = \begin{bmatrix} -1 & -1 \\ 2 & 3 \\ -3 & 2 \end{bmatrix}$, $\mathbf{x} = \begin{bmatrix} x_1 \\ x_2 \end{bmatrix}$, $\mathbf{y} = \begin{bmatrix} 3 \\ -1 \\ 2 \end{bmatrix}$. We

calculate $A^T A = \begin{bmatrix} -1 & -1 \\ 2 & 3 \\ -3 & 2 \end{bmatrix}^T \begin{bmatrix} -1 & -1 \\ 2 & 3 \\ -3 & 2 \end{bmatrix} = \begin{bmatrix} 14 & 1 \\ 1 & 14 \end{bmatrix}$ and $A^T \mathbf{y} = \begin{bmatrix} -1 & -1 \\ 2 & 3 \\ -3 & 2 \end{bmatrix}^T \begin{bmatrix} 3 \\ -1 \\ 2 \end{bmatrix} =$

$\begin{bmatrix} -11 \\ -2 \end{bmatrix}$. The normal equations are $\begin{bmatrix} 14 & 1 \\ 1 & 14 \end{bmatrix} \begin{bmatrix} x_1 \\ x_2 \end{bmatrix} = \begin{bmatrix} -11 \\ -2 \end{bmatrix}$. Row-reduce the corresponding augmented matrix

$$\begin{bmatrix} 14 & 1 & -11 \\ 1 & 14 & -2 \end{bmatrix} \sim \begin{bmatrix} 1 & 14 & -2 \\ 0 & -195 & 17 \end{bmatrix}$$

to obtain the least squares solution $x_1 = -\frac{152}{195}$ and $x_2 = -\frac{17}{195}$.

11. We solve the normal equations, $A^T A\mathbf{x} = A^T \mathbf{y}$. $A = \begin{bmatrix} 2 & -1 & -1 \\ -1 & 1 & 3 \\ 3 & -2 & -4 \end{bmatrix}$, $\mathbf{x} = \begin{bmatrix} x_1 \\ x_2 \\ x_3 \end{bmatrix}$, $\mathbf{y} = \begin{bmatrix} 1 \\ -1 \\ 3 \end{bmatrix}$. We

calculate $A^T A = \begin{bmatrix} 2 & -1 & -1 \\ -1 & 1 & 3 \\ 3 & -2 & -4 \end{bmatrix}^T \begin{bmatrix} 2 & -1 & -1 \\ -1 & 1 & 3 \\ 3 & -2 & -4 \end{bmatrix} =$

$\begin{bmatrix} 14 & -9 & -17 \\ -9 & 6 & 12 \\ -17 & 12 & 26 \end{bmatrix}$ and $A^T \mathbf{y} = \begin{bmatrix} 2 & -1 & -1 \\ -1 & 1 & 3 \\ 3 & -2 & -4 \end{bmatrix}^T \begin{bmatrix} 1 \\ -1 \\ 3 \end{bmatrix} = \begin{bmatrix} 12 \\ -8 \\ -16 \end{bmatrix}$. The normal equations are

$\begin{bmatrix} 14 & -9 & -17 \\ -9 & 6 & 12 \\ -17 & 12 & 26 \end{bmatrix} \begin{bmatrix} x_1 \\ x_2 \\ x_3 \end{bmatrix} = \begin{bmatrix} 12 \\ -8 \\ -16 \end{bmatrix}$. Row-reduce the corresponding augmented matrix

$$\begin{bmatrix} 14 & -9 & -17 & 12 \\ -9 & 6 & 12 & -8 \\ -17 & 12 & 26 & -16 \end{bmatrix} \sim \begin{bmatrix} -9 & 6 & 12 & -8 \\ 0 & \frac{1}{3} & \frac{5}{3} & -\frac{4}{9} \\ 0 & 0 & 0 & 0 \end{bmatrix}$$

We let $x_3 = t$ to obtain the least squares solutions $x_1 = -2t$, $x_2 = -5t - \frac{4}{3}$, and $x_3 = t$.

13. We consider a parabola of the form $f(x) = c_1 + c_2 x + c_3 x^2$. We evaluate at the given points

$$\begin{array}{lll} (0,1) \implies & c_1 & = 1 \\ (2,5) \implies & c_1 + 2c_2 + 4c_3 & = 5 \end{array}$$

This is equivalent to $A\mathbf{c} = \mathbf{y}$, where $A = \begin{bmatrix} 1 & 0 & 0 \\ 1 & 2 & 4 \end{bmatrix}$, $\mathbf{c} = \begin{bmatrix} c_1 \\ c_2 \\ c_3 \end{bmatrix}$, and $\mathbf{y} = \begin{bmatrix} 1 \\ 5 \end{bmatrix}$. We solve the

normal equations, $A^T A\mathbf{c} = A^T \mathbf{y}$; $A^T A = \begin{bmatrix} 1 & 0 & 0 \\ 1 & 2 & 4 \end{bmatrix}^T \begin{bmatrix} 1 & 0 & 0 \\ 1 & 2 & 4 \end{bmatrix} = \begin{bmatrix} 2 & 2 & 4 \\ 2 & 4 & 8 \\ 4 & 8 & 16 \end{bmatrix}$ and $A^T \mathbf{y} =$

$\begin{bmatrix} 1 & 0 & 0 \\ 1 & 2 & 4 \end{bmatrix}^T \begin{bmatrix} 1 \\ 5 \end{bmatrix} = \begin{bmatrix} 6 \\ 10 \\ 20 \end{bmatrix}$. The normal equations are $\begin{bmatrix} 2 & 2 & 4 \\ 2 & 4 & 8 \\ 4 & 8 & 16 \end{bmatrix} \begin{bmatrix} c_1 \\ c_2 \\ c_3 \end{bmatrix} = \begin{bmatrix} 6 \\ 10 \\ 20 \end{bmatrix}$. Row-

reduce the corresponding augmented matrix

$$\begin{bmatrix} 2 & 2 & 4 & 6 \\ 2 & 4 & 8 & 10 \\ 4 & 8 & 16 & 20 \end{bmatrix} \sim \begin{bmatrix} 2 & 2 & 4 & 6 \\ 0 & 2 & 4 & 4 \\ 0 & 0 & 0 & 0 \end{bmatrix}$$

We let $c_3 = t$ to obtain the least squares solutions $c_1 = 1$, $c_2 = 2 - 2t$, and $c_3 = t$. We have obtained infinitely many solutions since there are infinitely many parabolas that pass through two given points.

15. For example,

$$\begin{array}{rcl} x_1 & & & = & 0 \\ & x_2 & & = & 0 \\ x_1 & + & x_2 & = & 1 \end{array}$$

has no solution and a unique least squares solution $x_1 = \frac{1}{3}$, $x_2 = \frac{1}{3}$.

17. For example,

$$\begin{array}{rcl} x_1 & + & x_2 & = & 0 \\ 2x_1 & + & 2x_2 & = & 0 \\ 3x_1 & + & 3x_2 & = & 0 \\ 4x_1 & + & 4x_2 & = & 1 \end{array}$$

has no solution and infinitely many least squares solutions, since the columns of $A = \begin{bmatrix} 1 & 1 \\ 2 & 2 \\ 3 & 3 \\ 4 & 4 \end{bmatrix}$ are

linearly dependent.

19. For example,

$$\begin{array}{rcl} x_1 & & & = & 0 \\ & x_2 & & = & 0 \\ & & x_3 & = & 0 \end{array}$$

has a unique solution and a unique least squares solution.

21. False. A least squares solution can be found for any system. (See the solution to Exercise 18 for a counter-example.)

23. True. The least squares solutions $\mathbf{x}$ satisfy $A^T A\mathbf{x} = A\mathbf{y}$, and since A is $n \times m$, $\mathbf{x}$ must be in $\mathbf{R}^m$.

25. True, by Definition 8.28.

27. False. Since $A^T A\mathbf{x}_1 = A^T\mathbf{y}$ and $A^T A\mathbf{x}_2 = A^T\mathbf{y}$, $A^T A(\mathbf{x}_1 + \mathbf{x}_2) = A^T\mathbf{y} + A^T\mathbf{y} = 2A^T\mathbf{y} \neq A^T\mathbf{y}$, unless $A^T\mathbf{y} = \mathbf{0}$. For example, consider $x_1 + x_2 = 1$, then $A^T\mathbf{y} = \begin{bmatrix} 1 & 1 \end{bmatrix}^T [1] = \begin{bmatrix} 1 \\ 1 \end{bmatrix} \neq \mathbf{0}$.

29. If A is an $n \times m$ matrix with orthogonal columns, then $A^T A = I_m$, and the normal equations $A^T A\mathbf{x} = I_m\mathbf{x} = \mathbf{x} = A^T\mathbf{y}$ has the unique solution $\mathbf{x} = A^T\mathbf{y}$.

31. Since A is orthogonal, A is an $n \times n$ invertible matrix, with $A^{-1} = A^T$. Also, the least squares solution of $A\mathbf{x} = \mathbf{y}$ is $\mathbf{x} = A^{-1}\mathbf{y} = A^T\mathbf{y}$. Hence $\mathbf{x}$ is a linear combination of the columns of A^T, and thus a linear combination of the rows of A. So any least squares solution (there's only one) of $A\mathbf{x} = \mathbf{y}$ is a linear combination of the rows of A.

33. By Exercise 37 of Section 8.4, $\operatorname{rank}(A^T A) = \operatorname{rank}(A)$. Suppose A is an $n \times m$ matrix with linearly independent columns. Then $\operatorname{rank}(A) = m$, and thus $\operatorname{rank}(A^T A) = m$. Since $A^T A$ is an $m \times m$ matrix with rank m, we conclude that $A^T A$ is invertible. Now suppose $A^T A$ is invertible. Thus we have $\operatorname{rank}(A^T A) = m$. Hence $\operatorname{rank}(A) = m$, and A has linearly independent columns.

35. We consider a line of the form $y = c_1 + c_2 x$. We evaluate at the given points

$$\begin{array}{rcl} (-2, 1.3) & \Longrightarrow & c_1 & - & 2c_2 & = & 1.3 \\ (0, 1.8) & \Longrightarrow & c_1 & & & = & 1.8 \\ (1, 3) & \Longrightarrow & c_1 & + & c_2 & = & 3 \end{array}$$

This is equivalent to $A\mathbf{c} = \mathbf{y}$, where $A = \begin{bmatrix} 1 & -2 \\ 1 & 0 \\ 1 & 1 \end{bmatrix}$, $\mathbf{c} = \begin{bmatrix} c_1 \\ c_2 \end{bmatrix}$, and $\mathbf{y} = \begin{bmatrix} 1.3 \\ 1.8 \\ 3 \end{bmatrix}$. Since A

has linearly independent columns, the least squares solution of $A\mathbf{c} = \mathbf{y}$ is $\mathbf{c} = (A^T A)^{-1} A^T\mathbf{y} =$

$\begin{bmatrix} 3 & -1 \\ -1 & 5 \end{bmatrix}^{-1} \begin{bmatrix} 1 & 1 & 1 \\ -2 & 0 & 1 \end{bmatrix} \begin{bmatrix} 1.3 \\ 1.8 \\ 3 \end{bmatrix} \approx \begin{bmatrix} 2.2071 \\ 0.5214 \end{bmatrix}$. Thus the equation of the line that best fits the

data is $y \approx 2.2071 + 0.5214x$.

37. We consider a line of the form $y = c_1 + c_2 x$. We evaluate at the given points

$$\begin{array}{rcrcrcr} (-2, 2.0) & \implies & c_1 & - & 2c_2 & = & 2.0 \\ (-1, 1.7) & \implies & c_1 & - & c_2 & = & 1.7 \\ (1, 2.6) & \implies & c_1 & + & c_2 & = & 2.6 \\ (3, 2.1) & \implies & c_1 & + & 3c_2 & = & 2.1 \end{array}$$

This is equivalent to $A\mathbf{c} = \mathbf{y}$, where $A = \begin{bmatrix} 1 & -2 \\ 1 & -1 \\ 1 & 1 \\ 1 & 3 \end{bmatrix}$, $\mathbf{c} = \begin{bmatrix} c_1 \\ c_2 \end{bmatrix}$, and $\mathbf{y} = \begin{bmatrix} 2.0 \\ 1.7 \\ 2.6 \\ 2.1 \end{bmatrix}$. Since A

has linearly independent columns, the least squares solution of $A\mathbf{c} = \mathbf{y}$ is $\mathbf{c} = \left(A^T A\right)^{-1} A^T \mathbf{y} =$

$\begin{bmatrix} 4 & 1 \\ 1 & 15 \end{bmatrix}^{-1} \begin{bmatrix} 1 & 1 & 1 & 1 \\ -2 & -1 & 1 & 3 \end{bmatrix} \begin{bmatrix} 2.0 \\ 1.7 \\ 2.6 \\ 2.1 \end{bmatrix} \approx \begin{bmatrix} 2.081 \\ 0.07458 \end{bmatrix}$. Thus the equation of the line that best fits

the data is $y \approx 2.081 + 0.07458x$.

39. We consider a parabola of the form $y = c_1 + c_2 x + c_3 x^2$. We evaluate at the given points

$$\begin{array}{rcrcrcrcr} (-2, 3) & \implies & c_1 & - & 2c_2 & + & 4c_3 & = & 3 \\ (-1, 2) & \implies & c_1 & - & c_2 & + & c_3 & = & 2 \\ (1, 2.1) & \implies & c_1 & + & c_2 & + & c_3 & = & 2.1 \\ (2, 3.4) & \implies & c_1 & + & 2c_2 & + & 4c_3 & = & 3.4 \end{array}$$

This is equivalent to $A\mathbf{c} = \mathbf{y}$, where $A = \begin{bmatrix} 1 & -2 & 4 \\ 1 & -1 & 1 \\ 1 & 1 & 1 \\ 1 & 2 & 4 \end{bmatrix}$, $\mathbf{c} = \begin{bmatrix} c_1 \\ c_2 \\ c_3 \end{bmatrix}$, and $\mathbf{y} = \begin{bmatrix} 3 \\ 2 \\ 2.1 \\ 3.4 \end{bmatrix}$. Since

A has linearly independent columns, the least squares solution of $A\mathbf{c} = \mathbf{y}$ is $\mathbf{c} = \left(A^T A\right)^{-1} A^T \mathbf{y} =$

$\begin{bmatrix} 4 & 0 & 10 \\ 0 & 10 & 0 \\ 10 & 0 & 34 \end{bmatrix}^{-1} \begin{bmatrix} 1 & 1 & 1 & 1 \\ -2 & -1 & 1 & 2 \\ 4 & 1 & 1 & 4 \end{bmatrix} \begin{bmatrix} 3 \\ 2 \\ 2.1 \\ 3.4 \end{bmatrix} \approx \begin{bmatrix} 1.667 \\ 0.09 \\ 0.3833 \end{bmatrix}$. Thus the equation of the parabola

that best fits the data is $y \approx 1.667 + 0.09x + 0.3833x^2$.

41. We consider a parabola of the form $y = c_1 + c_2 x + c_3 x^2$. We evaluate at the given points

$$\begin{array}{rcrcrcrcr} (-2, 0) & \implies & c_1 & - & 2c_2 & + & 4c_3 & = & 0 \\ (-1, 1.5) & \implies & c_1 & - & c_2 & + & c_3 & = & 1.5 \\ (0, 2.5) & \implies & c_1 & & & & & = & 2.5 \\ (1, 1.3) & \implies & c_1 & + & c_2 & + & c_3 & = & 1.3 \\ (2, -0.2) & \implies & c_1 & + & 2c_2 & + & 4c_3 & = & -0.2 \end{array}$$

This is equivalent to $A\mathbf{c} = \mathbf{y}$, where $A = \begin{bmatrix} 1 & -2 & 4 \\ 1 & -1 & 1 \\ 1 & 0 & 0 \\ 1 & 1 & 1 \\ 1 & 2 & 4 \end{bmatrix}$, $\mathbf{c} = \begin{bmatrix} c_1 \\ c_2 \\ c_3 \end{bmatrix}$, and $\mathbf{y} = \begin{bmatrix} 0 \\ 1.5 \\ 2.5 \\ 1.3 \\ -0.2 \end{bmatrix}$. Since

A has linearly independent columns, the least squares solution of $A\mathbf{c} = \mathbf{y}$ is $\mathbf{c} = \left(A^T A\right)^{-1} A^T \mathbf{y} =$

$$\begin{bmatrix} 5 & 0 & 10 \\ 0 & 10 & 0 \\ 10 & 0 & 34 \end{bmatrix}^{-1} \begin{bmatrix} 1 & 1 & 1 & 1 & 1 \\ -2 & -1 & 0 & 1 & 2 \\ 4 & 1 & 0 & 1 & 4 \end{bmatrix} \begin{bmatrix} 0 \\ 1.5 \\ 2.5 \\ 1.3 \\ -0.2 \end{bmatrix} \approx \begin{bmatrix} 2.191 \\ -0.06 \\ -0.5857 \end{bmatrix}.$$ Thus the equation of the parabola

that best fits the data is $y \approx 2.191 - 0.06x - 0.5857x^2$.

43. Taking the natural logarithm we have $\ln(y) = \ln(a) + bx$. We evaluate at the given points

$$\begin{array}{rclcrcl}
(-1, 0.3) & \Longrightarrow & \ln(a) & - & b & = & \ln(0.3) \\
(0, 1.3) & \Longrightarrow & \ln(a) & & & = & \ln(1.3) \\
(1, 3.1) & \Longrightarrow & \ln(a) & + & b & = & \ln(3.1) \\
(2, 5.7) & \Longrightarrow & \ln(a) & + & 2b & = & \ln(5.7)
\end{array}$$

This is equivalent to $A\mathbf{x} = \mathbf{y}$, where $A = \begin{bmatrix} 1 & -1 \\ 1 & 0 \\ 1 & 1 \\ 1 & 2 \end{bmatrix}$, $\mathbf{x} = \begin{bmatrix} \ln(a) \\ b \end{bmatrix}$, and $\mathbf{y} = \begin{bmatrix} \ln(0.3) \\ \ln(1.3) \\ \ln(3.1) \\ \ln(5.7) \end{bmatrix}$. Since

A has linearly independent columns, the least squares solution of $A\mathbf{x} = \mathbf{y}$ is $\mathbf{x} = \left(A^T A\right)^{-1} A^T \mathbf{y} =$

$$\begin{bmatrix} 4 & 2 \\ 2 & 6 \end{bmatrix}^{-1} \begin{bmatrix} 1 & 1 & 1 & 1 \\ -1 & 0 & 1 & 2 \end{bmatrix} \begin{bmatrix} \ln(0.3) \\ \ln(1.3) \\ \ln(3.1) \\ \ln(5.7) \end{bmatrix} \approx \begin{bmatrix} -0.002553 \\ 0.9702 \end{bmatrix}.$$ Thus $\ln(a) \approx -0.002553 \quad \Rightarrow \quad a \approx$

0.9975 and $b \approx 0.9702$. Hence $y \approx 0.9975e^{0.9702x}$ best fits the data.

45. Taking the natural logarithm we have $\ln(y) = \ln(a) + bx$. We evaluate at the given points

$$\begin{array}{rclcrcl}
(2, 11.3) & \Longrightarrow & \ln(a) & + & 2b & = & \ln(11.3) \\
(4, 8.2) & \Longrightarrow & \ln(a) & + & 4b & = & \ln(8.2) \\
(5, 7.1) & \Longrightarrow & \ln(a) & + & 5b & = & \ln(7.1) \\
(7, 5.3) & \Longrightarrow & \ln(a) & + & 7b & = & \ln(5.3) \\
(10, 3.2) & \Longrightarrow & \ln(a) & + & 10b & = & \ln(3.2)
\end{array}$$

This is equivalent to $A\mathbf{x} = \mathbf{y}$, where $A = \begin{bmatrix} 1 & 2 \\ 1 & 4 \\ 1 & 5 \\ 1 & 7 \\ 1 & 10 \end{bmatrix}$, $\mathbf{x} = \begin{bmatrix} \ln(a) \\ b \end{bmatrix}$, and $\mathbf{y} = \begin{bmatrix} \ln(11.3) \\ \ln(8.2) \\ \ln(7.1) \\ \ln(5.3) \\ \ln(3.2) \end{bmatrix}$. Since

A has linearly independent columns, the least squares solution of $A\mathbf{x} = \mathbf{y}$ is $\mathbf{x} = \left(A^T A\right)^{-1} A^T \mathbf{y} =$

$$\begin{bmatrix} 5 & 28 \\ 28 & 194 \end{bmatrix}^{-1} \begin{bmatrix} 1 & 1 & 1 & 1 & 1 \\ 2 & 4 & 5 & 7 & 10 \end{bmatrix} \begin{bmatrix} \ln(11.3) \\ \ln(8.2) \\ \ln(7.1) \\ \ln(5.3) \\ \ln(3.2) \end{bmatrix} \approx \begin{bmatrix} 2.740 \\ -0.1564 \end{bmatrix}.$$ Thus $\ln(a) \approx 2.740 \quad \Rightarrow \quad a \approx$

15.49 and $b \approx -0.1564$. Hence $y \approx 15.49e^{-0.1564x}$ best fits the data.

47. Taking the natural logarithm we have $\ln(y) = \ln(a) + b\ln(x)$. We evaluate at the given points

$$\begin{array}{rclcrcl}
(2, 5.4) & \Longrightarrow & \ln(a) & + & \ln(2)b & = & \ln(5.4) \\
(4, 13.5) & \Longrightarrow & \ln(a) & + & \ln(4)b & = & \ln(13.5) \\
(5, 17.6) & \Longrightarrow & \ln(a) & + & \ln(5)b & = & \ln(17.6) \\
(7, 26.0) & \Longrightarrow & \ln(a) & + & \ln(7)b & = & \ln(26.0) \\
(9, 40.2) & \Longrightarrow & \ln(a) & + & \ln(9)b & = & \ln(40.2)
\end{array}$$

This is equivalent to $A\mathbf{x} = \mathbf{y}$, where $A = \begin{bmatrix} 1 & \ln(2) \\ 1 & \ln(4) \\ 1 & \ln(5) \\ 1 & \ln(7) \\ 1 & \ln(9) \end{bmatrix}$, $\mathbf{x} = \begin{bmatrix} \ln(a) \\ b \end{bmatrix}$, and $\mathbf{y} = \begin{bmatrix} \ln(5.4) \\ \ln(13.5) \\ \ln(17.6) \\ \ln(26.0) \\ \ln(40.2) \end{bmatrix}$. Since

A has linearly independent columns, the least squares solution of $A\mathbf{x} = \mathbf{y}$ is $\mathbf{x} = \left(A^T A\right)^{-1} A^T \mathbf{y} \approx$

$$\begin{bmatrix} 5.0 & 7.832 \\ 7.832 & 13.607 \end{bmatrix}^{-1} \begin{bmatrix} 1 & 1 & 1 & 1 & 1 \\ \ln 2 & \ln 4 & \ln 5 & \ln 7 & \ln 9 \end{bmatrix} \begin{bmatrix} \ln(5.4) \\ \ln(13.5) \\ \ln(17.6) \\ \ln(26.0) \\ \ln(40.2) \end{bmatrix}$$

$\approx \begin{bmatrix} 0.7761 \\ 1.306 \end{bmatrix}$. Thus $\ln(a) \approx 0.7761 \Rightarrow a \approx 2.173$ and $b \approx 1.306$. Hence $y \approx 2.173 x^{1.306}$ best fits the data.

49. Taking the natural logarithm we have $\ln(y) = \ln(a) + b\ln(x)$. We evaluate at the given points

$$\begin{array}{rclcrcrcl}
(2, 26.1) & \Longrightarrow & \ln(a) & + & \ln(2)\,b & = & \ln(26.1) \\
(3, 21.7) & \Longrightarrow & \ln(a) & + & \ln(3)\,b & = & \ln(21.7) \\
(5, 15.8) & \Longrightarrow & \ln(a) & + & \ln(5)\,b & = & \ln(15.8) \\
(7, 12.7) & \Longrightarrow & \ln(a) & + & \ln(7)\,b & = & \ln(12.7) \\
(10, 11.2) & \Longrightarrow & \ln(a) & + & \ln(10)\,b & = & \ln(11.2)
\end{array}$$

This is equivalent to $A\mathbf{x} = \mathbf{y}$, where

$$A = \begin{bmatrix} 1 & \ln(2) \\ 1 & \ln(3) \\ 1 & \ln(5) \\ 1 & \ln(7) \\ 1 & \ln(10) \end{bmatrix}, \quad \mathbf{x} = \begin{bmatrix} \ln(a) \\ b \end{bmatrix}, \text{ and } \mathbf{y} = \begin{bmatrix} \ln(26.1) \\ \ln(21.7) \\ \ln(15.8) \\ \ln(12.7) \\ \ln(11.2) \end{bmatrix}. \text{ Since } A \text{ has linearly independent columns,}$$

the least squares solution of $A\mathbf{x} = \mathbf{y}$ is

$$\mathbf{x} = \left(A^T A\right)^{-1} A^T \mathbf{y} \approx \begin{bmatrix} 5.0 & 7.650 \\ 7.650 & 13.37 \end{bmatrix}^{-1} \begin{bmatrix} 1 & 1 & 1 & 1 & 1 \\ \ln 2 & \ln 3 & \ln 5 & \ln 7 & \ln 10 \end{bmatrix} \begin{bmatrix} \ln(26.1) \\ \ln(21.7) \\ \ln(15.8) \\ \ln(12.7) \\ \ln(11.2) \end{bmatrix}$$

$\approx \begin{bmatrix} 3.651 \\ -0.5491 \end{bmatrix}$. Thus $\ln(a) \approx 3.651 \Rightarrow a \approx 38.51$ and $b \approx -0.5491$. Hence $y \approx 38.51 x^{-0.5491}$ best fits the data.

51. Using the data in Table 4, we have $A = \begin{bmatrix} 1 & 4.684 \\ 1 & 5.008 \\ 1 & 5.429 \end{bmatrix}$, $\mathbf{x} = \begin{bmatrix} a_1 \\ b \end{bmatrix}$, and $\mathbf{y} = \begin{bmatrix} 5.415 \\ 5.900 \\ 6.532 \end{bmatrix}$. Thus

$\hat{\mathbf{x}} = \left(A^T A\right)^{-1} A^T \mathbf{y} \approx \begin{bmatrix} -1.609 \\ 1.499 \end{bmatrix}$. Therefore $a_1 = -1.609$ and $b = 1.499$. Since $a_1 = \ln(a)$, $a = e^{-1.609} \approx 0.2001$. Thus our model is $p = 0.2001 d^{1.499}$.

53. We consider a quadratic polynomial of the form $f(t) = c_1 + c_2 t + c_3 t^2$. We evaluate at the given points

$$\begin{array}{rclcrcrcrcl}
(1, 2185) & \Longrightarrow & c_1 & + & c_2 & + & c_3 & = & 2185 \\
(2, 2140) & \Longrightarrow & c_1 & + & 2c_2 & + & 4c_3 & = & 2140 \\
(3, 2055) & \Longrightarrow & c_1 & + & 3c_2 & + & 9c_3 & = & 2055 \\
(4, 1943) & \Longrightarrow & c_1 & + & 4c_2 & + & 16c_3 & = & 1943
\end{array}$$

This is equivalent to $A\mathbf{c} = \mathbf{y}$, where $A = \begin{bmatrix} 1 & 1 & 1 \\ 1 & 2 & 4 \\ 1 & 3 & 9 \\ 1 & 4 & 16 \end{bmatrix}$, $\mathbf{c} = \begin{bmatrix} c_1 \\ c_2 \\ c_3 \end{bmatrix}$, and $\mathbf{y} = \begin{bmatrix} 2185 \\ 2140 \\ 2055 \\ 1943 \end{bmatrix}$. The least

squares solution of $A\mathbf{c} = \mathbf{y}$ is $\mathbf{c} = \left(A^T A\right)^{-1} A^T \mathbf{y} \approx \begin{bmatrix} 2199.8 \\ 2.65 \\ -16.75 \end{bmatrix}$. Thus the quadratic polynomial that

best fits the data is $f(t) \approx 2199.8 + 2.65t - 16.75t^2$. We predict when the anvil hits the ground by solving $f(t) \approx 2199.8 + 2.65t - 16.75t^2 = 0$, and obtain $t = 11.539$ seconds.

55. Taking the natural logarithm we have $\ln(y) = \ln(a) + bt$. We evaluate at the given points

$$
\begin{array}{rcl}
(2, 1.50) \implies \ln(a) + 2b &=& \ln(1.50) \\
(4, 0.97) \implies \ln(a) + 4b &=& \ln(0.97) \\
(6, 0.57) \implies \ln(a) + 6b &=& \ln(0.57) \\
(8, 0.41) \implies \ln(a) + 8b &=& \ln(0.41)
\end{array}
$$

This is equivalent to $A\mathbf{x} = \mathbf{y}$, where $A = \begin{bmatrix} 1 & 2 \\ 1 & 4 \\ 1 & 6 \\ 1 & 8 \end{bmatrix}$, $\mathbf{x} = \begin{bmatrix} \ln(a) \\ b \end{bmatrix}$, and $\mathbf{y} = \begin{bmatrix} \ln(1.50) \\ \ln(0.97) \\ \ln(0.57) \\ \ln(0.41) \end{bmatrix}$. Since A

has linearly independent columns, the least squares solution $\mathbf{x} = \left(A^T A\right)^{-1} A^T \mathbf{y} \approx \begin{bmatrix} 0.8360 \\ -0.2211 \end{bmatrix}$. Thus

$\ln(a) \approx 0.8360 \quad \Rightarrow \quad a \approx 2.307$ and $b \approx -0.2211$. Hence $y \approx 2.307 e^{-0.2211t}$ best fits the data. The initial size of the sample is determined by setting $t = 0$, and we obtain $y \approx 2.307$ grams. The amount present at $t = 15$ is determined by evaluating $y \approx 2.307 e^{-0.2211(15)} \approx 0.08370$ grams.

Chapter 9

Linear Transformations

9.1 Definition and Properties

1. $T(\mathbf{v}_2 - 2\mathbf{v}_1) = T(\mathbf{v}_2) - 2T(\mathbf{v}_1) = \begin{bmatrix} -3 \\ 1 \end{bmatrix} - 2 \begin{bmatrix} 1 \\ 2 \end{bmatrix} = \begin{bmatrix} -5 \\ -3 \end{bmatrix}.$

3. Write $2x^2 - 4x - 1 = c_1(x^2 + 1) + c_2(4x + 3) \quad \Rightarrow \quad c_1 = 2$ and $c_2 = -1$, so $2x^2 - 4x - 1 = 2(x^2 + 1) - (4x + 3)$. Thus $T(2x^2 - 4x - 1) = T(2(x^2 + 1) - (4x + 3)) = 2T(x^2 + 1) - T(4x + 3) = 2 \begin{bmatrix} -1 \\ 3 \end{bmatrix} - \begin{bmatrix} 2 \\ 1 \end{bmatrix} = \begin{bmatrix} -4 \\ 5 \end{bmatrix}.$

5. Let $a_1 x^2 + b_1 x + c_1$ and $a_2 x^2 + b_2 x + c_2$ belong to $\mathbf{P}^2$, and s and t scalars. Then
$T(s(a_1 x^2 + b_1 x + c_1) + t(a_2 x^2 + b_2 x + c_2)) = T((sa_1 + ta_2)x^2 + (sb_1 + tb_2)x + (sc_1 + tc_2))$
$= (sc_1 + tc_2)x^2 + (sb_1 + tb_2)x + (sa_1 + ta_2) = s(c_1 x^2 + b_1 x + a_1) + t(c_2 x^2 + b_2 x + a_2)$
$= sT(a_1 x^2 + b_1 x + c_1) + tT(a_2 x^2 + b_2 x + c_2)$. Hence by Theorem 9.2 T is a linear transformation.

7. Let $p_1(x)$ and $p_2(x)$ belong to $\mathbf{P}^n$, and c_1 and c_2 scalars. Then $T(c_1 p_1(x) + c_2 p_2(x))$
$= T((c_1 p_1 + c_2 p_2)(x)) = e^x(c_1 p_1 + c_2 p_2)(x) = e^x(c_1 p_1(x) + c_2 p_2(x)) = c_1(e^x p_1(x)) + c_2(e^x p_2(x)) = c_1 T(p_1(x)) + c_2 T(p_2(x))$. Hence by Theorem 9.2 T is a linear transformation.

9. Let $\begin{bmatrix} a_1 & b_1 \\ c_1 & d_1 \end{bmatrix}$ and $\begin{bmatrix} a_2 & b_2 \\ c_2 & d_2 \end{bmatrix}$ belong to $\mathbf{R}^{2\times 2}$, and s and t scalars. Then
$T\left(s \begin{bmatrix} a_1 & b_1 \\ c_1 & d_1 \end{bmatrix} + t \begin{bmatrix} a_2 & b_2 \\ c_2 & d_2 \end{bmatrix} \right) = T\left(\begin{bmatrix} sa_1 + ta_2 & sb_1 + tb_2 \\ sc_1 + tc_2 & sd_1 + td_2 \end{bmatrix} \right) =$
$\text{tr}\left(\begin{bmatrix} sa_1 + ta_2 & sb_1 + tb_2 \\ sc_1 + tc_2 & sd_1 + td_2 \end{bmatrix} \right) = (sa_1 + ta_2) + (sd_1 + td_2) = s(a_1 + d_1) + t(a_2 + d_2) =$
$s\,\text{tr}\left(\begin{bmatrix} a_1 & b_1 \\ c_1 & d_1 \end{bmatrix} \right) + t\,\text{tr}\left(\begin{bmatrix} a_2 & b_2 \\ c_2 & d_2 \end{bmatrix} \right) = sT\left(\begin{bmatrix} a_1 & b_1 \\ c_1 & d_1 \end{bmatrix} \right) + tT\left(\begin{bmatrix} a_2 & b_2 \\ c_2 & d_2 \end{bmatrix} \right).$
Hence by Theorem 9.2 T is a linear transformation.

11. Let $\mathbf{v}_1$ and $\mathbf{v}_2$ belong to $\mathbf{R}^n$, and c_1 and c_2 scalars. Then $T(c_1\mathbf{v}_1 + c_2\mathbf{v}_1) = -4(c_1\mathbf{v}_1 + c_2\mathbf{v}_1) = c_1(-4(\mathbf{v}_1)) + c_2(-4(\mathbf{v}_2)) = c_1 T(\mathbf{v}_1) + c_2 T(\mathbf{v}_2)$. Hence by Theorem 9.2 T is a linear transformation.

13. Let $\begin{bmatrix} a_1 \\ b_1 \end{bmatrix}$ and $\begin{bmatrix} a_2 \\ b_2 \end{bmatrix}$ belong to $\mathbf{R}^2$, and c_1 and c_2 scalars. Then
$T\left(c_1 \begin{bmatrix} a_1 \\ b_1 \end{bmatrix} + c_2 \begin{bmatrix} a_2 \\ b_2 \end{bmatrix} \right) = T\left(\begin{bmatrix} a_1 c_1 + a_2 c_2 \\ b_1 c_1 + b_2 c_2 \end{bmatrix} \right) = \begin{bmatrix} b_1 c_1 + b_2 c_2 \\ a_1 c_1 + a_2 c_2 \end{bmatrix} = c_1 \begin{bmatrix} b_1 \\ a_1 \end{bmatrix} + c_2 \begin{bmatrix} b_2 \\ a_2 \end{bmatrix} = c_1 T\left(\begin{bmatrix} a_1 \\ b_1 \end{bmatrix} \right) + c_2 T\left(\begin{bmatrix} a_2 \\ b_2 \end{bmatrix} \right).$
Hence by Theorem 9.2 T is a linear transformation.

15. Let $a_1x^2 + b_1x + c_1$ and $a_2x^2 + b_2x + c_2$ belong to $\mathbf{P}^2$, and s and t scalars. Then

$$T\left(s\left(a_1x^2 + b_1x + c_1\right) + t\left(a_2x^2 + b_2x + c_2\right)\right) = T\left((sa_1 + ta_2)x^2 + (sb_1 + tb_2)x + (sc_1 + tc_2)\right)$$

$$= \left[\begin{array}{c} (sa_1 + ta_2) - (sb_1 + tb_2) \\ (sb_1 + tb_2) + (sc_1 + tc_2) \end{array}\right] = s\left[\begin{array}{c} a_1 - b_1 \\ b_1 + c_1 \end{array}\right] + t\left[\begin{array}{c} a_2 - b_2 \\ b_2 + c_2 \end{array}\right]$$

$$= sT\left(a_1x^2 + b_1x + c_1\right) + tT\left(a_2x^2 + b_2x + c_2\right).$$

Hence by Theorem 9.2 T is a linear transformation.

17. Let A_1 and A_2 belong to $\mathbf{R}^{n \times n}$, and c_1 and c_2 scalars. Then $T\left(c_1A_1 + c_2A_2\right) = \left(c_1A_1 + c_2A_2\right)^T$
$= c_1A_1^T + c_2A_2^T = c_1T\left(A_1\right) + c_2T\left(A_2\right)$. Hence by Theorem 9.2 T is a linear transformation.

19. Let $A = \left[\begin{array}{ccc} a_{11} & \cdots & a_{1n} \\ \vdots & \ddots & \vdots \\ a_{n1} & \cdots & a_{nn} \end{array}\right]$ and $B = \left[\begin{array}{ccc} b_{11} & \cdots & b_{1n} \\ \vdots & \ddots & \vdots \\ b_{n1} & \cdots & b_{nn} \end{array}\right]$ belong to $\mathbf{R}^{n \times n}$, and s and t scalars. Then

$$T\left(s\left[\begin{array}{ccc} a_{11} & \cdots & a_{1n} \\ \vdots & \ddots & \vdots \\ a_{n1} & \cdots & a_{nn} \end{array}\right] + t\left[\begin{array}{ccc} b_{11} & \cdots & b_{1n} \\ \vdots & \ddots & \vdots \\ b_{n1} & \cdots & b_{nn} \end{array}\right]\right) =$$

$$T\left(\left[\begin{array}{ccc} sa_{11} + tb_{11} & \cdots & sa_{1n} + tb_{1n} \\ \vdots & \ddots & \vdots \\ sa_{n1} + tb_{n1} & \cdots & sa_{nn} + tb_{nn} \end{array}\right]\right) = \text{tr}\left(\left[\begin{array}{ccc} sa_{11} + tb_{11} & \cdots & sa_{1n} + tb_{1n} \\ \vdots & \ddots & \vdots \\ sa_{n1} + tb_{n1} & \cdots & sa_{nn} + tb_{nn} \end{array}\right]\right) = (sa_{11} + tb_{11})+$$

$\cdots + (sa_{nn} + tb_{nn}) = s\left(a_{11} + \cdots + a_{nn}\right) + t\left(b_{11} + \cdots + b_{nn}\right) = s\,\text{tr}\,(A) + t\,\text{tr}\,(B) = sT\,(A) + tT\,(B)$.
Hence by Theorem 9.2 T is a linear transformation.

21. Let $A = \left[\begin{array}{cc} 1 & 0 \\ 0 & 0 \\ 0 & 0 \end{array}\right]$. Then $T\,(2A) = \left(2\left[\begin{array}{cc} 1 & 0 \\ 0 & 0 \\ 0 & 0 \end{array}\right]\right)^T\left(2\left[\begin{array}{cc} 1 & 0 \\ 0 & 0 \\ 0 & 0 \end{array}\right]\right) = \left[\begin{array}{cc} 4 & 0 \\ 0 & 0 \end{array}\right]$, but $2T\,(A) =$

$2\left(\left[\begin{array}{cc} 1 & 0 \\ 0 & 0 \\ 0 & 0 \end{array}\right]^T\left[\begin{array}{cc} 1 & 0 \\ 0 & 0 \\ 0 & 0 \end{array}\right]\right) = \left[\begin{array}{cc} 2 & 0 \\ 0 & 0 \end{array}\right]$. Since $T\,(2A) \neq 2T\,(A)$, T is not a linear transformation.

23. $\ker\,(T)$ is the set of all polynomials $p\,(x) = ax + b$ such that $T\,(ax + b) = a - b = 0$. Thus $b = a$, and hence $\ker\,(T) = \{p\,(x) : p\,(x) = ax + a\}$. Since $T\,(0x - 1) = 0 - (-1) = 1$, and range$(T)$ is a subspace of $\mathbf{R}$, we conclude that range$(T) = \mathbf{R}$.

25. $\ker\,(T)$ is the set of all polynomials $p\,(x) = ax^2 + bx + c$ such that $T\left(ax^2 + bx + c\right) = \left[\begin{array}{cc} a & b \\ b & c \end{array}\right] = \left[\begin{array}{cc} 0 & 0 \\ 0 & 0 \end{array}\right]$. Thus $a = b = c = 0$, and $\ker\,(T) = \{p\,(x) : p\,(x) = 0\} = \{\mathbf{0}_{\mathbf{P}^2}\}$. The range of T consists of all matrices of the form $\left[\begin{array}{cc} a & b \\ b & c \end{array}\right] = a\left[\begin{array}{cc} 1 & 0 \\ 0 & 0 \end{array}\right] + b\left[\begin{array}{cc} 0 & 1 \\ 1 & 0 \end{array}\right] + c\left[\begin{array}{cc} 0 & 0 \\ 0 & 1 \end{array}\right]$, hence range$\,(T) = $ span $\left\{\left[\begin{array}{cc} 1 & 0 \\ 0 & 0 \end{array}\right], \left[\begin{array}{cc} 0 & 1 \\ 1 & 0 \end{array}\right], \left[\begin{array}{cc} 0 & 0 \\ 0 & 1 \end{array}\right]\right\}$.

27. Consider $f\,(x) = x^2 - 3x + 2$. Then $T\,(f) = \left[\begin{array}{c} f\,(1) \\ f\,(2) \end{array}\right] = \left[\begin{array}{c} 0 \\ 0 \end{array}\right]$. Since $\ker\,(T) \neq \{\mathbf{0}_{\mathbf{P}^2}\}$, T is not one-to-one. T is onto, since if $\left[\begin{array}{c} a \\ b \end{array}\right] \in \mathbf{R}^2$, then $T\,((b - a)\,x + (2a - b)) = \left[\begin{array}{c} a \\ b \end{array}\right]$.

29. Consider $f\,(x) = x - 1$, then $T\,(f) = f\,(1) = 0$. Since f is not the zero function, T is not one-to-one. If $a \in \mathbf{R}$, and $f\,(x) = a$, then $f \in C\,[0, 1]$ and $T\,(f) = f\,(1) = a$. Thus T is onto.

31. Let $V = \mathbf{R}$ and $W = \mathbf{R}^2$, and define $T(a) = \left(\begin{bmatrix} a \\ 0 \end{bmatrix} \right)$. Then T is a linear transformation which is one-to-one, but not onto.

33. Let $V = \mathbf{R}^2$ and $W = \mathbf{R}^2$, and define $T\left(\begin{bmatrix} a \\ b \end{bmatrix} \right) = \begin{bmatrix} a \\ 0 \end{bmatrix}$. Then T is a linear transformation that is neither one-to-one nor onto.

35. Let $V = \mathbf{R}^4$ and $W = \mathbf{R}^3$, and let $T\left(\begin{bmatrix} a \\ b \\ c \\ d \end{bmatrix} \right) = \begin{bmatrix} a \\ b \\ c \end{bmatrix}$. Then T is a linear transformation, $\ker(T) = \operatorname{span}\left\{ \begin{bmatrix} 0 \\ 0 \\ 0 \\ 1 \end{bmatrix} \right\}$, $\dim(\ker(T)) = 1$, and $\operatorname{range}(T) = \mathbf{R}^3$, $\dim(\operatorname{range}(T)) = 3$.

37. Let $V = \mathbf{R}^k$ and $W = \mathbf{R}$, and define $T(\mathbf{v}) = 0$. Then T is a linear transformation, and if $\{\mathbf{v}_1, \ldots, \mathbf{v}_k\}$ is linearly independent in V, then $\{T(\mathbf{v}_1), \ldots, T(\mathbf{v}_k)\} = \{\mathbf{0}, \ldots, \mathbf{0}\}$ is linearly dependent.

39. True, since $T(\mathbf{v}_1 - \mathbf{v}_2) = T(\mathbf{v}_1 + (-1)\mathbf{v}_2) = T(\mathbf{v}_1) + (-1)T(\mathbf{v}_2) = T(\mathbf{v}_1) - T(\mathbf{v}_2)$.

41. True, by Theorem 9.3.

43. False. For example, consider $T : \mathbf{R} \to \mathbf{R}$ defined by $T(x) = 0$. $\{1\}$ is a linearly independent set in $\mathbf{R}$, but $\{T(1)\} = \{0\}$ is linearly dependent.

45. True. $\dim(\mathbf{R}^{2\times 2}) = 4$, so by Theorem 9.7 $\dim(\operatorname{range}(T)) = \dim(\mathbf{R}^{2\times 2}) - \dim(\ker(T)) = 4 - \dim(\ker(T)) \le 4 - 0 = 4$. Since $\dim(\mathbf{P}^6) = 7 > 4 \ge \dim(\operatorname{range}(T))$, we conclude that $\operatorname{range}(T) \ne \mathbf{P}^6$, and T is not onto.

47. False. For example, consider $T : \mathbf{R} \to \mathbf{R}$ defined by $T(x) = x$. Let $\mathbf{w} = 1$, then the set of vectors $\mathbf{v}$ with $T(\mathbf{v}) = \mathbf{w} = 1$ consists of the single element set $\{1\}$, which is not a subspace.

49. By Theorem 9.7, $\dim(\operatorname{range}(T)) = \dim(\mathbf{R}^5) - \dim(\ker(T)) = 5 - 2 = 3$.

51. $T(\mathbf{v}_1 - \mathbf{v}_2) = T(\mathbf{v}_1) - T(\mathbf{v}_2) = \mathbf{0}$, so $\mathbf{v}_1 - \mathbf{v}_2$ belongs to $\ker(T)$.

53. Let $\mathbf{v}_1$ and $\mathbf{v}_2$ belong to V, and c_1 and c_2 scalars and define $T(\mathbf{v}) = T_1(\mathbf{v}) + T_2(\mathbf{v})$. Then

$$\begin{aligned}
T(c_1\mathbf{v}_1 + c_2\mathbf{v}_2) &= (T_1 + T_2)(c_1\mathbf{v}_1 + c_2\mathbf{v}_2) \\
&= T_1(c_1\mathbf{v}_1 + c_2\mathbf{v}_2) + T_2(c_1\mathbf{v}_1 + c_2\mathbf{v}_2) \\
&= (c_1 T_1(\mathbf{v}_1) + c_2 T_1(\mathbf{v}_2)) + (c_1 T_2(\mathbf{v}_1) + c_2 T_2(\mathbf{v}_2)) \\
&= (c_1 T_1(\mathbf{v}_1) + c_1 T_2(\mathbf{v}_1)) + (c_2 T_1(\mathbf{v}_2) + c_2 T_2(\mathbf{v}_2)) \\
&= c_1(T_1(\mathbf{v}_1) + T_2(\mathbf{v}_1)) + c_2(T_1(\mathbf{v}_2) + T_2(\mathbf{v}_2)) \\
&= c_1((T_1 + T_2)(\mathbf{v}_1)) + c_2((T_1 + T_2)(\mathbf{v}_2)) \\
&= c_1 T(\mathbf{v}_1) + c_2 T(\mathbf{v}_2).
\end{aligned}$$

Thus by Theorem 9.2 T is a linear transformation from V to W.

55. $T(\mathbf{0}_V) = T(\mathbf{0}_V - \mathbf{0}_V) = T(\mathbf{0}_V) - T(\mathbf{0}_V) = \mathbf{0}_W$.

57. Since $T(\mathbf{0}_V) = \mathbf{0}_W$, $\ker(T)$ contains the zero vector, $\mathbf{0}_V$. Suppose $\mathbf{v}_1$ and $\mathbf{v}_2$ belong to $\ker(T)$. Then $T(\mathbf{v}_1) = \mathbf{0}_W$ and $T(\mathbf{v}_2) = \mathbf{0}_W$. Thus, $T(\mathbf{v}_1 + \mathbf{v}_2) = T(\mathbf{v}_1) + T(\mathbf{v}_2) = \mathbf{0}_W + \mathbf{0}_W = \mathbf{0}_W$, and $\mathbf{v}_1 + \mathbf{v}_2$ belongs to $\ker(T)$. Let c be a scalar, then $T(c\mathbf{v}_1) = cT(\mathbf{v}_1) = c\mathbf{0}_W = \mathbf{0}_W$, so $c\mathbf{v}_1$ belongs to $\ker(T)$. Thus, $\ker(T)$ is a non-empty subset of V which is closed under vector addition and scalar multiplication, and so $\ker(T)$ is a subspace of V.

59. Suppose T is one-to-one. Since $T\left(\mathbf{0}_V\right) = \mathbf{0}_W$, $\mathbf{0}_V$ is the unique vector whose image is $\mathbf{0}_W$, and thus $\ker\left(T\right) = \{\mathbf{0}_V\}$. Now suppose $\ker\left(T\right) = \{\mathbf{0}_V\}$, and consider $T\left(\mathbf{v}_1\right) = T\left(\mathbf{v}_2\right)$ for some two vectors $\mathbf{v}_1$ and $\mathbf{v}_2$ in V. Then, since T is a linear transformation, $T\left(\mathbf{v}_1 - \mathbf{v}_2\right) = T\left(\mathbf{v}_1\right) - T\left(\mathbf{v}_2\right) = \mathbf{0}_W$. Therefore $\mathbf{v}_1 - \mathbf{v}_2$ belongs to $\ker\left(T\right) = \{\mathbf{0}_V\}$, and thus $\mathbf{v}_1 - \mathbf{v}_2 = \mathbf{0}_V$, and therefore $\mathbf{v}_1 = \mathbf{v}_2$. Hence, T is one-to-one.

61. Consider $c_1\mathbf{w}_1 + \cdots + c_m\mathbf{w}_m = \mathbf{0}_W$. Then $\mathbf{0}_W = c_1T\left(\mathbf{v}_1\right) + \cdots + c_mT\left(\mathbf{v}_m\right) = T\left(c_1\mathbf{v}_1 + \cdots + c_m\mathbf{v}_m\right)$. Hence $c_1\mathbf{v}_1 + \cdots + c_m\mathbf{v}_m$ is in $\ker\left(T\right)$. Since T is one-to-one, $\ker\left(T\right) = \{\mathbf{0}_V\}$, and thus $c_1\mathbf{v}_1 + \cdots + c_m\mathbf{v}_m = \mathbf{0}_V$. Since $\mathcal{V} = \{\mathbf{v}_1, \ldots, \mathbf{v}_m\}$ is linearly independent, each $c_i = 0$. Consequently, $\mathcal{W} = \{\mathbf{w}_1, \ldots, \mathbf{w}_m\}$ is linearly independent.

63. Let f and g belong to $C^1\left(a,b\right)$ and c_1 and c_2 scalars. Then $T\left(c_1f + c_2g\right) = \left(c_1f + c_2g\right)'\left(x\right) = c_1f'\left(x\right) + c_2g'\left(x\right) = c_1T\left(f\right) + c_2T\left(g\right)$. Thus, by Theorem 9.2, T is a linear transformation from $C^1\left(a,b\right)$ to $C\left(a,b\right)$.

65. Let p_1 and p_2 belong to $\mathbf{P}^4$ and c_1 and c_2 scalars. Then $T\left(c_1p_1 + c_2p_2\right) = \left(c_1p_1 + c_2p_2\right)''\left(x\right) = c_1p_1''\left(x\right) + c_2p_2''\left(x\right) = c_1T\left(p_1\right) + c_2T\left(p_2\right)$. Thus, by Theorem 9.2, T is a linear transformation from $\mathbf{P}^4$ to $\mathbf{P}^2$.

67. Let p_1 and p_2 belong to $\mathbf{P}^4$ and c_1 and c_2 scalars. Then $T\left(c_1p_1 + c_2p_2\right) = \left(x^2\left(c_1p_1 + c_2p_2\right)\left(x\right)\right)' = \left(c_1x^2p_1\left(x\right) + c_2x^2p_2\left(x\right)\right)' = \left(c_1x^2p_1\left(x\right)\right)' + \left(c_2x^2p_2\left(x\right)\right)' = c_1\left(x^2p_1\left(x\right)\right)' + c_2\left(x^2p_2\left(x\right)\right)' = c_1T\left(p_1\right) + c_2T\left(p_2\right)$. Thus, by Theorem 9.2, T is a linear transformation from $\mathbf{P}^4$ to $\mathbf{P}^5$.

9.2   Isomorphisms

1. $\dim\left(V\right) = \dim\left(\mathbf{R}^8\right) = 8$, and $\dim\left(W\right) = \dim\left(\mathbf{P}^9\right) = 10$. Since $\dim\left(V\right) \neq \dim\left(W\right)$, the vector spaces are not isomorphic.

3. $\dim\left(V\right) = \dim\left(\mathbf{R}^{3\times 6}\right) = 18$, and $\dim\left(W\right) = \dim\left(\mathbf{P}^{17}\right) = 18$. Since $\dim\left(V\right) = \dim\left(W\right)$, the vector spaces are isomorphic.

5. $\dim\left(V\right) = \dim\left(\mathbf{R}^{13}\right) = 13$, and $\dim\left(W\right) = \dim\left(\mathbf{C}\left[0,1\right]\right) = \infty$. Since $\dim\left(V\right) \neq \dim\left(W\right)$, the vector spaces are not isomorphic.

7. First check that T is a linear transformation. Let $\begin{bmatrix} a_1 \\ b_1 \\ c_1 \end{bmatrix}$ and $\begin{bmatrix} a_2 \\ b_2 \\ c_2 \end{bmatrix}$ be two vectors in $\mathbf{R}^3$, and s and t scalars. Then

$$T\left(s\begin{bmatrix} a_1 \\ b_1 \\ c_1 \end{bmatrix} + t\begin{bmatrix} a_2 \\ b_2 \\ c_2 \end{bmatrix}\right) = T\left(\begin{bmatrix} sa_1 + ta_2 \\ sb_1 + tb_2 \\ sc_1 + tc_2 \end{bmatrix}\right)$$
$$= \left(sc_1 + tc_2\right)x^2 + \left(sb_1 + tb_2\right)x + \left(sa_1 + ta_2\right)$$
$$= \left(\left(sc_1\right)x^2 + \left(sb_1\right)x + sa_1\right) + \left(\left(tc_2\right)x^2 + \left(tb_2\right)x + ta_2\right)$$
$$= s\left(c_1x^2 + b_1x + a_1\right) + t\left(c_2x^2 + b_2x + a_2\right)$$
$$= sT\left(\begin{bmatrix} a_1 \\ b_1 \\ c_1 \end{bmatrix}\right) + tT\left(\begin{bmatrix} a_2 \\ b_2 \\ c_2 \end{bmatrix}\right).$$

Hence T is a linear transformation. Suppose $T\left(\begin{matrix} a \\ b \\ c \end{matrix}\right) = cx^2 + bx + a = 0x^2 + 0x + 0$, the zero vector in $\mathbf{P}^2$. Then $c = b = a = 0$, and $\begin{bmatrix} a \\ b \\ c \end{bmatrix} = \begin{bmatrix} 0 \\ 0 \\ 0 \end{bmatrix}$, the zero vector in $\mathbf{R}^3$. Thus $\ker\left(T\right) = \{\mathbf{0}_{\mathbf{R}^3}\}$, and hence

T is one-to-one. By Theorem 9.3, $\dim(\text{range}(T)) = \dim(\mathbf{R}^3) - \dim(\ker(T)) = 3 - 0 = 3$. Hence, since $\dim(\mathbf{P}^2) = 3$, $\text{range}(T) = \mathbf{P}^2$, and T is onto. Therefore T is an isomorphism.

9. First check that T is a linear transformation. Let $p_1(x) = a_1x^3 + b_1x^2 + c_1x + d_1$ and $p_2(x) = a_2x^3 + b_2x^2 + c_2x + d_2$ be two vectors in $\mathbf{P}^3$, and s and t scalars. Then

$$T(sp_1(x) + tp_2(x)) = T\left(s\left(a_1x^3 + b_1x^2 + c_1x + d_1\right) + t\left(a_2x^3 + b_2x^2 + c_2x + d_2\right)\right) =$$
$$T\left((sa_1 + ta_2)x^3 + (sb_1 + tb_2)x^2 + (sc_1 + tc_2)x + (sd_1 + td_2)\right) =$$
$$\left[\begin{array}{cc} (sa_1 + ta_2) + (sb_1 + tb_2) + (sc_1 + tc_2) + (sd_1 + td_2) & (sa_1 + ta_2) + (sb_1 + tb_2) + (sc_1 + tc_2) \\ (sa_1 + ta_2) + (sb_1 + tb_2) & (sa_1 + ta_2) \end{array}\right]$$
$$= s\left[\begin{array}{cc} a_1 + b_1 + c_1 + d_1 & a_1 + b_1 + c_1 \\ a_1 + b_1 & a_1 \end{array}\right] + t\left[\begin{array}{cc} a_2 + b_2 + c_2 + d_2 & a_2 + b_2 + c_2 \\ a_2 + b_2 & a_2 \end{array}\right]$$
$$= sT(p_1(x)) + tT(p_2(x)).$$

Hence T is a linear transformation. Suppose $p(x) = ax^3 + bx^2 + cx + d$ is in $\mathbf{P}^3$, and $T(p(x)) = \left[\begin{array}{cc} a+b+c+d & a+b+c \\ a+b & a \end{array}\right] = \left[\begin{array}{cc} 0 & 0 \\ 0 & 0 \end{array}\right]$, the zero vector in $\mathbf{R}^{2\times2}$. Then

$$a + b + c + d = 0$$
$$a + b + c = 0$$
$$a + b = 0$$
$$a = 0$$

so $a = b = c = d = 0$. Thus $p(x) = 0$, $\ker(T) = \{\mathbf{0}_{\mathbf{P}^3}\}$, and hence T is one-to-one. By Theorem 9.3, $\dim(\text{range}(T)) = \dim(\mathbf{P}^3) - \dim(\ker(T)) = 4 - 0 = 4$. Hence, since $\dim(\mathbf{R}^{2\times2}) = 4$, $\text{range}(T) = \mathbf{R}^{2\times2}$, and T is onto. Therefore T is an isomorphism.

11. Suppose $T(ax + b) = \left[\begin{array}{c} a+b \\ b-a \end{array}\right] = \left[\begin{array}{c} 0 \\ 0 \end{array}\right]$. Then $a + b = 0$ and $b - a = 0$, and hence $a = b = 0$. Thus $\ker(T) = \{\mathbf{0}_{\mathbf{P}^1}\}$, and T is one-to-one. By Theorem 9.3, $\dim(\text{range}(T)) = \dim(\mathbf{P}^1) - \dim(\ker(T)) = 2 - 0 = 2$. Hence, since $\dim(\mathbf{R}^2) = 2$, $\text{range}(T) = \mathbf{R}^2$, and T is onto. Therefore T is an isomorphism.

13. Let $f(x) = \sin(\pi x)$. Then $T(f) = (f(1), f(2), \ldots) = (\sin(\pi), \sin(2\pi), \ldots) = (0, 0, \ldots)$, the zero vector in $\mathbf{R}^\infty$. Thus $\ker(T) \neq \{\mathbf{0}_{C(\mathbf{R})}\}$, and T is not one-to-one. Hence T is not an isomorphism.

15. $T^{-1}: \mathbf{R}^2 \to \mathbf{P}^1$. Suppose $T(ax + b) = \left[\begin{array}{c} 2b \\ a-b \end{array}\right] = \left[\begin{array}{c} c \\ d \end{array}\right]$. Then $2b = c$ and $a - b = d$, so $b = c/2$ and $a = c/2 + d$. Thus, $T^{-1}\left(\left[\begin{array}{c} c \\ d \end{array}\right]\right) = (c/2 + d)x + c/2$.

17. $T^{-1}: \mathbf{P}^2 \to \mathbf{P}^2$ is given by $T^{-1}(ax^2 + bx + c) = cx^2 - bx + a$. Thus $T = T^{-1}$.

19. Suppose $T\left(\left[\begin{array}{c} a_1 \\ a_2 \end{array}\right]\right) = \left[\begin{array}{c} a_1 \\ a_2 \\ 0 \end{array}\right] = \left[\begin{array}{c} 0 \\ 0 \\ 0 \end{array}\right]$. Then $a_1 = a_2 = 0$, so $\ker(T) = \{\mathbf{0}_{\mathbf{R}^2}\}$, and T is one-to-one. By Theorem 9.3, $\dim(\text{range}(T)) = \dim(\mathbf{R}^2) - \dim(\ker(T)) = 2 - 0 = 2$. Hence, since $\dim(S) = 2$, $\text{range}(T) = S$, and T is onto. Therefore T is an isomorphism.

21. Let $a_n x^{2n} + \cdots + a_1 x^2 + a_0$ and $b_n x^{2n} + \cdots + b_1 x^2 + b_0$ belong to $\mathbf{P}_e$, and c_1 and c_2 scalars. Then

$$T\left(c_1\left(a_n x^{2n} + \cdots + a_1 x^2 + a_0\right) + c_2\left(b_n x^{2n} + \cdots + b_1 x^2 + b_0\right)\right)$$
$$= T\left(\left(c_1 a_n + c_2 b_n\right) x^{2n} + \cdots + \left(c_1 a_1 + c_2 b_1\right) x^2 + \left(c_1 a_0 + c_2 b_0\right)\right)$$
$$= \left(c_1 a_n + c_2 b_n\right) x^n + \cdots + \left(c_1 a_1 + c_2 b_1\right) x + \left(c_1 a_0 + c_2 b_0\right)$$
$$= c_1\left(a_n x^n + \cdots + a_1 x + a_0\right) + c_2\left(b_n x^n + \cdots + b_1 x + b_0\right)$$
$$= c_1 T\left(a_n x^{2n} + \cdots + a_1 x^2 + a_0\right) + c_2 T\left(b_n x^{2n} + \cdots + b_1 x^2 + b_0\right).$$

Thus T is a linear transformation. Suppose $T\left(a_n x^{2n} + \cdots + a_1 x^2 + a_0\right) = a_n x^n + \cdots + a_1 x + a_0 = 0$. Then $a_n = a_{n-1} = \cdots = a_0 = 0$, $\ker(T) = \{\mathbf{0}_{\mathbf{P}_e}\}$, and T is one-to-one. Suppose $p(x) = a_n x^n + \cdots + a_1 x + a_0$ is in $\mathbf{P}$. Then $q(x) = a_n x^{2n} + \cdots + a_1 x^2 + a_0$ is in $\mathbf{P}_e$, and $T(q) = T\left(a_n x^{2n} + \cdots + a_1 x^2 + a_0\right) = a_n x^n + \cdots + a_1 x + a_0 = p$. Hence T is onto, and therefore T is an isomorphism.

23. Define the linear transformation
$$T\left(\begin{bmatrix} a \\ b \\ c \\ d \\ e \end{bmatrix}\right) = ax^4 + bx^3 + cx^2 + dx + e. \text{ If } T\left(\begin{bmatrix} a \\ b \\ c \\ d \\ e \end{bmatrix}\right) = ax^4 + bx^3 + cx^2 + dx + e = 0,$$
then $a = b = c = d = e = 0$, and $\ker(T) = \{\mathbf{0}_{\mathbf{R}^5}\}$, and T is one-to-one. By Theorem 9.3, $\dim(\text{range}(T)) = \dim(\mathbf{R}^5) - \dim(\ker(T)) = 5 - 0 = 5$. Hence, since $\dim(\mathbf{P}^4) = 5$, $\text{range}(T) = \mathbf{P}^4$, and T is onto. Therefore T is an isomorphism.

25. Define the linear transformation $T\left(\begin{bmatrix} a & b \\ c & d \end{bmatrix}\right) = ax^3 + bx^2 + cx + d$. If $T\left(\begin{bmatrix} a & b \\ c & d \end{bmatrix}\right) = ax^3 + bx^2 + cx + d = 0$, then $a = b = c = d = 0$, and $\ker(T) = \{\mathbf{0}_{\mathbf{R}^{2 \times 2}}\}$, and T is one-to-one. By Theorem 9.3, $\dim(\text{range}(T)) = \dim(\mathbf{R}^{2 \times 2}) - \dim(\ker(T)) = 4 - 0 = 4$. Hence, since $\dim(\mathbf{P}^3) = 4$, $\text{range}(T) = \mathbf{P}^3$, and T is onto. Therefore T is an isomorphism.

27. Let S be the subspace of $\mathbf{R}^{2 \times 3}$ defined by
$$S = \text{span}\left\{\begin{bmatrix} 1 & 0 & 0 \\ 0 & 0 & 0 \end{bmatrix}, \begin{bmatrix} 0 & 1 & 0 \\ 0 & 0 & 0 \end{bmatrix}, \begin{bmatrix} 0 & 0 & 0 \\ 1 & 0 & 0 \end{bmatrix}, \begin{bmatrix} 0 & 0 & 0 \\ 0 & 1 & 0 \end{bmatrix}\right\}.$$
Define $T : \mathbf{P}^3 \to S$ by $T\left(ax^3 + bx^2 + cx + d\right) = \begin{bmatrix} a & b & 0 \\ c & d & 0 \end{bmatrix}$. Suppose
$$T\left(ax^3 + bx^2 + cx + d\right) = \begin{bmatrix} a & b & 0 \\ c & d & 0 \end{bmatrix} = \begin{bmatrix} 0 & 0 & 0 \\ 0 & 0 & 0 \end{bmatrix}, \text{ then } a = b = c = d = 0, \ker(T) = \{\mathbf{0}_{\mathbf{P}^3}\},$$
and T is one-to-one. By Theorem 9.3, $\dim(\text{range}(T)) = \dim(\mathbf{P}^3) - \dim(\ker(T)) = 4 - 0 = 4$. Since $\dim(S) = 4$, $\text{range}(T) = S$, and so T is onto. Therefore T is an isomorphism, and S is isomorphic to $\mathbf{P}^3$.

29. Let $S = \{(a_1, a_2, \ldots) \in \mathbf{R}^\infty : \text{ there exists } N \text{ such that } a_n = 0 \text{ for all } n > N\}$. Then S is a proper subspace of $\mathbf{R}^\infty$. Define the linear transformation $T : S \to \mathbf{P}$ by $T(a_1, \ldots, a_n, 0, 0, \ldots) = a_1 + a_2 x + \cdots + a_n x^{n-1}$. If $T(a_1, \ldots, a_n, 0, 0, \ldots) = a_1 + a_2 x + \cdots + a_n x^{n-1} = 0$, then $a_1 = a_2 = \cdots = a_n = 0$, and thus $(a_1, \ldots, a_n, 0, 0, \ldots) = (0, \ldots, 0, \ldots)$. So $\ker(T) = \{\mathbf{0}_S\}$, and T is one-to-one. If $p(x) = a_0 + \cdots + a_n x^n$ is any element in $\mathbf{P}$, then $(a_0, \ldots, a_n, 0, 0, \ldots)$ belongs to S, and $T(a_0, \ldots, a_n, 0, 0, \ldots) = a_0 + \cdots + a_n x^n = p(x)$. Thus T is onto, and hence T is an isomorphism from S to $\mathbf{P}$.

31. False. For example, $T : \mathbf{R} \to \mathbf{R}$ defined by $T(x) = 0$ is a linear transformation, but not one-to-one, and hence not an isomorphism.

33. True. Since T is one-to-one, we have $\ker(T) = \{\mathbf{0}\}$ so that $\dim(\ker(T)) = 0$. Thus by Theorem 9.7, $\dim(\mathbf{R}^n) = \dim(\text{range}(T))$. Since $\text{range}(T)$ is a subspace of $\mathbf{R}^n$, it follows that $\text{range}(T) = \mathbf{R}^n$. Thus T is onto and hence an isomorphism.

35. False. For example, $\mathbf{R}$ is isomorphic to itself, but $T_1(x) = x$ and $T_2(x) = 2x$ are both isomorphisms from $\mathbf{R}$ to $\mathbf{R}$.

37. False. For example, let $T_1(x) = x$ and $T_2(x) = -x$ be isomorphisms from $\mathbf{R}$ to $\mathbf{R}$. Then $(T_1 + T_2)(x) = T_1(x) + T_2(x) = x - x = 0$, so $T_1 + T_2$ is not one-to-one, and hence not an isomorphism.

39. False. For example, let $T\left(a_n x^n + a_{n-1} x^{n-1} + \cdots + a_1 x + a_0\right) = a_n x^{n-1} + a_{n-1} x^{n-2} + \cdots + a_2 x + a_1$. Then range $(T) = \mathbf{P}$, but T is not one-to-one. Indeed, $T(1) = 0$, so ker $(T) \neq \{\mathbf{0_P}\}$. Thus, T is not an isomorphism.

41. False. Consider the constant function $f(x) = 1$, which belongs to $C^\infty(a, b)$. Then $T(f) = f'(x) = 0$, so f belongs to ker (T), and since f is not the zero vector, T is not one-to-one. Therefore T is not an isomorphism.

43. False. Consider $p(x) = 1$. Then $T(p) = xp'(x) = x(0) = 0$, so $p(x)$ belongs to ker (T), and since $p(x)$ is not the zero vector, T is not one-to-one. Therefore T is not an isomorphism.

45. Suppose the linear transformation $T : V \to W$ is an isomorphism. For each $\mathbf{w}$ in W there exists a vector $\mathbf{v}$ in V such that $T(\mathbf{v}) = \mathbf{w}$, since T is onto. Moreover, $\mathbf{v}$ is unique, since T is one-to-one. Thus we may define $T^{-1} : W \to V$ by $T^{-1}(\mathbf{w}) = \mathbf{v}$, where $\mathbf{v}$ is the unique vector in V such that $T(\mathbf{v}) = \mathbf{w}$. Hence T has an inverse. Now suppose $T : V \to W$ is a linear transformation which has an inverse. Hence there exists $T^{-1} : W \to V$ such that $T^{-1}(\mathbf{w}) = \mathbf{v}$ where $T(\mathbf{v}) = \mathbf{w}$. If $T(\mathbf{v_1}) = T(\mathbf{v_2})$, then $\mathbf{v_1} = T^{-1}(T(\mathbf{v_1})) = T^{-1}(T(\mathbf{v_2})) = \mathbf{v_2}$, so T is one-to-one. If $\mathbf{w}$ is in W, then $T(T^{-1}(\mathbf{w})) = \mathbf{w}$, so T is onto. Therefore T is an isomorphism.

47. Let $\mathbf{w_1}$ and $\mathbf{w_2}$ belong to W, and c_1 and c_2 scalars. Then $T^{-1}(\mathbf{w_1}) = \mathbf{v_1}$ and $T^{-1}(\mathbf{w_2}) = \mathbf{v_2}$ where $T(\mathbf{v_1}) = \mathbf{w_1}$ and $T(\mathbf{v_2}) = \mathbf{w_2}$. Now $T(c_1\mathbf{v_1} + c_2\mathbf{v_2}) = c_1 T(\mathbf{v_1}) + c_2 T(\mathbf{v_2}) = c_1\mathbf{w_1} + c_2\mathbf{w_2}$, so $T^{-1}(c_1\mathbf{w_1} + c_2\mathbf{w_2}) = c_1\mathbf{v_1} + c_2\mathbf{v_2} = c_1 T^{-1}(\mathbf{w_1}) + c_2 T^{-1}(\mathbf{w_2})$, and T^{-1} is a linear transformation. Suppose $T^{-1}(\mathbf{w}) = \mathbf{0}_V$. Then $\mathbf{w} = T(\mathbf{0}_V) = \mathbf{0}_W$, and hence T is one-to-one. Let $\mathbf{v}$ belong to V, then $T^{-1}(T(\mathbf{v})) = \mathbf{v}$, so T^{-1} is onto. Thus T^{-1} is an isomorphism.

49. We may consider the mapping $\widetilde{T} : V \to \text{range}(T)$ by $\widetilde{T}(\mathbf{v}) = T(\mathbf{v})$ for all $\mathbf{v}$ in V. Since T is a one-to-one linear transformation, so is the mapping $\widetilde{T}$. Moreover, range $\left(\widetilde{T}\right) = \text{range}(T)$, so $\widetilde{T}$ is onto. Thus, $\widetilde{T}$ is an isomorphism, and V and range (T) are isomorphic.

9.3   The Matrix of a Linear Transformation

1. $\mathbf{v} = (-4) \begin{bmatrix} 3 \\ 2 \end{bmatrix} + (1) \begin{bmatrix} 1 \\ 4 \end{bmatrix} = \begin{bmatrix} -11 \\ -4 \end{bmatrix}$

3. $\mathbf{v} = (-1)\left(x^2 - x + 3\right) + (0)\left(3x^2 + 4\right) + (3)(-5x - 2) = -x^2 - 14x - 9$

5. Set $\mathbf{v} = \begin{bmatrix} a \\ b \end{bmatrix}_\mathcal{G}$, then $\begin{bmatrix} 8 \\ 9 \end{bmatrix} = a\begin{bmatrix} 2 \\ 0 \end{bmatrix} + b\begin{bmatrix} 0 \\ 3 \end{bmatrix} = \begin{bmatrix} 2a \\ 3b \end{bmatrix}$. So $a = 4$ and $b = 3$, and $\mathbf{v}_\mathcal{G} = \begin{bmatrix} 4 \\ 3 \end{bmatrix}_\mathcal{G}$.

7. Set $\mathbf{v} = \begin{bmatrix} a \\ b \\ c \end{bmatrix}_\mathcal{G}$, then $12x^2 - 15x + 30 = a\left(4x^2\right) + b(3x) + c(5) = (4a)x^2 + (3b)x + 5c$. So $4a =$

   $12 \quad \Rightarrow \quad a = 3, 3b = -15 \quad \Rightarrow \quad b = -5$, and $5c = 30 \quad \Rightarrow \quad c = 6$. Thus $\mathbf{v}_\mathcal{G} = \begin{bmatrix} 3 \\ -5 \\ 6 \end{bmatrix}_\mathcal{G}$.

9. Set $\mathbf{v} = \begin{bmatrix} a \\ b \end{bmatrix}_\mathcal{G}$, then $\begin{bmatrix} 5 \\ -5 \end{bmatrix} = a\begin{bmatrix} 1 \\ 2 \end{bmatrix} + b\begin{bmatrix} 3 \\ 1 \end{bmatrix} = \begin{bmatrix} a + 3b \\ 2a + b \end{bmatrix}$. We solve the system

$$a + 3b = 5$$
$$2a + b = -5$$

and obtain $a = -4$ and $b = 3$. Thus $\mathbf{v} = \begin{bmatrix} -4 \\ 3 \end{bmatrix}_{\mathcal{G}}$.

11. Set $\mathbf{v} = \begin{bmatrix} a \\ b \\ c \end{bmatrix}_{\mathcal{G}}$, then $9x + 1 = a\left(x^2 - 1\right) + b\left(2x + 1\right) + c\left(3x^2 - 1\right) = (a + 3c)\,x^2 + 2bx + (b - a - c)$. We solve the system

$$a + 3c = 0$$
$$2b = 9$$
$$b - a - c = 1$$

and obtain $a = \frac{21}{4}$, $b = \frac{9}{2}$, and $c = -\frac{7}{4}$. Thus $\mathbf{v} = \begin{bmatrix} \frac{21}{4} \\ \frac{9}{2} \\ -\frac{7}{4} \end{bmatrix}_{\mathcal{G}}$.

13. $T\left(\mathbf{v}_{\mathcal{G}}\right) = A\mathbf{v}_{\mathcal{G}} = \begin{bmatrix} 1 & 3 \\ 2 & -1 \end{bmatrix}\begin{bmatrix} 4 \\ -3 \end{bmatrix}_{\mathcal{G}} = \begin{bmatrix} -5 \\ 11 \end{bmatrix}_{\mathcal{Q}} = (-5)\begin{bmatrix} 1 \\ 1 \end{bmatrix} + (11)\begin{bmatrix} 2 \\ 3 \end{bmatrix} = \begin{bmatrix} 17 \\ 28 \end{bmatrix}$

15. $T\left(\mathbf{v}_{\mathcal{G}}\right) = A\mathbf{v}_{\mathcal{G}} = \begin{bmatrix} 1 & 1 & 2 \\ 0 & 1 & 3 \\ 0 & 1 & 1 \end{bmatrix}\begin{bmatrix} 1 \\ 3 \\ 4 \end{bmatrix}_{\mathcal{G}} = \begin{bmatrix} 12 \\ 15 \\ 7 \end{bmatrix}_{\mathcal{Q}} = (12)\left(x^2 - 2x\right) + (15)\left(x^2 + x + 4\right) + (7)\left(3x + 1\right) = 27x^2 + 12x + 67$

17. $T\left(\mathbf{v}_{\mathcal{G}}\right) = A\mathbf{v}_{\mathcal{G}} = \begin{bmatrix} 0 & 4 & 3 \\ 2 & -1 & -3 \\ 2 & -2 & -1 \end{bmatrix}\begin{bmatrix} 2 \\ -3 \\ 1 \end{bmatrix}_{\mathcal{G}} = \begin{bmatrix} -9 \\ 4 \\ 9 \end{bmatrix}_{\mathcal{Q}} = (-9)\sin(x) + (4)\cos(x) + (9)\,e^{-x} = 4\cos x - 9\sin x + 9e^{-x}$

19. $T\left(\mathbf{e}_1\right) = T\left(\begin{bmatrix} 1 \\ 0 \end{bmatrix}\right) = (0)\,x - 1 = -1 \quad \Rightarrow \quad \left[T\left(\mathbf{e}_1\right)\right]_{\mathcal{Q}} = \begin{bmatrix} 0 \\ -1 \end{bmatrix}_{\mathcal{Q}}$ and $T\left(\mathbf{e}_2\right) = T\left(\begin{bmatrix} 0 \\ 1 \end{bmatrix}\right) = (1)\,x - 0 = x \quad \Rightarrow \quad \left[T\left(\mathbf{e}_2\right)\right]_{\mathcal{Q}} = \begin{bmatrix} 1 \\ 0 \end{bmatrix}_{\mathcal{Q}}$. Therefore $A = \begin{bmatrix} 0 & 1 \\ -1 & 0 \end{bmatrix}$.

21. $T\left(x^2\right) = \left(x^2\right)' = 2x \quad \Rightarrow \quad \left[T\left(x^2\right)\right]_{\mathcal{Q}} = \begin{bmatrix} 2 \\ 0 \end{bmatrix}_{\mathcal{Q}}$, $T(x) = (x)' = 1 \quad \Rightarrow \quad \left[T(x)\right]_{\mathcal{Q}} = \begin{bmatrix} 0 \\ 1 \end{bmatrix}_{\mathcal{Q}}$, and $T(1) = (1)' = 0 \quad \Rightarrow \quad \left[T(1)\right]_{\mathcal{Q}} = \begin{bmatrix} 0 \\ 0 \end{bmatrix}_{\mathcal{Q}}$. Therefore $A = \begin{bmatrix} 2 & 0 & 0 \\ 0 & 1 & 0 \end{bmatrix}$.

23. $T\left(\mathbf{e}_1\right) = T\left(\begin{bmatrix} 1 \\ 0 \end{bmatrix}\right) = \begin{bmatrix} 0 - 1 \\ 1 + 2\,(0) \end{bmatrix} = \begin{bmatrix} -1 \\ 1 \end{bmatrix}$. Set $\begin{bmatrix} -1 \\ 1 \end{bmatrix} = \begin{bmatrix} a \\ b \end{bmatrix}_{\mathcal{Q}}$, then $\begin{bmatrix} -1 \\ 1 \end{bmatrix} = a\begin{bmatrix} 1 \\ 1 \end{bmatrix} + b\begin{bmatrix} 1 \\ 2 \end{bmatrix} = \begin{bmatrix} a + b \\ a + 2b \end{bmatrix}$. Solve the system

$$a + b = -1$$
$$a + 2b = 1$$

to obtain $a = -3$ and $b = 2$. Thus $\left[T\left(\mathbf{e}_1\right)\right]_{\mathcal{Q}} = \begin{bmatrix} -3 \\ 2 \end{bmatrix}_{\mathcal{Q}}$. $T\left(\mathbf{e}_2\right) = T\left(\begin{bmatrix} 0 \\ 1 \end{bmatrix}\right) = \begin{bmatrix} 1 - 0 \\ 0 + 2\,(1) \end{bmatrix} = \begin{bmatrix} 1 \\ 2 \end{bmatrix}$. Set $\begin{bmatrix} 1 \\ 2 \end{bmatrix} = \begin{bmatrix} a \\ b \end{bmatrix}_{\mathcal{Q}}$, then $\begin{bmatrix} 1 \\ 2 \end{bmatrix} = a\begin{bmatrix} 1 \\ 1 \end{bmatrix} + b\begin{bmatrix} 1 \\ 2 \end{bmatrix} = \begin{bmatrix} a + b \\ a + 2b \end{bmatrix}$. Solve the system

$$a + b = 1$$
$$a + 2b = 2$$

to obtain $a = 0$ and $b = 1$. Thus $[T(\mathbf{e}_2)]_Q = \begin{bmatrix} 0 \\ 1 \end{bmatrix}_Q$. Therefore $A = \begin{bmatrix} -3 & 0 \\ 2 & 1 \end{bmatrix}$.

25. $T\left(\begin{bmatrix} 2 \\ 1 \end{bmatrix}\right) = (-1)x + 2 + 1 = -x + 3$. Set $-x + 3 = \begin{bmatrix} a \\ b \end{bmatrix}_Q$, then $-x + 3 = a(5x + 3) + b(2x + 1) = (5a + 2b)x + (3a + b)$. Solve the system

$$5a + 2b = -1$$
$$3a + b = 3$$

to obtain $a = 7$ and $b = -18$. Thus $\left[T\left(\begin{bmatrix} 2 \\ 1 \end{bmatrix}\right)\right]_Q = \begin{bmatrix} 7 \\ -18 \end{bmatrix}_Q$. $T\left(\begin{bmatrix} 3 \\ 2 \end{bmatrix}\right) = (-2)x + 3 + 2 =$

$-2x + 5$. Set $-2x + 5 = \begin{bmatrix} a \\ b \end{bmatrix}_Q$, then $-2x + 5 = a(5x + 3) + b(2x + 1) = (5a + 2b)x + (3a + b)$. Solve

the system

$$5a + 2b = -2$$
$$3a + b = 5$$

to obtain $a = 12$ and $b = -31$. Thus $\left[T\left(\begin{bmatrix} 3 \\ 2 \end{bmatrix}\right)\right]_Q = \begin{bmatrix} 12 \\ -31 \end{bmatrix}_Q$. Therefore $A = \begin{bmatrix} 7 & 12 \\ -18 & -31 \end{bmatrix}$.

27. (a) $T(\mathbf{g}_1) = \begin{bmatrix} a \\ d \end{bmatrix}_Q = a\mathbf{q}_1 + d\mathbf{q}_2 = (a/2)(2\mathbf{q}_1) + d\mathbf{q}_2 = \begin{bmatrix} a/2 \\ d \end{bmatrix}_R$

$T(\mathbf{g}_2) = \begin{bmatrix} b \\ e \end{bmatrix}_Q = b\mathbf{q}_1 + e\mathbf{q}_2 = (b/2)(2\mathbf{q}_1) + e\mathbf{q}_2 = \begin{bmatrix} b/2 \\ e \end{bmatrix}_R$

$T(\mathbf{g}_3) = \begin{bmatrix} c \\ f \end{bmatrix}_Q = c\mathbf{q}_1 + f\mathbf{q}_2 = (c/2)(2\mathbf{q}_1) + f\mathbf{q}_2 = \begin{bmatrix} c/2 \\ f \end{bmatrix}_R$

Thus the matrix of T with respect to the bases $\mathcal{H}$ and $\mathcal{R}$ is $\begin{bmatrix} a/2 & b/2 & c/2 \\ d & e & f \end{bmatrix}$.

(b) $T(3\mathbf{g}_1) = 3T(\mathbf{g}_1) = 3\begin{bmatrix} a \\ d \end{bmatrix}_Q = 3(a\mathbf{q}_1 + d\mathbf{q}_2) = (3a)\mathbf{q}_1 + (3d)\mathbf{q}_2 = \begin{bmatrix} 3a \\ 3d \end{bmatrix}_R$

$T(\mathbf{g}_2) = \begin{bmatrix} b \\ e \end{bmatrix}_Q = b\mathbf{q}_1 + e\mathbf{q}_2 = \begin{bmatrix} b \\ e \end{bmatrix}_R$

$T(\mathbf{g}_3) = \begin{bmatrix} c \\ f \end{bmatrix}_Q = c\mathbf{q}_1 + f\mathbf{q}_2 = \begin{bmatrix} c \\ f \end{bmatrix}_R$

Thus the matrix of T with respect to the bases $\mathcal{H}$ and $\mathcal{R}$ is $\begin{bmatrix} 3a & b & c \\ 3d & e & f \end{bmatrix}$.

29. (a) $T(\mathbf{g}_3) = \begin{bmatrix} c \\ f \end{bmatrix}_Q = c\mathbf{q}_1 + f\mathbf{q}_2 = \begin{bmatrix} c \\ f \end{bmatrix}_R$

$T(\mathbf{g}_1) = \begin{bmatrix} a \\ d \end{bmatrix}_Q = a\mathbf{q}_1 + d\mathbf{q}_2 = \begin{bmatrix} a \\ d \end{bmatrix}_R$

$T(\mathbf{g}_2) = \begin{bmatrix} b \\ e \end{bmatrix}_Q = b\mathbf{q}_1 + e\mathbf{q}_2 = \begin{bmatrix} b \\ e \end{bmatrix}_R$

Thus the matrix of T with respect to the bases $\mathcal{H}$ and $\mathcal{R}$ is $\begin{bmatrix} c & a & b \\ f & d & e \end{bmatrix}$.

(b) $T(\mathbf{g}_1) = \begin{bmatrix} a \\ d \end{bmatrix}_Q = a\mathbf{q}_1 + d\mathbf{q}_2 = d\mathbf{q}_2 + a\mathbf{q}_1 = \begin{bmatrix} d \\ a \end{bmatrix}_R$

$T(\mathbf{g}_3) = \begin{bmatrix} c \\ f \end{bmatrix}_Q = c\mathbf{q}_1 + f\mathbf{q}_2 = f\mathbf{q}_2 + c\mathbf{q}_1 = \begin{bmatrix} f \\ c \end{bmatrix}_R$

$$T\left(\mathbf{g}_2\right) = \left[\begin{array}{c} b \\ e \end{array}\right]_{\mathcal{Q}} = b\mathbf{q}_1 + e\mathbf{q}_2 = e\mathbf{q}_2 + b\mathbf{q}_1 = \left[\begin{array}{c} e \\ b \end{array}\right]_{\mathcal{R}}.$$

Thus the matrix of T with respect to the bases $\mathcal{H}$ and $\mathcal{R}$ is $\left[\begin{array}{ccc} d & f & e \\ a & c & b \end{array}\right]$.

31. Write $x = (1)\,x + (0)\,1 = (0)\,1 + (1)\,x = \left[\begin{array}{c} 0 \\ 1 \end{array}\right]_{\mathcal{Q}}$, so $T^{-1}\left(x\right) = T^{-1}\left(\left[\begin{array}{c} 0 \\ 1 \end{array}\right]_{\mathcal{Q}}\right) = A^{-1}\left[\begin{array}{c} 0 \\ 1 \end{array}\right]_{\mathcal{Q}} = $
$\left[\begin{array}{cc} 1 & 3 \\ 0 & 1 \end{array}\right]^{-1}\left[\begin{array}{c} 0 \\ 1 \end{array}\right]_{\mathcal{Q}} = \left[\begin{array}{c} -3 \\ 1 \end{array}\right]_{\mathcal{G}} = (-3)\left(x+1\right) + (1)\left(x-1\right) = -2x-4.$

33. Let $x + 1 = a\left(x\right) + b\left(2x+1\right) = \left(a+2b\right)x + b$, and solve the system

$$\begin{aligned} a + 2b &= 1 \\ b &= 1 \end{aligned}$$

to obtain $a = -1$, and $b = 1$. Thus $x + 1 = \left[\begin{array}{c} -1 \\ 1 \end{array}\right]_{\mathcal{Q}}$, and $T^{-1}\left(x+1\right) = T^{-1}\left(\left[\begin{array}{c} -1 \\ 1 \end{array}\right]_{\mathcal{Q}}\right) = $
$A^{-1}\left[\begin{array}{c} -1 \\ 1 \end{array}\right]_{\mathcal{Q}} = \left[\begin{array}{cc} 2 & 1 \\ 3 & 2 \end{array}\right]^{-1}\left[\begin{array}{c} -1 \\ 1 \end{array}\right]_{\mathcal{Q}} = \left[\begin{array}{c} -3 \\ 5 \end{array}\right]_{\mathcal{G}} = (-3)\left[\begin{array}{c} 1 \\ 3 \end{array}\right] + (5)\left[\begin{array}{c} 2 \\ 1 \end{array}\right] = \left[\begin{array}{c} 7 \\ -4 \end{array}\right].$

35. For example, let $\mathbf{v} = 2x^2 - 3$, and $\mathcal{G} = \left\{x^2, x, 1\right\}$. Then $(2)\,x^2 + (0)\,x + (-3)\,1 = 2x^2 - 3 = \mathbf{v}$, so
$\mathbf{v}_{\mathcal{G}} = \left[\begin{array}{c} 2 \\ 0 \\ -3 \end{array}\right].$

37. For example, let $\mathcal{G} = \left\{\left[\begin{array}{c} -7/3 \\ 0 \end{array}\right], \left[\begin{array}{c} 0 \\ 5/4 \end{array}\right]\right\}$. Then $(-3)\left[\begin{array}{c} -7/3 \\ 0 \end{array}\right] + (4)\left[\begin{array}{c} 0 \\ 5/4 \end{array}\right] = \left[\begin{array}{c} 7 \\ 5 \end{array}\right]$, so $\left[\begin{array}{c} 7 \\ 5 \end{array}\right] = $
$\left[\begin{array}{c} -3 \\ 4 \end{array}\right]_{\mathcal{G}}.$

39. Let $V = \mathbf{R}^3$ and $W = \mathbf{R}^2$, and let $\mathcal{G} = \left\{\left[\begin{array}{c} 1 \\ 0 \\ 0 \end{array}\right], \left[\begin{array}{c} 0 \\ 1 \\ 0 \end{array}\right], \left[\begin{array}{c} 0 \\ 0 \\ 1 \end{array}\right]\right\}$ be the basis for V, and $\mathcal{Q} = $
$\left\{\left[\begin{array}{c} 1 \\ 0 \end{array}\right], \left[\begin{array}{c} 0 \\ 1 \end{array}\right]\right\}$ be the basis for W. Define $T\left(\mathbf{v}\right) = A\mathbf{v}$, where $A = \left[\begin{array}{ccc} 2 & 1 & 2 \\ 0 & 3 & 1 \end{array}\right]$. Then $T\left(\left[\begin{array}{c} 1 \\ 0 \\ 0 \end{array}\right]\right) = $
$\left[\begin{array}{ccc} 2 & 1 & 2 \\ 0 & 3 & 1 \end{array}\right]\left[\begin{array}{c} 1 \\ 0 \\ 0 \end{array}\right] = \left[\begin{array}{c} 2 \\ 0 \end{array}\right] = \left[\begin{array}{c} 2 \\ 0 \end{array}\right]_{\mathcal{Q}},$
$T\left(\left[\begin{array}{c} 0 \\ 1 \\ 0 \end{array}\right]\right) = \left[\begin{array}{ccc} 2 & 1 & 2 \\ 0 & 3 & 1 \end{array}\right]\left[\begin{array}{c} 0 \\ 1 \\ 0 \end{array}\right] = \left[\begin{array}{c} 1 \\ 3 \end{array}\right] = \left[\begin{array}{c} 1 \\ 3 \end{array}\right]_{\mathcal{Q}},$ and
$T\left(\left[\begin{array}{c} 0 \\ 0 \\ 1 \end{array}\right]\right) = \left[\begin{array}{ccc} 2 & 1 & 2 \\ 0 & 3 & 1 \end{array}\right]\left[\begin{array}{c} 0 \\ 0 \\ 1 \end{array}\right] = \left[\begin{array}{c} 2 \\ 1 \end{array}\right] = \left[\begin{array}{c} 2 \\ 1 \end{array}\right]_{\mathcal{Q}}.$ So the matrix representing T with respect to
the bases $\mathcal{G}$ and $\mathcal{Q}$ is $\left[\begin{array}{ccc} 2 & 1 & 2 \\ 0 & 3 & 1 \end{array}\right] = A.$

41. False. For instance, $T : \mathbf{R}^2 \to \mathbf{R}^3$ with respect to any choice of bases will be represented by a 3×2 matrix.

43. True. Since $\left[\begin{array}{c} a \\ b \end{array}\right]_{\mathcal{G}}$ has two components, the basis $\mathcal{G} = \left\{\mathbf{g}_1, \mathbf{g}_2\right\}$ has two vectors, and $\left[\begin{array}{c} 2a \\ -3b \end{array}\right]_{\mathcal{G}}$ is the coordinate vector of $(2a)\,\mathbf{g}_1 + (-3b)\,\mathbf{g}_2.$

45. True. Let $\mathcal{G} = \{\mathbf{g}_1, \ldots, \mathbf{g}_m\}$ be a basis of V, and write $\mathbf{0}_V = \begin{bmatrix} c_1 \\ \vdots \\ c_m \end{bmatrix}_\mathcal{G} = c_1\mathbf{g}_1 + \cdots + c_m\mathbf{g}_m$. Since $\mathcal{G}$

is a basis, $\{\mathbf{g}_1, \ldots, \mathbf{g}_m\}$ is linearly independent, each $c_i = 0$, so $\mathbf{0}_V = \begin{bmatrix} 0 \\ \vdots \\ 0 \end{bmatrix}_\mathcal{G}$.

47. Let $\mathcal{G} = \{\mathbf{g}_1, \ldots, \mathbf{g}_m\}$ be a basis of V. We first note the general result that $[c\mathbf{v}]_\mathcal{G} = c[\mathbf{v}]_\mathcal{G}$ and $[\mathbf{v}_1 + \mathbf{v}_2]_\mathcal{G} = [\mathbf{v}_1]_\mathcal{G} + [\mathbf{v}_2]_G$, which follows from $c\mathbf{v} = c(a_1\mathbf{g}_1 + \cdots + a_m\mathbf{g}_m) = (ca_1)\mathbf{g}_1 + \cdots + (ca_m)\mathbf{g}_m$, and $\mathbf{v}_1 + \mathbf{v}_2 = (a_1\mathbf{g}_1 + \cdots + a_m\mathbf{g}_m) + (b_1\mathbf{g}_1 + \cdots + b_m\mathbf{g}_m) = (a_1 + b_1)\mathbf{g}_1 + \cdots + (a_m + b_m)\mathbf{g}_m$. Now suppose that $\{[\mathbf{v}_1]_\mathcal{G}, \ldots, [\mathbf{v}_k]_\mathcal{G}\}$ is linearly independent, and consider $c_1\mathbf{v}_1 + \cdots + c_k\mathbf{v}_k = \mathbf{0}_V$. Then $[c_1\mathbf{v}_1 + \cdots + c_k\mathbf{v}_k]_\mathcal{G} = [\mathbf{0}_V]_\mathcal{G} \Rightarrow c_1[\mathbf{v}_1]_\mathcal{G} + \cdots + c_k[\mathbf{v}_k]_\mathcal{G} = \mathbf{0}$, and hence each $c_i = 0$ since $\{[\mathbf{v}_1]_\mathcal{G}, \ldots, [\mathbf{v}_k]_\mathcal{G}\}$ is linearly independent. Thus $\{\mathbf{v}_1, \ldots, \mathbf{v}_k\}$ is linearly independent. Suppose now that $\{[\mathbf{v}_1]_\mathcal{G}, \ldots, [\mathbf{v}_k]_\mathcal{G}\}$ is linearly dependent. There exist c_i, with at least one $c_i \neq 0$, such that $c_1[\mathbf{v}_1]_\mathcal{G} + \cdots + c_k[\mathbf{v}_k]_\mathcal{G} = [c_1\mathbf{v}_1 + \cdots + c_k\mathbf{v}_k]_\mathcal{G} = \mathbf{0}$. Hence $c_1\mathbf{v}_1 + \cdots + c_k\mathbf{v}_k = (0)\mathbf{g}_1 + \cdots + (0)\mathbf{g}_m = \mathbf{0}_V$. But this shows that $\{\mathbf{v}_1, \ldots, \mathbf{v}_k\}$ is linearly dependent. Consequently, if $\{\mathbf{v}_1, \ldots, \mathbf{v}_k\}$ is linearly independent, then $\{[\mathbf{v}_1]_\mathcal{G}, \ldots, [\mathbf{v}_k]_\mathcal{G}\}$ is linearly independent.

49. Suppose $\mathbf{v} = c_1\mathbf{v}_1 + \cdots + c_k\mathbf{v}_k$. Then $\mathbf{v}_\mathcal{G} = [c_1\mathbf{v}_1 + \cdots + c_k\mathbf{v}_k]_\mathcal{G} = c_1[\mathbf{v}_1]_\mathcal{G} + \cdots + c_k[\mathbf{v}_k]_\mathcal{G}$, which follows from the general results that $[c\mathbf{v}]_\mathcal{G} = c[\mathbf{v}]_\mathcal{G}$ and $[\mathbf{v}_1 + \mathbf{v}_2]_\mathcal{G} = [\mathbf{v}_1]_\mathcal{G} + [\mathbf{v}_2]_G$ as established in Exercise 46. Thus, if $\mathbf{v}$ is a linear combination of $\mathbf{v}_1, \ldots, \mathbf{v}_k$, then $\mathbf{v}_\mathcal{G}$ is a linear combination of $[\mathbf{v}_1]_\mathcal{G}, \ldots, [\mathbf{v}_k]_\mathcal{G}$. Now suppose $\mathbf{v}_\mathcal{G} = c_1[\mathbf{v}_1]_\mathcal{G} + \cdots + c_k[\mathbf{v}_k]_\mathcal{G} = [c_1\mathbf{v}_1 + \cdots + c_k\mathbf{v}_k]_\mathcal{G}$. Then by Exercise 47, $\mathbf{v} = c_1\mathbf{v}_1 + \cdots + c_k\mathbf{v}_k$. Thus, if $\mathbf{v}_\mathcal{G}$ is a linear combination of $[\mathbf{v}_1]_\mathcal{G}, \ldots, [\mathbf{v}_k]_\mathcal{G}$, then $\mathbf{v}$ is a linear combination of $\mathbf{v}_1, \ldots, \mathbf{v}_k$.

51. (a) $[T^2(\mathbf{v})]_\mathcal{G} = [T(T(\mathbf{v}))]_\mathcal{G} = A[T(\mathbf{v})]_\mathcal{G} = A(A\mathbf{v}_\mathcal{G}) = A^2\mathbf{v}_\mathcal{G}$, so A^2 is the matrix of the linear transformation T^2.

(b) By induction on n. If $n = 1$, then $[T^1(\mathbf{v})]_\mathcal{G} = [T(\mathbf{v})]_\mathcal{G} = A\mathbf{v}_\mathcal{G} = A^1\mathbf{v}_\mathcal{G}$, and the result holds. Suppose that $[T^{n-1}(\mathbf{v})]_\mathcal{G} = A^{n-1}\mathbf{v}_\mathcal{G}$. Then $[T^n(\mathbf{v})]_\mathcal{G} = [T(T^{n-1}\mathbf{v})]_\mathcal{G} = A[T^{n-1}(\mathbf{v})]_\mathcal{G} = A(A^{n-1}\mathbf{v}_\mathcal{G}) = A^n\mathbf{v}_\mathcal{G}$. Therefore, by induction, A^n is the matrix of the transformation T^n for all n.

9.4 Similarity

1. Set $2x - 1 = s_{11}(x) + s_{21}(1)$, to obtain $s_{11} = 2$ and $s_{21} = -1$. Set $5x + 4 = s_{21}(x) + s_{22}(1)$, to obtain $s_{21} = 5$ and $s_{22} = 4$. Thus $S = \begin{bmatrix} 2 & 5 \\ -1 & 4 \end{bmatrix}$.

3. Set $\begin{bmatrix} 3 & 2 \\ 1 & 0 \end{bmatrix} = s_{11}\begin{bmatrix} 1 & 0 \\ 0 & 0 \end{bmatrix} + s_{21}\begin{bmatrix} 0 & 1 \\ 0 & 0 \end{bmatrix} + s_{31}\begin{bmatrix} 0 & 0 \\ 1 & 0 \end{bmatrix} + s_{41}\begin{bmatrix} 0 & 0 \\ 0 & 1 \end{bmatrix}$, to obtain $s_{11} = 3$, $s_{21} = 2$, $s_{31} = 1$, and $s_{41} = 0$. Set $\begin{bmatrix} 4 & 0 \\ 0 & 2 \end{bmatrix} = s_{12}\begin{bmatrix} 1 & 0 \\ 0 & 0 \end{bmatrix} + s_{22}\begin{bmatrix} 0 & 1 \\ 0 & 0 \end{bmatrix} + s_{32}\begin{bmatrix} 0 & 0 \\ 1 & 0 \end{bmatrix} + s_{42}\begin{bmatrix} 0 & 0 \\ 0 & 1 \end{bmatrix}$, to obtain $s_{12} = 4$, $s_{22} = 0$, $s_{32} = 0$, and $s_{42} = 2$. Set $\begin{bmatrix} 1 & 7 \\ 5 & 1 \end{bmatrix} = s_{13}\begin{bmatrix} 1 & 0 \\ 0 & 0 \end{bmatrix} + s_{23}\begin{bmatrix} 0 & 1 \\ 0 & 0 \end{bmatrix} + s_{33}\begin{bmatrix} 0 & 0 \\ 1 & 0 \end{bmatrix} + s_{43}\begin{bmatrix} 0 & 0 \\ 0 & 1 \end{bmatrix}$, to obtain $s_{13} = 1$, $s_{23} = 7$, $s_{33} = 5$, and $s_{43} = 1$. Set $\begin{bmatrix} 0 & 6 \\ 2 & 3 \end{bmatrix} = s_{14}\begin{bmatrix} 1 & 0 \\ 0 & 0 \end{bmatrix} + s_{24}\begin{bmatrix} 0 & 1 \\ 0 & 0 \end{bmatrix} + s_{34}\begin{bmatrix} 0 & 0 \\ 1 & 0 \end{bmatrix} + s_{44}\begin{bmatrix} 0 & 0 \\ 0 & 1 \end{bmatrix}$, to obtain $s_{14} = 0$, $s_{24} = 6$, $s_{34} = 2$, and

$s_{44} = 3$. Thus $S = \begin{bmatrix} 3 & 4 & 1 & 0 \\ 2 & 0 & 7 & 6 \\ 1 & 0 & 5 & 2 \\ 0 & 2 & 1 & 3 \end{bmatrix}$.

5. Set $x - 2 = s_{11}(x) + s_{21}(1) + s_{31}(x^2)$, to obtain $s_{11} = 1$, $s_{21} = -2$, and $s_{31} = 0$. Set $x^2 + 9x = s_{12}(x) + s_{22}(1) + s_{32}(x^2)$, to obtain $s_{12} = 9$, $s_{22} = 0$, and $s_{32} = 1$. Set $x^2 - x - 1 = s_{13}(x) + s_{23}(1) + s_{33}(x^2)$, to obtain $s_{13} = -1$, $s_{23} = -1$, and $s_{33} = 1$. Thus $S = \begin{bmatrix} 1 & 9 & -1 \\ -2 & 0 & -1 \\ 0 & 1 & 1 \end{bmatrix}$.

7. Set $7x + 4 = s_{11}(2x + 1) + s_{21}(5x + 3) = (2s_{11} + 5s_{21})x + (s_{11} + 3s_{21})$, and solve the system

$$2s_{11} + 5s_{21} = 7$$
$$s_{11} + 3s_{21} = 4$$

to obtain $s_{11} = 1$, and $s_{21} = 1$. Set $3x + 2 = s_{12}(2x + 1) + s_{22}(5x + 3) = (2s_{12} + 5s_{22})x + (s_{12} + 3s_{22})$, and solve the system

$$2s_{12} + 5s_{22} = 3$$
$$s_{12} + 3s_{22} = 2$$

to obtain $s_{12} = -1$, and $s_{22} = 1$. Thus $S = \begin{bmatrix} 1 & -1 \\ 1 & 1 \end{bmatrix}$.

9. $A = S^{-1}BS = \begin{bmatrix} 1 & 3 \\ 2 & 7 \end{bmatrix}^{-1} \begin{bmatrix} 2 & 3 \\ -4 & 1 \end{bmatrix} \begin{bmatrix} 1 & 3 \\ 2 & 7 \end{bmatrix} = \begin{bmatrix} 62 & 204 \\ -18 & -59 \end{bmatrix}$

11. $A = S^{-1}BS = \begin{bmatrix} 1 & 0 & 0 \\ 2 & 1 & 0 \\ 1 & 3 & 1 \end{bmatrix}^{-1} \begin{bmatrix} 1 & 0 & 2 \\ 0 & 1 & 1 \\ 1 & 2 & -1 \end{bmatrix} \begin{bmatrix} 1 & 0 & 0 \\ 2 & 1 & 0 \\ 1 & 3 & 1 \end{bmatrix} = \begin{bmatrix} 3 & 6 & 2 \\ -3 & -8 & -3 \\ 10 & 17 & 6 \end{bmatrix}$

13. Since $3x + 1 = (3)(x) + (1)(1)$ and $2x + 1 = (2)(x) + (1)(1)$, $S = \begin{bmatrix} 3 & 2 \\ 1 & 1 \end{bmatrix}$. Thus $A = S^{-1}BS = \begin{bmatrix} 3 & 2 \\ 1 & 1 \end{bmatrix}^{-1} \begin{bmatrix} 1 & 2 \\ 1 & 1 \end{bmatrix} \begin{bmatrix} 3 & 2 \\ 1 & 1 \end{bmatrix} = \begin{bmatrix} -3 & -2 \\ 7 & 5 \end{bmatrix}$.

15. Set $2x - 1 = s_{11}(x + 3) + s_{21}(2x + 5) = (s_{11} + 2s_{21})x + (3s_{11} + 5s_{21})$, and solve the system

$$s_{11} + 2s_{21} = 2$$
$$3s_{11} + 5s_{21} = -1$$

to obtain $s_{11} = -12$, and $s_{21} = 7$. Set $-3x + 2 = s_{12}(x + 3) + s_{22}(2x + 5) = (s_{12} + 2s_{22})x + (3s_{12} + 5s_{22})$, and solve the system

$$s_{12} + 2s_{22} = -3$$
$$3s_{12} + 5s_{22} = 2$$

to obtain $s_{12} = 19$, and $s_{22} = -11$. Thus $S = \begin{bmatrix} -12 & 19 \\ 7 & -11 \end{bmatrix}$. So

$A = S^{-1}BS = \begin{bmatrix} -12 & 19 \\ 7 & -11 \end{bmatrix}^{-1} \begin{bmatrix} 2 & 2 \\ 4 & 1 \end{bmatrix} \begin{bmatrix} -12 & 19 \\ 7 & -11 \end{bmatrix} = \begin{bmatrix} -889 & 1411 \\ -562 & 892 \end{bmatrix}$.

17. $\det(A) = \det\left(\begin{bmatrix} 1 & 3 \\ 2 & 5 \end{bmatrix}\right) = -1$ and $\det(B) = \det\left(\begin{bmatrix} 2 & 1 \\ 3 & 2 \end{bmatrix}\right) = 1$. Since $\det(A) \neq \det(B)$, A and B are not similar matrices.

19. Let $S = \begin{bmatrix} a_{11} & a_{12} & a_{13} \\ a_{21} & a_{22} & a_{23} \\ a_{31} & a_{32} & a_{33} \end{bmatrix}$, and set $SA = BS$ to obtain the system of equations

$$a_{12} - a_{21} - 2a_{31} = 0$$
$$a_{13} - 4a_{12} - a_{11} - a_{22} - 2a_{32} = 0$$
$$3a_{11} - 3a_{12} + a_{13} - a_{23} - 2a_{33} = 0$$
$$a_{21} - a_{11} + a_{22} - a_{31} = 0$$
$$a_{23} - a_{21} - 3a_{22} - a_{12} - a_{32} = 0$$
$$3a_{21} - a_{13} - 3a_{22} + 2a_{23} - a_{33} = 0$$
$$2a_{31} - a_{21} + a_{32} = 0$$
$$a_{33} - a_{31} - 2a_{32} - a_{22} = 0$$
$$3a_{31} - a_{23} - 3a_{32} + 3a_{33} = 0$$

One solution, obtained by letting $a_{22} = a_{32} = 0$ and $a_{33} = 1$, is $S = \begin{bmatrix} 1 & 4 & 17 \\ 2 & 0 & 6 \\ 1 & 0 & 1 \end{bmatrix}$. Then

$$S^{-1}BS = \begin{bmatrix} 1 & 4 & 17 \\ 2 & 0 & 6 \\ 1 & 0 & 1 \end{bmatrix}^{-1} \begin{bmatrix} 1 & 1 & 2 \\ 1 & 0 & 1 \\ 0 & 1 & -1 \end{bmatrix} \begin{bmatrix} 1 & 4 & 17 \\ 2 & 0 & 6 \\ 1 & 0 & 1 \end{bmatrix} = \begin{bmatrix} 1 & -1 & 3 \\ 1 & -3 & -3 \\ 0 & 1 & 2 \end{bmatrix} = A,$$

so A and B are similar matrices.

21. For example, let $V = \mathbf{R}^2$, and let $\mathcal{G} = \left\{ \begin{bmatrix} 3 \\ 2 \end{bmatrix}, \begin{bmatrix} 4 \\ 3 \end{bmatrix} \right\}$ and $\mathcal{H} = \left\{ \begin{bmatrix} 1 \\ 0 \end{bmatrix}, \begin{bmatrix} 0 \\ 1 \end{bmatrix} \right\}$. Since $\begin{bmatrix} 3 \\ 2 \end{bmatrix} =$ $(3)\begin{bmatrix} 1 \\ 0 \end{bmatrix} + (2)\begin{bmatrix} 0 \\ 1 \end{bmatrix}$ and $\begin{bmatrix} 4 \\ 3 \end{bmatrix} = (4)\begin{bmatrix} 1 \\ 0 \end{bmatrix} + (3)\begin{bmatrix} 0 \\ 1 \end{bmatrix}$, $S = \begin{bmatrix} 3 & 4 \\ 2 & 3 \end{bmatrix}$.

23. Let $B = \begin{bmatrix} 1 & 2 \\ 3 & 4 \end{bmatrix}$, and $A = S^{-1}BS = \begin{bmatrix} 5 & 2 \\ 8 & 3 \end{bmatrix}^{-1} \begin{bmatrix} 1 & 2 \\ 3 & 4 \end{bmatrix} \begin{bmatrix} 5 & 2 \\ 8 & 3 \end{bmatrix} = \begin{bmatrix} 31 & 12 \\ -67 & -26 \end{bmatrix}$, then A and B are similar matrices related by S.

25. True, by Definition 9.19.

27. False. For example, $A = \begin{bmatrix} 1 & 0 \\ 0 & 1 \end{bmatrix}$ and $B = \begin{bmatrix} 2 & 0 \\ 0 & 2 \end{bmatrix}$ both have rank 2, but A and B are not similar, since they have different determinants.

29. False. $A = \begin{bmatrix} 1 & 0 \\ 0 & 1 \end{bmatrix}$ is not similar to $B = \begin{bmatrix} 0 & 0 \\ 0 & 0 \end{bmatrix}$, and B is not similar to $C = \begin{bmatrix} 1 & 0 \\ 0 & 1 \end{bmatrix}$, but A is similar to C.

31. False. For example, let $A = \begin{bmatrix} 1 & 0 \\ 0 & 0 \end{bmatrix}$, $B = \begin{bmatrix} 2 & 0 \\ 0 & 0 \end{bmatrix}$, and $S = \begin{bmatrix} 0 & 0 \\ 0 & 1 \end{bmatrix}$. Then $SA = \begin{bmatrix} 0 & 0 \\ 0 & 1 \end{bmatrix}\begin{bmatrix} 1 & 0 \\ 0 & 0 \end{bmatrix} = \begin{bmatrix} 0 & 0 \\ 0 & 0 \end{bmatrix}$ and $BS = \begin{bmatrix} 2 & 0 \\ 0 & 0 \end{bmatrix}\begin{bmatrix} 0 & 0 \\ 0 & 1 \end{bmatrix} = \begin{bmatrix} 0 & 0 \\ 0 & 0 \end{bmatrix}$, but A and B are not similar, since they have different eigenvalues.

33. False. Let $A = \begin{bmatrix} 1 & 0 \\ 0 & 0 \end{bmatrix}$ and $B = \begin{bmatrix} 2 & 1 \\ -2 & -1 \end{bmatrix}$. Then $A = S^{-1}BS$, where $S = \begin{bmatrix} 1 & -1 \\ -1 & 2 \end{bmatrix}$. But null$(A) = $ span $\left\{ \begin{bmatrix} 0 \\ 1 \end{bmatrix} \right\}$ and null$(B) = $ span $\left\{ \begin{bmatrix} 1 \\ -2 \end{bmatrix} \right\}$, so null$(A) \neq $ null(B).

35. We have
$$A = S_1^{-1}BS_1 = S_1^{-1}\left(S_2^{-1}CS_2\right)S_1 = (S_2S_1)^{-1}C(S_2S_1)$$
Therefore if $D = S_2S_1$ then $A = D^{-1}CD$.

37. Since A and B are diagonalizable with the same eigenvalues including multiplicities, both A and B are similar to the same diagonal matrix D, with $A = P^{-1}DP$ and $B = Q^{-1}DQ$ for invertible matrices P and Q. Thus $QBQ^{-1} = D$, and hence $A = P^{-1}DP = P^{-1}\left(QBQ^{-1}\right)P = \left(Q^{-1}P\right)^{-1}B\left(Q^{-1}P\right)$. Therefore A is similar to B, with similarity matrix $Q^{-1}P$.

39. If A is similar to B, then $A = S^{-1}BS$ for some invertible matrix S. Then $A^T = \left(S^{-1}BS\right)^T = S^T B^T \left(S^{-1}\right)^T = U^{-1}B^T U$, where $U = \left(S^{-1}\right)^T = \left(S^T\right)^{-1}$, so $U^{-1} = S^T$. Therefore A^T and B^T are similar, with similarity matrix $U = \left(S^{-1}\right)^T = \left(S^T\right)^{-1}$.

41. $\det(A) = \det\left(\begin{bmatrix} 1 & -2 & 4 \\ 5 & 1 & 2 \\ 0 & 1 & -3 \end{bmatrix}\right) = -15$ and $\det(B) = \det\left(\begin{bmatrix} 1 & -1 & 1 \\ 5 & 0 & 0 \\ 0 & 1 & 2 \end{bmatrix}\right) = 15$. Since $\det(A) \neq \det(B)$, A and B are not similar matrices.

43. Let $S = \begin{bmatrix} 17 & 6 & -16 & 39 \\ 0 & 21 & -21 & -42 \\ -10 & -6 & 23 & 24 \\ -28 & 0 & 14 & 0 \end{bmatrix}$. Then $S^{-1}BS = \begin{bmatrix} 1 & 0 & 1 & 3 \\ -1 & 2 & 4 & 1 \\ 2 & 3 & -1 & 0 \\ 0 & 2 & -2 & -2 \end{bmatrix} = A$, so A and B are similar matrices.

Chapter 10

Inner Product Spaces

10.1 Inner Products

1. $\langle \mathbf{u}, \mathbf{v} \rangle = t_1 u_1 v_1 + t_2 u_2 v_2 + t_3 u_3 v_3 = (2)(1)(3) + (3)(2)(4) + (1)(1)(2) = 32$.

3. $\langle p, q \rangle = p(-1)q(-1) + p(0)q(0) + p(2)q(2) = (3(-1)+2)(-(-1)+1) + (3(0)+2)(-(0)+1) + (3(2)+2)(-(2)+1) = -2 + 2 - 8 = -8$.

5. $\langle f, g \rangle = \int_{-1}^{1} (x+3)(x^2)\, dx = \int_{-1}^{1} (x^3 + 3x^2)\, dx = 2$.

7. $\langle A, B \rangle = \operatorname{tr}(A^T B) = \operatorname{tr}\left(\begin{bmatrix} 2 & -1 \\ 3 & 4 \end{bmatrix}^T \begin{bmatrix} 5 & 2 \\ -3 & -2 \end{bmatrix} \right) = \operatorname{tr}\left(\begin{bmatrix} 1 & -2 \\ -17 & -10 \end{bmatrix} \right) = -9$.

9. $\langle \mathbf{u}, \mathbf{v} \rangle = t_1 u_1 v_1 + t_2 u_2 v_2 + t_3 u_3 v_3 = (3)(1)(2) + (1)(0)(1) + (a)(-1)(2) = 6 - 2a = 0 \Rightarrow a = 3$.

11. $\langle p, q \rangle = p(-1)q(-1) + p(a)q(a) + p(2)q(2) = ((-1)+2)(-3(-1)+1) + ((a)+2)(-3(a)+1) + ((2)+2)(-3(2)+1) = -3a^2 - 5a - 14 = 0$. This equation has no real solutions, so no possible value of a will make p and q orthogonal.

13. $\langle f, g \rangle = \int_{-1}^{1} (2x)(x+b)\, dx = \frac{4}{3} \neq 0$, so f and g are not orthogonal for any value of b.

15. Let $\mathbf{u} = \begin{bmatrix} 1 \\ -3 \\ 2 \end{bmatrix}$, then $\|\mathbf{u}\| = \sqrt{\langle \mathbf{u}, \mathbf{u} \rangle} = \sqrt{(2)(1)(1) + (3)(-3)(-3) + (1)(2)(2)} = \sqrt{33}$.

17. Let $p(x) = 3x - 5$, then $\|p\| = \sqrt{\langle p, p \rangle} = \sqrt{p(-2)p(-2) + p(1)p(1) + p(4)p(4)}$
$= \sqrt{(-11)^2 + (-2)^2 + (7)^2} = \sqrt{174}$.

19. Let $f(x) = x^3$, then $\|f\| = \sqrt{\langle f, f \rangle} = \sqrt{\int_{-1}^{1} (x^3)(x^3)\, dx} = \sqrt{\frac{2}{7}} = \frac{1}{7}\sqrt{14}$.

21. $\|A\| = \sqrt{\langle A, A \rangle} = \sqrt{\operatorname{tr}(A^T A)} =$
$\left(\operatorname{tr}\left(\begin{bmatrix} 3 & -1 \\ 2 & 0 \end{bmatrix}^T \begin{bmatrix} 3 & -1 \\ 2 & 0 \end{bmatrix} \right) \right)^{1/2} = \left(\operatorname{tr}\left(\begin{bmatrix} 13 & -3 \\ -3 & 1 \end{bmatrix} \right) \right)^{1/2} = \sqrt{14}$.

23. $\operatorname{proj}_{\mathbf{u}} \mathbf{v} = \frac{\langle \mathbf{u}, \mathbf{v} \rangle}{\langle \mathbf{u}, \mathbf{u} \rangle} \mathbf{u} = \frac{(2)(1)(3) + (3)(2)(4) + (1)(1)(2)}{(2)(1)(1) + (3)(2)(2) + (1)(1)(1)} \begin{bmatrix} 1 \\ 2 \\ 1 \end{bmatrix} = \frac{32}{15} \begin{bmatrix} 1 \\ 2 \\ 1 \end{bmatrix} = \begin{bmatrix} \frac{32}{15} \\ \frac{64}{15} \\ \frac{32}{15} \end{bmatrix}$.

25. $\text{proj}_p q = \dfrac{\langle p, q \rangle}{\langle p, p \rangle} p = \dfrac{p(-1)q(-1) + p(0)q(0) + p(2)q(2)}{p(-1)p(-1) + p(0)p(0) + p(2)p(2)} (3x+2) =$
$\dfrac{(-1)(2) + (2)(1) + (8)(-1)}{(-1)^2 + (2)^2 + (8)^2} (3x+2) = -\dfrac{8}{69}(3x+2) = -\dfrac{8}{23}x - \dfrac{16}{69}.$

27. $\text{proj}_f g = \dfrac{\langle f, g \rangle}{\langle f, f \rangle} f = \dfrac{\int_{-1}^{1} (x)(x^2)\, dx}{\int_{-1}^{1} (x)(x)\, dx} (x) = \dfrac{0}{\frac{2}{3}} (x) = 0.$

29. $\text{proj}_A B = \dfrac{\langle A, B \rangle}{\langle A, A \rangle} A = \dfrac{\text{tr}(A^T B)}{\text{tr}(A^T A)} A = \dfrac{\text{tr}\left(\begin{bmatrix} 2 & -1 \\ 1 & 0 \end{bmatrix}^T \begin{bmatrix} 2 & 3 \\ 0 & -2 \end{bmatrix} \right)}{\text{tr}\left(\begin{bmatrix} 2 & -1 \\ 1 & 0 \end{bmatrix}^T \begin{bmatrix} 2 & -1 \\ 1 & 0 \end{bmatrix} \right)} \begin{bmatrix} 2 & -1 \\ 1 & 0 \end{bmatrix} =$

$\dfrac{\text{tr}\left(\begin{bmatrix} 4 & 4 \\ -2 & -3 \end{bmatrix} \right)}{\text{tr}\left(\begin{bmatrix} 5 & -2 \\ -2 & 1 \end{bmatrix} \right)} \begin{bmatrix} 2 & -1 \\ 1 & 0 \end{bmatrix} = \dfrac{1}{6} \begin{bmatrix} 2 & -1 \\ 1 & 0 \end{bmatrix} = \begin{bmatrix} \frac{1}{3} & -\frac{1}{6} \\ \frac{1}{6} & 0 \end{bmatrix}.$

31. For example, let $\mathbf{u} = \begin{bmatrix} 1 \\ 0 \end{bmatrix}$ and $\mathbf{v} = \begin{bmatrix} 0 \\ 1 \end{bmatrix}$. Then $\{\mathbf{u}, \mathbf{v}\}$ is an orthogonal basis for $\mathbf{R}^2$ with respect to the given inner product.

33. For example, let $t_1 = t_2 = 1$ and $\mathbf{u} = \begin{bmatrix} 6/\sqrt{13} \\ -4/\sqrt{13} \end{bmatrix}$.

35. For example, let $A = \begin{bmatrix} 1 & 0 & 0 \\ 0 & 2 & 0 \\ 0 & 0 & 3 \end{bmatrix}$. Then $\langle \mathbf{u}, \mathbf{v} \rangle = \mathbf{u}^T \begin{bmatrix} 1 & 0 & 0 \\ 0 & 2 & 0 \\ 0 & 0 & 3 \end{bmatrix} \mathbf{v} =$

$\begin{bmatrix} u_1 \\ u_2 \\ u_3 \end{bmatrix}^T \begin{bmatrix} 1 & 0 & 0 \\ 0 & 2 & 0 \\ 0 & 0 & 3 \end{bmatrix} \begin{bmatrix} v_1 \\ v_2 \\ v_3 \end{bmatrix} = u_1 v_1 + 2u_2 v_2 + 3u_3 v_3$ is an inner product, as in Example 2, with weights $t_1 = 1$, $t_2 = 2$, and $t_3 = 3$.

37. For example, let $w(x) = \cos(x)$. Define $\langle p, q \rangle = \int_0^1 p(x)q(x)w(x)\, dx$, then this satisfies the requirements of Definition 10.1, and hence is an inner product on $\mathbf{P}^2$.

39. For example, let $\langle p, q \rangle = 0$ for all p and q in $\mathbf{P}^2$. Then (a)-(c) Definition 10.1 are satisfied, but (d) fails, as the constant polynomial $p(x) = 1 \neq 0$, but $\langle p, p \rangle = 0$.

41. True, since $\langle 2\mathbf{u}, -4\mathbf{v} \rangle = (2)\langle \mathbf{u}, -4\mathbf{v} \rangle = (2)(-4)\langle \mathbf{u}, \mathbf{v} \rangle = (-8)\langle \mathbf{u}, \mathbf{v} \rangle = (-8)(3) = -24.$

43. True. By Theorem 10.4, since $\mathbf{u}$ and $\mathbf{v}$ are orthogonal, $\|\mathbf{u} + \mathbf{v}\|^2 = \|\mathbf{u}\|^2 + \|\mathbf{v}\|^2 = 3^2 + 4^2 = 25$, so $\|\mathbf{u} + \mathbf{v}\| = \sqrt{25} = 5$.

45. True. Since $\{\mathbf{u}, \mathbf{v}\}$ is an orthogonal set, $\langle \mathbf{u}, \mathbf{v} \rangle = 0$. Thus $\langle c_1 \mathbf{u}, c_2 \mathbf{v} \rangle = c_1 c_2 \langle \mathbf{u}, \mathbf{v} \rangle = c_1 c_2 (0) = 0$, so $\{c_1 \mathbf{u}, c_2 \mathbf{v}\}$ is an orthogonal set.

47. False. For example, let $\mathbf{u} = \begin{bmatrix} 1 \\ 0 \end{bmatrix}$ and $\mathbf{v} = \begin{bmatrix} 0 \\ 1 \end{bmatrix}$ in $\mathbf{R}^2$ with the standard inner product $\langle \mathbf{u}, \mathbf{v} \rangle = u_1 v_1 + u_2 v_2$. Then $\|\mathbf{u}\| - \|\mathbf{v}\| = 1 - 1 = 0$, but $\|\mathbf{u} - \mathbf{v}\| = \left\| \begin{bmatrix} 1 \\ 0 \end{bmatrix} - \begin{bmatrix} 0 \\ 1 \end{bmatrix} \right\| = \left\| \begin{bmatrix} 1 \\ -1 \end{bmatrix} \right\| = \sqrt{2}$, so $\|\mathbf{u} - \mathbf{v}\| > \|\mathbf{u}\| - \|\mathbf{v}\|$ in this case.

49. False. Let $p(x) = x^2 - x$, $x_0 = 0$ and $x_1 = 1$. Then $\langle p, p \rangle = p(0)^2 + p(1)^2 = 0^2 + 0^2 = 0$, but $p \neq \mathbf{0}$, so Definition 10.1 (d) fails.

51. We need to establish properties (b) and (c). Evaluate $\langle \mathbf{u} + \mathbf{v}, \mathbf{w} \rangle = t_1 (u_1 + v_1) w_1 + t_2 (u_2 + v_2) w_2 + \cdots + t_n (u_n + v_n) w_n = (t_1 u_1 w_1 + t_2 u_2 w_2 + \cdots + t_n u_n w_n) + (t_1 v_1 w_1 + t_2 v_2 w_2 + \cdots + t_n v_n w_n) = \langle \mathbf{u}, \mathbf{w} \rangle + \langle \mathbf{v}, \mathbf{w} \rangle$, so property (b) holds. Evaluate $\langle c\mathbf{u}, \mathbf{v} \rangle = t_1 (cu_1) v_1 + t_2 (cu_2) v_2 + \cdots + t_n (cu_n) v_n = c (t_1 u_1 v_1 + t_2 u_2 v_2 + \cdots + t_n u_n v_n) = c \langle \mathbf{u}, \mathbf{v} \rangle$, so $\langle c\mathbf{u}, \mathbf{v} \rangle = c \langle \mathbf{u}, \mathbf{v} \rangle = c \langle \mathbf{v}, \mathbf{u} \rangle = \langle c\mathbf{v}, \mathbf{u} \rangle = \langle \mathbf{u}, c\mathbf{v} \rangle$, and property (c) holds. Thus the weighted dot product of Example 1 is an inner product.

53. We establish properties (a)-(d). Evaluate
$$\langle p, q \rangle = t (x_0) p (x_0) q (x_0) + \cdots + t (x_n) p (x_n) q (x_n)$$
$$= t (x_0) q (x_0) p (x_0) + \cdots + t (x_n) q (x_n) p (x_n) = \langle q, p \rangle$$
so property (a) holds. Evaluate
$$\langle p + q, r \rangle = t (x_0) (p (x_0) + q (x_0)) r (x_0) + \cdots + t (x_n) (p (x_n) + q (x_n)) r (x_n)$$
$$= (t (x_0) p (x_0) r (x_0) + \cdots + t (x_n) p (x_n) r (x_n)) +$$
$$(t (x_0) q (x_0) r (x_0) + \cdots + t (x_n) q (x_n) r (x_n))$$
$$= \langle p, r \rangle + \langle q, r \rangle$$
so property (b) holds. Evaluate
$$\langle cp, q \rangle = t (x_0) (cp (x_0)) q (x_0) + \cdots + t (x_n) (cp (x_n)) q (x_n)$$
$$= c (t (x_0) p (x_0) q (x_0) + \cdots + t (x_n) p (x_n) q (x_n)) = c \langle p, q \rangle$$
so $\langle cp, q \rangle = c \langle p, q \rangle = c \langle q, p \rangle = \langle cq, p \rangle = \langle p, cq \rangle$, and property (c) holds. Evaluate $\langle p, p \rangle = t (x_0) p (x_0) p (x_0) + \cdots + t (x_n) p (x_n) p (x_n) = t (x_0) (p (x_0))^2 + \cdots + t (x_n) (p (x_n))^2 \geq 0$ since each $t (x_i) > 0$. If $\langle p, p \rangle = 0$, then $t (x_0) (p (x_0))^2 + \cdots + t (x_n) (p (x_n))^2 = 0$, so each $p (x_i) = 0$ since each $t (x_i) > 0$. Since p is a polynomial in $\mathbf{P}^n$, $\deg (p) \leq n$, and p vanishes at $n + 1$ distinct points, it follows from the Fundamental Theorem of Algebra that $p = 0$, and property (d) holds. Hence $\langle p, q \rangle$ defines an inner product.

55. We establish properties (a)-(d). We have $\langle f, g \rangle = \frac{1}{\pi} \int_{-\pi}^{\pi} f (x) g (x) \, dx = \frac{1}{\pi} \int_{-\pi}^{\pi} g (x) f (x) \, dx = \langle g, f \rangle$, so property (a) holds. Next we have
$$\langle f + g, h \rangle = \frac{1}{\pi} \int_{-\pi}^{\pi} (f (x) + g (x)) h (x) \, dx$$
$$= \frac{1}{\pi} \int_{-\pi}^{\pi} f (x) h (x) \, dx + \frac{1}{\pi} \int_{-\pi}^{\pi} g (x) h (x) \, dx = \langle f, h \rangle + \langle g, h \rangle$$
so property (b) holds. We also have $\langle cf, g \rangle = \frac{1}{\pi} \int_{-\pi}^{\pi} (cf (x)) g (x) \, dx = c \frac{1}{\pi} \int_{-\pi}^{\pi} f (x) g (x) \, dx = c \langle f, g \rangle$, so $\langle cf, g \rangle = c \langle f, g \rangle = c \langle g, f \rangle = \langle cg, f \rangle = \langle f, cg \rangle$, and property (c) holds. Finally, we have $\langle f, f \rangle = \frac{1}{\pi} \int_{-\pi}^{\pi} (f (x))^2 \, dx \geq 0$ since $(f (x))^2 \geq 0$ for x in $[-1, 1]$. If $\langle f, f \rangle = 0$, then $\frac{1}{\pi} \int_{-\pi}^{\pi} (f (x))^2 \, dx = 0$, and since $(f (x))^2 \geq 0$ we must have $f (x) = 0$ for all x in $[-1, 1]$. (A rigorous justification of this claim requires arguments from advanced calculus, but the claim is plausible.) Thus property (d) holds. Hence $\langle f, g \rangle$ defines an inner product.

57. We establish properties (a)-(d). Evaluate $\langle A, B \rangle = \operatorname{tr} (A^T B) = \operatorname{tr} ((A^T B)^T) = \operatorname{tr} (B^T A) = \langle B, A \rangle$, and property (a) holds. Evaluate $\langle A + B, C \rangle = \operatorname{tr} ((A + B)^T C) = \operatorname{tr} ((A^T + B^T) C) = \operatorname{tr} (A^T C + B^T C) = \operatorname{tr} (A^T C) + \operatorname{tr} (B^T C) = \langle A, C \rangle + \langle B, C \rangle$, and property (b) holds. Evaluate $\langle cA, B \rangle = \operatorname{tr} ((cA)^T B) = \operatorname{tr} (c (A^T B)) = c \operatorname{tr} (A^T B) = c \langle A, B \rangle$, so $\langle cA, B \rangle = c \langle A, B \rangle = c \langle B, A \rangle = \langle cB, A \rangle = \langle A, cB \rangle$, and property (c) holds. Let $A = [\ \mathbf{a}_1 \quad \mathbf{a}_2 \quad \mathbf{a}_3\ ]$, then $A^T A = \begin{bmatrix} \mathbf{a}_1^T \mathbf{a}_1 & \mathbf{a}_1^T \mathbf{a}_2 & \mathbf{a}_1^T \mathbf{a}_3 \\ \mathbf{a}_2^T \mathbf{a}_1 & \mathbf{a}_2^T \mathbf{a}_2 & \mathbf{a}_2^T \mathbf{a}_3 \\ \mathbf{a}_3^T \mathbf{a}_1 & \mathbf{a}_3^T \mathbf{a}_2 & \mathbf{a}_3^T \mathbf{a}_3 \end{bmatrix}$, so $\langle A, A \rangle = \operatorname{tr} (A^T A) = \mathbf{a}_1^T \mathbf{a}_1 + \mathbf{a}_2^T \mathbf{a}_2 + \mathbf{a}_3^T \mathbf{a}_3 = \|\mathbf{a}_1\|^2 + \|\mathbf{a}_2\|^2 + \|\mathbf{a}_3\|^3 \geq 0$.

If $\langle A, A \rangle = 0$, then each $\|\mathbf{a}_i\| = 0$, so each $\mathbf{a}_i = \mathbf{0}$, and thus $A = [\begin{array}{ccc} \mathbf{0} & \mathbf{0} & \mathbf{0} \end{array}]$, and property (d) holds. Hence $\langle A, B \rangle$ defines an inner product.

59. $\|c\mathbf{v}\|^2 = \langle c\mathbf{v}, c\mathbf{v} \rangle = c^2 \langle \mathbf{v}, \mathbf{v} \rangle = c^2 \|\mathbf{v}\|^2$, so $\|c\mathbf{v}\| = |c| \|\mathbf{v}\|$.

61. By induction on k. If $k = 1$, then $\langle c_1\mathbf{u}_1, \mathbf{w} \rangle = c_1 \langle \mathbf{u}_1, \mathbf{w} \rangle$ by property (c) of Definition 10.1. Suppose $\langle c_1\mathbf{u}_1 + \cdots + c_{k-1}\mathbf{u}_{k-1}, \mathbf{w} \rangle = c_1 \langle \mathbf{u}_1, \mathbf{w} \rangle + \cdots + c_{k-1} \langle \mathbf{u}_{k-1}, \mathbf{w} \rangle$. Then

$$\langle c_1\mathbf{u}_1 + \cdots + c_{k-1}\mathbf{u}_{k-1} + c_k\mathbf{u}_k, \mathbf{w} \rangle$$
$$= \langle (c_1\mathbf{u}_1 + \cdots + c_{k-1}\mathbf{u}_{k-1}) + c_k\mathbf{u}_k, \mathbf{w} \rangle$$
$$= \langle c_1\mathbf{u}_1 + \cdots + c_{k-1}\mathbf{u}_{k-1}, \mathbf{w} \rangle + \langle c_k\mathbf{u}_k, \mathbf{w} \rangle, \text{ by property (b)}$$
$$= (c_1 \langle \mathbf{u}_1, \mathbf{w} \rangle + \cdots + c_{k-1} \langle \mathbf{u}_{k-1}, \mathbf{w} \rangle) + c_k \langle \mathbf{u}_k, \mathbf{w} \rangle, \text{ by induction and property (c)}$$
$$= c_1 \langle \mathbf{u}_1, \mathbf{w} \rangle + \cdots + c_k \langle \mathbf{u}_k, \mathbf{w} \rangle.$$

Hence, by induction, the statement holds for all positive integers k.

63. $\|\mathbf{w}\| = \left\| \dfrac{1}{\|\mathbf{v}\|}\mathbf{v} \right\| = \left| \dfrac{1}{\|\mathbf{v}\|} \right| \|\mathbf{v}\| = \dfrac{1}{\|\mathbf{v}\|} \|\mathbf{v}\| = 1$.

65. Let $\mathbf{u}_1$ and $\mathbf{u}_2$ belong to V, and let c_1 and c_2 be scalars. Then $T_{\mathbf{v}}(c_1\mathbf{u}_1 + c_2\mathbf{u}_2) = \langle c_1\mathbf{u}_1 + c_2\mathbf{u}_2, \mathbf{v} \rangle = c_1 \langle \mathbf{u}_1, \mathbf{v} \rangle + c_2 \langle \mathbf{u}_2, \mathbf{v} \rangle = c_1 T(\mathbf{u}_1) + c_2 T(\mathbf{u}_2)$, and hence T is a linear transformation.

67. Since $\mathbf{u}$ and $\mathbf{v}$ are orthogonal, so are $\mathbf{u}$ and $(-\mathbf{v})$, and by the Pythagorean Theorem $\|\mathbf{u} - \mathbf{v}\|^2 = \|\mathbf{u} + (-\mathbf{v})\|^2 = \|\mathbf{u}\|^2 + \|-\mathbf{v}\|^2 = \|\mathbf{u}\|^2 + \|\mathbf{v}\|^2$. Thus, the distance between $\mathbf{u}$ and $\mathbf{v}$ is $\|\mathbf{u} - \mathbf{v}\| = \sqrt{\|\mathbf{u}\|^2 + \|\mathbf{v}\|^2}$.

69. Since $\langle \mathbf{0}, \mathbf{s} \rangle = 0$ for all $\mathbf{s} \in S$, $\mathbf{0} \in S^\perp$. Suppose $\mathbf{s}_1$ and $\mathbf{s}_2$ are in $S^\perp$, and $\mathbf{s} \in S$. Then $\langle \mathbf{s}_1 + \mathbf{s}_2, \mathbf{s} \rangle = \langle \mathbf{s}_1, \mathbf{s} \rangle + \langle \mathbf{s}_2, \mathbf{s} \rangle = 0 + 0 = 0$, so $\mathbf{s}_1 + \mathbf{s}_2 \in S^\perp$. Also, $\langle c\mathbf{s}_1, \mathbf{s} \rangle = c \langle \mathbf{s}_1, \mathbf{s} \rangle = c(0) = 0$, so $c\mathbf{s}_1 \in S^\perp$. Since $S^\perp$ is a non-empty subset closed under vector addition and scalar multiplication, $S^\perp$ is a subspace.

71. Since $\mathbf{s} \in S$ and $\mathbf{s}^\perp \in S^\perp$, $\langle \mathbf{s}^\perp, \mathbf{s} \rangle = 0$, and $\mathbf{s}^\perp$ and $\mathbf{s}$ are orthogonal, and thus $(-\mathbf{s}^\perp)$ and $\mathbf{s}$ are also orthogonal. By the Pythagorean Theorem, $\|\mathbf{s} \pm \mathbf{s}^\perp\|^2 = \|\mathbf{s}\|^2 + \|\mathbf{s}^\perp\|^2$.

10.2   The Gram-Schmidt Process Revisited

1. $\|\mathbf{v}_1\| = \sqrt{\langle \mathbf{v}_1, \mathbf{v}_1 \rangle} = \sqrt{2(4)^2 + 3(5)^2 + (1)^2} = 6\sqrt{3}$, so $\mathbf{u}_1 = \dfrac{1}{\|\mathbf{v}_1\|}\mathbf{v}_1 = \dfrac{1}{6\sqrt{3}}\begin{bmatrix} 4 \\ 5 \\ 1 \end{bmatrix} = \begin{bmatrix} \frac{2}{9}\sqrt{3} \\ \frac{5}{18}\sqrt{3} \\ \frac{1}{18}\sqrt{3} \end{bmatrix}$.

$\|\mathbf{v}_2\| = \sqrt{\langle \mathbf{v}_2, \mathbf{v}_2 \rangle} = \sqrt{2(-3)^2 + 3(2)^2 + (-6)^2} = \sqrt{66}$, so $\mathbf{u}_2 = \dfrac{1}{\|\mathbf{v}_2\|}\mathbf{v}_2 = \dfrac{1}{\sqrt{66}}\begin{bmatrix} -3 \\ 2 \\ -6 \end{bmatrix} = \begin{bmatrix} -\frac{1}{22}\sqrt{66} \\ \frac{1}{33}\sqrt{66} \\ -\frac{1}{11}\sqrt{66} \end{bmatrix}$.

$\|\mathbf{v}_3\| = \sqrt{\langle \mathbf{v}_3, \mathbf{v}_3 \rangle} = \sqrt{2(16)^2 + 3(-7)^2 + (-23)^2} = 6\sqrt{33}$, so $\mathbf{u}_3 = \dfrac{1}{\|\mathbf{v}_3\|}\mathbf{v}_3 =$

$\dfrac{1}{6\sqrt{33}}\begin{bmatrix} 16 \\ -7 \\ -23 \end{bmatrix} = \begin{bmatrix} \frac{8}{99}\sqrt{33} \\ -\frac{7}{198}\sqrt{33} \\ -\frac{23}{198}\sqrt{33} \end{bmatrix}$. So $\{\mathbf{u}_1, \mathbf{u}_2, \mathbf{u}_3\}$ is the orthonormal set derived from $\{\mathbf{v}_1, \mathbf{v}_2, \mathbf{v}_3\}$.

3. Let $\mathbf{v}_1 = \begin{bmatrix} 2 \\ -1 \\ 1 \end{bmatrix}$, $\mathbf{v}_2 = \begin{bmatrix} 2 \\ 1 \\ a \end{bmatrix}$, and $\mathbf{v}_3 = \begin{bmatrix} 1 \\ 2 \\ 2 \end{bmatrix}$. Then $\langle \mathbf{v}_1, \mathbf{v}_2 \rangle = (2)(2)(2) + (3)(-1)(1) + (1)(1)(a) = a + 5 = 0 \Rightarrow a = -5$. We check $\langle \mathbf{v}_1, \mathbf{v}_3 \rangle = (2)(2)(1) + (3)(-1)(2) + (1)(1)(2) = 0$ and

$\langle \mathbf{v}_2, \mathbf{v}_3 \rangle = (2)(2)(1) + (3)(1)(2) + (1)(-5)(2) = 0$. To normalize, we determine $\|\mathbf{v}_1\| = \sqrt{\langle \mathbf{v}_1, \mathbf{v}_1 \rangle} = \sqrt{(2)(2)^2 + (3)(-1)^2 + (1)(1)^2} = 2\sqrt{3}$, $\|\mathbf{v}_2\| = \sqrt{\langle \mathbf{v}_2, \mathbf{v}_2 \rangle} = \sqrt{(2)(2)^2 + (3)(1)^2 + (1)(-5)^2} = 6$,

and $\|\mathbf{v}_3\| = \sqrt{\langle \mathbf{v}_3, \mathbf{v}_3 \rangle} = \sqrt{(2)(1)^2 + (3)(2)^2 + (1)(2)^2} = 3\sqrt{2}$. So $\mathbf{u}_1 = \dfrac{1}{\|\mathbf{v}_1\|}\mathbf{v}_1 = \dfrac{1}{2\sqrt{3}}\begin{bmatrix} 2 \\ -1 \\ 1 \end{bmatrix} =$

$\begin{bmatrix} \frac{1}{3}\sqrt{3} \\ -\frac{1}{6}\sqrt{3} \\ \frac{1}{6}\sqrt{3} \end{bmatrix}$, $\mathbf{u}_2 = \dfrac{1}{\|\mathbf{v}_2\|}\mathbf{v}_2 = \dfrac{1}{6}\begin{bmatrix} 2 \\ 1 \\ -5 \end{bmatrix} = \begin{bmatrix} \frac{1}{3} \\ \frac{1}{6} \\ -\frac{5}{6} \end{bmatrix}$, and $\mathbf{u}_3 = \dfrac{1}{\|\mathbf{v}_3\|}\mathbf{v}_3 = \dfrac{1}{3\sqrt{2}}\begin{bmatrix} 1 \\ 2 \\ 2 \end{bmatrix} = \begin{bmatrix} \frac{1}{6}\sqrt{2} \\ \frac{1}{3}\sqrt{2} \\ \frac{1}{3}\sqrt{2} \end{bmatrix}$. So

the corresponding orthonormal set is $\left\{ \begin{bmatrix} \frac{1}{3}\sqrt{3} \\ -\frac{1}{6}\sqrt{3} \\ \frac{1}{6}\sqrt{3} \end{bmatrix}, \begin{bmatrix} \frac{1}{3} \\ \frac{1}{6} \\ -\frac{5}{6} \end{bmatrix}, \begin{bmatrix} \frac{1}{6}\sqrt{2} \\ \frac{1}{3}\sqrt{2} \\ \frac{1}{3}\sqrt{2} \end{bmatrix} \right\}$.

5. Let $p_1(x) = x^2 + x$, $p_2(x) = x^2 + ax - 2$, and $p_3(x) = x^2 - 2x$. Then $\langle p_1, p_2 \rangle = p_1(-1)p_2(-1) + p_1(0)p_2(0) + p_1(2)p_2(2) = (0)(-1-a) + (0)(-2) + (6)(2+2a) = 12a + 12 = 0 \Rightarrow a = -1$. We check $\langle p_1, p_3 \rangle = p_1(-1)p_3(-1) + p_1(0)p_3(0) + p_1(2)p_3(2) = (0)(0) + (0)(-2) + (6)(0) = 0$, and $\langle p_2, p_3 \rangle = p_2(-1)p_3(-1) + p_2(0)p_3(0) + p_2(2)p_3(2) = (0)(3) + (-2)(0) + (0)(0) = 0$. To normalize, we determine $\|p_1\| = \sqrt{\langle p_1, p_1 \rangle} = \sqrt{(0)^2 + (0)^2 + (6)^2} = 6$, $\|p_2\| = \sqrt{\langle p_2, p_2 \rangle} = \sqrt{(0)^2 + (-2)^2 + (0)^2} = 2$, and $\|p_3\| = \sqrt{\langle p_3, p_3 \rangle} = \sqrt{(3)^2 + (0)^2 + (0)^2} = 3$. So $q_1(x) = \frac{1}{\|p_1\|}p_1(x) = \frac{1}{6}(x^2 + x) = \frac{1}{6}x^2 + \frac{1}{6}x$, $q_2(x) = \frac{1}{\|p_2\|}p_2(x) = \frac{1}{2}(x^2 - x - 2) = \frac{1}{2}x^2 - \frac{1}{2}x - 1$, and $q_3(x) = \frac{1}{\|p_3\|}p_3(x) = \frac{1}{3}(x^2 - 2x) = \frac{1}{3}x^2 - \frac{2}{3}x$. So the corresponding orthonormal set is $\left\{ \frac{1}{6}x^2 + \frac{1}{6}x, \frac{1}{2}x^2 - \frac{1}{2}x - 1, \frac{1}{3}x^2 - \frac{2}{3}x \right\}$.

7. $s_1 = \dfrac{\langle \mathbf{v}_1, \mathbf{v} \rangle}{\|\mathbf{v}_1\|^2} = \dfrac{2(4)(1) + 3(5)(34) + (1)(22)}{2(4)^2 + 3(5)^2 + (1)^2} = 5$,

$s_2 = \dfrac{\langle \mathbf{v}_2, \mathbf{v} \rangle}{\|\mathbf{v}_2\|^2} = \dfrac{2(-3)(1) + 3(2)(34) + (-6)(22)}{2(-3)^2 + 3(2)^2 + (-6)^2} = 1$, and

$s_3 = \dfrac{\langle \mathbf{v}_3, \mathbf{v} \rangle}{\|\mathbf{v}_3\|^2} = \dfrac{2(16)(1) + 3(-7)(34) + (-23)(22)}{2(16)^2 + 3(-7)^2 + (-23)^2} = -1$.

Hence $\mathbf{v} = (5)\begin{bmatrix} 4 \\ 5 \\ 1 \end{bmatrix} + (1)\begin{bmatrix} -3 \\ 2 \\ -6 \end{bmatrix} + (-1)\begin{bmatrix} 16 \\ -7 \\ -23 \end{bmatrix}$.

9. $\text{proj}_S \mathbf{v} = \dfrac{\langle \mathbf{v}_1, \mathbf{v} \rangle}{\|\mathbf{v}_1\|^2}\mathbf{v}_1 + \dfrac{\langle \mathbf{v}_2, \mathbf{v} \rangle}{\|\mathbf{v}_2\|^2}\mathbf{v}_2 = \dfrac{2(4)(1) + 3(5)(0) + (1)(-1)}{2(4)^2 + 3(5)^2 + (1)^2}\begin{bmatrix} 4 \\ 5 \\ 1 \end{bmatrix} +$

$\dfrac{2(-3)(1) + 3(2)(0) + (-6)(-1)}{2(-3)^2 + 3(2)^2 + (-6)^2}\begin{bmatrix} -3 \\ 2 \\ -6 \end{bmatrix} = \left(\frac{7}{108}\right)\begin{bmatrix} 4 \\ 5 \\ 1 \end{bmatrix} + (0)\begin{bmatrix} -3 \\ 2 \\ -6 \end{bmatrix} = \begin{bmatrix} \frac{7}{27} \\ \frac{35}{108} \\ \frac{7}{108} \end{bmatrix}$.

11. Since $\{1/\sqrt{2}, \cos(x), \sin(x)\}$ is orthonormal,

$\text{proj}_S f = \langle f_1, f \rangle f_1 + \langle f_2, f \rangle f_2 + \langle f_3, f \rangle f_3 = \left(\frac{1}{\pi} \int_{-\pi}^{\pi} (1/\sqrt{2})(x)\, dx \right)(1/\sqrt{2}) +$

$\left(\frac{1}{\pi} \int_{-\pi}^{\pi} (\cos(x))(x)\, dx \right)(\cos(x)) + \left(\frac{1}{\pi} \int_{-\pi}^{\pi} (\sin(x))(x)\, dx \right)(\sin(x)) =$

$(0)(1/\sqrt{2}) + (0)(\cos(x)) + (2)(\sin(x)) = 2\sin(x)$.

13. Let $\mathbf{s}_1 = \begin{bmatrix} 1 \\ -1 \\ 0 \end{bmatrix}$ and $\mathbf{s}_2 = \begin{bmatrix} 2 \\ 0 \\ 1 \end{bmatrix}$. Set $\mathbf{v}_1 = \mathbf{s}_1 = \begin{bmatrix} 1 \\ -1 \\ 0 \end{bmatrix}$, and $S_1 = \text{span}\{\mathbf{s}_1\}$. Let $\mathbf{v}_2 = \mathbf{s}_2 -$

$$\text{proj}_{S_1}\mathbf{s}_2 = \mathbf{s}_2 - \frac{\langle \mathbf{s}_1, \mathbf{s}_2 \rangle}{\|\mathbf{s}_1\|^2}\mathbf{s}_1 = \begin{bmatrix} 2 \\ 0 \\ 1 \end{bmatrix} - \frac{2(1)(2) + 3(-1)(0) + (0)(1)}{2(1)^2 + 3(-1)^2 + (0)^2}\begin{bmatrix} 1 \\ -1 \\ 0 \end{bmatrix} = \begin{bmatrix} 2 \\ 0 \\ 1 \end{bmatrix} - \frac{4}{5}\begin{bmatrix} 1 \\ -1 \\ 0 \end{bmatrix} =$$

$\begin{bmatrix} \frac{6}{5} \\ \frac{4}{5} \\ 1 \end{bmatrix}$. Then $\{\mathbf{v}_1, \mathbf{v}_2\}$ is an orthogonal basis for $S = \text{span}\{\mathbf{s}_1, \mathbf{s}_2\}$.

15. Let $f_1(x) = 1$, and $f_2(x) = x^2$. Set $g_1(x) = f_1(x) = 1$, and $S_1 = \text{span}\{g_1\}$. Let $g_2(x) = f_2(x) - (\text{proj}_{S_1}f_2)(x) = f_2(x) - \frac{\langle g_1, f_2 \rangle}{\|g_1\|^2}g_1(x) = x^2 - \frac{\int_0^1 (1)(x^2)\,dx}{\int_0^1 (1)(1)\,dx}(1) = x^2 - \frac{\frac{1}{3}}{1}(1) = x^2 - \frac{1}{3}$. Then $\{1, x^2 - \frac{1}{3}\}$ is an orthogonal basis for $S = \text{span}\{f_1, f_2\}$.

17. Let $p_1(x) = x$ and $p_2(x) = 1$. Set $q_1(x) = p_1(x) = x$, and $S_1 = \text{span}\{q_1\}$. Let $q_2(x) = p_2(x) - (\text{proj}_{S_1}p_2)(x) = p_2(x) - \frac{\langle q_1, p_2 \rangle}{\|q_1\|^2}q_1(x) = (1) - \frac{q_1(-1)p_2(-1) + q_1(0)p_2(0) + q_1(2)p_2(2)}{(q_1(-1))^2 + (q_1(0))^2 + (q_1(2))^2}(x) = 1 - \frac{(-1)(1) + (0)(1) + (2)(1)}{(-1)^2 + (0)^2 + (2)^2}(x) = -\frac{1}{5}x + 1$. Then $\{x, -\frac{1}{5}x + 1\}$ is an orthogonal basis for $S = \text{span}\{p_1, p_2\}$.

19. For example, let $\mathbf{u}_2 = \begin{bmatrix} 4 \\ -3 \end{bmatrix}$, then $\langle \mathbf{u}_1, \mathbf{u}_2 \rangle = 3(1)(4) + 2(2)(-3) = 0$, so $\{\mathbf{u}_1, \mathbf{u}_2\}$ is an orthogonal basis for $\mathbf{R}^2$.

21. For example, let $p_2(x) = 5x - 3$, then $\langle p_1, p_2 \rangle = \int_0^1 (3x + 1)(5x - 3)\,dx = 0$, so $\langle p_1, p_2 \rangle$ is an orthogonal basis for $\mathbf{P}^1$.

23. For example, $f_1(x) = 0$, $f_2(x) = 1$, and $f_3(x) = \cos(x)$. Then $\{0, 1, \cos(x)\}$ is orthogonal, but cannot be made orthonormal, since $\|f_1\| = 0$.

25. False, since if $|c_1| \neq 1$, then $\|c_1\mathbf{v}_1\| = |c_1|\,\|\mathbf{v}_1\| = |c_1|(1) = |c_1| \neq 1$, and hence $\{c_1\mathbf{v}_1, c_2\mathbf{v}_2, c_3\mathbf{v}_3\}$ is not orthonormal.

27. True, by Theorem 10.15, with S the span of the linearly independent set $\mathcal{G}$, provided $\mathcal{G}$ is a finite set. (If $\mathcal{G} = \{\mathbf{s}_1, \mathbf{s}_2, \ldots\}$, then the result still holds, as Theorem 10.15 can be extended to such sets. However, if $\mathcal{G}$ is an arbitrary linearly independent set, then we cannot conclude the result based on Theorem 10.15.)

29. False. For example, let $V = \mathbf{R}^2$, $\mathbf{v} = \begin{bmatrix} 1 \\ 0 \end{bmatrix}$, $\mathbf{u} = \begin{bmatrix} 0 \\ 1 \end{bmatrix}$, and $S = \text{span}\{\mathbf{u}\}$. Then, with the standard inner product, $\text{proj}_S\mathbf{v} = \frac{\langle \mathbf{u}, \mathbf{v} \rangle}{\langle \mathbf{u}, \mathbf{u} \rangle}\mathbf{u} = \frac{0}{1}\mathbf{u} = \mathbf{0}$.

31. True. If one of the vectors $\mathbf{s}_i = \mathbf{0}$, then the corresponding vector $\mathbf{v}_i = \mathbf{s}_i - \text{proj}_{S_{i-1}}\mathbf{s}_i = \mathbf{0} - \mathbf{0} = \mathbf{0}$. Otherwise, there will be a vector $\mathbf{s}_i \in \text{span}\{\mathbf{s}_1, \ldots, \mathbf{s}_{i-1}\}$. And since $S_{i-1} = \text{span}\{\mathbf{v}_1, \ldots, \mathbf{v}_{i-1}\} = \text{span}\{\mathbf{s}_1, \ldots, \mathbf{s}_{i-1}\}$, $\mathbf{s}_i$ belongs to S_{i-1}. Thus $\mathbf{v}_i = \mathbf{s}_i - \text{proj}_{S_{i-1}}\mathbf{s}_i = \mathbf{s}_i - \mathbf{s}_i = \mathbf{0}$.

33. Consider $c_1\mathbf{v}_1 + \cdots + c_m\mathbf{v}_m = \mathbf{0}$, and take the inner product with respect to $\mathbf{v}_i$. Then $0 = \langle \mathbf{0}, \mathbf{v}_i \rangle = \langle c_1\mathbf{v}_1 + \cdots + c_m\mathbf{v}_m, \mathbf{v}_i \rangle = c_1\langle \mathbf{v}_1, \mathbf{v}_i \rangle + \cdots + c_i\langle \mathbf{v}_i, \mathbf{v}_i \rangle + \cdots + c_m\langle \mathbf{v}_m, \mathbf{v}_i \rangle = c_1(0) + \cdots + c_i\|\mathbf{v}_i\|^2 + \cdots + c_m(0) = c_i\|\mathbf{v}_i\|^2$. Since $\mathbf{v}_i \neq \mathbf{0}$, we conclude $c_i = 0$. Since every $c_i = 0$, we conclude that $\{\mathbf{v}_1, \ldots, \mathbf{v}_m\}$ is linearly independent.

35. Since the orthonormal basis $\{\mathbf{v}_1, \ldots, \mathbf{v}_k\}$ is also orthogonal, by Theorem 10.11, every vector $\mathbf{v}$ can be written as $\mathbf{v} = s_1\mathbf{v}_1 + \cdots + s_k\mathbf{v}_k$ where each $s_i = \frac{\langle \mathbf{v}_i, \mathbf{v} \rangle}{\langle \mathbf{v}_i, \mathbf{v}_i \rangle} = \frac{\langle \mathbf{v}_i, \mathbf{v} \rangle}{\|\mathbf{v}_i\|^2}$. Since $\{\mathbf{v}_1, \ldots, \mathbf{v}_k\}$ is orthonormal, each $\|\mathbf{v}_i\| = 1$, so each $s_i = \langle \mathbf{v}_i, \mathbf{v} \rangle$. Thus, $\mathbf{v} = \langle \mathbf{v}_1, \mathbf{v} \rangle\mathbf{v}_1 + \cdots + \langle \mathbf{v}_k, \mathbf{v} \rangle\mathbf{v}_k$.

37. Since $\text{proj}_S \mathbf{u} \in S$, $\text{proj}_S \left(\text{proj}_S \mathbf{u} \right) = \text{proj}_S \mathbf{u}$ by Theorem 10.14 part (c).

39. (a) $\text{proj}_S \mathbf{u} = \mathbf{u}$, by Theorem 10.14, part (c).

   (b) Since $\mathbf{u} - \text{proj}_{S^\perp} \mathbf{u}$ is orthogonal to $S^\perp$, $\mathbf{u} - \text{proj}_{S^\perp} \mathbf{u} \in \left(S^\perp \right)^\perp = S$. Since $\mathbf{u} \in S$, $\text{proj}_{S^\perp} \mathbf{u} = \mathbf{u} - \left(\mathbf{u} - \text{proj}_{S^\perp} \mathbf{u} \right) \in S$. But $\text{proj}_{S^\perp} \mathbf{u} \in S^\perp$, by Theorem 10.14, part (a). Hence $\text{proj}_{S^\perp} \mathbf{u} \in S \cap S^\perp = \{\mathbf{0}\}$, thus $\text{proj}_{S^\perp} \mathbf{u} = \mathbf{0}$.

41. By Theorem 10.12, for any $\mathbf{v}$ in V, $\mathbf{v} = \langle \mathbf{v}_1, \mathbf{v} \rangle \mathbf{v}_1 + \cdots + \langle \mathbf{v}_k, \mathbf{v} \rangle \mathbf{v}_k$. Thus

$$
\begin{aligned}
\|\mathbf{v}\|^2 &= \langle \mathbf{v}, \mathbf{v} \rangle \\
&= \langle \langle \mathbf{v}_1, \mathbf{v} \rangle \mathbf{v}_1 + \cdots + \langle \mathbf{v}_k, \mathbf{v} \rangle \mathbf{v}_k, \mathbf{v} \rangle \\
&= \langle \mathbf{v}_1, \mathbf{v} \rangle \langle \mathbf{v}_1, \mathbf{v} \rangle + \cdots + \langle \mathbf{v}_k, \mathbf{v} \rangle \langle \mathbf{v}_k, \mathbf{v} \rangle \\
&= \left(\langle \mathbf{v}_1, \mathbf{v} \rangle \right)^2 + \cdots + \left(\langle \mathbf{v}_k, \mathbf{v} \rangle \right)^2.
\end{aligned}
$$

10.3   Applications of Inner Products

1. We evaluate $\frac{\langle \mathbf{a}_1, \mathbf{y} \rangle}{\langle \mathbf{a}_1, \mathbf{a}_1 \rangle} = \frac{1(1)(0)+2(1)(2)+2(1)(3)+1(1)(5)}{1(1^2)+2(1^2)+2(1^2)+1(1^2)} = \frac{5}{2}$ and $\frac{\langle \mathbf{a}_2, \mathbf{y} \rangle}{\langle \mathbf{a}_2, \mathbf{a}_2 \rangle} = \frac{1(-2)(0)+2(-1)(2)+2(1)(3)+1(2)(5)}{1(-2)^2+2(-1)^2+2(1)^2+1(2)^2} = 1$, and obtain the weighted least-squares line $y = \frac{5}{2} + x$.

3. The slope of ℓ_1 would be greater than the slope of ℓ_2. The slope of ℓ_1 would reflect all the data points equally, including the endpoints which correspond to a line of larger slope. The slope of ℓ_2 concentrates the weight on the middle points, which fall on a line with smaller slope.

5. The resulting line will be the same. Tripling the weights will triple the inner product, but these factors will cancel in the ratios $\frac{\langle \mathbf{a}_1, \mathbf{y} \rangle}{\langle \mathbf{a}_1, \mathbf{a}_1 \rangle}$ and $\frac{\langle \mathbf{a}_2, \mathbf{y} \rangle}{\langle \mathbf{a}_2, \mathbf{a}_2 \rangle}$.

7. $a_0 = \frac{1}{2\pi} \int_{-\pi/2}^{\pi/2} 1 \, dx = \frac{1}{2}$, $a_1 = \frac{1}{\pi} \int_{-\pi/2}^{\pi/2} 1 \cos(x) \, dx = \frac{2}{\pi}$, and $a_2 = \frac{1}{\pi} \int_{-\pi/2}^{\pi/2} 1 \cos(2x) \, dx = 0$. $b_1 = b_2 = 0$ since f is an even function. Thus $f_2(x) = \frac{1}{2} + \frac{2}{\pi} \cos(x)$.

9. Evaluate

$$
\begin{aligned}
a_0 &= \frac{1}{2\pi} \left(\int_0^\pi 1 \, dx \right) = \frac{1}{2}, \\
a_1 &= \frac{1}{\pi} \left(\int_0^\pi 1 \cos(x) \, dx \right) = 0, \\
a_2 &= \frac{1}{\pi} \left(\int_0^\pi 1 \cos(2x) \, dx \right) = 0, \\
b_1 &= \frac{1}{\pi} \left(\int_0^\pi 1 \sin(x) \, dx \right) = \frac{2}{\pi}, \\
b_2 &= \frac{1}{\pi} \left(\int_0^\pi 1 \sin(2x) \, dx \right) = 0.
\end{aligned}
$$

Thus $f_2(x) = \frac{1}{2} + \frac{2}{\pi} \sin(x)$.

11. Evaluate

$$a_0 = \frac{1}{2\pi} \left(\int_{-\pi}^{\pi} (x+1)\, dx \right) = 1,$$

$$a_1 = \frac{1}{\pi} \left(\int_{-\pi}^{\pi} (x+1) \cos(x)\, dx \right) = 0,$$

$$a_2 = \frac{1}{\pi} \left(\int_{-\pi}^{\pi} (x+1) \cos(2x)\, dx \right) = 0,$$

$$b_1 = \frac{1}{\pi} \left(\int_{-\pi}^{\pi} (x+1) \sin(x)\, dx \right) = 2,$$

$$b_2 = \frac{1}{\pi} \left(\int_{-\pi}^{\pi} (x+1) \sin(2x)\, dx \right) = -1.$$

Thus $f_2(x) = 1 + 2\sin(x) - \sin(2x)$.

13. Evaluate

$$a_0 = \frac{1}{2\pi} \left(\int_{-\pi}^{\pi} (x^2)\, dx \right) = \frac{1}{3}\pi^2,$$

$$a_1 = \frac{1}{\pi} \left(\int_{-\pi}^{\pi} (x^2) \cos(x)\, dx \right) = -4,$$

$$a_2 = \frac{1}{\pi} \left(\int_{-\pi}^{\pi} (x^2) \cos(2x)\, dx \right) = 1,$$

$$b_1 = \frac{1}{\pi} \left(\int_{-\pi}^{\pi} (x^2) \sin(x)\, dx \right) = 0,$$

$$b_2 = \frac{1}{\pi} \left(\int_{-\pi}^{\pi} (x^2) \sin(2x)\, dx \right) = 0.$$

Thus $f_2(x) = \frac{1}{3}\pi^2 - 4\cos(x) + \cos(2x)$.

15. Since $\cos(2x)$ and $\sin(3x)$ are basis functions, we have $a_2 = 1$ and $b_3 = -1$, with all other coefficients zero.

17. Rewrite $f(x) = 1 + \sin^2(4x) = 1 + \frac{1}{2}(1 - \cos(8x)) = \frac{3}{2} - \frac{1}{2}\cos(8x)$. Since 1 and $\cos(8x)$ are basis functions, we have $a_0 = \frac{3}{2}$ and $a_8 = -\frac{1}{2}$, with all other Fourier coefficients zero.

19. Evaluate

$$c_0 = \frac{1}{2}(f(0) + f(\pi)) = \frac{1}{2}(1+2) = \frac{3}{2},$$

$$c_1 = \frac{2}{2}(f(0)\cos(0) + f(\pi)\cos(\pi)) = \frac{2}{2}(1\cos(0) + 2\cos(\pi)) = -1,$$

$$d_1 = \frac{2}{2}(f(0)\sin(0) + f(\pi)\sin(\pi)) = \frac{2}{2}(1\sin(0) + 2\sin(\pi)) = 0.$$

Thus $g_1(x) = \frac{3}{2} - \cos(x)$.

21. Evaluate

$$c_0 = \frac{1}{4}\left(f\left(-\frac{\pi}{2}\right) + f(0) + f\left(\frac{\pi}{2}\right) + f(\pi)\right) = \frac{1}{4}(0 + 1 + 3 - 2) = \frac{1}{2},$$

$$c_1 = \frac{2}{4}\left(f\left(-\frac{\pi}{2}\right)\cos\left(-\frac{\pi}{2}\right) + f(0)\cos(0) + f\left(\frac{\pi}{2}\right)\cos\left(\frac{\pi}{2}\right) + f(\pi)\cos(\pi)\right)$$

$$= \frac{2}{4}\left(0\cos\left(-\frac{\pi}{2}\right) + 1\cos(0) + 3\cos\left(\frac{\pi}{2}\right) - 2\cos(\pi)\right) = \frac{3}{2},$$

$$d_1 = \frac{2}{4}\left(f\left(-\frac{\pi}{2}\right)\sin\left(-\frac{\pi}{2}\right) + f(0)\sin(0) + f\left(\frac{\pi}{2}\right)\sin\left(\frac{\pi}{2}\right) + f(\pi)\sin(\pi)\right)$$

$$= \frac{2}{4}\left(0\sin\left(-\frac{\pi}{2}\right) + 1\sin(0) + 3\sin\left(\frac{\pi}{2}\right) - 2\sin(\pi)\right) = \frac{3}{2}.$$

Thus $g_1(x) = \frac{1}{2} + \frac{3}{2}\cos(x) + \frac{3}{2}\sin(x)$.

23. For example, consider the data set $\{(-1, -1), (0, 1), (1, -1)\}$. This set has ordinary least squares regression line $y = 0$, and weighted least squares regression line with triple the weight on the right-most point $y = -\frac{2}{5} - \frac{1}{2}x$.

25. For example, let $f(x) = 1$. Then, $a_0 = 1$, and all other Fourier coefficients are zero.

27. For example, let $f(x) = x+1$. Then $a_0 = \frac{1}{2\pi}\int_{-\pi}^{\pi}(x+1)\,dx = 1$. For $k \geq 1$, $a_k = \frac{1}{\pi}\int_{-\pi}^{\pi}(x+1)\cos(kx)\,dx = \frac{1}{\pi}\int_{-\pi}^{\pi}x\cos(kx)\,dx + \frac{1}{\pi}\int_{-\pi}^{\pi}\cos(kx)\,dx = 0$, since the first term is the integral of an odd function, and for the second term the anti-derivative of $\cos(kx)$ is $\frac{1}{k}\sin(kx)$, which vanishes at the endpoints.

29. True. Otherwise property (d) of Definition 10.1 of the inner product does not hold.

31. False. Weighted least squares can be applied to any data set, although the computations are more involved.

33. True. To determine the Fourier coefficients, we need to know $f(x)$, and hence the values of f at the required points.

35. For all $k \geq 1$, $\langle 1, \sin(kx)\rangle = \frac{1}{\pi}\int_{-\pi}^{\pi}(1)(\sin(kx))\,dx = \frac{1}{\pi}\int_{-\pi}^{\pi}\sin(kx)\,dx = \frac{-1}{2k}\cos(kx)\big|_{x=-\pi}^{x=\pi} = \frac{-1}{2k}(\cos(\pi k) - \cos(-\pi k)) = \frac{-1}{2k}(\cos(\pi k) - \cos(\pi k)) = 0$.

37. $\|\sin(kx)\|^2 = \langle\sin(kx), \sin(kx)\rangle = \frac{1}{\pi}\int_{-\pi}^{\pi}(\sin(kx))(\sin(kx))\,dx = \frac{1}{\pi}\int_{-\pi}^{\pi}\sin^2(kx)\,dx = \frac{1}{2\pi}\int_{-\pi}^{\pi}(1 - \cos(2kx))\,dx = \frac{1}{2\pi}\int_{-\pi}^{\pi}1\,dx - \frac{1}{2\pi}\int_{-\pi}^{\pi}\cos(2kx)\,dx = 1 + \frac{1}{4k\pi}\sin(2kx)\big|_{x=-\pi}^{x=\pi} = 1 + \frac{1}{4k\pi}(\sin(2k\pi) - \sin(-2k\pi)) = 1 + 0 = 1$, so $\|\sin(kx)\| = 1$ for all $k \geq 1$.

39. $\frac{1}{\pi}\int_{-\pi}^{\pi}x\cos(kx)\,dx = \left(\frac{1}{k\pi}x\sin(kx)\right)\big|_{x=-\pi}^{x=\pi} - \frac{1}{k\pi}\int_{-\pi}^{\pi}\sin(kx)\,dx = \left(\frac{1}{k\pi}\pi\sin(k\pi) - \frac{1}{k\pi}(-\pi)\sin(-k\pi)\right) + \frac{1}{k^2\pi}\cos(kx)\big|_{x=-\pi}^{x=\pi} = 0 + \frac{1}{k^2\pi}(\cos(k\pi) - \cos(-k\pi)) = \frac{1}{k^2\pi}(\cos(k\pi) - \cos(k\pi)) = 0$ for all $k \geq 1$.

41. Apply the identities:

$$\cos\left(\frac{2jk\pi}{n} - k\pi\right) = \cos\left(\frac{2jk\pi}{n}\right)\cos(k\pi) + \sin\left(\frac{2jk\pi}{n}\right)\sin(k\pi)$$

$$= \cos\left(\frac{2jk\pi}{n}\right)(-1)^k + \sin\left(\frac{2jk\pi}{n}\right)(0)$$

$$= (-1)^k\cos\left(\frac{2jk\pi}{n}\right),$$

so

$$c_k = \frac{2}{n} \sum_{j=1}^{n} f\left(\frac{2jk}{n} - \pi\right) \cos\left(\frac{2jk\pi}{n} - k\pi\right)$$

$$= \frac{2}{n} \sum_{j=1}^{n} f\left(\frac{2jk}{n} - \pi\right) (-1)^k \cos\left(\frac{2jk\pi}{n}\right)$$

$$= \frac{2}{n} \sum_{j=1}^{n} (-1)^k f\left(\frac{2jk}{n} - \pi\right) \cos\left(\frac{2jk\pi}{n}\right)$$

And

$$\sin\left(\frac{2jk\pi}{n} - k\pi\right) = \sin\left(\frac{2jk\pi}{n}\right) \cos\left(k\pi\right) - \sin\left(k\pi\right) \cos\left(\frac{2jk\pi}{n}\right)$$

$$= \sin\left(\frac{2jk\pi}{n}\right) (-1)^k - (0) \cos\left(\frac{2jk\pi}{n}\right)$$

$$= (-1)^k \sin\left(\frac{2jk\pi}{n}\right),$$

so

$$d_k = \frac{2}{n} \sum_{j=1}^{n} f\left(\frac{2jk}{n} - \pi\right) \sin\left(\frac{2jk\pi}{n} - k\pi\right)$$

$$= \frac{2}{n} \sum_{j=1}^{n} f\left(\frac{2jk}{n} - \pi\right) (-1)^k \sin\left(\frac{2jk\pi}{n}\right)$$

$$= \frac{2}{n} \sum_{j=1}^{n} (-1)^k f\left(\frac{2jk}{n} - \pi\right) \sin\left(\frac{2jk\pi}{n}\right).$$

43. We compute the following:

n	c_i	d_i
0	2.7125	*
1	0.5445	−0.06768
2	0.05	−0.025
3	0.1555	0.03232
4	0.075	0
5	0.1555	−0.03232

Thus $g_5(x) \approx 2.7125 + 0.5445 \cos(x) + 0.05 \cos(2x) + 0.1555 \cos(3x) + 0.075 \cos(4x) + 0.1555 \cos(5x) - 0.06768 \sin(x) - 0.025 \sin(2x) + 0.03232 \sin(3x) - 0.03232 \sin(5x).$

45. The Fourier coefficients are $a_0 = \frac{\sinh(\pi)}{\pi} \approx 3.6761$, $a_1 = -\frac{\sinh(\pi)}{\pi} \approx -3.6761$, $a_2 = \frac{2\sinh(\pi)}{5\pi} \approx 1.4704$, $a_3 = -\frac{\sinh(\pi)}{5\pi} \approx -0.7352$, $b_1 = \frac{\sinh(\pi)}{\pi} \approx 3.6761$, $b_2 = -\frac{4\sinh(\pi)}{5\pi} \approx -2.9409$, and $b_3 = \frac{3\sinh(\pi)}{5\pi} \approx 2.2057$. Therefore $f_3(x) \approx 3.6761 - 3.6761 \cos(x) + 1.4704 \cos(2x) - 0.7352 \cos(3x) + 3.6761 \sin(x) - 2.9409 \sin(2x) + 2.2057 \sin(3x).$

To find g_3, we start by evaluating $f(x)$ at the 50 evenly distributed points $x = \frac{2\pi}{50} - \pi, \frac{4\pi}{50} - \pi, \frac{6\pi}{50} - \pi, \ldots, \pi$. Using a computer algebra system, we then apply the formulas in equation (7) to find that $c_0 \approx 3.9119$, $c_1 \approx -4.1477$, $c_2 \approx 1.9421$, $c_3 \approx -1.2069$, $d_1 \approx 3.6664$, $d_2 \approx -2.9215$, and $d_3 \approx 2.1766$. Hence we find that $g_3(x) \approx 3.9119 - 4.1477 \cos(x) + 1.9421 \cos(2x) - 1.2069 \cos(3x) + 3.6664 \sin(x) - 2.9215 \sin(2x) + 2.1766 \sin(3x).$

Note: The problem is stated incorrectly in the text, where is calls for the 50 points to have values $0.02, 0.04, \ldots, 1.0$. The solution given here is the appropriate one.

Chapter 11

Additional Topics and Applications

11.1 Quadratic Forms

1. $Q(\mathbf{x}_0) = Q\left(\begin{bmatrix} 3 \\ 1 \end{bmatrix}\right) = (3)^2 - 5(1)^2 + 6(3)(1) = 22.$

3. $Q(\mathbf{x}_0) = Q\left(\begin{bmatrix} 0 \\ 3 \\ 1 \end{bmatrix}\right) = 3(0)^2 + (3)^2 - (1)^2 + 6(0)(1) = 8.$

5. $Q(\mathbf{x}) = Q\left(\begin{bmatrix} x_1 \\ x_2 \end{bmatrix}\right) = 4x_1^2 + x_2^2.$

7. $Q(\mathbf{x}) = Q\left(\begin{bmatrix} x_1 \\ x_2 \end{bmatrix}\right) = x_1^2 + 2x_2^2 + 6x_1x_2.$

9. $Q(\mathbf{x}) = Q\left(\begin{bmatrix} x_1 \\ x_2 \\ x_3 \end{bmatrix}\right) = x_1^2 + 3x_2^2 - 2x_3^2.$

11. $Q(\mathbf{x}) = Q\left(\begin{bmatrix} x_1 \\ x_2 \\ x_3 \\ x_4 \end{bmatrix}\right) = 2x_1^2 + x_2^2 + 2x_3^2 + 3x_4^2.$

13. $A = \begin{bmatrix} 1 & 3 \\ 3 & -5 \end{bmatrix}.$

15. $A = \begin{bmatrix} 3 & 0 & 3 \\ 0 & 1 & 0 \\ 3 & 0 & -1 \end{bmatrix}.$

17. $A = \begin{bmatrix} 5 & 0 & 3 \\ 0 & -1 & -6 \\ 3 & -6 & 3 \end{bmatrix}.$

19. $\det(\lambda I_2 - A) = \det\left(\lambda\begin{bmatrix} 1 & 0 \\ 0 & 1 \end{bmatrix} - \begin{bmatrix} 1 & 2 \\ 2 & 1 \end{bmatrix}\right) = \lambda^2 - 2\lambda - 3 = (\lambda+1)(\lambda-3) = 0 \quad \Rightarrow \quad \lambda_1 = -1 < 0$ and $\lambda_2 = 3 > 0$. Since A has both a positive and negative eigenvalue, by Theorem 11.4 (c), A is indefinite.

21. $\det(\lambda I_2 - A) = \det\left(\lambda\begin{bmatrix} 1 & 0 \\ 0 & 1 \end{bmatrix} - \begin{bmatrix} 2 & 2 \\ 2 & -1 \end{bmatrix}\right) = \lambda^2 - \lambda - 6 = (\lambda + 2)(\lambda - 3) = 0 \Rightarrow \lambda_1 = -2 <$
0 and $\lambda_2 = 3 > 0$. Since A has both a positive and negative eigenvalue, by Theorem 11.4 (c), A is indefinite.

23. $\det(\lambda I_3 - A) = \det\left(\lambda\begin{bmatrix} 1 & 0 & 0 \\ 0 & 1 & 0 \\ 0 & 0 & 1 \end{bmatrix} - \begin{bmatrix} 0 & 1 & 0 \\ 1 & 0 & 0 \\ 0 & 0 & 1 \end{bmatrix}\right) = \lambda^3 - \lambda^2 - \lambda + 1 = (\lambda + 1)(\lambda - 1)^2 = 0 \Rightarrow$
$\lambda_1 = -1 < 0$ and $\lambda_2 = 1 > 0$. Since A has both a positive and negative eigenvalue, by Theorem 11.4 (c), A is indefinite.

25. $\det(\lambda I_4 - A) = \det\left(\lambda\begin{bmatrix} 1 & 0 & 0 & 0 \\ 0 & 1 & 0 & 0 \\ 0 & 0 & 1 & 0 \\ 0 & 0 & 0 & 1 \end{bmatrix} - \begin{bmatrix} 0 & 0 & 1 & 0 \\ 0 & 1 & 0 & 0 \\ 1 & 0 & 0 & 0 \\ 0 & 0 & 0 & 1 \end{bmatrix}\right) = \lambda^4 - 2\lambda^3 + 2\lambda - 1 = (\lambda + 1)(\lambda - 1)^3 =$
$0 \Rightarrow \lambda_1 = -1 < 0$ and $\lambda_2 = 1 > 0$. Since A has both a positive and negative eigenvalue, by Theorem 11.4 (c), A is indefinite.

27. For example, let $Q(\mathbf{x}) = Q\left(\begin{bmatrix} x_1 \\ x_2 \end{bmatrix}\right) = x_1^2 + x_2^2$, and $c = -1$.

29. For example, let $Q(\mathbf{x}) = Q\left(\begin{bmatrix} x_1 \\ x_2 \end{bmatrix}\right) = x_1^2 - x_2^2$, and $c = 0$. Then $Q(\mathbf{x}) = x_1^2 - x_2^2 = 0 \Rightarrow$
$(x_1 - x_2)(x_1 + x_2) = 0$, and the graph consists of the intersecting lines $x_1 - x_2 = 0$ and $x_1 + x_2 = 0$.

31. The only quadratic form which is also a linear transformation is $Q(\mathbf{x}) = 0$ for all $\mathbf{x}$.

33. For example, let $Q(\mathbf{x}) = Q\left(\begin{bmatrix} x_1 \\ x_2 \\ x_3 \end{bmatrix}\right) = x_1^2 + x_2^2$.

35. True, by Definition 11.1.

37. False. For example, if $Q(\mathbf{x}) = Q\left(\begin{bmatrix} x_1 \\ x_2 \end{bmatrix}\right) = x_1 x_2$, then $Q\left(\begin{bmatrix} 1 \\ 0 \end{bmatrix}\right) = 0$, $Q\left(\begin{bmatrix} 0 \\ 1 \end{bmatrix}\right) = 0$, but
$Q\left(\begin{bmatrix} 1 \\ 0 \end{bmatrix} + \begin{bmatrix} 0 \\ 1 \end{bmatrix}\right) = Q\left(\begin{bmatrix} 1 \\ 1 \end{bmatrix}\right) = 1$.

39. True. If $Q_1(\mathbf{x}) = \mathbf{x}^T A_1 \mathbf{x}$ and $Q_2(\mathbf{x}) = \mathbf{x}^T A_2 \mathbf{x}$, then $Q_1(\mathbf{x}) + Q_2(\mathbf{x}) = \mathbf{x}^T A_1 \mathbf{x} + \mathbf{x}^T A_2 \mathbf{x} = \mathbf{x}^T (A_1 + A_2)\mathbf{x}$. Since A_1 and A_2 are both symmetric, $A_1 + A_2$ is also symmetric, and thus the matrix of a quadratic form $Q = Q_1 + Q_2$.

41. The quadratic form $Q: \mathbf{R} \to \mathbf{R}$ will have the form $Q(\mathbf{x}) = Q(x) = \mathbf{x}^T A\mathbf{x} = xax = ax^2$. Thus $Q(x)$ cannot have both positive and negative values, and hence Q cannot be indefinite.

43. Let $A = I_n$, then $Q(\mathbf{x}) = \mathbf{x}^T I \mathbf{x} = \mathbf{x}^T \mathbf{x} = \|\mathbf{x}\|^2$, so the identity matrix I_n is the matrix of $Q(\mathbf{x}) = \|\mathbf{x}\|^2$.

45. $Q(\mathbf{0}) = \mathbf{0}^T (A\mathbf{0}) = \mathbf{0}^T \mathbf{0} = 0$.

11.2   Positive Definite Matrices

1. $A_1 = [3]$, $A_2 = \begin{bmatrix} 3 & 5 \\ 5 & 7 \end{bmatrix}$.

3. $A_1 = [1]$, $A_2 = \begin{bmatrix} 1 & 4 \\ 4 & 0 \end{bmatrix}$, $A_3 = \begin{bmatrix} 1 & 4 & -3 \\ 4 & 0 & 2 \\ -3 & 2 & 5 \end{bmatrix}$.

5. $A_1 = [2]$, $A_2 = \begin{bmatrix} 2 & 1 \\ 1 & 3 \end{bmatrix}$, $A_3 = \begin{bmatrix} 2 & 1 & 0 \\ 1 & 3 & 4 \\ 0 & 4 & 0 \end{bmatrix}$, $A_4 = \begin{bmatrix} 2 & 1 & 0 & -1 \\ 1 & 3 & 4 & 1 \\ 0 & 4 & 0 & -2 \\ -1 & 1 & -2 & 1 \end{bmatrix}$.

7. $\det(A_1) = \det([2]) = 2 > 0$, $\det(A_2) = \det\left(\begin{bmatrix} 2 & 1 \\ 1 & -2 \end{bmatrix}\right) = -5 < 0$, so A is not positive definite.

9. $\det(A_1) = \det([1]) = 1 > 0$, $\det(A_2) = \det\left(\begin{bmatrix} 1 & 2 \\ 2 & 5 \end{bmatrix}\right) = 1 > 0$, $\det(A_3) =$

$\det\left(\begin{bmatrix} 1 & 2 & -1 \\ 2 & 5 & 1 \\ -1 & 1 & 11 \end{bmatrix}\right) = 1 > 0$, so A is positive definite.

11. $\det(A_1) = \det([1]) = 1 > 0$, $\det(A_2) = \det\left(\begin{bmatrix} 1 & 1 \\ 1 & 2 \end{bmatrix}\right) = 1 > 0$, $\det(A_3) =$

$\det\left(\begin{bmatrix} 1 & 1 & 2 \\ 1 & 2 & 1 \\ 2 & 1 & 6 \end{bmatrix}\right) = 1 > 0$, $\det(A_4) = \det\left(\begin{bmatrix} 1 & 1 & 2 & 1 \\ 1 & 2 & 1 & 0 \\ 2 & 1 & 6 & 3 \\ 1 & 0 & 3 & 6 \end{bmatrix}\right) = 4 > 0$, so A is positive definite.

13. $\det(A_1) = \det([1]) = 1 > 0$, $\det(A_2) = \det\left(\begin{bmatrix} 1 & -2 \\ -2 & 5 \end{bmatrix}\right) = 1 > 0$, so $\begin{bmatrix} 1 & -2 \\ -2 & 5 \end{bmatrix}$ is positive definite.

$$\begin{bmatrix} 1 & -2 \\ -2 & 5 \end{bmatrix} \overset{2R_1+R_2 \Rightarrow R_2}{\sim} \begin{bmatrix} 1 & -2 \\ 0 & 1 \end{bmatrix} \Rightarrow L = \begin{bmatrix} 1 & \bullet \\ -2 & 1 \end{bmatrix}$$

Thus $L = \begin{bmatrix} 1 & 0 \\ -2 & 1 \end{bmatrix}$, $U = \begin{bmatrix} 1 & -2 \\ 0 & 1 \end{bmatrix}$, and $A = LU$.

15. $\det([1]) = 1 > 0$, $\det\left(\begin{bmatrix} 1 & -2 \\ -2 & 5 \end{bmatrix}\right) = 1 > 0$, and $\det\left(\begin{bmatrix} 1 & -2 & 2 \\ -2 & 5 & -5 \\ 2 & -5 & 6 \end{bmatrix}\right) = 1 > 0$, so $\begin{bmatrix} 1 & -2 & 2 \\ -2 & 5 & -5 \\ 2 & -5 & 6 \end{bmatrix}$

is positive definite.

$$\begin{bmatrix} 1 & -2 & 2 \\ -2 & 5 & -5 \\ 2 & -5 & 6 \end{bmatrix} \overset{\substack{2R_1+R_2 \Rightarrow R_2 \\ -2R_1+R_3 \Rightarrow R_3}}{\sim} \begin{bmatrix} 1 & -2 & 2 \\ 0 & 1 & -1 \\ 0 & -1 & 2 \end{bmatrix} \Rightarrow L = \begin{bmatrix} 1 & \bullet & \bullet \\ -2 & 1 & \bullet \\ 2 & 0 & 1 \end{bmatrix}$$

$$\overset{R_2+R_3 \Rightarrow R_3}{\sim} \begin{bmatrix} 1 & -2 & 2 \\ 0 & 1 & -1 \\ 0 & 0 & 1 \end{bmatrix} \Rightarrow L = \begin{bmatrix} 1 & \bullet & \bullet \\ -2 & 1 & \bullet \\ 2 & -1 & 1 \end{bmatrix}$$

Thus $L = \begin{bmatrix} 1 & 0 & 0 \\ -2 & 1 & 0 \\ 2 & -1 & 1 \end{bmatrix}$, $U = \begin{bmatrix} 1 & -2 & 2 \\ 0 & 1 & -1 \\ 0 & 0 & 1 \end{bmatrix}$, and $A = LU$.

17. $\det([1]) = 1 > 0$ and $\det\left(\begin{bmatrix} 1 & -2 \\ -2 & 8 \end{bmatrix}\right) = 4 > 0$, so $\begin{bmatrix} 1 & -2 \\ -2 & 8 \end{bmatrix}$ is positive definite.

$$\begin{bmatrix} 1 & -2 \\ -2 & 8 \end{bmatrix} \overset{2R_1+R_2 \Rightarrow R_2}{\sim} \begin{bmatrix} 1 & -2 \\ 0 & 4 \end{bmatrix} \Rightarrow L = \begin{bmatrix} 1 & \bullet \\ -2 & 1 \end{bmatrix}$$

Thus $L = \begin{bmatrix} 1 & 0 \\ -2 & 1 \end{bmatrix}$, $D = \begin{bmatrix} 1 & 0 \\ 0 & 4 \end{bmatrix}$ and $U = \begin{bmatrix} 1 & -2 \\ 0 & 1 \end{bmatrix}$, and $A = LDU$.

19. $\det([1]) = 1 > 0$, $\det\left(\begin{bmatrix} 1 & 3 \\ 3 & 10 \end{bmatrix}\right) = 1 > 0$, and $\det\left(\begin{bmatrix} 1 & 3 & 2 \\ 3 & 10 & 8 \\ 2 & 8 & 9 \end{bmatrix}\right) = 1 > 0$, so $\begin{bmatrix} 1 & 3 & 2 \\ 3 & 10 & 8 \\ 2 & 8 & 9 \end{bmatrix}$ is

positive definite.

$$\begin{bmatrix} 1 & 3 & 2 \\ 3 & 10 & 8 \\ 2 & 8 & 9 \end{bmatrix} \overset{\underset{\sim}{-3R_1+R_2 \Rightarrow R_2}}{-2R_1+R_3 \Rightarrow R_3} \begin{bmatrix} 1 & 3 & 2 \\ 0 & 1 & 2 \\ 0 & 2 & 5 \end{bmatrix} \Rightarrow L = \begin{bmatrix} 1 & \bullet & \bullet \\ 3 & 1 & \bullet \\ 2 & 0 & 1 \end{bmatrix}$$

$$\overset{-2R_1+R_3 \Rightarrow R_3}{\underset{\sim}{}} \begin{bmatrix} 1 & 3 & 2 \\ 0 & 1 & 2 \\ 0 & 0 & 1 \end{bmatrix} \Rightarrow L = \begin{bmatrix} 1 & \bullet & \bullet \\ 3 & 1 & \bullet \\ 2 & 2 & 1 \end{bmatrix}$$

Thus $L = \begin{bmatrix} 1 & 0 & 0 \\ 3 & 1 & 0 \\ 2 & 2 & 1 \end{bmatrix}$, $D = \begin{bmatrix} 1 & 0 & 0 \\ 0 & 1 & 0 \\ 0 & 0 & 1 \end{bmatrix}$, $U = \begin{bmatrix} 1 & 3 & 2 \\ 0 & 1 & 2 \\ 0 & 0 & 1 \end{bmatrix}$, and $A = LDU$.

21. $\det([1]) = 1 > 0$ and $\det\left(\begin{bmatrix} 1 & 2 \\ 2 & 5 \end{bmatrix}\right) = 1 > 0$, so $\begin{bmatrix} 1 & 2 \\ 2 & 5 \end{bmatrix}$ is positive definite.

$$\begin{bmatrix} 1 & 2 \\ 2 & 5 \end{bmatrix} \overset{-2R_1+R_2 \Rightarrow R_2}{\underset{\sim}{}} \begin{bmatrix} 1 & 2 \\ 0 & 1 \end{bmatrix} \Rightarrow L = \begin{bmatrix} 1 & \bullet \\ 2 & 1 \end{bmatrix}$$

Thus $L = \begin{bmatrix} 1 & 0 \\ 2 & 1 \end{bmatrix}$, $D = \begin{bmatrix} 1 & 0 \\ 0 & 1 \end{bmatrix}$, and $U = \begin{bmatrix} 1 & 3 \\ 0 & 1 \end{bmatrix}$. Hence $L_c = LD^{1/2} =$

$\begin{bmatrix} 1 & 0 \\ 2 & 1 \end{bmatrix} \begin{bmatrix} 1 & 0 \\ 0 & 1 \end{bmatrix}^{1/2} = \begin{bmatrix} 1 & 0 \\ 2 & 1 \end{bmatrix} \begin{bmatrix} 1 & 0 \\ 0 & 1 \end{bmatrix} = \begin{bmatrix} 1 & 0 \\ 2 & 1 \end{bmatrix}$, and $A = L_c L_c^T$ is the Cholesky decomposition.

23. $\det([1]) = 1 > 0$, $\det\left(\begin{bmatrix} 1 & 2 \\ 2 & 5 \end{bmatrix}\right) = 1 > 0$, and $\det\left(\begin{bmatrix} 1 & 2 & 3 \\ 2 & 5 & 6 \\ 3 & 6 & 10 \end{bmatrix}\right) = 1 > 0$, so $\begin{bmatrix} 1 & 2 & 3 \\ 2 & 5 & 6 \\ 3 & 6 & 10 \end{bmatrix}$ is

positive definite.

$$\begin{bmatrix} 1 & 2 & 3 \\ 2 & 5 & 6 \\ 3 & 6 & 10 \end{bmatrix} \overset{\underset{\sim}{-2R_1+R_2 \Rightarrow R_2}}{-3R_1+R_3 \Rightarrow R_3} \begin{bmatrix} 1 & 2 & 3 \\ 0 & 1 & 0 \\ 0 & 0 & 1 \end{bmatrix} \Rightarrow L = \begin{bmatrix} 1 & \bullet & \bullet \\ 2 & 1 & \bullet \\ 3 & 0 & 1 \end{bmatrix}$$

Thus $L = \begin{bmatrix} 1 & 0 & 0 \\ 2 & 1 & 0 \\ 3 & 0 & 1 \end{bmatrix}$, $D = \begin{bmatrix} 1 & 0 & 0 \\ 0 & 1 & 0 \\ 0 & 0 & 1 \end{bmatrix}$, and $U = \begin{bmatrix} 1 & 2 & 3 \\ 0 & 1 & 0 \\ 0 & 0 & 1 \end{bmatrix}$. Hence $L_c = LD^{1/2}$

$= \begin{bmatrix} 1 & 0 & 0 \\ 2 & 1 & 0 \\ 3 & 0 & 1 \end{bmatrix} \begin{bmatrix} 1 & 0 & 0 \\ 0 & 1 & 0 \\ 0 & 0 & 1 \end{bmatrix}^{1/2} = \begin{bmatrix} 1 & 0 & 0 \\ 2 & 1 & 0 \\ 3 & 0 & 1 \end{bmatrix} \begin{bmatrix} 1 & 0 & 0 \\ 0 & 1 & 0 \\ 0 & 0 & 1 \end{bmatrix} = \begin{bmatrix} 1 & 0 & 0 \\ 2 & 1 & 0 \\ 3 & 0 & 1 \end{bmatrix}$, and $A = L_c L_c^T$ is the

Cholesky decomposition.

25. For example, $A = \begin{bmatrix} -1 & 0 \\ 0 & -1 \end{bmatrix}$.

27. For example, $A = \begin{bmatrix} -1 & 0 & 0 \\ 0 & -1 & 0 \\ 0 & 0 & -1 \end{bmatrix}$.

29. For example, $A = \begin{bmatrix} -1 & 0 & 0 \\ 0 & -1 & 0 \\ 0 & 0 & 1 \end{bmatrix}$.

31. False. For example, $A = \begin{bmatrix} -1 & 0 \\ 0 & -1 \end{bmatrix}$ is not positive definite, but $\det(A) = 1 > 0$.

33. False. AB is symmetric if $AB = (AB)^T = B^T A^T = BA$, so we would need A and B to commute, in order for AB to be symmetric. For example, let $A = \begin{bmatrix} 1 & 0 \\ 0 & 2 \end{bmatrix}$ and $B = \begin{bmatrix} 3 & 2 \\ 2 & 4 \end{bmatrix}$.

35. True. Since A is positive definite, $\det(A) > 0$ so A^{-1} exists and A^{-1} is also symmetric. Furthermore $\mathbf{x}^T A^{-1} \mathbf{x} = \mathbf{x}^T I_n A^{-1} \mathbf{x} = \mathbf{x}^T \left(A^{-1} A \right) A^{-1} \mathbf{x} = \left(\mathbf{x}^T \left(A^{-1} \right)^T \right) A \left(A^{-1} \mathbf{x} \right) = \left(A^{-1} \mathbf{x} \right)^T A \left(A^{-1} \mathbf{x} \right) > 0$ if $A^{-1} \mathbf{x} \neq \mathbf{0}$, since A is positive definite. But $A^{-1} \mathbf{x} \neq \mathbf{0}$ if and only if $\mathbf{x} \neq \mathbf{0}$, and hence A^{-1} is positive definite.

37. Since $A = L_1 D_1 U_1 = L_2 D_2 U_2$, then $L_2^{-1} L_1 D_1 = D_2 U_2 U_1^{-1}$. Since the matrix $L_2^{-1} L_1 D_1$ is a product of lower triangular matrices, $L_2^{-1} L_1 D_1$ is a lower triangular matrix. And since $D_2 U_2 U_1^{-1}$ is a product of upper triangular matrices, $D_2 U_2 U_1^{-1}$ is an upper triangular matrix. A lower and upper triangular matrix are equal only if they are both diagonal, so let D be this common diagonal matrix, $D = L_2^{-1} L_1 D_1 = D_2 U_2 U_1^{-1}$. Thus $L_2^{-1} L_1 = D D_1^{-1}$ is a diagonal matrix, as a product of diagonal matrices. And since L_1 and L_2 have 1's on the diagonal, and L_2^{-1} also has 1's on the diagonal, so $L_2^{-1} L_1$ has 1's on the diagonal. Hence $L_2^{-1} L_1 = I_n$, and $L_2 = L_1$. Now we observe that $D_1 = L_2^{-1} L_1 D_1 = D_2 U_2 U_1^{-1}$, so $U_2 U_1^{-1} = D_2^{-1} D_1$. Apply the similar argument, noting that $U_2^{-1} U_1$ is an upper triangular matrix with 1's on the diagonal, and equal to a diagonal matrix, to conclude that $U_2^{-1} U_1 = I_n$, and hence $U_2 = U_1$. Finally, $D_2^{-1} D_1 = U_2 U_1^{-1} = I_n$, so $D_2 = D_1$. Therefore the LDU factorization of an invertible matrix A is unique.

11.3   Constrained Optimization

1. $\max\{q_1, q_2\} = \max\{2, -3\} = 2$ and $\min\{q_1, q_2\} = \min\{2, -3\} = -3$, so by Theorem 11.11 the maximum and minimum values of $Q(\mathbf{x})$ subject to $\|\mathbf{x}\| = 1$ are 2 and -3, respectively.

3. $\max\{q_1, q_2, q_3\} = \max\{3, -3, -5\} = 3$ and $\min\{q_1, q_2, q_3\} = \min\{3, -3, -5\} = -5$, so by Theorem 11.11 the maximum and minimum values of $Q(\mathbf{x})$ subject to $\|\mathbf{x}\| = 1$ are 3 and -5, respectively.

5. $\max\{q_1, q_2, q_3, q_4\} = \max\{1, -1, -4, 2\} = 2$ and $\min\{q_1, q_2, q_3, q_4\} = \min\{1, -1, -4, 2\} = -4$, so by Theorem 11.11 the maximum and minimum values of $Q(\mathbf{x})$ subject to $\|\mathbf{x}\| = 1$ are 2 and -4, respectively.

7. $Q(\mathbf{x}) = \mathbf{x}^T \begin{bmatrix} 4 & 2 \\ 2 & 1 \end{bmatrix} \mathbf{x}$, $\det(\lambda I_2 - A) = \det\left(\lambda \begin{bmatrix} 1 & 0 \\ 0 & 1 \end{bmatrix} - \begin{bmatrix} 4 & 2 \\ 2 & 1 \end{bmatrix} \right) = \lambda^2 - 5\lambda = \lambda(\lambda - 5) = 0 \Rightarrow$ $\lambda_1 = 0 < \lambda_2 = 5$. By Theorem 11.12, subject to $\|\mathbf{x}\| = 1$, the maximum value of $Q(\mathbf{x})$ is 5 and the minimum value of $Q(\mathbf{x})$ is 0.

9. $Q(\mathbf{x}) = \mathbf{x}^T \begin{bmatrix} 0 & 0 & 1 \\ 0 & 1 & 0 \\ 1 & 0 & 0 \end{bmatrix} \mathbf{x}$, $\det(\lambda I_3 - A) = \det\left(\lambda \begin{bmatrix} 1 & 0 & 0 \\ 0 & 1 & 0 \\ 0 & 0 & 1 \end{bmatrix} - \begin{bmatrix} 0 & 0 & 1 \\ 0 & 1 & 0 \\ 1 & 0 & 0 \end{bmatrix} \right) = \lambda^3 - \lambda^2 - \lambda + 1 = (\lambda + 1)(\lambda - 1)^2 = 0 \Rightarrow \lambda_1 = -1 < \lambda_2 = \lambda_3 = 1$. By Theorem 11.12, subject to $\|\mathbf{x}\| = 1$, the maximum value of $Q(\mathbf{x})$ is 1 and the minimum value of $Q(\mathbf{x})$ is -1.

11. $Q(\mathbf{x}) = \mathbf{x}^T \begin{bmatrix} 1 & 1 & 1 \\ 1 & 1 & 1 \\ 1 & 1 & 1 \end{bmatrix} \mathbf{x}$, $\det(\lambda I_3 - A) = \det\left(\lambda \begin{bmatrix} 1 & 0 & 0 \\ 0 & 1 & 0 \\ 0 & 0 & 1 \end{bmatrix} - \begin{bmatrix} 1 & 1 & 1 \\ 1 & 1 & 1 \\ 1 & 1 & 1 \end{bmatrix} \right) = \lambda^3 - 3\lambda^2 = \lambda^2(\lambda - 3) = 0 \Rightarrow \lambda_1 = \lambda_2 = 0 < \lambda_3 = 3$. By Theorem 11.12, subject to $\|\mathbf{x}\| = 1$, the maximum value of $Q(\mathbf{x})$ is 3 and the minimum value of $Q(\mathbf{x})$ is 0.

13. $Q(\mathbf{x}) = \mathbf{x}^T \begin{bmatrix} 1 & 2 \\ 2 & 4 \end{bmatrix} \mathbf{x}$, $\det(\lambda I_2 - A) = \det\left(\lambda \begin{bmatrix} 1 & 0 \\ 0 & 1 \end{bmatrix} - \begin{bmatrix} 1 & 2 \\ 2 & 4 \end{bmatrix} \right) = \lambda^2 - 5\lambda = \lambda(\lambda - 5) = 0 \Rightarrow$ $\lambda_1 = 0 < \lambda_2 = 5$. Thus, subject to $\|\mathbf{x}\| = c = 2$, the maximum value of $Q(\mathbf{x})$ is $c^2(5) = (2)^2(5) = 20$, and the minimum value of $Q(\mathbf{x})$ is $c^2(0) = (2)^2 0 = 0$.

15. $Q(\mathbf{x}) = \mathbf{x}^T \begin{bmatrix} 1 & 0 & 0 \\ 0 & 0 & 1 \\ 0 & 1 & 0 \end{bmatrix} \mathbf{x}$, $\det(\lambda I_3 - A) = \det\left(\lambda \begin{bmatrix} 1 & 0 & 0 \\ 0 & 1 & 0 \\ 0 & 0 & 1 \end{bmatrix} - \begin{bmatrix} 1 & 0 & 0 \\ 0 & 0 & 1 \\ 0 & 1 & 0 \end{bmatrix}\right) = \lambda^3 - \lambda^2 - \lambda + 1 =$
$(\lambda + 1)(\lambda - 1)^2 = 0 \Rightarrow \lambda_1 = -1 < \lambda_2 = \lambda_3 = 1$. Thus, subject to $\|\mathbf{x}\| = c = 10$, the maximum value of $Q(\mathbf{x})$ is $c^2(1) = (10)^2(1) = 100$, and the minimum value of $Q(\mathbf{x})$ is $c^2(-1) = (10)^2(-1) = -100$.

17. Let $w_1 = x_1/5$ and $w_2 = x_2/2$, so $x_1 = 5w_1$ and $x_2 = 2w_2$. Then $Q(\mathbf{x}) = 4(5w_1)^2 + (2w_2)^2 + 4(5w_1)(2w_2) = 100w_1^2 + 40w_1 w_2 + 4w_2^2 = \mathbf{w}^T \begin{bmatrix} 100 & 20 \\ 20 & 4 \end{bmatrix} \mathbf{w}$. So,

$\det\left(\lambda \begin{bmatrix} 1 & 0 \\ 0 & 1 \end{bmatrix} - \begin{bmatrix} 100 & 20 \\ 20 & 4 \end{bmatrix}\right) = \lambda^2 - 104\lambda = \lambda(\lambda - 104) = 0 \Rightarrow$
$\lambda_1 = 0 < \lambda_2 = 104$. Thus, subject to $4x_1^2 + 25x_2^2 = 100 \Leftrightarrow \|\mathbf{w}\| = 1$, the maximum value of $Q(\mathbf{x})$ is 104, and the minimum value of $Q(\mathbf{x})$ is 0.

19. Let $w_1 = x_1$ and $w_2 = x_2/3$, so $x_1 = w_1$ and $x_2 = 3w_2$. Then $Q(\mathbf{x}) = 4(w_1)^2 + 4(3w_2)^2 + 6(w_1)(3w_2) = 4w_1^2 + 18w_1 w_2 + 36w_2^2 = \mathbf{w}^T \begin{bmatrix} 2 & 9 \\ 9 & 36 \end{bmatrix} \mathbf{w}$. So,

$\det\left(\lambda \begin{bmatrix} 1 & 0 \\ 0 & 1 \end{bmatrix} - \begin{bmatrix} 2 & 9 \\ 9 & 36 \end{bmatrix}\right) = \lambda^2 - 38\lambda - 9 = 0 \Rightarrow$
$\lambda_1 = 19 - \sqrt{370} < \lambda_2 = 19 + \sqrt{370}$. Thus, subject to $9x_1^2 + x_2^2 = 9 \Leftrightarrow \|\mathbf{w}\| = 1$, the maximum value of $Q(\mathbf{x})$ is $19 + \sqrt{370}$, and the minimum value of $Q(\mathbf{x})$ is $19 - \sqrt{370}$.

21. $Q(\mathbf{x}) = \mathbf{x}^T \begin{bmatrix} 4 & 0 & 0 \\ 0 & 1 & 0 \\ 0 & 0 & 3 \end{bmatrix} \mathbf{x}$, A has eigenvalues $\lambda_1 = 1 < \lambda_2 = 3 < \lambda_3 = 4$. By Theorem 11.13, subject to $\|\mathbf{x}\| = 1$ and $\mathbf{x} \cdot \mathbf{u}_3 = 0$, the maximum value of $Q(\mathbf{x})$ is $\lambda_2 = 3$, and the minimum value of $Q(\mathbf{x})$ is $\lambda_1 = 1$.

23. $Q(\mathbf{x}) = \mathbf{x}^T \begin{bmatrix} 1 & 0 & 0 \\ 0 & 4 & 2 \\ 0 & 2 & 1 \end{bmatrix} \mathbf{x}$, $\det(\lambda I_3 - A) = \det\left(\lambda \begin{bmatrix} 1 & 0 & 0 \\ 0 & 1 & 0 \\ 0 & 0 & 1 \end{bmatrix} - \begin{bmatrix} 1 & 0 & 0 \\ 0 & 4 & 2 \\ 0 & 2 & 1 \end{bmatrix}\right) = \lambda^3 - 6\lambda^2 + 5\lambda =$
$\lambda(\lambda - 1)(\lambda - 5) = 0 \Rightarrow \lambda_1 = 0 < \lambda_2 = 1 < \lambda_3 = 5$. By Theorem 11.13, subject to $\|\mathbf{x}\| = 1$ and $\mathbf{x} \cdot \mathbf{u}_3 = 0$, the maximum value of $Q(\mathbf{x})$ is $\lambda_2 = 1$, and the minimum value of $Q(\mathbf{x})$ is $\lambda_1 = 0$.

25. For example, let $Q(\mathbf{x}) = Q\left(\begin{bmatrix} x_1 \\ x_2 \end{bmatrix}\right) = x_1^2 + 5x_2^2$.

27. For example, let $Q(\mathbf{x}) = Q\left(\begin{bmatrix} x_1 \\ x_2 \end{bmatrix}\right) = x_1^2 + 6x_2^2$.

29. For example, let $Q(\mathbf{x}) = Q\left(\begin{bmatrix} x_1 \\ x_2 \end{bmatrix}\right) = -\frac{1}{9}x_1^2 + \frac{4}{9}x_2^2$.

31. True, by Theorem 11.12.

33. False. If $Q(\mathbf{x}) = 0$, then $m = M = 0$.

35. False. If the minimum value of $Q(\mathbf{x})$ subject to $\|\mathbf{x}\| = 1$ is $m = 1$, then the minimum value of $Q(\mathbf{x})$ subject to $\|\mathbf{x}\| = 2$ is $2^2(1) = 4$.

37. (a) Each $q_k \leq q_i$, so $Q(\mathbf{x}) = q_1 x_1^2 + \cdots + q_n x_n^2 \leq q_i x_1^2 + \cdots + q_i x_n^2 = q_i(x_1^2 + \cdots + x_n^2) = q_i \|\mathbf{x}\|^2 = q_i(1)^2 = q_i$. And $Q(\mathbf{e}_i) = q_1(0)^2 + \cdots + q_i(1)^2 + \cdots + q_n(0)^2 = q_i$. Thus, the maximum value of $Q(\mathbf{x})$ is q_i, and attained at $\mathbf{x} = \mathbf{e}_i$.

(b) Each $q_j \geq q_i$, so $Q(\mathbf{x}) = q_1 x_1^2 + \cdots + q_n x_n^2 \geq q_j x_1^2 + \cdots + q_j x_n^2 = q_j \left(x_1^2 + \cdots + x_n^2 \right) = q_j \|\mathbf{x}\|^2 = q_j (1)^2 = q_j$. And $Q(\mathbf{e}_j) = q_1 (0)^2 + \cdots + q_j (1)^2 + \cdots + q_n (0)^2 = q_j$. Thus, the minimum value of $Q(\mathbf{x})$ is q_j, and attained at $\mathbf{x} = \mathbf{e}_j$.

39. $Q(c\mathbf{x}) = (c\mathbf{x})^T A (c\mathbf{x}) = c^2 \mathbf{x}^T A \mathbf{x} = c^2 Q(\mathbf{x})$.

11.4 Complex Vector Spaces

1. (a) $\mathbf{u} - \mathbf{v} = (2 + 3i, -1, 3 + 4i) - (2, 3 - i, 1 + 5i) = (3i, -4 + i, 2 - i)$.

    (b) $\mathbf{w} + 3\mathbf{v} = (2 + i, 2 - i, 4 - 3i) + 3(2, 3 - i, 1 + 5i) = (8 + i, 11 - 4i, 7 + 12i)$.

    (c) $-\mathbf{u} + 2i\mathbf{w} - 5\mathbf{v} = -(2 + 3i, -1, 3 + 4i) + 2i(2 + i, 2 - i, 4 - 3i) - 5(2, 3 - i, 1 + 5i)$
    $= (-14 + i, -12 + 9i, -2 - 21i)$.

3. $\mathbf{u} + i\mathbf{v} = c\mathbf{w} \Leftrightarrow (2 + 3i, -1, 3 + 4i) + i(2, 3 - i, 1 + 5i) = c(2 + i, 2 - i, 4 - 3i) \Leftrightarrow$
$(2 + 5i, 3i, -2 + 5i) = (c(2 + i), c(2 - i), c(4 - 3i))$. Thus we need $2 + 5i = c(2 + i)$, $3i = c(2 - i)$, and $-2 - 5i = c(4 - 3i)$. From $2 + 5i = c(2 + i)$ we obtain $c = \frac{2 + 5i}{2 + i} = \frac{9}{5} + \frac{8}{5}i$. We check $\left(\frac{9}{5} + \frac{8}{5}i\right)(2 - i) = \frac{26}{5} + \frac{7}{5}i \neq 3i$, and conclude that c does not exist.

5. By Exercise 4, $\{\mathbf{u}, \mathbf{v}, \mathbf{w}\}$ are linearly independent, hence form a basis for $\mathbf{C}^3$. Therefore $(-5 + 2i, -3, -5 + i)$ is in the span of $\{\mathbf{u}, \mathbf{v}, \mathbf{w}\}$. Indeed, $(-5 + 2i, -3, -5 + i) = \left(\frac{45}{401} + \frac{1595}{1604}i\right)\mathbf{u} + \left(-\frac{595}{802} - \frac{141}{1604}i\right)\mathbf{v} + \left(-\frac{267}{802} + \frac{147}{1604}i\right)\mathbf{w}$.

7. (a) $\langle \mathbf{u}, \mathbf{v} \rangle = u_1 \overline{v_1} + u_2 \overline{v_2} + u_3 \overline{v_3} = (2 + 3i)\overline{2} + (-1)\overline{(3 - i)} + (3 + 4i)\overline{(1 + 5i)} = (2 + 3i)2 + (-1)(3 + i) + (3 + 4i)(1 - 5i) = 24 - 6i$.

    (b) $\langle i\mathbf{v}, -2\mathbf{w} \rangle = i\overline{(-2)}\langle \mathbf{v}, \mathbf{w} \rangle = -2i(v_1 \overline{w_1} + v_2 \overline{w_2} + v_3 \overline{w_3}) = -2i$
    $\left(2\overline{(2 + i)} + (3 - i)\overline{(2 - i)} + (1 + 5i)\overline{(4 - 3i)} \right) =$
    $-2i(2(2 - i) + (3 - i)(2 + i) + (1 + 5i)(4 + 3i)) = -2i(22i) = 44$.

    (c) $\|\mathbf{w}\| = \sqrt{\langle \mathbf{w}, \mathbf{w} \rangle} = \sqrt{w_1 \overline{w_1} + w_2 \overline{w_2} + w_3 \overline{w_3}} = \sqrt{|w_1|^2 + |w_2|^2 + |w_3|^2} = \sqrt{|2 + i|^2 + |2 - i|^2 + |4 - 3i|^2} = \sqrt{5 + 5 + 25} = \sqrt{35}$.

9. Let $\dfrac{\mathbf{u}}{\|\mathbf{u}\|} = \dfrac{(2 + 3i, -1, 3 + 4i)}{\|(2 + 3i, -1, 3 + 4i)\|} = \dfrac{(2 + 3i, -1, 3 + 4i)}{\sqrt{39}} = \left(\frac{2}{39}\sqrt{39} + \frac{1}{13}\sqrt{39}i, -\frac{1}{39}\sqrt{39}, \frac{1}{13}\sqrt{39} + \frac{4}{39}\sqrt{39}i\right)$, and $\dfrac{\mathbf{v}}{\|\mathbf{v}\|} = \dfrac{(2, 3 - i, 1 + 5i)}{\|(2, 3 - i, 1 + 5i)\|} = \dfrac{(2, 3 - i, 1 + 5i)}{2\sqrt{10}} = \left(\frac{1}{10}\sqrt{10}, \frac{3}{20}\sqrt{10} - \frac{1}{20}\sqrt{10}i, \frac{1}{20}\sqrt{10} + \frac{1}{4}\sqrt{10}i\right)$.

11. (a) $A - iC = \begin{bmatrix} 2 + i & 3 \\ 1 - i & 2 + 3i \end{bmatrix} - i \begin{bmatrix} 0 & 3 + i \\ -4i & 1 + i \end{bmatrix} =$
$\begin{bmatrix} 2 + i & 3 \\ 1 - i & 2 + 3i \end{bmatrix} + \begin{bmatrix} 0 & 1 - 3i \\ -4 & 1 - i \end{bmatrix} = \begin{bmatrix} 2 + i & 4 - 3i \\ -3 - i & 3 + 2i \end{bmatrix}$.

    (b) $2B - A - 4iC = 2 \begin{bmatrix} -i & 4 \\ 2 + 2i & 1 + 4i \end{bmatrix} - \begin{bmatrix} 2 + i & 3 \\ 1 - i & 2 + 3i \end{bmatrix} - 4i \begin{bmatrix} 0 & 3 + i \\ -4i & 1 + i \end{bmatrix}$
    $= \begin{bmatrix} -2i & 8 \\ 4 + 4i & 2 + 8i \end{bmatrix} - \begin{bmatrix} 2 + i & 3 \\ 1 - i & 2 + 3i \end{bmatrix} + \begin{bmatrix} 0 & 4 - 12i \\ -16 & 4 - 4i \end{bmatrix}$
    $= \begin{bmatrix} -2 - 3i & 9 - 12i \\ -13 + 5i & 4 + i \end{bmatrix}$.

13. We rewrite $A - cB = iC$ as $cB = A - iC$, hence $c \begin{bmatrix} -i & 4 \\ 2+2i & 1+4i \end{bmatrix} = \begin{bmatrix} 2+i & 3 \\ 1-i & 2+3i \end{bmatrix} - i \begin{bmatrix} 0 & 3+i \\ -4i & 1+i \end{bmatrix} = \begin{bmatrix} 2+i & 4-3i \\ -3-i & 3+2i \end{bmatrix}$. From $c(-i) = 2+i$, we determine $c = -1+2i$. We check $c(4) = (-1+2i)(4) = -4+8i \neq 4-3i$, hence no constant c exists such that $A - cB = iC$.

15. To determine if $\begin{bmatrix} 3+2i & -3-7i \\ 4+8i & 5+2i \end{bmatrix}$ is in the span of $\{A, B, C\}$, we set $\begin{bmatrix} 3+2i & -3-7i \\ 4+8i & 5+2i \end{bmatrix}$ $= c_1 A + c_2 B + c_3 C$ and obtain the system

$$(2+i)c_1 + (-i)c_2 + (0)c_3 = 3+2i$$
$$(3)c_1 + (4)c_2 + (3+i)c_3 = -3-7i$$
$$(1-i)c_1 + (2+2i)c_2 + (-4i)c_3 = 4+8i$$
$$(2+3i)c_1 + (1+4i)c_2 + (1+i)c_3 = 5+2i$$

Use Gaussian elimination

$$\begin{bmatrix} 2+i & -i & 0 & 3+2i \\ 3 & 4 & 3+i & -3-7i \\ 1-i & 2+2i & -4i & 4+8i \\ 2+3i & 1+4i & 1+i & 5+2i \end{bmatrix} \begin{matrix} -\frac{3}{2+i}R_1 + R_2 \Rightarrow R_2 \\ -\frac{1-i}{2+i}R_1 + R_3 \Rightarrow R_3 \\ -\frac{2+3i}{2+i}R_1 + R_4 \Rightarrow R_4 \\ \sim \end{matrix}$$

$$\begin{bmatrix} 2+i & -i & 0 & 3+2i \\ 0 & \frac{23}{5}+\frac{6}{5}i & 3+i & -\frac{39}{5}-\frac{38}{5}i \\ 0 & \frac{13}{5}+\frac{11}{5}i & -4i & \frac{11}{5}+\frac{47}{5}i \\ 0 & \frac{1}{5}+\frac{27}{5}i & 1+i & \frac{12}{5}-\frac{16}{5}i \end{bmatrix} \begin{matrix} -\frac{13+11i}{23+6i}R_2 + R_3 \Rightarrow R_3 \\ -\frac{1+27i}{23+6i}R_2 + R_4 \Rightarrow R_4 \\ \sim \end{matrix}$$

$$\begin{bmatrix} 2+i & -i & 0 & 3+2i \\ 0 & \frac{23}{5}+\frac{6}{5}i & 3+i & -\frac{39}{5}-\frac{38}{5}i \\ 0 & 0 & -\frac{184}{113}-\frac{630}{113}i & \frac{552}{113}+\frac{1890}{113}i \\ 0 & 0 & \frac{125}{113}-\frac{293}{113}i & -\frac{375}{113}+\frac{879}{113}i \end{bmatrix} \begin{matrix} -\frac{125-293i}{-184-630i}R_3 + R_4 \Rightarrow R_4 \\ \sim \end{matrix}$$

$$\begin{bmatrix} 2+i & -i & 0 & 3+2i \\ 0 & \frac{23}{5}+\frac{6}{5}i & 3+i & -\frac{39}{5}-\frac{38}{5}i \\ 0 & 0 & -\frac{184}{113}-\frac{630}{113}i & \frac{552}{113}+\frac{1890}{113}i \\ 0 & 0 & 0 & 0 \end{bmatrix}$$

Thus $c_3 = \frac{\frac{552}{113}+\frac{1890}{113}i}{-\frac{184}{113}-\frac{630}{113}i} = -3$, $c_2 = \frac{-\frac{39}{5}-\frac{38}{5}i-(-3)(3+i)}{\frac{23}{5}+\frac{6}{5}i} = -i$, and $c_1 = \frac{3+2i-(-i)(-i)}{2+i} = 2$, and $\begin{bmatrix} 3+2i & -3-7i \\ 4+8i & 5+2i \end{bmatrix}$ is in the span of $\{A, B, C\}$.

17. (a) $\langle A, C \rangle = a_{11}\overline{c_{11}} + a_{12}\overline{c_{12}} + a_{21}\overline{c_{21}} + a_{22}\overline{c_{22}} = (2+i)\overline{(0)} + (3)\overline{(3+i)} + (1-i)\overline{(-4i)} + (2+3i)\overline{(1+i)} = (2+i)(0) + (3)(3-i) + (1-i)(4i) + (2+3i)(1-i) = 18+2i$

(b) $\langle iB, -2A \rangle = i\overline{(-2)}\langle B, A \rangle = i(-2)(b_{11}\overline{a_{11}} + b_{12}\overline{a_{12}} + b_{21}\overline{a_{21}} + b_{22}\overline{a_{22}}) = i(-2)\left(-i\overline{(2+i)} + 4\overline{(3)} + (2+2i)\overline{(1-i)} + (1+4i)\overline{(2+3i)}\right) = i(-2)(25+7i) = 14-50i$

(c) $\|B\| = \sqrt{\langle B, B \rangle} = \sqrt{b_{11}\overline{b_{11}} + b_{12}\overline{b_{12}} + b_{21}\overline{b_{21}} + b_{22}\overline{b_{22}}} = \sqrt{|b_{11}|^2 + |b_{12}|^2 + |b_{21}|^2 + |b_{22}|^2} = \sqrt{|-i|^2 + |4|^2 + |2+2i|^2 + |1+4i|^2} = \sqrt{42}$

19. $\|A\| = \sqrt{\langle A, A \rangle} = \sqrt{|2+i|^2 + |3|^2 + |1-i|^2 + |2+3i|^2} = \sqrt{29}$, so $\frac{1}{\|A\|}A = \frac{1}{\sqrt{29}}\begin{bmatrix} 2+i & 3 \\ 1-i & 2+3i \end{bmatrix} = \begin{bmatrix} \frac{2}{29}\sqrt{29} + \frac{1}{29}\sqrt{29}i & \frac{3}{29}\sqrt{29} \\ \frac{1}{29}\sqrt{29} - \frac{1}{29}\sqrt{29}i & \frac{2}{29}\sqrt{29} + \frac{329}{29}\sqrt{29}i \end{bmatrix}$. $\|C\| = \sqrt{\langle C, C \rangle} =$

$$\sqrt{|0|^2 + |3+i|^2 + |-4i|^2 + |1+i|^2} = 2\sqrt{7}, \text{ so}$$

$$\frac{1}{\|C\|}C = \frac{1}{2\sqrt{7}}\begin{bmatrix} 0 & 3+i \\ -4i & 1+i \end{bmatrix} = \begin{bmatrix} 0 & \frac{3}{14}\sqrt{7} + \frac{1}{14}\sqrt{7}i \\ -\frac{2}{7}\sqrt{7}i & \frac{1}{14}\sqrt{7} + \frac{1}{14}\sqrt{7}i \end{bmatrix}.$$

21. (a) $h_1(x) + (4-i)h_2(x) = (1+ix) + (4-i)(i-x) = (2+4i) - (4-2i)x$

   (b) $ih_1(x) - h_2(x) + 3h_3(x) = i(1+ix) - (i-x) + 3(3-(1+i)x) = 9 - (3+3i)x$

23. We set $h_3(x) + h_2(x) = ch_1(x)$ and obtain $(3-(1+i)x) + (i-x) = c(1+ix) \Rightarrow (3+i) - (2+i)x = c + (ci)x$. Equate the constant terms to obtain $3+i = c$. We check the x coefficient, $(ci) = (3+i)(i) = -1 + 3i \neq -(2+i)$. Therefore no constant c exists such that $h_3(x) + h_2(x) = ch_1(x)$.

25. Set

$$(2+i) + (3-2i)x = c_1 h_1(x) + c_2 h_2(x) + c_3 h_3(x)$$
$$= c_1(1+ix) + c_2(i-x) + c_3(3-(1+i)x)$$
$$= (c_1 + ic_2 + 3c_3) + (ic_1 - c_2 - (1+i)c_3)x$$

Equating coefficients, we obtain the system

$$c_1 + ic_2 + 3c_3 = 2+i$$
$$ic_1 - c_2 - (1+i)c_3 = 3-2i$$

Use Gaussian elimination

$$\begin{bmatrix} 1 & i & 3 & 2+i \\ i & -1 & -1-i & 3-2i \end{bmatrix} \quad -iR_1 + R_2 \Rightarrow R_2 \quad \begin{bmatrix} 1 & i & 3 & 2+i \\ 0 & 0 & -1-4i & 4-4i \end{bmatrix}$$

There exists a solution, so $(2+i) + (3-2i)x$ is in the span of $\{h_1(x), h_2(x), h_3(x)\}$. In particular, we have $c_3 = \frac{4-4i}{-1-4i} = \frac{12}{17} + \frac{20}{17}i$, we can set $c_2 = 0$, and then $c_1 = (2+i) - 3\left(\frac{12}{17} + \frac{20}{17}i\right) = -\frac{2}{17} - \frac{43}{17}i$. So $(2+i) + (3-2i)x = \left(-\frac{2}{17} - \frac{43}{17}i\right)(1+ix) + \left(\frac{12}{17} + \frac{20}{17}i\right)(3-(1+i)x)$.

27. (a) $\langle h_1, h_3 \rangle = \int_0^1 h_1(x)\overline{h_3(x)}\,dx = \int_0^1 (1+ix)\overline{(3-(1+i)x)}\,dx = \int_0^1 (1+ix)(3-(1-i)x)\,dx = \frac{13}{6} + \frac{5}{3}i$

   (b) $\langle ih_2, -2h_3 \rangle = i\overline{(-2)}\langle h_2, h_3 \rangle = i\overline{(-2)}\int_0^1 h_2(x)\overline{h_3(x)}\,dx = i\overline{(-2)}\int_0^1 (i-x)\overline{(3-(1+i)x)}\,dx = i\overline{(-2)}\int_0^1 (i-x)(3-(1-i)x)\,dx = i(-2)\left(-\frac{5}{3} + \frac{13}{6}i\right) = \frac{13}{3} + \frac{10}{3}i$

   (c) $\|h_1\| = \sqrt{\langle h_1, h_1 \rangle} = \sqrt{\int_0^1 h_1(x)\overline{h_1(x)}\,dx} = \sqrt{\int_0^1 (1+ix)\overline{(1+ix)}\,dx} = \sqrt{\int_0^1 (1+ix)(1-ix)\,dx} = \sqrt{\frac{4}{3}} = \frac{2}{3}\sqrt{3}$

29. From Exercise 27 (c), $\|h_1\| = \frac{2}{3}\sqrt{3}$, thus $\frac{1}{\|h_1\|}h_1(x) = \frac{1}{\frac{2}{3}\sqrt{3}}(1+ix) = \frac{1}{2}\sqrt{3} + \frac{1}{2}\sqrt{3}ix$. $\|h_2\| = \sqrt{\langle h_2, h_2 \rangle} = \sqrt{\int_0^1 (i-x)\overline{(i-x)}\,dx} = \sqrt{\int_0^1 (i-x)(-i-x)\,dx} = \frac{2}{3}\sqrt{3}$. Thus $\frac{1}{\|h_2\|}h_2(x) = \frac{1}{\frac{2}{3}\sqrt{3}}(i-x) = \frac{1}{2}\sqrt{3}i - \frac{1}{2}\sqrt{3}x$.

31. For example, if $\mathbf{u} = 1$ and $\mathbf{v} = i$, then $\|\mathbf{u}\|^2 = \|\mathbf{v}\|^2 = 1$ and $\|\mathbf{u} + \mathbf{v}\|^2 = 2$, but $\langle \mathbf{u}, \mathbf{v} \rangle = \bar{i} = -i \neq 0$.

33. For example, define $\langle \mathbf{u}, \mathbf{v} \rangle = 2u_1\overline{v_1} + \cdots + 2u_n\overline{v_n}$.

35. For example, let $V = \mathbf{C}$, and $S = \mathbf{R}$.

37. True, by Definition 11.14 (a) and (b).

39. True. By Definition 11.15 (d), $\langle \mathbf{v}, \mathbf{v} \rangle \geq 0$, and since $\mathbf{v} \neq \mathbf{0}$, $\langle \mathbf{v}, \mathbf{v} \rangle \neq 0$. Hence $\|\mathbf{v}\| = \sqrt{\langle \mathbf{v}, \mathbf{v} \rangle} > 0$.

41. False. By Definition 11.15 (a), $\langle \mathbf{u}, \mathbf{v} \rangle = \overline{\langle \mathbf{v}, \mathbf{u} \rangle}$, so if $\langle \mathbf{u}, \mathbf{v} \rangle \notin \mathbf{R}$, then $\langle \mathbf{u}, \mathbf{v} \rangle \neq \langle \mathbf{v}, \mathbf{u} \rangle$. For example, in $\mathbf{C}$, $\langle 1, i \rangle = 1\overline{(i)} = -i$, and $\langle i, 1 \rangle = i\overline{(1)} = i\,(1) = i$.

43. We verify the remaining parts of Definition 11.14:

(2). For $A = \begin{bmatrix} a_{11} & a_{12} \\ a_{21} & a_{22} \end{bmatrix}$ and $c \in \mathbf{C}$, $cA = c\begin{bmatrix} a_{11} & a_{12} \\ a_{21} & a_{22} \end{bmatrix} = \begin{bmatrix} ca_{11} & ca_{12} \\ ca_{21} & ca_{22} \end{bmatrix} \in \mathbf{C}^{2 \times 2}$.

(4). For $A = \begin{bmatrix} a_{11} & a_{12} \\ a_{21} & a_{22} \end{bmatrix}$, define $-A = \begin{bmatrix} -a_{11} & -a_{12} \\ -a_{21} & -a_{22} \end{bmatrix}$. Then $A + (-A) =$

$\begin{bmatrix} a_{11} & a_{12} \\ a_{21} & a_{22} \end{bmatrix} + \begin{bmatrix} -a_{11} & -a_{12} \\ -a_{21} & -a_{22} \end{bmatrix} = \begin{bmatrix} 0 & 0 \\ 0 & 0 \end{bmatrix} = \mathbf{0}$.

Let $A = \begin{bmatrix} a_{11} & a_{12} \\ a_{21} & a_{22} \end{bmatrix}$, $B = \begin{bmatrix} b_{11} & b_{12} \\ b_{21} & b_{22} \end{bmatrix}$, and $C = \begin{bmatrix} c_{11} & c_{12} \\ c_{21} & c_{22} \end{bmatrix}$, and s_1 and s_2 complex scalars.

(5a). $A + B = \begin{bmatrix} a_{11} & a_{12} \\ a_{21} & a_{22} \end{bmatrix} + \begin{bmatrix} b_{11} & b_{12} \\ b_{21} & b_{22} \end{bmatrix} = \begin{bmatrix} a_{11} + b_{11} & a_{12} + b_{12} \\ a_{21} + b_{21} & a_{22} + b_{22} \end{bmatrix} =$

$\begin{bmatrix} b_{11} + a_{11} & b_{12} + a_{12} \\ b_{21} + a_{21} & b_{22} + a_{22} \end{bmatrix} = \begin{bmatrix} b_{11} & b_{12} \\ b_{21} & b_{22} \end{bmatrix} + \begin{bmatrix} a_{11} & a_{12} \\ a_{21} & a_{22} \end{bmatrix} = B + A$.

(5b). $(A + B) + C = \left(\begin{bmatrix} a_{11} & a_{12} \\ a_{21} & a_{22} \end{bmatrix} + \begin{bmatrix} b_{11} & b_{12} \\ b_{21} & b_{22} \end{bmatrix} \right) + \begin{bmatrix} c_{11} & c_{12} \\ c_{21} & c_{22} \end{bmatrix} =$

$\begin{bmatrix} a_{11} + b_{11} & a_{12} + b_{12} \\ a_{21} + b_{21} & a_{22} + b_{22} \end{bmatrix} + \begin{bmatrix} c_{11} & c_{12} \\ c_{21} & c_{22} \end{bmatrix} = \begin{bmatrix} (a_{11} + b_{11}) + c_{11} & (a_{12} + b_{12}) + c_{12} \\ (a_{21} + b_{21}) + c_{21} & (a_{22} + b_{22}) + c_{22} \end{bmatrix} =$

$\begin{bmatrix} a_{11} + (b_{11} + c_{11}) & a_{12} + (b_{12} + c_{12}) \\ a_{21} + (b_{21} + c_{21}) & a_{22} + (b_{22} + c_{22}) \end{bmatrix} = \begin{bmatrix} a_{11} & a_{12} \\ a_{21} & a_{22} \end{bmatrix} + \begin{bmatrix} b_{11} + c_{11} & b_{12} + c_{12} \\ b_{21} + c_{21} & b_{22} + c_{22} \end{bmatrix} =$

$\begin{bmatrix} a_{11} & a_{12} \\ a_{21} & a_{22} \end{bmatrix} + \left(\begin{bmatrix} b_{11} & b_{12} \\ b_{21} & b_{22} \end{bmatrix} + \begin{bmatrix} c_{11} & c_{12} \\ c_{21} & c_{22} \end{bmatrix} \right) = A + (B + C)$

(5c). $s_1 (A + B) = s_1 \left(\begin{bmatrix} a_{11} & a_{12} \\ a_{21} & a_{22} \end{bmatrix} + \begin{bmatrix} b_{11} & b_{12} \\ b_{21} & b_{22} \end{bmatrix} \right) = s_1 \left(\begin{bmatrix} a_{11} + b_{11} & a_{12} + b_{12} \\ a_{21} + b_{21} & a_{22} + b_{22} \end{bmatrix} \right) =$

$\begin{bmatrix} s_1(a_{11} + b_{11}) & s_1(a_{12} + b_{12}) \\ s_1(a_{21} + b_{21}) & s_1(a_{22} + b_{22}) \end{bmatrix} = \begin{bmatrix} s_1 a_{11} + s_1 b_{11} & s_1 a_{12} + s_1 b_{12} \\ s_1 a_{21} + s_1 b_{21} & s_1 a_{22} + s_1 b_{22} \end{bmatrix} = \begin{bmatrix} s_1 a_{11} & s_1 a_{12} \\ s_1 a_{21} & s_1 a_{22} \end{bmatrix} +$

$\begin{bmatrix} s_1 b_{11} & s_1 b_{12} \\ s_1 b_{21} & s_1 b_{22} \end{bmatrix} = s_1 \begin{bmatrix} a_{11} & a_{12} \\ a_{21} & a_{22} \end{bmatrix} + s_1 \begin{bmatrix} b_{11} & b_{12} \\ b_{21} & b_{22} \end{bmatrix} = s_1 A + s_2 B$

(5d). $(s_1 + s_2) A = (s_1 + s_2) \begin{bmatrix} a_{11} & a_{12} \\ a_{21} & a_{22} \end{bmatrix} = \begin{bmatrix} (s_1 + s_2) a_{11} & (s_1 + s_2) a_{12} \\ (s_1 + s_2) a_{21} & (s_1 + s_2) a_{22} \end{bmatrix} =$

$\begin{bmatrix} s_1 a_{11} + s_2 a_{11} & s_1 a_{12} + s_2 a_{12} \\ s_1 a_{21} + s_2 a_{21} & s_1 a_{22} + s_2 a_{22} \end{bmatrix} = \begin{bmatrix} s_1 a_{11} & s_1 a_{12} \\ s_1 a_{21} & s_1 a_{22} \end{bmatrix} + \begin{bmatrix} s_2 a_{11} & s_2 a_{12} \\ s_2 a_{21} & s_2 a_{22} \end{bmatrix} =$

$s_1 \begin{bmatrix} a_{11} & a_{12} \\ a_{21} & a_{22} \end{bmatrix} + s_2 \begin{bmatrix} a_{11} & a_{12} \\ a_{21} & a_{22} \end{bmatrix} = s_1 A + s_2 A$.

(5e). $(s_1 s_2) A = (s_1 s_2) \begin{bmatrix} a_{11} & a_{12} \\ a_{21} & a_{22} \end{bmatrix} = \begin{bmatrix} (s_1 s_2) a_{11} & (s_1 s_2) a_{12} \\ (s_1 s_2) a_{21} & (s_1 s_2) a_{22} \end{bmatrix} =$

$\begin{bmatrix} s_1 (s_2 a_{11}) & s_1 (s_2 a_{12}) \\ s_1 (s_2 a_{21}) & s_1 (s_2 a_{22}) \end{bmatrix} = s_1 \begin{bmatrix} s_2 a_{11} & s_2 a_{12} \\ s_2 a_{21} & s_2 a_{22} \end{bmatrix} = s_1 \left(s_2 \begin{bmatrix} a_{11} & a_{12} \\ a_{21} & a_{22} \end{bmatrix} \right) = s_1 (s_2 A)$.

(5f). $1 \cdot A = 1 \begin{bmatrix} a_{11} & a_{12} \\ a_{21} & a_{22} \end{bmatrix} = \begin{bmatrix} 1(a_{11}) & 1(a_{12}) \\ 1(a_{21}) & 1(a_{22}) \end{bmatrix} = \begin{bmatrix} a_{11} & a_{12} \\ a_{21} & a_{22} \end{bmatrix} = A$.

45. Definition 11.14 (2) fails, as $i \begin{bmatrix} 1 & 0 \\ 0 & 1 \end{bmatrix} = \begin{bmatrix} i & 0 \\ 0 & i \end{bmatrix} \notin \mathbf{R}^{2 \times 2}$.

47. (a) $\langle \mathbf{0}, \mathbf{u} \rangle = \langle 0\mathbf{0}, \mathbf{u} \rangle = 0 \langle \mathbf{0}, \mathbf{u} \rangle = 0$, $\langle \mathbf{u}, \mathbf{0} \rangle = \overline{\langle \mathbf{0}, \mathbf{u} \rangle} = \overline{0} = 0$.

(b) $\langle \mathbf{u}, \mathbf{v} + \mathbf{w} \rangle = \overline{\langle \mathbf{v} + \mathbf{w}, \mathbf{u} \rangle} = \overline{\langle \mathbf{v}, \mathbf{u} \rangle + \langle \mathbf{w}, \mathbf{u} \rangle} = \overline{\langle \mathbf{v}, \mathbf{u} \rangle} + \overline{\langle \mathbf{w}, \mathbf{u} \rangle} = \langle \mathbf{u}, \mathbf{v} \rangle + \langle \mathbf{u}, \mathbf{w} \rangle$.

(c) $\langle \mathbf{u}, c\mathbf{v} \rangle = \overline{\langle c\mathbf{v}, \mathbf{u} \rangle} = \overline{c \langle \mathbf{v}, \mathbf{u} \rangle} = \overline{c}\,\overline{\langle \mathbf{v}, \mathbf{u} \rangle} = \overline{c} \langle \mathbf{u}, \mathbf{v} \rangle$

49. By property (d) of Definition 11.15, $\langle \mathbf{u}, \mathbf{u} \rangle \geq 0$, and since square roots are non-negative, $\|\mathbf{u}\| = \sqrt{\langle \mathbf{u}, \mathbf{u} \rangle} \geq 0$. If $\|\mathbf{u}\| = \sqrt{\langle \mathbf{u}, \mathbf{u} \rangle} = 0$, then $\langle \mathbf{u}, \mathbf{u} \rangle = 0$, and hence $\mathbf{u} = \mathbf{0}$ by property (d) of Definition 11.15.

51. If $\mathbf{v} = \mathbf{0}$, then both sides of $|\mathbf{u} \cdot \mathbf{v}| \leq \|\mathbf{u}\| \|\mathbf{v}\|$ are 0, so $|\mathbf{u} \cdot \mathbf{v}| \leq \|\mathbf{u}\| \|\mathbf{v}\|$ holds. Now suppose $\mathbf{v} \neq \mathbf{0}$, and consider

$$0 \leq \left\| \mathbf{u} - \frac{\langle \mathbf{u}, \mathbf{v} \rangle}{\|\mathbf{v}\|^2} \mathbf{v} \right\|^2 = \left\langle \mathbf{u} - \frac{\langle \mathbf{u}, \mathbf{v} \rangle}{\|\mathbf{v}\|^2} \mathbf{v}, \mathbf{u} - \frac{\langle \mathbf{u}, \mathbf{v} \rangle}{\|\mathbf{v}\|^2} \mathbf{v} \right\rangle$$

$$= \langle \mathbf{u}, \mathbf{u} \rangle - \frac{\langle \mathbf{u}, \mathbf{v} \rangle}{\|\mathbf{v}\|^2} \langle \mathbf{v}, \mathbf{u} \rangle - \frac{\overline{\langle \mathbf{u}, \mathbf{v} \rangle}}{\|\mathbf{v}\|^2} \langle \mathbf{u}, \mathbf{v} \rangle + \frac{|\langle \mathbf{u}, \mathbf{v} \rangle|^2}{\|\mathbf{v}\|^4} \langle \mathbf{v}, \mathbf{v} \rangle$$

$$= \|\mathbf{u}\|^2 - \frac{\langle \mathbf{u}, \mathbf{v} \rangle}{\|\mathbf{v}\|^2} \overline{\langle \mathbf{u}, \mathbf{v} \rangle} - \frac{\overline{\langle \mathbf{u}, \mathbf{v} \rangle}}{\|\mathbf{v}\|^2} \langle \mathbf{u}, \mathbf{v} \rangle + \frac{|\langle \mathbf{u}, \mathbf{v} \rangle|^2}{\|\mathbf{v}\|^4} \|\mathbf{v}\|^2$$

$$= \|\mathbf{u}\|^2 - 2\frac{|\langle \mathbf{u}, \mathbf{v} \rangle|^2}{\|\mathbf{v}\|^2} + \frac{|\langle \mathbf{u}, \mathbf{v} \rangle|^2}{\|\mathbf{v}\|^2}$$

$$= \|\mathbf{u}\|^2 - \frac{|\langle \mathbf{u}, \mathbf{v} \rangle|^2}{\|\mathbf{v}\|^2} = \frac{1}{\|\mathbf{v}\|^2} \left(\|\mathbf{u}\|^2 \|\mathbf{v}\|^2 - |\langle \mathbf{u}, \mathbf{v} \rangle|^2 \right)$$

Hence $\|\mathbf{u}\|^2 \|\mathbf{v}\|^2 - |\langle \mathbf{u}, \mathbf{v} \rangle|^2 \geq 0$, so $|\langle \mathbf{u}, \mathbf{v} \rangle|^2 \leq \|\mathbf{u}\|^2 \|\mathbf{v}\|^2$. Take the square root to obtain $|\langle \mathbf{u}, \mathbf{v} \rangle| \leq \|\mathbf{u}\| \|\mathbf{v}\|$.

11.5   Hermitian Matrices

1. $A^* = \begin{bmatrix} 1+i & 3i \\ 2-i & 1+4i \end{bmatrix}^* = \overline{\begin{bmatrix} 1+i & 3i \\ 2-i & 1+4i \end{bmatrix}^T} = \overline{\begin{bmatrix} 1+i & 2-i \\ 3i & 1+4i \end{bmatrix}} = \begin{bmatrix} 1-i & 2+i \\ -3i & 1-4i \end{bmatrix}$

3. $A^* = \begin{bmatrix} 3+i & 5i & 1-i \\ 1-4i & -8 & 6+i \\ 2+2i & 0 & -7i \end{bmatrix}^* = \overline{\begin{bmatrix} 3+i & 5i & 1-i \\ 1-4i & -8 & 6+i \\ 2+2i & 0 & -7i \end{bmatrix}^T} = \overline{\begin{bmatrix} 3+i & 1-4i & 2+2i \\ 5i & -8 & 0 \\ 1-i & 6+i & -7i \end{bmatrix}}$

$= \begin{bmatrix} 3-i & 1+4i & 2-2i \\ -5i & -8 & 0 \\ 1+i & 6-i & 7i \end{bmatrix}$

5. $A^* = \begin{bmatrix} 1 & -2i & 3 & 4i \\ 2i & 5 & -6i & 1+i \\ 3 & 6i & 7 & 3+2i \\ -4i & 1-i & 3-2i & 11 \end{bmatrix}^* = \overline{\begin{bmatrix} 1 & -2i & 3 & 4i \\ 2i & 5 & -6i & 1+i \\ 3 & 6i & 7 & 3+2i \\ -4i & 1-i & 3-2i & 11 \end{bmatrix}^T} =$

$\overline{\begin{bmatrix} 1 & 2i & 3 & -4i \\ -2i & 5 & 6i & 1-i \\ 3 & -6i & 7 & 3-2i \\ 4i & 1+i & 3+2i & 11 \end{bmatrix}} = \begin{bmatrix} 1 & -2i & 3 & 4i \\ 2i & 5 & -6i & 1+i \\ 3 & 6i & 7 & 3+2i \\ -4i & 1-i & 3-2i & 11 \end{bmatrix}$

7. $A^* = \begin{bmatrix} 1+i & 3i \\ 3i & 2 \end{bmatrix}^* = \begin{bmatrix} 1-i & -3i \\ -3i & 2 \end{bmatrix} \neq A$, so A is not Hermitian.

9. $A^* = \begin{bmatrix} 3 & 5i & 1-i \\ -5i & -5 & 0 \\ 1+i & 0 & 7 \end{bmatrix}^* = \begin{bmatrix} 3 & 5i & 1-i \\ -5i & -5 & 0 \\ 1+i & 0 & 7 \end{bmatrix} = A$, so A is Hermitian.

11. $A^* = \begin{bmatrix} 1 & -2i & 3 & 4i \\ 2i & 5 & -6i & 1+i \\ 3 & 6i & 7 & 3+2i \\ -4i & 1-i & 3-2i & 11 \end{bmatrix}^* = \begin{bmatrix} 1 & -2i & 3 & 4i \\ 2i & 5 & -6i & 1+i \\ 3 & 6i & 7 & 3+2i \\ -4i & 1-i & 3-2i & 11 \end{bmatrix} = A$, so A is Hermitian.

13. $A^*A = \begin{bmatrix} 1 & 2-5i \\ 2+5i & 3 \end{bmatrix}^* \begin{bmatrix} 1 & 2-5i \\ 2+5i & 3 \end{bmatrix} = \begin{bmatrix} 1 & 2-5i \\ 2+5i & 3 \end{bmatrix} \begin{bmatrix} 1 & 2-5i \\ 2+5i & 3 \end{bmatrix}$

$= \begin{bmatrix} 30 & 8-20i \\ 8+20i & 38 \end{bmatrix}$ and $AA^* = \begin{bmatrix} 1 & 2-5i \\ 2+5i & 3 \end{bmatrix} \begin{bmatrix} 1 & 2-5i \\ 2+5i & 3 \end{bmatrix}^* =$

$\begin{bmatrix} 1 & 2-5i \\ 2+5i & 3 \end{bmatrix} \begin{bmatrix} 1 & 2-5i \\ 2+5i & 3 \end{bmatrix} = \begin{bmatrix} 30 & 8-20i \\ 8+20i & 38 \end{bmatrix}$. Since $A^*A = AA^*$, A is normal.

15. $A^*A = \begin{bmatrix} -i & -i \\ i & -i \end{bmatrix}^* \begin{bmatrix} -i & -i \\ i & -i \end{bmatrix} = \begin{bmatrix} i & -i \\ i & i \end{bmatrix} \begin{bmatrix} -i & -i \\ i & -i \end{bmatrix} = \begin{bmatrix} 2 & 0 \\ 0 & 2 \end{bmatrix}$ and AA^*

$= \begin{bmatrix} -i & -i \\ i & -i \end{bmatrix} \begin{bmatrix} -i & -i \\ i & -i \end{bmatrix}^* = \begin{bmatrix} -i & -i \\ i & -i \end{bmatrix} \begin{bmatrix} i & -i \\ i & i \end{bmatrix} = \begin{bmatrix} 2 & 0 \\ 0 & 2 \end{bmatrix}$. Since $A^*A = AA^*$, A is normal.

17. $A^*A = \begin{bmatrix} -1 & -i & 1-i \\ i & 5 & -2i \\ 1+i & 2i & 0 \end{bmatrix}^* \begin{bmatrix} -1 & -i & 1-i \\ i & 5 & -2i \\ 1+i & 2i & 0 \end{bmatrix} =$

$\begin{bmatrix} -1 & -i & 1-i \\ i & 5 & -2i \\ 1+i & 2i & 0 \end{bmatrix} \begin{bmatrix} -1 & -i & 1-i \\ i & 5 & -2i \\ 1+i & 2i & 0 \end{bmatrix} = \begin{bmatrix} 4 & 2-2i & -3+i \\ 2+2i & 30 & 1-9i \\ -3-i & 1+9i & 6 \end{bmatrix}$ and AA^*

$= \begin{bmatrix} -1 & -i & 1-i \\ i & 5 & -2i \\ 1+i & 2i & 0 \end{bmatrix} \begin{bmatrix} -1 & -i & 1-i \\ i & 5 & -2i \\ 1+i & 2i & 0 \end{bmatrix}^* =$

$\begin{bmatrix} -1 & -i & 1-i \\ i & 5 & -2i \\ 1+i & 2i & 0 \end{bmatrix} \begin{bmatrix} -1 & -i & 1-i \\ i & 5 & -2i \\ 1+i & 2i & 0 \end{bmatrix} = \begin{bmatrix} 4 & 2-2i & -3+i \\ 2+2i & 30 & 1-9i \\ -3-i & 1+9i & 6 \end{bmatrix}$. Since $A^*A = AA^*$, A is normal.

19. For example, $A = \begin{bmatrix} 0 & i & i \\ i & 0 & i \\ i & i & 0 \end{bmatrix}$.

21. For example, $A = \begin{bmatrix} i & 0 & 0 \\ 0 & i & 0 \\ 0 & 0 & i \end{bmatrix}$.

23. True, by Theorem 11.19.

25. False. For example, $A = \begin{bmatrix} i & i \\ i & i \end{bmatrix}$ has complex entries, and $A = A^T$, but $A \neq A^*$, so A is not Hermitian.

27. True, since $(A+B)^* = \overline{(A+B)^T} = \overline{A^T + B^T} = \overline{A^T} + \overline{B^T} = A^* + B^* = A + B$, so $A+B$ is Hermitian.

29. True, by Exercise 37.

31. If A has real entries, then A^T has real entries, and so $A^* = \overline{A^T} = A^T$.

33. Using Exercise 32, $(A^*)^* = \left(\overline{(A^*)}\right)^T = \overline{(A^*)^T} = \overline{\left(\overline{(\overline{A})^T}\right)^T} = \overline{(\overline{A})} = A$.

35. $(AB)^* = \left(\overline{(AB)}\right)^T = (\bar{A}\bar{B})^T = (\overline{B})^T(\overline{A})^T = B^*A^*$.

37. Let $A = [\ \mathbf{a}_1 \ \cdots \ \mathbf{a}_n\ ]$, and suppose A is unitary. Then $A^{-1} = A^*$, so $A^*A = I_n$. Thus $I_n =$
$\begin{bmatrix} (\overline{\mathbf{a}_1})^T \\ \vdots \\ (\overline{\mathbf{a}_n})^T \end{bmatrix} [\ \mathbf{a}_1 \ \cdots \ \mathbf{a}_n\ ]$. Therefore, if $i \neq j$, $0 = \overline{\mathbf{a}_i} \cdot \mathbf{a}_j = \langle \overline{\mathbf{a}_i}, \mathbf{a}_j \rangle = \overline{\langle \mathbf{a}_j, \overline{\mathbf{a}_i} \rangle} \quad \Rightarrow \quad \langle \mathbf{a}_j, \overline{\mathbf{a}_i} \rangle = 0$,
which shows that the columns of A are orthogonal. And if $i = j$, $1 = \overline{\mathbf{a}_i} \cdot \mathbf{a}_i = \langle \overline{\mathbf{a}_i}, \mathbf{a}_i \rangle = \overline{\langle \mathbf{a}_i, \overline{\mathbf{a}_i} \rangle} \quad \Rightarrow$
$\langle \mathbf{a}_i, \overline{\mathbf{a}_i} \rangle = 1 \quad \Rightarrow \quad \|\mathbf{a}_i\| = 1$, and hence the columns of A are orthonormal. Conversely, suppose the
columns are orthonormal, then $A^*A = \begin{bmatrix} (\overline{\mathbf{a}_1})^T \\ \vdots \\ (\overline{\mathbf{a}_n})^T \end{bmatrix} [\ \mathbf{a}_1 \ \cdots \ \mathbf{a}_n\ ] = \overline{\begin{bmatrix} (\mathbf{a}_1)^T \\ \vdots \\ (\mathbf{a}_n)^T \end{bmatrix}} \ \overline{[\ \mathbf{a}_1 \ \cdots \ \mathbf{a}_n\ ]} =$
$\overline{I_n} = I_n$, since the i,j entry of this product is given by $\mathbf{a}_i \cdot \overline{\mathbf{a}_j} = \langle \mathbf{a}_i, \mathbf{a}_j \rangle$ which is 1 if $i = j$, and 0
otherwise. Hence $A^{-1} = A^*$, and A is unitary.

39. Suppose $A = \begin{bmatrix} a_{11} & a_{12} & \cdots & a_{1n} \\ 0 & a_{22} & \cdots & a_{2n} \\ \vdots & \vdots & \ddots & \vdots \\ 0 & 0 & \cdots & a_{nn} \end{bmatrix}$ is upper triangular and normal. Then $A^*A = AA^*$. Now

$A^*A = \begin{bmatrix} \overline{a_{11}} & 0 & \cdots & 0 \\ \overline{a_{12}} & \overline{a_{22}} & \cdots & 0 \\ \vdots & \vdots & \ddots & \vdots \\ \overline{a_{1n}} & \overline{a_{2n}} & \cdots & \overline{a_{nn}} \end{bmatrix} \begin{bmatrix} a_{11} & a_{12} & \cdots & a_{1n} \\ 0 & a_{22} & \cdots & a_{2n} \\ \vdots & \vdots & \ddots & \vdots \\ 0 & 0 & \cdots & a_{nn} \end{bmatrix}$, so $(A^*A)_{11} = |a_{11}|^2$. And

$AA^* = \begin{bmatrix} a_{11} & a_{12} & \cdots & a_{1n} \\ 0 & a_{22} & \cdots & a_{2n} \\ \vdots & \vdots & \ddots & \vdots \\ 0 & 0 & \cdots & a_{nn} \end{bmatrix} \begin{bmatrix} \overline{a_{11}} & 0 & \cdots & 0 \\ \overline{a_{12}} & \overline{a_{22}} & \cdots & 0 \\ \vdots & \vdots & \ddots & \vdots \\ \overline{a_{1n}} & \overline{a_{2n}} & \cdots & \overline{a_{nn}} \end{bmatrix}$, so $(AA^*)_{11} = |a_{11}|^2 + \cdots + |a_{1n}|^2$. Thus

$|a_{11}|^2 = |a_{11}|^2 + \cdots + |a_{1n}|^2$, and hence each $a_{1j} = 0$ for $j > 1$, and $A = \begin{bmatrix} a_{11} & 0 & \cdots & 0 \\ 0 & a_{22} & \cdots & a_{2n} \\ \vdots & \vdots & \ddots & \vdots \\ 0 & 0 & \cdots & a_{nn} \end{bmatrix}$.

Now consider $(A^*A)_{22}$, using $a_{12} = 0$, to obtain $(A^*A)_{22} = |a_{22}|^2$. But $(AA^*)_{22} = |a_{22}|^2 + \cdots + |a_{1n}|^2$,
so $|a_{22}|^2 = |a_{22}|^2 + \cdots + |a_{1n}|^2$, and hence $a_{2j} = 0$ for $j > 2$. Continue in this way to obtain $a_{ij} = 0$
for $j > i$, and therefore A is diagonal. If A is lower triangular and normal, then A^* is upper triangular
and normal. By the above, A^* is diagonal, and hence A is diagonal.